DIGITAL COMMUNICATIONS BY SATELLITE

DIGITAL COMMUNICATIONS BY SATELLITE

Modulation, Multiple Access and Coding

VIJAY K. BHARGAVA
Concordia University
Montréal

DAVID HACCOUN
Ecole Polytechnique de Montréal
Montréal

ROBERT MATYAS
Canadian Astronautics Ltd.
Ottawa

PETER P. NUSPL
INTELSAT
Washington, D. C.

A WILEY-INTERSCIENCE PUBLICATION
JOHN WILEY & SONS

New York · Chichester · Brisbane · Toronto · Singapore

Library of Congress Cataloging in Publication Data:

Main entry under title:

Digital communications by satellite.

"A Wiley-Interscience publication."
Includes bibliographical references and index.
1. Digital communications.   2. Artificial
satellites in telecommunication. I. Bhargava,
Vijay K., 1948-
TK5103.7.D53   621.38′0422   81-10276
ISBN 0-471-08316-X   AACR2

Printed in the United States of America

10 9 8 7 6 5 4 3

To

Our Parents and

Yolande
Alexandre-Ram

Lyson
Nathalie, Laurent

Ann

Maria
Tina, Tony, Randy, Nicky

Preface

The application of digital methods to satellite communications requires the utilization of *modulation, multiple-access* and *coding techniques* which are powerful, reliable and practical. Although detailed treatment of these techniques may be found in various engineering journals, the literature is widely scattered and is frequently written for experts only. This book brings together theories, tradeoffs and implications for system design of modulation, multiple-access and coding techniques that can be applied in the area of digital communications by satellite. The material presented is suitable for senior undergraduate and graduate students and may be used as a self-study text or reference by researchers and practicing engineers in this field. The prerequisite knowledge is a typical undergraduate background in elementary probability theory, some introductory random process theory and a first course in communications theory.

The material is divided into three main parts. Part 1 covers modulation techniques (Chapters 2–6); Part 2 presents multiple-access techniques (Chapters 7–10); and in the third part, coding is treated in Chapters 11–14.

Chapter 1 introduces digital communications by satellites. Link equations are presented, discussed briefly and illustrated. The chapter outlines the important areas of modulation, multiple access and coding and concludes with a perspective on satellite communications.

The basic framework for evaluating the error-rate performance of modulation techniques is described in Chapter 2. These techniques include binary and quaternary phase-shift keying (BPSK and QPSK, respectively), offset-keyed QPSK, frequency-shift keying and a specific form, which is known either as minimum shift keying (MSK) or fast frequency-shift keying (FFSK). In Chapter 3 the generation and detection of various modulation techniques are described in a number of ways to provide the reader with some insight into the underlying properties.

In Chapter 4 performance degradations due to channel nonlinearities, channel filtering (band limitation) and interference are reviewed. A comparison is presented of the performance of several digital modulation techniques over satellite channels. Carrier and clock recovery are aspects of synchronization that directly affect modem implementation and performance. Recovery techniques applicable to PSK and FFSK signals are discussed in Chapter 5.

Chapter 6 presents multi-*h* phase-coded modulation techniques which offer performance advantages relative to PSK systems. These techniques have coding inherent in the modulation structure and are representative of recent developments in modulation processing.

Multiple access refers to the technique by which a number of Earth stations form communication links through one or more satellite radio-frequency (RF) channels. Frequency-division multiple access (FDMA), time-division multiple access (TDMA), and code-division multiple access (CDMA) are three basic forms of multiple access. The chapters on multiple access reflect communication systems design and are broad surveys of research, developments and implementations which are still maturing.

Chapter 7 deals with FDMA, currently the most common form of multiple access, and presents FDMA principles, methods and descriptions of systems; it includes a design example for a single-channel-per-carrier (SCPC) system.

TDMA is an area of widespread interest for future satellite systems and is described in Chapter 8. TDMA synchronization methods are catalogued and a broad overview of experimental, operational and planned TDMA systems is presented. The chapter contains a detailed example of TDMA system design.

In addition to multiple-access capability, CDMA, described in Chapter 9, is useful in providing protection against jamming, low probability of intercept and privacy. Chapter 10 is a brief survey of the advancing technology of packet communications, specifically as applied to satellite systems.

Part 3 presents fundamentals of error-control coding (block and convolutional, forward-error correction and automatic repeat request) and its application to digital satellite links. This part explains both basic and advanced material from publications generally considered to be difficult. Chapter 11 presents the fundamental coding problem, channel capacity and coding concepts, together with ARQ techniques applied to satellite links. Chapter 12 is devoted to convolutional encoding and to the powerful decoding techniques of sequential and Viterbi decoding. The chapter is augmented by an extensive tabulation of the best known convolutional codes suitable for these decoding techniques.

Chapter 13 treats encoding and decoding techniques for block codes, emphasizing cyclic and related codes such as the BCH and Reed–Solomon codes. The chapter includes examples of the applications of block codes to digital communications. A concise unified treatment of threshold decoding for both block and convolutional codes is given in Chapter 14.

The concluding chapter deals with related technologies and trends which influence the main themes of this book: modulation, multiple access and coding.

Appendices include descriptions of digital speech interpolation (DSI) techniques and demand assignment (DA). There is a set of definitions which are germane to the book and which are not usually collected in one source. The notation is summarized in the Glossary at the end of the book.

Included at the end of most chapters are problems which range from simple numerical examples to those designed to enhance and augment the material presented.

The book can serve as a text for a short course in digital satellite communications for practicing engineers, and we have used preliminary versions in such courses. It could also serve for a one-semester course in digital satellite communications for senior students who have an adequate background in communications. Other courses could be organized as illustrated in the following list:

Digital Communications Engineering	Multiple-Access Systems	Error-Control Techniques
1	1	1
2.1 to 2.5		
3.1 to 3.5		
4.1 to 4.3	4.4, 4.5	11
5.1, 5.2, 5.5	7	
11.1 to 11.3	8	12
12.1 to 12.4, 12.8	9	
13.1 to 13.6	10	13
14.1, 14.3	11.2	
		14
15	15	15

This book is the result of joint, collaborative work first begun for an intensive short course given at Concordia University, Montreal, in September 1979 to an international audience of senior engineers involved in digital communications. Chapters 1 and 15 were prepared by Peter Nuspl with major contributions from each of us. Chapters 2, 3, 4, 5, and 6 were written by Robert Matyas; Chapters 7, 8 and 10 by Peter Nuspl; Chapters 11 and 12 by David Haccoun; and Chapters 9, 13 and 14 by Vijay Bhargava. The overall coordination in the preparation of the manuscript was accomplished by Vijay Bhargava.

This book was written while Robert Matyas and Peter Nuspl were associated with the Communications Research Centre, Ottawa.

VIJAY K. BHARGAVA
DAVID HACCOUN
ROBERT MATYAS
PETER P. NUSPL

Montreal
Montreal
Ottawa
Washington, D. C.

August 1981

Acknowledgments

We gratefully acknowledge the kind permission from R. C. Kirby of Comité Consultatif International des Radio Communications to adapt Figures 1.8, 1.9, 1.10, 7.2, 7.3, 10.2, E.1 and E.2; Appendix E on digital speech interpolation techniques; and some definitions relating to satellite communications from CCIR documents. We thank L. C. Palmer for the use of course notes relating to trellis detection of MSK. We wish to thank H. C. Chan, J. Conan, N. G. Davies, H. H. Hoc, L. N. Lee, G. E. Séguin and D. P. Taylor for reading parts of the manuscript and for suggesting numerous improvements; A. Antoniou, Y. Gervais, R. Langlois, B. Lavigueur, and M. N. S. Swamy for encouraging this work; and Concordia University and Ecole Polytechnique de Montréal for their support. V. K. Bhargava and David Haccoun wish to thank Natural Sciences and Engineering Research Council of Canada and le Programme de Formation de Chercheurs et d'Action Concertée-Québec for supporting the research that led to many of the new results presented. R. Matyas and P. Nuspl acknowledge the extensive experience acquired in work assignments at the Communications Research Centre of the Department of Communications, Canada.

We also thank Lina D'Iorio, Sharon Speevak, Marie Berryman, and Monica Etwaroo for typing the manuscript, N. Tata for doing the drawings, B. Nguyen, and A. Benyamin-Seeyar for his help in proofreading the manuscript.

V.K.B.
D.H.
R.M.
P.P.N.

Contents

PART 3   CODING

Digital Communications by Satellite

We begin with a discussion of the title of this book. **Communications**, in its most general meaning, is the transfer or interchange of information. This book is really about telecommunications, which is communications over distances using technological means; we limit the treatment to technical matters as distinct from human communications. **Digital** means that the information is expressed directly as digits; "digital" also conveys the idea that the information is discrete, in distinction from analog or continuous. By definition, a **satellite** is a secondary body in orbit about a primary body; clearly, man-made Earth-orbiting satellites are the specific interest in this book. Satellites are the means by which truly long-distance, wide-area and flexible communications are achieved.

This chapter begins with a presentation of satellite characteristics and a general discussion of digital communications by satellite. A brief section on link calculations describes the major parameters and their tradeoffs in practices of the present era. A number of factors are reviewed for their impact on the selection of a modulation technique. Some concepts and methods of multiple access are introduced and compared. Error-control techniques are characterized, with a brief discussion of advantages and disadvantages.

1.1 SATELLITE CHARACTERISTICS

This section gives some definitions and facts that are pertinent to satellites. There are short discussions on why satellites are useful in communications and on some systems implications.

Some Definitions and Facts

Many detailed definitions are contained in the Glossary. Directly relevant terms are introduced and defined here. A **geosynchronous** satellite is in Earth

1

orbit and has a period equal to that of the Earth's rotations, a sidereal day of 23 h 56 min 4.09 s (mean value). A **geostationary** satellite is geosynchronous with an equatorial (in the plane of the equator), circular and direct (same direction of rotation as Earth) orbit. A small inclination from the equatorial plane and small ellipticity are usually present in real orbits. The terms "geosynchronous" and "geostationary" are often used interchangeably in the literature. "Synchronous" and "geosynchronous" are equivalent for Earth (geos) satellites, as are "stationary" and "geostationary."

The period of an elliptical orbit is given by

$$\text{period} = 2\pi\sqrt{\frac{A^3}{\mu}} \qquad (1.1)$$

where A = semimajor axis of the ellipse
μ = gravitational constant

For Earth $\mu = 3.99 \times 10^5$ km^3/s^2 (mean value). For a circular orbit to have a period of one sidereal day, the altitude h of a geostationary satellite above the Earth's surface is calculated as 35 800 km.

As a result of many complex forces, a geostationary satellite moves about in a relatively large volume of space; a sphere of radius 60 km can be envisioned. Furthermore, the motion has a number of drift components with different rates and there are necessarily random perturbations. At infrequent but predictable events called station-keeping, the satellite is deliberately accelerated by on-board thrusters. In general, not every communications station can follow these variations in the orbit and this fact has a major impact on some forms of multiple access, which is simultaneous use of a channel by several transmitters.

Communications satellites of the present era include several generations, yet can be simply characterized as to their communications functions. Figure 1.1 illustrates the essential features of a communications satellite. A receive antenna collects electromagnetic energy; a desired bandwidth is filtered and amplified. For each frequency band, the signal is power-amplified and radiated through the transmit antenna. That there are many complexities and many tradeoffs should be evident. What is outlined here is a **repeater**, which is a frequency-translating amplifier; this is called a transponder in most of the literature.

It will be evident that the satellite power amplifier is a key element and a problem area. Thus far such amplifiers use travelling-wave tubes (TWT) for their low weight and relatively high efficiency. These TWT amplifiers (TWTA) are nonlinear and are often operated near saturation. Weight and power limitations in a satellite dictate that wideband channels (20 to 250 MHz) are used (or planned) and that there are 4 to 24 such channels per satellite.

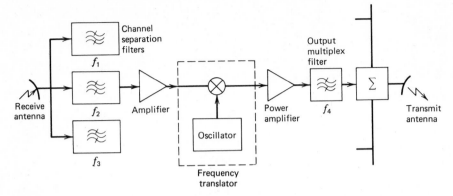

Figure 1.1 Typical satellite repeater, showing receiver amplification, frequency translation and power amplification.

The reader is referred to some introductory reading on orbits and orbit parameters [1], to some characteristics of communications satellite systems [2] and to an overview of satellite communications [3].

Why Satellites?

Satellites and communications systems based on satellites have experienced phenomenal growth since their beginning in 1965. Yet there remain much scope and many opportunities for satellite technologies and their applications. Some basic factors can be readily identified as unique and critical as to why satellites are used; and there are compelling reasons for not using satellites. The emphasis is on geostationary satellites used for commercial telecommunications and the following discussion is about the three topics: (*a*) network topology, (*b*) long distances and (*c*) costs.

Network Topology

The most obvious and unique characteristic of satellites is that they are at the center of a *star topology*, as illustrated in Figure 1.2*a*. All signals in the network pass through the satellite. This is in sharp contrast to terrestrial communications systems, in which the natural topologies are *linear* (a chain of repeater stations in radio systems), as in Figure 1.2*b*, or the topology is a *loop*, as in Figure 1.2*c*. A telephone subscriber loop is a simple example and many computer networks are connected in loop form. Figure 1.2*d* illustrates the very powerful topology of the *mesh*, such as used in a network of telephone switching centers.

The *wide-area coverage* of satellites has led to applications in transoceanic, regional and national networks. Such coverage and topology make distribution and broadcasting of TV and radio programmes attractive and practical. Data

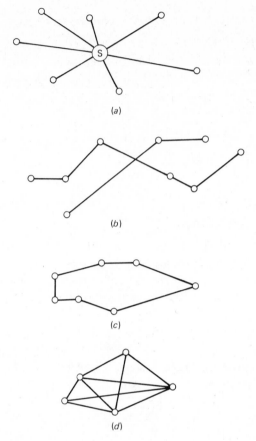

Figure 1.2 Some topologies in networks. (*a*) Star or hub; (*b*) linear or chain; (*c*) loop; (*d*) mesh, shown fully connected.

could also be distributed and broadcasted; such services are usually called point-to-multipoint. News, weather, stock information and computer networks are examples.

Of course, telephone and data services are now provided by satellite networks, which take advantage of the *connectivity*. A **circuit** is defined as a two-way connection between users, such as in telephony; a **half-circuit** is a one-way connection, such as in broadcasting to receive-only stations. Since the star topology with a satellite can be readily viewed as a mesh, there is with a satellite the possibility of a fully interconnected mesh, even for a large number of stations. Furthermore, it is possible to connect stations by various assignment schemes, up to fully demand-assigned circuits, making a switching system. Connectivity, assignments and other features are discussed in the following chapters and appendices.

Long Distances

It is evident that satellites imply truly *long-distance* communications. For a geostationary satellite, the distances between satellite and earth stations are 36 000 to 41 000 km. A formula for this distance is

$$D_r = \sqrt{(h+r)^2 + r^2 - 2(h+r)r\cos\theta} \quad \text{m} \quad \text{(range)} \quad (1.2)$$

where geometry and variables are defined in Figure 1.3. As a result of these great distances, the propagation paths have high losses (>200 dB at 14/12 GHz); fortunately, in the bands used the losses are quite steady and stable, requiring fade margins of 3 to 7 dB. This contrasts with much lower losses for terrestrial links, which are more under the engineer's control, but with fade margins of 30 to 40 dB. A **fade margin** is the additional effective power required for a microwave signal to allow for losses during fading, mostly attributable to rain. Formulas for path loss and fade margin are given in Section 1.3.

For the satellite portion of circuits, we will use the terminology **uplink** to mean the transmission or path from a station to a satellite, **downlink** for the direction satellite to earth station. A **single hop** consists of an uplink and a downlink; where there is no possibility of confusion, this is called a **link**. A **double hop** involves two single hops, often with a large station in the middle to perform a network function. There is a significant delay in transmissions to the geostationary orbit: approximately $\frac{1}{8}$ s in each direction, up and down; $\frac{1}{4}$ s for a single hop; $\frac{1}{2}$ s for a double hop. A **satellite loop** is a single hop back to the same station. The geometry for a loop is shown in Figure 1.3 and the transmission delay for a loop is given by

$$d = \frac{2D_r}{c} \quad \text{s} \quad \text{(loop delay)} \quad (1.3)$$

where c is the propagation velocity in free space ($\simeq 3 \times 10^8$ m/s). The

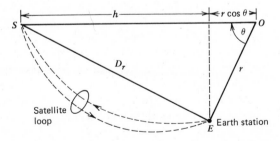

Figure 1.3 Geocentric geometry, showing the plane containing the satellite and an earth station. O is the geocentric origin; earth radius is r; θ is the angle between OS and OE.

minimum delay occurs for a station directly below the satellite and the maximum occurs for stations on the horizon as viewed from the satellite.

$$d_{min} = \frac{2h}{c} \qquad \text{at } \cos\theta = 1 \qquad (1.4a)$$

$$d_{max} = \frac{2h}{c}\sqrt{1 + \frac{2r}{h}} \qquad \text{at } \cos\theta = \frac{r}{r+h} \qquad (1.4b)$$

A single-hop delay would be well within these bounds, because two stations would be separated and would likely have elevation angles higher than 3°. The **elevation angle** for a station is the angle between the satellite direction and the local horizontal plane.

Costs

Costs are very important factors in why satellites and satellite communications are used and not used. However, this book does not include economics of satellite communications systems. Brief discussions on some qualitative factors are given here and some quantitative data are presented in Chapter 15.

A satellite is often the only way, or the *only practical way*, for some specific service to be provided. For example, a reliable, high-performance data or voice circuit for a small island in the Pacific, a community in the Canadian North or a ship on the Atlantic can readily be provided by satellite. *Broadcast* services over wide areas are also economically attractive. That satellite circuits are *distance insensitive* (approximately) and *terrain independent* with respect to cost and performance are important characteristics. Some very *special services* can be provided via satellites and compatible earth stations: weather maps, communications during disaster relief, precise time-transfer and long-baseline interferometry. See Chapter 8 for definitions and references.

Some very careful *cost–benefit* analyses have been done by many organizations. In many cases, satellite systems have proven cost-effective when compared to microwave relay or undersea cables. It is important but difficult to assess these advancing satellite technologies against established technologies (cables, microwave, radio) and against other new technologies (fibre optics, lasers).

Another factor is that satellite systems are proven to be valid *alternatives*, even if not always unique or cost-effective. For example, alternate routing by satellite, spare and growth capacity, and peak offloading from terrestrial systems are used by the designer and operator to enhance the performance or to minimize the risk to the network. Hence even though initial or lifetime costs are high, satellite systems have their role in communications.

Systems Implications

The factors discussed above tend to have the following systems implications. Satellites now have travelling-wave-tube amplifiers (TWTAs) with 20 to 200 W of possible output power each and there are 4 to 24 in one satellite. These satellites require high prime power from large solar sails, since the sun is the only large source of power presently available for commercial satellites. Large and complex antenna systems are used on the satellite. Receivers in the satellite and earth stations require low-noise amplifiers (LNAs). Of course, to date, satellites cannot be repaired, so there must be sufficient on-board component redundancy and switchover capability.

Transmitters in earth stations have high-power amplifiers (HPAs) which are large (100 to 600 W), bulky and costly. But single voice circuits are achieved with 1 W. The HPAs are also often operated near saturation and are nonlinear, too. Earth station antennas are usually parabolic reflectors with diameters from 2 m (TV receive only, telephony), 5 m (multichannel telephony and medium-speed data) to 30 m for multipurpose, high-capacity stations. Given the narrow beamwidths of large antennas and the motion of the satellite, there is a need for such stations to track the satellite.

Networking aspects are very significant, since many users generally acquire service from the same satellite channel. There exist interesting problems in intermodulation and interference. In satellite design, link budgets are carefully assessed to fractions of a decibel; examples are given in Section 1.3.

1.2 DIGITAL COMMUNICATIONS

Next, some reasons for digital communications are presented and concepts of digital communications by satellite are introduced.

Why Digital?

As part of this introduction, some of the reasons for using digital communications are recalled. Of course, the emphasis is on satellite systems and not all arguments apply in every case. The factors are categorized as follows and briefly discussed: (*a*) compatibility, (*b*) flexibility and (*c*) economy.

Compatibility

An increasingly significant factor is the accepted and growing use of *digital modulation techniques* in terrestrial telecommunications (2400 bps to 1.544 Mbps and recently much higher). Voice traffic is still predominant, with data traffic at about 10% and gaining. There are now systems using encoded voice at 64 and 56 kbps; some at 40 and 32 kbps are under development and in field trials. Such satellite services have been available for several years. *Voice processing* (source encoding) can be done at the source or close to the source. For example, a 3.4-kHz voice signal can be processed by linear predictive encoding to produce a 2400-bps digital stream, with good quality on reconstruction. Such source coding and its emerging viability are important to selection of digital methods in communications. Also, as in voice services, instrumentation and to some degree in TV, there is a strong trend to digital.

Digital modulation and digital buffers are essential for compatibility with *multiple-access* techniques such as TDMA and CDMA, treated in Chapters 8 and 9, respectively. Digital formatting also permits *privacy encoding* and is necessary in systems using *encryption* for security. Digital methods are readily applied when carrier *energy dispersal* and *band spreading* techniques are needed.

We also note that some satellite networks (private, specialized) do not require compatability with exterior signals, so decisions on the forms of modulation, multiple access and coding can be based on performance or cost factors only. Some private networks and national networks use FM and FSK modulation; many use digital modulation.

Flexibility

Probably the most significant reason for a choice of digital modulation for terrestrial networks is the ability to *regenerate* a digital signal. Although corrupted by noise, the bit stream is received and demodulated with few errors; then the bit stream modulates a carrier and is transmitted down the chain of repeaters. Significantly, no such regeneration is used in commercial communications satellites, but this will soon come. Regeneration in the satellite will have the effect of separating the uplink and downlink.

Flexibility is available because *"a bit is a bit"*, in that the subsystems process each in the same manner. Multiplexing, switching, storing, regeneration and other functions are possible, usually with bits, words, packets or blocks. The information content is not always relevant.

A very significant feature is the possibility for *error control*, giving the user the potential for low error rates and high reliability. Improved performance can also mean better fidelity, range, resolution, purity and fast response time, sometimes not available from analog systems. Some new services are possible only because of the flexibility of digital implementations.

Economy

Digital communications systems are taking full advantage of *integrated circuit technology*, in particular microprocessors, memory and MSI (medium-scale integration) implementations. Digital systems design has become sophisticated and automated; there is a trend toward *aggregation of functions* such as modems and codecs, converters and filters. Such digital subsystems are expected to be more *economical to manufacture* once the initial high cost of design is paid. When installed and running, experience has shown that operations and maintenance of digital systems are no more difficult than for analog systems; for some, these practical aspects are simplified and there are improved monitor, alarm and control mechanisms. A strong argument for digital systems is the *increased capacity* to be obtained in some satellite services.

There are definite economic disincentives to using communications satellites. Satellites are *technology intensive* and the related technologies are known to a relatively small community. Research, development and field trials are often long and *costly*. Production preparations can be expensive. There is now *insufficient volume*, with a few exceptions, to take advantage of scale. Of course, staff must be trained to install, operate and maintain this new equipment. Also, in this high technology, as in the computer industry, equipment becomes obsolete quickly.

Digital Systems Concepts Using Satellites

A satellite communications system consists of one or more links (half-circuits) and a network is a collection of such links operating under some form of control. A **satellite communications link** consists of a space segment (a satellite or spacecraft) and an earth segment (one, two or more Earth stations). These definitions include deep-space links, telemetry, tracking and control links and, of course, Earth–satellite–Earth links and networks. We specialize to systems illustrated in Figure 1.4, in which the important elements of the satellite are the repeater and the antenna subsystems. The Earth stations consist of several subsystems: RF transmitter and receiver, IF transmitting side and IF receiving side, and baseband equipment at each end of the link. Figure 1.5 illustrates a simplified Earth station. Modulation (MOD) and demodulation (DEMOD) equipment is usually called a **modem**; coding and decoding equipment has the short name **codec**. Most communications links have all the elements shown in Figure 1.5, but it is possible that only the sending or receiving side are of interest. For example, a terminal* may be used to send only, such as environmental data destined for a data bank. On the other hand, a central information source may operate in a broadcast mode, such as to provide news material to a number of receivers.

The concepts of signal-to-noise ratio (SNR), carrier power-to-noise power (C/N) and carrier-to-noise density (C/N_0) ratios are well known and apply to satellite uplinks and downlinks in analog or digital modulation. C/N_0 link calculations are presented below.

The concept of energy per bit needs a brief introduction. A binary information digit **(bit)** is a measure of information and the energy required to convey that bit, E_b, is a fundamental quantity. It is therefore not surprising to be interested in the ratio of this energy per bit and the noise density. The first is what we pay for and the second is what we must combat. E_b/N_0 is the basis of comparison among modems, codecs and other processors of digital information. E_b/N_0 is dimensionless, since E_b has the unit joules and N_0 has units of power per hertz, which is (joules/second)×second.

---

*The terms "station" and "terminal" are used somewhat interchangeably in the literature; some authors reserve "terminal" for mobile equipment, and others use it to designate IF and baseband equipment in stations.

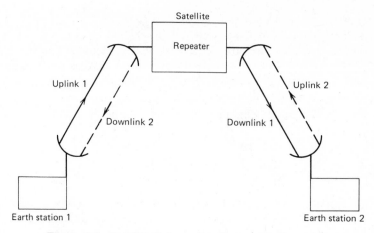

Figure 1.4   Satellite links connecting two Earth stations.

1.3   LINK CALCULATIONS

Pertinent to an understanding of the following chapters, the following link
equations are presented, discussed briefly and illustrated. Notation in this
book is consistent with AIAA and IEEE recommendations of 1977, but these
are not standards [4]. It is convenient to express link equations in decibels. A
decibel (dB) is properly used only as a measure of power or ratios of powers.

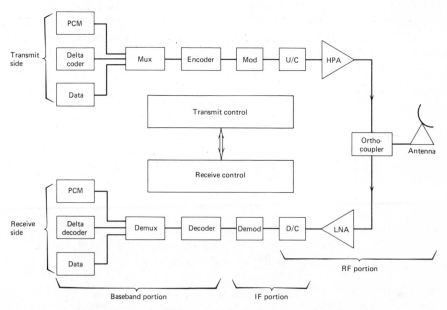

Figure 1.5   Simplified configuration for an Earth station.

But it is customary throughout the literature to use dB for other quantities. Any quantity X with unit u is converted to dB by

$$X\,u \rightarrow 10\log_{10} X \qquad \mathrm{dB}{-}u$$

The merits of this approach are the change of scale and the conversion of products to sums, since $\log AB = \log A + \log B$. In the other direction, y dB is converted to its numeric value by

$$y\,\mathrm{dB} \rightarrow 10^{y/10}$$

The term "dB" is *not* a unit but a scale function, $10\log(\cdot)$. In products and ratios, units *multiply and divide* (combine and cancel) when collecting units. For example,

$$\text{cost of trip} = \frac{40\ \mathrm{km}}{\mathrm{h}} \times \frac{12\ell}{100\ \mathrm{km}} \times \frac{\$0.25}{\ell} \times 5\ \mathrm{h}$$

$$= \$6$$

Expressed in dB, units *add and subtract*:

$$\text{cost of trip} = 10\log 40\ \mathrm{dBkm} - 0\ \mathrm{dBh}$$

$$+ 10\log 12\ \mathrm{dB}\ell - 20\ \mathrm{dBkm}$$

$$+ 10\log(0.25)\ \mathrm{dB}\$ - 0\ \mathrm{dB}\ell$$

$$+ 10\log 5\ \mathrm{dBh}$$

$$= 0.778\ \mathrm{dB}\$$$

$$= \$6$$

Statements such as $6 = 0.778$ dB are often made.

Uplink Equation

The uplink equation appears in two forms, depending on the perspective and available information. Where satellite receiver characteristics are given, (1.5) is used; where Earth station transmit parameters are given, (1.6) is convenient. C/N_0 has the unit hertz, (Hz) meaning $1/\text{second}$.

$$\left.\frac{C}{N_0}\right)_u = \phi_s - G_1 - \mathrm{BO}_i + \left.\frac{G}{T}\right)_s - k - L_{r,u} \qquad \mathrm{dB\text{-}Hz} \qquad (1.5)$$

where u designates the uplink
s designates the satellite

$\phi_s =$ flux density at TWTA saturation, dBW/m^2
$G_1 =$ gain of 1 m^2 antenna at f_u (up frequency) dB;
BO$_i =$ input backoff, dB

$\left. \dfrac{G}{T} \right)_s =$ figure of merit for satellite, dB/K

  $k =$ Boltzmann's constant

   $k = 1.380 \times 10^{-23}$ J/K, -228.6 dBW/K-Hz

$L_{r,u} =$ uplink margin for rainfall attenuation, dB

$$\left. \frac{C}{N_0} \right)_u = \text{e.i.r.p.})_e - L_{p,u} + \left. \frac{G}{T} \right)_s - k - L_{r,u} \qquad \text{dB-Hz} \qquad (1.6)$$

where e.i.r.p.)$_e =$ effective isotropic radiated transmitted power, Earth station, dBW

  $L_{p,u} =$ free-space path loss, uplink, dB

$G/T)_s$ is calculated as

$$\left. \frac{G_r}{T} \right)_s = G_r - T \qquad \text{dB/K} \qquad (1.7)$$

where $G_r =$ antenna gain of the receiver, dBi

  $T =$ (receiver) noise temperature, dB-K

The additional subscripts s for satellite and e for Earth station are commonly used. G is the antenna gain measured at the input of a low-noise amplifier, expressed in dB relative to an isotropic radiator, hence the notation dBi. T is the receiving system noise temperature referred to the input of a low-noise amplifier, expressed in dB relative to 1 kelvin (K).*

For a parabolic reflector in the antenna system, the gain G is calculated as

$$G = 10 \log \left(\eta \frac{\pi^2 D^2}{\lambda^2} \right) \qquad \text{dBi} \qquad (1.8)$$

where $\eta =$ efficiency of the antenna (50–60%)

  $D =$ antenna diameter, m

  $\lambda =$ wavelength, m ($\lambda = c/f$)

The reasons for directional antennas are obvious: to concentrate the radiated energy; in uplinks to avoid interference with other orbit positions; and in downlinks to form global, regional or spot beams for the required coverage.

---

*Thermodynamic temperature is called kelvin, instead of "degree kelvin," and has the symbol K without the symbol ° (degree).

The receiver noise temperature T is given by the formula

$$T = T_r + \left(\frac{\ell-1}{\ell}\right)T_w + \frac{T_g}{\ell} + \frac{T_{sky}}{a\ell} + \left(\frac{a-1}{a\ell}\right)T_{rain} \qquad K \qquad (1.9a)$$

where T_r = noise temperature of receiver and following stages, K
$\quad\ell$ = waveguide loss factor
$\quad T_w$ = waveguide temperature, K
$\quad T_{sky}$ = sky contribution to T, K
$\quad a$ = rain absorption factor
$\quad T_{rain}$ = effective temperature of rain, K

$$T_r = T_{LNA} + \sum_{m=2}^{N} \frac{T_m}{g_m - 1} \qquad K \qquad (1.9b)$$

where T_{LNA} = receiver (LNA) noise temperature, K
$\quad T_m$ = noise temperature of following mth stage, K
$\quad g_{m-1}$ = gain before the mth stage

In the absence of rain, for so-called clear weather conditions, set $a = 1$. This equation is for Earth station receivers; for the satellite, the effective temperature is usually measured. We comment that G/T is called a figure of merit in a receiver system; however, note that it is frequency dependent through G.

Effective isotropic radiated power has the symbol **e.i.r.p.** (and very often EIRP); it is calculated as

$$e.i.r.p. = P_t - L_f + G_t \qquad dBW \qquad (1.10)$$

where P_t = power of the transmitter, dBW
$\quad L_f$ = losses due to feed cable, branching filters, dB
$\quad G_t$ = antenna gain of the transmitter, dBi

Path loss L_p is for free-space conditions. From the subsatellite point on Earth to a geostationary satellite

$$L_p = \frac{4\pi D_r^2}{\lambda^2} \qquad (1.11)$$

where D_r is given by (1.2). It is necessary to take into account the actual distance (range) to stations.

C/N is readily calculated.

$$\left.\frac{C}{N}\right)_u = \left.\frac{C}{N_0}\right)_u - B_s \qquad dB \qquad (1.12)$$

where B_s is the receiver noise bandwidth at the satellite, dB-Hz (defined in Section 4.4).

Table 1.1   Example of an Uplink Budget[a]

Flux density at saturation	ϕ_s	-81.0	dBW/m^2
Gain of 1 m^2	$-G_1$	-44.5	dBi
Input backoff	$-BO_i$	-0.0	dB
Satellite figure of merit	$+\dfrac{G}{T}\Big)_s$	$+1.9$	dB/K
Boltzmann's constant	$-k$	$+228.6$	dB/K-Hz
Rain loss, dry weather	$-L_{r,u}$	0.0	dB
Carrier-to-noise density	$\dfrac{C}{N_0}\Big)_{u,s}$	105.0	dB-Hz

[a]Frequency-dependent terms are calculated for $f_u = 14$ GHz.

The saturating flux density for the satellite and the e.i.r.p. for the Earth station are key design parameters. Once they are fixed, we note that C/N_0 and C/N are not directly dependent on the operating frequency f_u: in (1.5) G_1 and $(G/T)_s$ have f^2 terms but they cancel; in (1.6) $L_{p,u}$ and $(G/T)_s$ do the same. The rain losses introduce slight frequency dependence. However, e.i.r.p. does depend on the antenna gain; hence for a given available power P_t, the dependence is as f^2. Also, for a fixed e.i.r.p., a tradeoff between P_t and G_t can be made, subject to available power, pointing accuracies and reflector sizes. At a given frequency, this means that a larger reflector will reduce the required transmitter power.

As an example and for use in later chapters, Table 1.1 shows an uplink budget for $C/N_0)_u$ under the conditions of saturation of the satellite TWTA and no rain attenuation.

Downlink Equation

For the downlink, the equation for $C/N_0)_d$ due to thermal noise is given by

$$\frac{C}{N_0}\bigg)_d = \text{e.i.r.p.})_s - BO_o - L_{p,d} + \frac{G}{T}\bigg)_e - k - L_{r,d} \qquad \text{dB-Hz} \qquad (1.13)$$

where d designates the downlink
     e designates the Earth station
e.i.r.p.)$_s$ = effective power at the satellite, dBW
   BO$_o$ = output backoff, dB

Other terms are defined in (1.5) and (1.6). There are also other losses, such as due to pointing of antennas in static conditions, in tracking conditions and under wind loads; these losses can be subtracted from (1.13) or from (1.8). Once again, for a fixed satellite e.i.r.p., $C/N_0)_d$ is independent of frequency, except for rain losses. The **output backoff** is expressed in dB from the saturated

Table 1.2 Example of a Downlink Budget[a]

Satellite radiated power	e.i.r.p.)$_s$	+46.5	dBW
Output backoff	$-\text{BO}_o$	−0.0	dB
Free-space path loss	$-L_{p,d}$	−205.8	dB
Earth station figure of merit	$+\dfrac{G}{T}\biggr)_e$	+20.7	dB/K
Boltzmann's constant	$-k$	+228.6	dBW/K-Hz
Rain loss, dry weather	$-L_{r,d}$	0.0	dB
	$\dfrac{C}{N_0}\biggr)_{d,s}$	90.0	dB-Hz

[a]Frequency-dependent terms are calculated for $f_d = 12$ GHz.

output point relative to saturated power. Since the power transfer function is nonlinear (discussed at length in later chapters), BO_o is a nonlinear function of BO_i.

Table 1.2 illustrates a typical downlink budget for an example used later; this is also for saturation and no rain. The numerical results of Table 1.1 and Table 1.2 are for a small Earth station for use at 14/12 GHz with ANIK-B, a Canadian domestic satellite.

Intermodulation

Some details of intermodulation are discussed in Chapter 7. As a brief introduction, **intermodulation** is the production by nonlinear amplifiers of signal-dependent, unwanted outputs. When few carriers are present, it is necessary to analyse the intermodulation terms individually. When large numbers of carriers are present, intermodulation produces an average effect that resembles noise to the wanted signals. In the repeater bandwidth B, the carrier-to-intermodulation noise ratio C/N_{im} is found by analysis, computation, simulation or actual measurements. The equivalent carrier-to-intermod noise density is then expressed as

$$\frac{C}{N_0}\biggr)_{im} = \frac{C}{N}\biggr)_{im} + B_e \qquad \text{dB-Hz} \qquad (1.14)$$

where B_e is the receiver noise bandwidth at the Earth station, in dB-Hz. For sufficient backoff, $C/N)_{im}$ and $C/N_0)_{im}$ are often high enough so that intermodulation effects are negligible.

Available C/N_0 and E_b/N_0

Under the assumption that the uplink, downlink and intermodulation noises are independent, which is only approximately true for satellite repeaters, the total $C/N_0)_t$ is given by

$$\frac{C}{N_0}\bigg)_t = \left[\left(\frac{C}{N_0}\right)_u^{-1} + \left(\frac{C}{N_0}\right)_d^{-1} + \left(\frac{C}{N_0}\right)_{im}^{-1}\right]^{-1} \quad \text{Hz} \quad (1.15)$$

It is necessary to make this calculation in numerical, absolute values, not in dB.

Some engineers include rain margins $L_{r,u}$ and $L_{r,d}$ in the link equations, as in (1.5) or (1.6) and (1.13), but not other margins. Engineers sometimes prefer to take account of all margins together, usually at the receiver. The total available E_b/N_0 is given by

$$\frac{E_b}{N_0}\bigg)_t = \frac{C}{N_0}\bigg)_t - rR \quad \text{dB} \quad (1.16)$$

where R = transmission rate, dB-bps = dB-Hz
$\quad\quad r$ = coding rate ($r=1$ for no coding)

E_b/N_0 is dimensionless, expressed in dB. The coding rate r is introduced for completeness when error-control coding is used.

Required E_b/N_0 and C/N_0

Usually, service requirements are stated at some capacity R bps and some performance requirements. The usual method of specifying performance in digital systems includes the bit-error rate (BER), more correctly called the probability of bit error P (bit error), P_b. As discussed in the chapters on modulation, demodulation and coding, there are relations between P_b and E_b/N_0. The required E_b/N_0 is obtained from the curves or can be calculated for the case of some elementary relations. These relations are for ideal processing, in the sense that only additive, white Gaussian noise is considered and also with the assumption that the mathematical operations can be ideally implemented.

The required $C/N_0)_{req}$ is then calculated.

$$\frac{C}{N_0}\bigg)_{req} = \frac{E_b}{N_0}\bigg)_t - G_c + rR + M_i + M_f + M_{sys} \quad \text{dB-Hz} \quad (1.17)$$

where G_c = coding gain, dB
$\quad\quad M_i$ = margin for modem implementation, dB
$\quad\quad M_f$ = margin for rain fade, dB [set $L_{r,u} = L_{r,d}$ in (1.5), (1.6) and (1.13)]
$\quad\quad M_{sys}$ = margin for system

G_c is the actual coding gain for a coder/decoder; it depends strongly on r and somewhat on the type of code selected, as discussed in Chapter 11. Several margins are added to account for various losses and imperfections.

Earth Station Parameters

The Earth station e.i.r.p. is calculated as follows:

$$\text{e.i.r.p.})_e = \phi_s - G_1 - \text{BO}_i + L_{p,u} \quad \text{dBW} \quad (1.18)$$

All terms have been defined previously. This equation is the relation between (1.5) and (1.6). The power for the HPA is calculated from (1.10), but with additional terms for pointing losses and HPA backoff.

$$P_{\text{HPA}} = \text{e.i.r.p.})_e - G_t + L_f + L_{\text{ptg}} + \text{BO}_{o,\text{HPA}} \quad \text{dBW} \quad (1.19)$$

where $\text{BO}_{o,\text{HPA}} =$ output backoff for HPA, dB
       $L_{\text{ptg}} =$ antenna pointing losses, dB

This gives an approximate size of the HPA for an uplink with BO_i of the TWTA and BO_o for the HPA. Table 1.3 shows a typical calculation for a small Earth station operating with ANIK B at 14 GHz.

Availability and Rain Margins

The foregoing link equations include the margins for attenuation due to precipitation; rain is the major contributor. There remains the problem of finding the margins for a given system specification. **Availability** is expressed as a percentage of time over which a link has the specified C/N_0 or some other performance criterion. A useful relationship between rain rate and specific attenuation due to rain is aR^b, expressed in dB/km. For each link, coefficients a and b are found from tables in Ref. 5. The effective path length for rain can

Table 1.3 Typical Calculation of HPA Power

Flux density at saturation	ϕ_s	-81.0	dBW/m^2
Gain of 1 m^2	$-G_1$	-44.5	dB$_i$
Input backoff	$-\text{BO}_i$	-10.0	dB
Free-space path loss	$+L_{p,u}$	$+207.3$	dB
Radiated power, earth station	$\text{e.i.r.p.})_e$	$+71.8$	dBW
Antenna gain, transmitter	$-G_t$	-52.3	
Losses in waveguide	$+L_f$	$+0.7$	
Antenna pointing loss	$+L_{\text{ptg}}$	$+0.8$	
Output backoff, HPA	$+\text{BO}_{o,\text{HPA}}$	$+3.0$	
Power for HPA	P_{HPA}	24.0	dBW
	P_{HPA}	$=251$	W

be obtained from an empirical equation [6], which is a good fit to path-length curves in Figure 1 of CCIR Report 564-1 [7].

$$\ell(R, \alpha) = \left[7.41 \times 10^{-3} R^{0.766} + (0.232 - 1.80 \times 10^{-4} R) \sin \alpha \right]^{-1} \quad \text{km}$$

(1.20a)

where R = rain rate, mm/h
      α = elevation angle

A fade margin calculation for 99.9% availability is illustrated:

$$
\begin{aligned}
A(0.1\%) &= \left[a_1 R(0.1\%)^{b_1} + a_2 R(0.1\%)^{b_2} \right] \ell(R, \alpha) \\
&= \left[0.0196 \times (7)^{1.150} + 0.0297 \times (7)^{1.129} \right] \times 7.6 \\
&= 3.4 \text{ dB}
\end{aligned}
$$

(1.20b)

For this example the frequencies are 14 and 12 GHz; the rain rate for 0.1% in southern Ontario is 7 mm/h; 0°C rain temperature is used; and the elevation angle for the station at Ottawa is 25°.

Small-Signal Gains

The foregoing link equations include the nonlinear effects of the TWTA through inclusion of the backoff terms. During conditions when the TWTA is operated near its linear region, small-signal power gain G_{ss} and small-signal voltage gain g_{ss} are convenient models for quick calculations.

$$\text{output power} = G_{ss} \cdot (\text{input power})$$

(1.21)

$$\text{output voltage} = g_{ss} \cdot (\text{input voltage})$$

(1.22)

1.4 DIGITAL MODULATION AND DEMODULATION

In Figure 1.5 a simplified Earth station is depicted. An external input signal is passed through a baseband processor which can digitize the signal, if necessary, and which combines it with any other inputs to form a digital data stream. The analog-to-digital conversion can be accomplished by a pulse-code modulation (PCM) or delta encoder. Channel encoders are introduced below. The emphasis here is on the relationships among the source processor, modulation and channel coding.

Modems

In Figure 1.5 the modulator (MOD) transforms digital information into a form suitable for transmission. Conventionally, the modulator output spectrum is centred about some intermediate frequency (IF) and has a bandwidth related to the transmission rate R. In the Earth station context, this IF is usually 70 or 140 MHz. The radio frequency (RF) can also be modulated directly if the frequency is consistent with circuit limitations. The IF and RF waveforms are called carriers. The type of modulation used depends upon many criteria, and these are examined in subsequent chapters. In a digital modulator, input bits are used to modify the amplitude, frequency or phase of the carrier either individually or in combination. This mapping determines the power and bandwidth efficiency of the modulation technique.

The modulator output at the IF is shifted to the station output RF by means of an up-converter (U/C) by linear translation. The U/C output is amplified by the HPA according to the power requirement in the uplink. For low power, solid-state amplifiers are becoming available; for medium power, the HPA can be a TWTA; for high power, a TWT driver and a klystron amplifier are used. The station output RF can be in one of many frequency bands; for commercial communications the uplinks are currently at 6 GHz and 14 GHz.

The antenna is connected by means of an orthocoupler* to a low-noise amplifier (LNA) which amplifies the wideband signal received from the satellite. Corresponding downlinks for commercial communications are at 4 and 12 GHz. The remaining blocks in the receiver perform the complementary operation of the transmitter units; the demodulator (DEMOD), controller and decoder are examined in later chapters.

System Aspects and Constraints

The digital communications modem is part of an overall system. Which particular modem is used in a given situation depends on many factors:

1 Bandwidth and power limitations
2 Spectral spreading
3 Resistance to interference
4 Circuit complexity
5 Implementation and availability

The weight and size of a satellite are significant factors in arriving at a cost. It follows that bandwidth and power requirements have an impact on the components and subsystems selected and consequently the cost of the total system. An efficient system design minimizes both bandwidth and power utilization in some optimum mix.

Spectral spreading caused by the TWTA nonlinearity is important and refers to the increase in occupied bandwidth of a signal arising from nonlinear

---

*An orthocoupler allows the simultaneous use of the antenna for transmission and reception of the signal.

processing. This spreading requires filters in the satellite and can cause interference to adjacent repeater bands. A further complication arises from the fact that bandlimited signals cannot have truly constant amplitude, although some filtered modulations are nearly so. The TWTA will distort a signal by means of amplitude-to-phase (AM/PM) and amplitude-to-amplitude (AM/AM) conversions. AM and PM refer to amplitude and phase modulation, respectively.

Signals are subjected to various kinds of interferences, including adjacent channel interference (ACI) and co-channel interference (CCI). ACI within a repeater limits the packing density (proximity of carriers). CCI results in problems with spurious tones or jamming (deliberate interference).

Performance improvement can be attained by increased circuit complexity. In situations where size, power and cost limitations are not severe, increased circuit complexity may be warranted. However, in applications where portability is important, limits are placed on subsystems. An interesting area in modem implementation is the impact of custom large-scale integration (LSI). These microcircuits, called chips, will eventually assist in implementing more sophisticated stations in economic formats.

Although some modem techniques may appear to be attractive, their availability must be evaluated. Most satellite digital modems are not yet off-the-shelf items. A satellite system now employing BPSK (binary phase-shift keying) cannot be suddenly changed over to one using eight-phase PSK. Before a modem is used, sufficient evaluation will be required to verify its performance under a variety of operating conditions. A particular technique can only be accepted if it has a viable implementation. Also, to be usable in operational environments, maintainability must be addressed.

The following are other modem characteristics that are relevant to the system design:

Continuous or Burst Mode   Does the modem have to acquire the signal quickly?

Frequency Range   What constraints are imposed by oscillator instabilities and satellite Doppler shift?

Modem Types   Are a variety of modem types to be used?

Modem Sensitivities   Is it sensitive to environmental factors, such as impulse noise?

Compatibility with Existing Equipment   Is the modem to be retrofitted to other equipment? What constraints are imposed by a retrofit?

Definition of Bandwidth Occupancy

Available E_b/N_0

Implementation Margins

Source and Channel Coding Considerations

In later chapters various modulation techniques are compared; certain assumptions are implicit as to the characteristics of the input to the modulator. If these are not assured, modem performance could degrade or the modem could fail to operate. It is required that the input data not have long strings of ones or zeros. Otherwise, two problems arise:

1 The transmitted spectrum could consist of high-level components at line frequencies, which might contravene regulations or agreements.
2 The lack of sufficient transitions would impede clock recovery by the demodulator.

It is also required that the input data be non-return-to-zero (NRZ). NRZ data have a constant level over a bit duration. Other schemes, which include Miller and Manchester coding, have additional transitions in the data which assist demodulator recovery circuits but at the expense of increased transmission bandwidth.

Finally, the input data must be essentially random. The modem might easily have anomalous behaviour for repeated patterns of data bits.

Another consideration is the type of signal source. If the source is voice, the manner of digitization can be used to advantage. Consider the case where the processor is a delta encoder. This device tries to follow the voice waveform by transmitting a "1" if the signal increases and a "0" if the signal decreases. In coherent demodulation schemes such as coherent phase-shift keying (CPSK), the data are recovered with an ambiguity; the data bits may all be inverted. In the case of the recovered delta-coded voice, the analog voice waveform could be inverted. Fortunately, the ear is not sensitive to this inversion. In other situations, data inversion is unacceptable. Here, the PSK modem requires a kind of coding, called differential coding, to resolve the ambiguity. A penalty in signal-to-noise ratio (SNR) is paid. At low values of SNR corresponding to an error rate of about 10^{-2}, delta-coded voice is intelligible and the ability to operate without differential coding is important.

Where an FEC system is used, there are two modem considerations. If a differential codec is used, errors can occur in pairs. Therefore, the FEC must be able to correct at least two errors. Also, the placement of the differential codec, which can be between the FEC codec and modem at each side, or it can be put prior to the FEC encoder and after the FEC decoder, affects the effective coding gain of the FEC. If the differential processing is outside the FEC and the FEC can handle the ambiguity, then it is required to handle only half of the number of errors of the preceding case.

1.5 MULTIPLE ACCESS

In this section multiple-access concepts are introduced, briefly discussed and compared. Chapters 7 to 10 treat each generic type in some detail.

Multiplexing and Demultiplexing

Concepts of multiplexing have already been introduced and are well known. As an introduction to multiple access, multiplexing is a starting place of comparison and contrast. In telecommunications, to **multiplex** is to combine orthogonally individual signals into a single signal. The types of multiplex are:

Frequency-Division Multiplexing   Signals occupying nonoverlapping frequency bands are added and a specific signal can be recovered by filtering. These are asynchronous techniques. This is FDM.

Time-Division Multiplexing   Signals are compressed into high-speed bursts and combined in nonoverlapping time slots. Recovery of a specific signal is by selection of its time slot and hence requires timing references. These are synchronous operations. This is TDM.

Code-Division Multiplexing   Signals can be given their own unique signatures (identifying characteristics or codes) before combining them in the frequency–time domain. Demultiplexing is by cross-correlation with a known reference signal and signals are usually digital. This is CDM.

Other forms of multiplexing are spatial multiplexing, which is the physical separation of the signal energy (e.g., wires, cables, directional beams); polarization multiplexing, where signals are kept distinct by the type of their electromagnetic radiations (horizontal, verticle polarization); and some others. These forms of multiplexing are used in satellite communications but do not lead to types of multiple access.

Types of Multiple Access

Multiple access is a variant of multiplexing which is specific to satellite communications. It may be defined as any technique by which a number of Earth stations from communications links through one or more satellite RF channels. Multiple access is multiplexing of RF signals in the satellite channel; it is cooperative shared use by multiusers.

Multiple access is necessary to take advantage of the star topology and the wide-area coverage of very wide bandwidth repeaters. Single uplinks seldom have the requirement for the capacity offered by the satellite channel. Also, economic factors are compelling: there is a high product of cost×bandwidth× power×coverage; each repeater must be used effectively. There are three basic froms of multiple access, illustrated in Figure 1.6.

Frequency-Division Multiple Access (FDMA)   This is FDM applied to a satellite repeater. Each uplink RF carrier occupies a defined frequency allocation and is assigned a specific location within a multicarrier repeater. Receiving Earth stations select a desired carrier by RF or IF filtering.

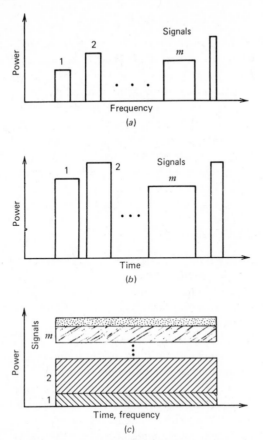

Figure 1.6 Diagramatic presentation of (*a*) FDMA, (*b*) TDMA and (*c*) CDMA. Note the labels on the *x*-axis.

Time-Division Multiple Access (TDMA) This is TDM applied to a satellite repeater. Each uplink burst of transmission is timed so as to pass through the repeater such that bursts are interleaved and do not overlap in time. Receiving stations are synchronized to the network and select the desired bursts for demodulation.

Code-Division Multiple Access (CDMA) Each uplink occupies the same frequency band at the same time, but each signal has its own structure to provide orthogonality. A receiver must have knowledge of this structure and be synchronized before reception can occur.

There are many hybrids and variants of these multiple-access techniques. A type of CDMA is called spread-spectrum multiple access (SSMA) and is a useful form of multiple access. However, not all spread-spectrum systems are multiple-access systems.

Time-Random Multiple Access (TRMA)

TDMA establishes circuits between stations, and in more advanced forms there is circuit switching. All forms of TDMA include some form of control and there is synchronization (provision of timing) of each Earth station to the satellite network. When TDMA is not synchronized or not controlled at all, the class of random-access systems results; many variations were developed in parallel with TDMA and some are discussed in Chapter 10.

Such random access of satellite channels is very useful in packet switching or in applications where direct point-to-point connections are not needed or not possible. For consistency with the other abbreviations, we call this time-random multiple access (TRMA); it is also known as broadcast multiple access and has come to be widely known as packet satellite networks (PSN).

Space-Division Multiple Access (SDMA)

As a point of clarification, when spatial multiplexing is used in satellites, directional beams in the satellite receiver or in the transmitter make possible the reuse of the same frequency band in other repeaters. This is called space-division multiple access (SDMA); it is not multiple access of a repeater but of the *whole satellite*.

It is useful to reflect on what is not multiple access. In a satellite system for distribution of TV, switched distribution does not rate as multiple access, even if the switching is fast, such as for video conferencing. A party-line use of a voice or data service is not multiple access. In general, multiple use or multiple users do not necessarily imply multiple access.

Figure 1.7 illustrates the relative capacities for the four forms of multiple access. In each case, regions of possible operations are shown, since the choice

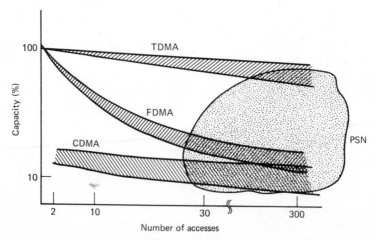

Figure 1.7 Comparison of throughput for a uniform network, versus number of accesses.

of modulation, coding, filtering, performance, control and many other factors will affect the capacity for a given network. TDMA capacity drops off linearly and is most efficient in its use of the nonlinear satellite channel. FDMA drops off more rapidly due to the need for back off to reduce intermodulation noise. CDMA appears very inefficient, primarily because of the self-interference effects of the broadband signals, but CDMA systems do have their advantages. Packet satellite networks (PSN) exhibit their best applications when large numbers of users require intermittent service or when stringent network control is not desirable or not feasible. Elaborations on these are given in the chapters on these multiple-access methods.

Advantages and Disadvantages

FDMA is now predominant in every operating satellite system. It is based on extensive experience with FDM in terrestrial systems which was available when satellite communications started. It is relatively simple to implement but not as simple as random access. FDMA compared to TDMA and CDMA has up to now had lower initial cost and is less demanding of repeater and Earth station parameters. Although FDMA has design adaptibility, FDMA systems exhibit inflexibility to changing requirements over the life of the system. FDM/FM/FDMA and FDM/PSK/FDMA systems have proven themselves in trunking applications with high capacities. Single-channel-per carrier (SCPC), a class of FDMA, is well suited to thin-route applications requiring modest flexibility in providing voice and data services.

TDMA is now attracting the most attention for future satellite systems. For large systems, significantly higher capacities are achieved, as illustrated in Figure 1.7, in comparison to FDMA and CDMA. Flexibility and performance are very attractive, since TDMA systems can be reconfigured, programmed for large changes and demand assigned in real time. TDMA systems must use digital modulation and have higher instantaneous transmission rates; TDMA has the advantages of digital systems, the more notable being compatibility with advancing technologies and the capability of graceful growth. Without doubt, in its first 15 years, TDMA has had much higher initial costs than did FDMA systems. As is demonstrated in Chapter 8, the industry is finding suitable applications and cost-effective TDMA systems for them.

CDMA systems have been around since the first communications satellites, originally for antijamming features and not only for multiple access. In the 1960s, the applications were in military systems requiring security, resistance to interference and low detectability. As is evident in Chapter 9, CDMA techniques are not as efficient as FDMA or TDMA; Figure 1.7 illustrates the point. In the 1970s, there was a rediscovery of CDMA principles, made more attractive by new devices and processing methods, and there are new applications being considered.

Random-access methods for satellites commenced with the Aloha system [8], with many new types now existing and being designed. Random access has its biggest advantage in its simplicity, there being no or little coordination by stations in the network. Random access is a natural choice for packet-switched systems, wherein a burst is treated much like a letter in the mail. This randomness is also the major reason for the lower throughput (defined below), which is the main topic of systems design at present. It is not surprising that there are strong ties between random access and TDMA; several papers have pointed out hybrid and transitional concepts.

Typical System Comparisons

Figure 1.7 compares the throughput against number of accessing stations in a uniform network. Another form of comparison is the capacity measured in total number of telephone channels, as illustrated in Figure 1.8 for a single

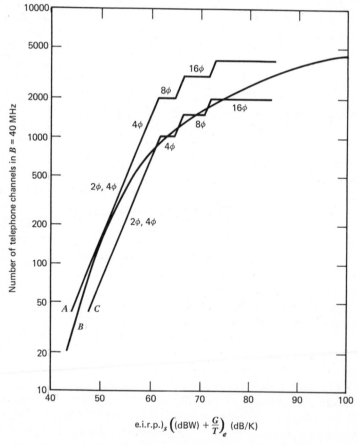

Figure 1.8 Capacities for single access: phases in PSK. Curves: *A*, TDM/PCM/DSI/CPSK; *B*, FDM/FM; *C*; TDM/PCM/CPSK. (Adapted from CCIR, Report 708, Vol. IV, 1978.)

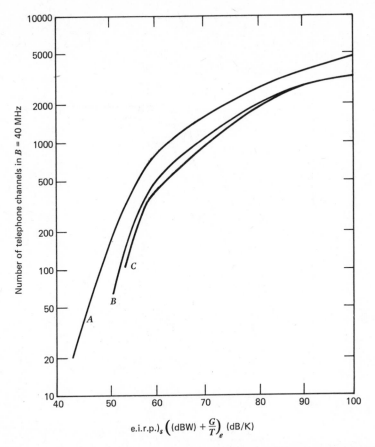

Figure 1.9 *FDMA capacities*. Curves: *A*, FDM/FM, 1 carrier; *B*; FDM/FM/FDMA, 5 carriers; *C*, FDM/FM/FDMA, 10 carriers. (Adapted from CCIR, Report 708, Vol. IV, 1978.)

access, in Figure 1.9 for FDMA and in Figure 1.10 for FDMA and TDMA. These figures are adapted with permission from a CCIR report [9], which presents the detailed assumptions of the comparisons. The assumptions include 10 accessing Earth stations, each with 600 one-way telephony channels—this is a high-capacity network.

1.6 ERROR CONTROL

In this section we introduce the general problem of error control for data transmission through a geostationary satellite. Satellite and terrestrial data channels have different characteristics with respect to propagation delay and noise distribution. For a satellite link, error-control techniques different from

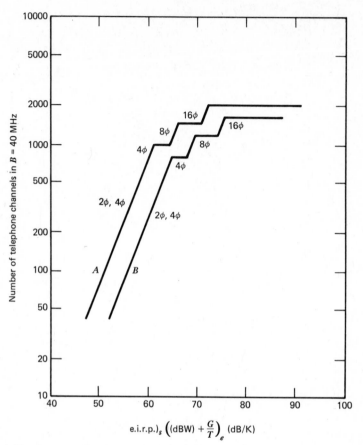

Figure 1.10 Capacities for multiple access: phases in PSK. Curves: A, TDM/PCM/CPSK/TDMA, 10 stations, 10 bursts; B, TDM/PCM/CPSK/FDMA, 10 stations, 10 carriers. (Adapted from CCIR, Report 708, Vol. IV, 1978.)

those used in terrestrial systems may have to be considered. The two principal characteristics of a satellite channel are:

1   The primary additive disturbance is adequately modelled by white, Gaussian noise for which there is independence from symbol to symbol, provided the signalling rate is small enough so that intersymbol interference is negligible.

2   The satellite channel is well described by the discrete, memoryless, binary symmetric channel (assuming hard-quantized detection), which has a bit-error rate much smaller than 10^{-2}.

Error control techniques are divided into three categories: forward-error-control (FEC) schemes, error detection with retransmission (automatic repeat request, ARQ), and hybrid systems that employ both FEC and ARQ techniques.

Forward-Acting Error Control

In a digital communication system using FEC, illustrated in Figure 1.11, the data source generates binary information digits (bits) at the rate R bps. These bits are encoded for the purpose of error protection, where encoding means the addition of redundant bits. The encoder output is a binary sequence at rate R_s bps. The subscript s for encoder output is used in this and other sections on coding. The coding rate r is

$$r = \frac{R}{R_s} \tag{1.23}$$

Since $r < 1$, $R_s > R$ and the transmission rate is higher than the data rate delivered by the source. Equivalently, the introduction of error-control coding requires more capacity; this can be in the form of wider bandwidth in FDMA systems, longer bursts in TDMA systems or a higher "chip" rate in CDMA systems.

The encoded sequence is suitably modulated and transmitted over the noisy channel. At the receiver after demodulation, the decoder attempts to reconstruct the original input to the source and deliver them to the data sink at the rate R bps. The purpose of the decoding operation is the removal of any

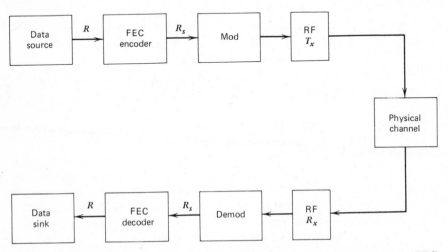

Figure 1.11 Block diagram of a digital link using forward-error correction (FEC). Note the absence of a reverse (feedback) channel.

channel-introduced errors. The advantages of an FEC system are:

1 No reverse channel is required.
2 A constant information throughput efficiency* is attained.
3 A constant overall delay may be obtained (whenever the decoder operates with a constant decoding lag).

The decoding lag (delay) in the decoding operations is frequently constant, although it may be highly variable in some cases, for example in sequential decoding. A constant decoding lag, which will ensure real-time processing, is an important feature for terminal equipment beyond the satellite link, when such equipment derives timing and synchronization from the bit stream. However, where the decoding delay is variable, these variations can be smoothed out by using a buffer at the decoder output. A decoding delay may also require an input buffer to store the incoming data waiting to be decoded. The necessary size of the input and output buffers will depend on the variations, and an overflow of these buffers has different consequences. An insufficiently large output buffer will not maintain a constant output data rate, whereas an input buffer overflow will result in data loss, or even in complete communication breakdown.

Whenever a reverse channel is not available or undesirable, for example in data broadcast applications, FEC must be used. Although readily available in most terrestrial systems, a reverse channel (feedback channel) may be a costly proposition for satellite applications.

The disadvantages of FEC include the following:

1 A relatively moderate throughput efficiency is obtained; this parameter decreases substantially when powerful codes must be used.
2 The selection of an appropriate error-correcting code and its decoding algorithm might be a difficult task; this is particularly true when high reliability of the data is required.
3 The reliability of the received data is very sensitive to any degradation in the channel transmission conditions.

To make an intelligent selection of the coding and decoding schemes, detailed knowledge of the error statistics of the channel is required. Within a block length (block coding) or a constraint length (convolutional coding), most channels exhibit a mixture of independent errors and burst error patterns. For these channels the selection of a code is a rather difficult task. A code suitable for independent errors will fail for long bursts of errors, and a code designed for a burst channel does not work properly when random errors occur between

---

*The **throughput efficiency** is defined as the ratio of the number of information bits delivered to the total number of bits transmitted. In FEC systems, it remains constant at the coding rate r.

bursts. Some codes correct both kinds of errors. Therefore, FEC for compound (tandem) channels, such as in terrestrial links, will result in occasional decoding failures (undetected, or detected but uncorrectable errors). The user must be able to tolerate these failures. To reduce the number of failures, the more powerful codes must be chosen; that is, the amount of redundancy in the codes must be high. Naturally, this lowers the coding rate and increases the required capacity. If this is obtained through wide bandwidth and higher signalling speed, the decoding problem becomes further complicated.

Channel degradation increases the undetected or detected errors in the decoder output, because it increases the occurrence of error patterns. Consequently, during periods of high noise or interference in the channel, an FEC system will deliver unreliable or garbled data to the user.

Another point of interest is the cost of an FEC error-control system. This cost is related to the coding rate r. Error correction requires more redundant symbols than error detection; the greater the redundancy, the higher are the encoder and decoder costs. Furthermore, the longer the constraint length or block length of the code, the larger are the amount of decoder storage and decoding time delay.

Error-Detection and Retransmission (ARQ) Systems

A digital communications system employing simple ARQ is shown in Figure 1.12. In these systems the data to be transmitted are organized in blocks or packets of length N bits, divided into K information bits and $N-K$ bits for control and overhead. Block coding is used with a sufficient number of redundant bits to achieve the required error-detection capability. Basically, no error correction is performed by the decoder, but whenever an error is detected in a block, a retransmission of that block is requested through a reverse channel. The transmitter is informed of whether a block has been correctly or incorrectly received by an acknowledgement (ACK) or nonacknowledgement (NACK) control signal sent over the reverse channel. If an ACK signal is received, a new block is transmitted, whereas if the control is NACK, the same block is transmitted again. Buffer storage must therefore be provided to keep a copy of each transmitted block until the corresponding ACK is received. With this procedure a block is delivered to the user only after it appears to be error free.

Clearly, then, the important measures of performance are the undetected error probability P_u which is typically very small ($\ll 10^{-10}$), and the throughput efficiency of the system.

The principal advantages of an ARQ error control are:

1 Very low undetected error probability
2 Effectiveness on any channel
3 Encoder/decoder simplicity

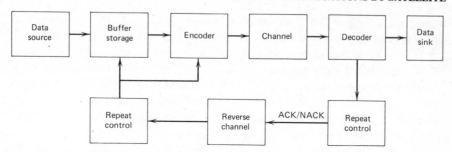

Figure 1.12 Block diagram of a digital link using ARQ, emphasizing the reverse channel.

The low undetected error probability is the consequence of the power of the error-detecting codes. Moreover, all information blocks delivered to the user can be accepted with equally high confidence, even during periods of high noise, high interference or other poor channel quality.

The information throughput of an ARQ system depends strongly on the number of requested retransmissions, that is, on channel quality, and to a lesser degree on the actual error-detecting code utilized.

The selection of a code is far simpler for ARQ systems than it is for FEC systems. In fact, several error-detecting block codes have been standardized for use in ARQ systems with a minimum amount of hardware [10]. Because a code used for error detection is not very sensitive to the actual error patterns, it can detect the vast majority of the error patterns; it does not matter very much how errors occur on the channel. Consequently, unlike FEC systems, the use of ARQ error control is effective on most channels.

The cost of an ARQ system is substantially lower than that of an FEC system. This is due to the fact that an error pattern need not be determined as in FEC systems: only the presence of an error pattern has to be detected, which is a simpler operation. However, storage in an ARQ system is required at the transmitter; this storage may present problems for long transmission delays such as those encountered in satellite links.

The disadvantages of ARQ systems include the following:

1 A return channel is required.
2 A variable decoding delay is possible.
3 The data source must be controllable and/or buffering must be provided.

One of the most serious disadvantages of ARQ systems is the requirement for a reverse channel. For some communications systems, such as a telephone data system, a reverse channel is readily available. However, for some other systems, providing a reverse channel may be a prohibitive proposition, or even an impossibility.

The occurrence of retransmissions induces a decoding delay, which is measured as the time between the first arrival of a block at the decoder and its delivery to the user. A multiple transmission of the same block increases the decoding delay and reduces the throughput efficiency. In addition to the decoding delay, the round-trip propagation delay may be important and further lowers the information throughput. For a satellite channel, the round-trip or two-hop delay is of the order of 500 ms. The use of brute-force ARQ systems on such channels will drastically reduce the throughput.

Finally, during retransmission periods the data source may have to be interrupted to prevent an undue accumulation and possible loss of the information bits at the encoder input. Therefore, provisions must be made either to control the source and/or to provide buffer storage for the blocks waiting to be transmitted. Depending on the nature of the data source, the controllability and buffering requirements may present a problem. It is not practical for real-time services, such as digital voice.

Hybrid ARQ and FEC

From the discussions on FEC and ARQ techniques, we observe that ARQ techniques provide very good error performance independently of the channel quality. However, as the channel becomes noisier, the requests for retransmissions increase, leading to reduced throughput. On the other hand, FEC techniques provide a constant throughput regardless of the channel quality, but the error performance will fall when the channel degrades.

In situations where the channel error rate is too high to guaranty sufficient throughput using ARQ and where the required system error performance is too severe to be achieved by FEC alone, a combination of FEC and ARQ systems may be attractive. A hybrid FEC/ARQ system consists of an FEC system contained within an ARQ system. In principle, such systems would combine the advantages of both techniques. The inner FEC system corrects many channel errors and hence reduces the requests for retransmissions, while the outer ARQ system provides a very low probability of undetected error.

The overall coding rate of such a hybrid system is the product of the coding rates of the error-correcting code and the error-detecting code. In well-designed hybrid schemes the reduction of the coding rate introduced by the FEC system is more than compensated by the reduced retransmission requests, yielding a larger throughput than does ARQ alone.

Hybrid FEC/ARQ techniques are well suited to data transmission systems that demand both high throughput and very low undetected error probability: satellite data links, satellite and terrestrial data links in tandem. Further discussion and examples of applications of hybrid FEC/ARQ systems over satellite channels are given in Chapter 11.

1.7 A PERSPECTIVE ON SATELLITE COMMUNICATIONS

Readers may be interested in the following perspective on satellite communications. See also Ref. 11.

Era One—Repeaters in Orbit

We are now in era one, which can be identified as having repeaters in orbit. The majority of commercial satellites are geostationary and have 12, 16 or 24 frequency-translating, amplifying repeaters, all of similar design. There are several generations in this era, having some of the following characteristics: 6/4-GHz or 14/12-GHz bands; spin- or three-axis stabilized; global, spot and multiple beams.

There are a number of families (living or conceived!) which span the era and coexist:

> INTELSAT I, II, III, IV and IV-A, V
> MOLNIYA I and II; ORBITA; STATIONAR and STATIONAR-T
> ANIK A, B, C, and D
> WESTAR; SATCOM; COMSTAR; SBS
> BS of Japan

and their contemporary experimental satellites:

> ATS 1, 3, 5 and 6; the LES series
> SYMPHONIE A and B
> CTS (HERMES); OTS; SIRIO
> CS of Japan

Each family has one, two, or three members, such as ANIK-A F1, F2, F3; some are retired or deceased and only a few satellites had launch failures or difficulties.

Era Two—Processors in Orbit

Era two might be characterized as having processors in orbit, with the first generation to have satellite-switched TDMA. Channelized repeaters already bridge the two eras; remodulation, regeneration, specialized antenna systems (flying spot beam) and antenna farms will be on board in following generations. There will be intersatellite links at high bit rates and we can expect tailored designs for special services.

The family of Advanced WESTAR is an example of this era that satellite communications is entering.

REFERENCES

1 CCIR, *Recommendations and Reports of the CCIR, 1978*, Vol. IV, "General Considerations Relating to the Choice of Orbit Parameters in the Fixed Satellite Service", Report 206-3, Geneva, 1978, pp. 8–10.

2 CCIR, *Recommendations and Reports of the CCIR, 1978*, Vol. IV, "Characteristics of Some Typical Experimental and Operational Communication Satellite Systems", Report 207-4, Geneva, 1978, pp. 10–18.

3 W. L. Pritchard, "Satellite Communication—An Overview of the Problems and Programs", *Proc. IEEE*, Vol. 65, Mar. 1977, pp. 294–307. Also in ref. 11.

4 *Abbreviations, Acronyms, and Symbols Common to Communication and Broadcast Satellite Systems*, compiled by Communication and Broadcast Satellite Systems Committee of the IEEE Aerospace and Electronic Systems Society, Aug. 1977.

5 R. L. Olsen, D. V. Rogers and D. B. Hodge, "The aR^b Relationship in the Calculation of Rain Attenuation", *IEEE Trans. Antennas Propag.*, Vol. AP-26, Mar. 1978, pp. 318–329.

6 W. L. Nowland, R. L. Olsen and I. P. Shkarofsky, "Theoretical Relationship between Rain Depolarization and Attenuation", *Electron. Lett.*, Vol. 13, Oct. 1977, pp. 676–678.

7 CCIR, *Recommendations and Reports of the CCIR, 1978*, Vol. V, "Propagation Data Required for Space Telecommunication Systems", Report 564-1, Geneva, 1978, pp. 219–239.

8 N. Abramson, "Packet Switching with Satellites", in *Proc. AFIPS Conf.*, Vol. 42, June 1973.

9 CCIR, *Recommendations and Reports of the CCIR, 1978*, Vol. IV, "Methods of Modulation and Multiple Access", Report 708, Geneva, 1978, pp. 109–135.

10 J. E. McNamara, "Technical Aspects of Data Communications", Digital Equipment Corporation, Bedford, Mass., 1977.

11 H. L. Van Trees (Ed.), *Satellite Communications*, IEEE Press, New York, 1979. Also distributed by Wiley, New York.

PROBLEMS

1 Calculate the approximate distance between Earth and Moon.

2 Verify (1.2), (1.4a) and (1.4b).

3 **(a)** State assumptions on available hardware at 4 GHz and performance requirements.

 (b) Calculate the receiver temperature.

 (c) Calculate the required antenna size for reception of 1 Mbps.

 (d) What other factors should be considered?

4 **(a)** For the downlink of Problem 3, calculate the probability of a word error in a 6-bit telex service.

 (b) What is the performance expressed in error-free seconds (EFS)?

5 **(a)** For 99.98% availability, calculate the required fade margin using (1.20) for a 6/4-GHz link. (Use London–New York through INTELSAT IV.)

 (b) For the same assumptions, calculate the rain losses and the total C/N_0.

 (c) Compare to dry weather conditions and to the result for part (a).

PART 1

Modulation Techniques

CHAPTER 2

Error-Rate Performance of Digital Modulation Techniques

The commonly used measure of performance for digital communications systems is the relationship between bit-error rate and signal-to-noise ratio (SNR). The SNR is usually expressed as the ratio of the energy per bit to the noise spectral density (E_b/N_0). Generally speaking, the desired modulation technique will be the one that satisfies the various system constraints and requires the least value of E_b/N_0 for a specified threshold error rate.

In this chapter a framework is given for evaluating the error-rate performance of modulation techniques. The source of interference is assumed to be additive white Gaussian noise. The performances of the more common signal formats that might be proposed for or are currently used in digital satellite communications systems are reviewed. These techniques include binary and four-phase or quaternary phase-shift keying (BPSK and QPSK, respectively), offset-keyed QPSK (OK-QPSK), frequency-shift keying (FSK) and a specific form of FSK which is known as either minimum-shift keying (MSK) or fast-frequency-shift keying (FFSK). In addition, results for M-ary PSK (MPSK) will be provided.

2.1 BASIS OF ERROR-RATE PERFORMANCE EVALUATION

A digital communications system transmits waveforms which represent numerical values. These values may be based upon an analog source. For example, a voice waveform may be sampled and the sample values quantized to one of a number of discrete values. These discrete values are encoded according to some rule, with a series of binary digits representing the encoded value. The function of the digital system is to reproduce the series of binary digits. Note the distinction between the analog and digital systems. In an analog system waveforms are reproduced and the difference between the transmitted and recovered waveforms is compared. A fidelity criterion such as percent distortion is used to compare various analog systems. The digital system reproduces discrete values. If the transmitted values are compared with the received values, the number of incorrectly detected values is a figure of merit for the system. The fewer errors made, the better the system is. Hence in digital systems the error rate (number of errors/number of values sent) is the measure of performance.

A digital modulator produces an output signal which is usually sinusoidal. This sinusoid may be changed in amplitude, phase or frequency, depending upon which source symbol is at the input to the modulator.

Digital systems can be compared on the basis of ideal error rate. In comparing digital satellite communications systems the ideal error rate is used to ascertain the best performance attainable. Various implementation degradations can then be determined either by analysis, computer simulation or experiment to give the realizable performance. The system designer can then select Earth station parameters so as to provide sufficient margin to assure this performance.

Inherent in the term "ideal error-rate conditions" are the following assumptions:

1   No phase or frequency uncertainties are introduced by Earth station or satellite oscillators.
2   No error-correcting codec is used.
3   Carrier and clock recovery is perfect, so that there is no recovered carrier phase or clock jitter.
4   The only source of errors is the addition of random Gaussian noise by the channel.

Prior to discussing specific modulations, a framework for determining performance which is common to these modulations is discussed briefly.

2.2   MAXIMUM LIKELIHOOD DETECTION

Figure 2.1 depicts the ideal system block diagram. Every T seconds the data source produces a message symbol. The transmitter produces one of a given set of waveforms (say M in total) corresponding to each symbol. The waveform

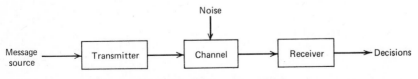

Figure 2.1   Ideal system model.

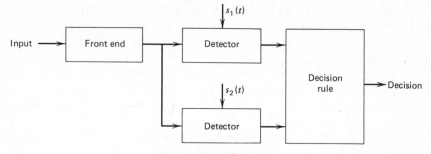

Figure 2.2   Receiver.

itself is arbitrary. In communications systems it is usually a pulse of carrier (i.e., many cycles of a sinusoid). The channel adds noise which is assumed flat over the signal spectrum with double-sided noise power spectral density $N_0/2$ W/Hz. The receiver uses an established decision rule to estimate which source symbol was actually sent.

Figure 2.2 depicts the receiver. The signal at the receiver input is given by

$$r(t) = s(t) + n(t)$$

where $s(t) =$ transmitted waveform
$\quad n(t) =$ noise added

The received signal corrupted by noise is compared in some manner with the set of possible waveforms that could have been sent. A decision circuit produces an estimate of the transmitted waveform.

The maximum likelihood receiver is a detector/decision configuration which, as the term implies, determines to which of the possible set of waveforms transmitted the received waveform is closest. In other words, given an observed waveform, it maximizes the likelihood that a certain waveform was sent.

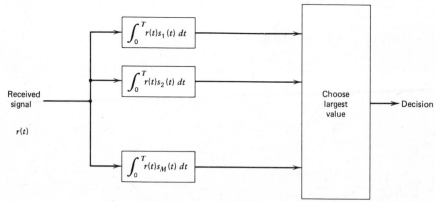

Figure 2.3   Optimum receiver.

It can be shown [1] that the optimum receiver is as depicted in Figure 2.3. It consists of M cross-correlators whose outputs are integrated over the symbol duration. The outputs of the cross-correlators are compared and the largest output indicates which of the M waveforms was the likeliest one sent.

2.3 MATCHED FILTER

The optimum receiver consists of a bank of cross-correlators wherein the received waveform is compared with locally generated versions of the transmitted waveforms. Another approach is to view the process in terms of a filtering operation.

The output of a filter is the convolution integral

$$u(t)=\int_0^t h(t-\tau)r(\tau)\,d\tau$$

where at $t=T$,

$$u(T)=\int_0^T h(T-\tau)r(\tau)\,d\tau$$

Here

$$h(t)=\text{filter impulse response}$$

$$r(t)=\text{received waveform}$$

The term $u(T)$ is equivalent to the integrated output of the cross-correlator if

$$h(T-\tau)=s(\tau)$$

namely

$$u(T)=\int_0^T s(\tau)r(\tau)\,d\tau$$

as shown in Figure 2.4.

This filter is then a time inverse of the correlation signal $s(t)$ and is called a matched filter because it is best matched to the incoming signal in terms of optimum reception. Therefore, in the optimum receiver a series of matched filters could be substituted for the correlators.

One-Shot Receiver

Assume that a pulse is transmitted and it is not influenced by adjacent pulses. The processing of this pulse is performed by the so-called one-shot receiver. Consider the case where the pulse is a sinusoidal pulse of duration T seconds.

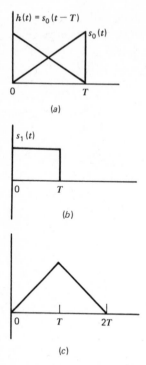

Figure 2.4 Matched filter convolution operation. (*a*) Waveform and matched impulse response; (*b*) rectangular pulse; (*c*) matched filter output for $s_1(t)$.

The matched filter has a rectangular envelope impulse response which is the time inverse of the tone burst. The filter output is then a triangular pulse which reaches a maximum at $t = T$. It is precisely this output at $t = T$ that is used to form an estimate of the signal transmitted.

A matched filter as described (i.e., a tone-burst type of impulse response) requires a properly shaped bandpass filter. An alternative approach is to use a correlator with the incoming signal multiplied by a local sinusoidal reference phase-locked to the transmitter's reference followed by a bit integrator as shown in Figure 2.5. Normally, the expression for the received signal would include an arbitrary carrier phase angle θ. Because the local reference is assumed to be phase-locked, this angle can be set to zero. Unless stated otherwise, phase lock will be assumed. The establishment of carrier phase synchronization is not trivial and is a major modem design problem considered in Chapter 5.

Consider the binary case where one of two waveforms $s_1(t)$ and $s_2(t)$ is sent. It will be assumed that these waveforms are of equal energy, that is,

$$E_b = \int_0^T [s_1(t)]^2 \, dt = \int_0^T [s_2(t)]^2 \, dt$$

where each waveform is of duration T seconds.

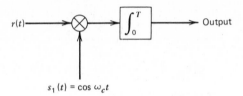

Figure 2.5 Correlator for tone burst.

Assume that a correlator exists at the receiver for each waveform. The correlator outputs assuming that $s_1(t)$ was sent, are given by

$$u_1 = \int_0^T r(t)s_1(t)\,dt = \int_0^T [s_1(t)]^2\,dt + \int_0^T n(t)s_1(t)\,dt$$

$$u_2 = \int_0^T r(t)s_2(t)\,dt = \int_0^T s_1(t)s_2(t)\,dt + \int_0^T n(t)s_2(t)\,dt$$

where $n(t)$ is Gaussian noise. The noise has an autocorrelation given by

$$\mathcal{R}(\tau) = \frac{N_0}{2}\delta(\tau)$$

where $\delta(\tau)$ is the Dirac delta function.
   If $s_1(t)$ was sent, then

$$D = u_1 - u_2$$

should be greater than zero. If not, an error will be made. Let

$$\rho = \text{correlation coefficient}$$

$$= \frac{1}{E_b}\int_0^T s_1(t)s_2(t)\,dt$$

Then

$$D = u_1 - u_2$$

$$= \int_0^T [s_1(t)]^2\,dt - \int_0^T s_1(t)s_2(t)\,dt + \int_0^T n(t)[s_1(t) - s_2(t)]\,dt$$

$$= E_b - \rho E_b + N$$

$$= E_b(1 - \rho) + N \tag{2.1}$$

where

$$N = \int_0^T n(t)[s_1(t) - s_2(t)]\,dt$$

is a noise term. It can be shown that N is Gaussian with zero mean and variance

$$\sigma^2 = E[N^2] = \int_0^T \frac{N_0}{2} [s_1(t) - s_2(t)]^2 \, dt$$

$$= E_b(1-\rho)N_0$$

where $E[\cdot]$ denotes the expected value.

The correlator difference given in (2.1) can be shown to be Gaussian with mean value $E_b(1-\rho)$ and variance σ^2. The probability density function (pdf) of a Gaussian random variable in general is given by

$$p(x) = \frac{1}{\left(\sqrt{2\pi}\right)\sigma_1} \exp\left[-\frac{(x-\bar{m})^2}{2\sigma_1^2} \right]$$

where $\bar{m}$ = mean value
σ_1^2 = variance

The probability of error, that is, the probability that D is less than zero when $s_1(t)$ is sent, is given by

$$P_b = \int_{-\infty}^0 p(t) \, dt$$

$$= \frac{1}{\left(\sqrt{2\pi}\right)\sigma} \int_{-\infty}^0 \exp\left[\frac{-[t - E_b(1-\rho)]^2}{2\sigma^2} \right] dt$$

$$= \frac{1}{2} \mathrm{erfc}\left[\sqrt{\frac{E_b(1-\rho)}{2N_0}} \right] \tag{2.2}$$

where erfc $(\cdot)$ is the complementary error function given by

$$\mathrm{erfc}(x) = \frac{2}{\sqrt{\pi}} \int_x^\infty e^{-y^2} \, dy$$

For large x,

$$\mathrm{erfc}(x) \simeq \frac{1}{x\sqrt{\pi}} \exp(-x^2)$$

The result in (2.2) also applies when $s_2(t)$ is sent and so represents the total error expression. The larger the argument of erfc, the smaller is the probability of error.

A function often used is the Q function, which is related to the erfc function by

$$\text{erfc}(x) = 2Q\left(\sqrt{2}\,x\right)$$

where

$$Q(x) = \frac{1}{\sqrt{2\pi}} \int_x^{\infty} e^{-y^2/2}\, dy$$

The integrals for these expressions require numerical evaluation. Approximations to the integrals are given in Wozencraft and Jacobs [2] (see also Appendix C).

To repeat, the expression for the error probability is

$$P_b = \frac{1}{2}\,\text{erfc}\left[\sqrt{\frac{E_b(1-\rho)}{2N_0}}\,\right]$$

Certain cases are of particular interest.

CASE 1   ANTIPODAL SIGNALLING   For antipodal signalling, $s_1(t) = -s_2(t)$ and $\rho = -1$. In this case the pulse shapes are identical and only the signs are different. Here

$$P_b = \frac{1}{2}\,\text{erfc}\left[\sqrt{\frac{E_b}{N_0}}\,\right]$$

CASE 2   ORTHOGONAL SIGNALLING   Orthogonality refers to the similarity between waveforms and indicates that the waveforms are unique, having independent information in each. For this case

$$\rho = \frac{1}{E_b} \int_0^T s_1(t)s_2(t)\, dt = 0$$

and

$$P_b = \frac{1}{2}\,\text{erfc}\left[\sqrt{\frac{E_b}{2N_0}}\,\right]$$

Note that in this expression the signal-to-noise term is $\frac{1}{2}$ or 3 dB less than for the antipodal case. In general, the correlation coefficient lies between $-1 \leq \rho \leq 1$ and best results are obtained for $\rho = -1$.

Implications of Unmatched Filtering

The results given for the probability of error apply to matched filter detection. The use of matched filters is not always possible because of practical implementation reasons. If a suboptimum or unmatched filter is used, it can be expected that a decrease in the filter output at $t = T$ will occur, thus reducing the signal-to-noise ratio term and thereby increasing the probability of error. For a rectangular pulse, an integrate-and-dump filter is optimum. The integrate-and-dump filter integrates the input rectangular signal from 0 to T, its output is sampled and it is then immediately quenched so as to avoid residual energy from affecting a decision on the next pulse. A single-pole RC filter could be used instead but with a lower effective SNR.

2.4 IDEAL ERROR-RATE PERFORMANCE

The probability of error expressions derived from a statistical analysis are not restricted to NRZ pulses and can be applied to specific modulation techniques.

Binary Phase-Shift Keying (BPSK)

BPSK transmission entails sending a sinusoid which has one of two phase values over each data interval

$$s(t) = \sqrt{2P}\cos[\omega_c t + \phi_k], \qquad \phi_k = 0 \quad \text{or} \quad \pi$$

$$= \left(\sqrt{2P}\right) a_k \cos \omega_c t$$

where $a_k = \pm 1$.

In terms of the previous terminology,

$$s_1(t) = \sqrt{2P}\cos \omega_c t$$
$$s_2(t) = -s_1(t)$$

where the power in the continuous waveform is $P = E_b / T$. BPSK is a form of antipodal signalling and has an error rate given by

$$P_{\text{BPSK}} = \frac{1}{2}\text{erfc}\left[\sqrt{\frac{E_b}{N_0}}\right]$$

Quadrature Phase-Shift Keying (QPSK or $4-\phi$ PSK)

QPSK is a technique that provides the same error-rate performance as antipodal signalling but also halves the bandwidth occupancy as compared with BPSK. QPSK transmission uses two input bits at a time. These input bits could be from parallel input channels or be consecutive even and odd bits from a serial input. Each bit can be a logical 0 or a 1, and therefore four combinations corresponding to four output carrier phases are possible. The modulator output is given by

$$s(t)=\sqrt{2P}\cos(\omega_c t-\phi_k)$$

where ω_c is the angular carrier frequency and ϕ_k is one of four possible phases. If standard trigonometric identities are used, then

$$s(t)=\sqrt{2P}\left[\cos\phi_k\cos\omega_c t+\sin\phi_k\sin\omega_c t\right]$$

The two terms on the right of the expression for $s(t)$ consist of a constant carrier term ($\cos\omega_c t$ or $\sin\omega_c t$), each multiplied by a factor related to the information bits. By selecting $\phi_k=\pm45°,\pm135°$ $s(t)$ becomes

$$s(t)=\sqrt{P}\left[\pm\cos\omega_c t\pm\sin\omega_c t\right]$$

This expression indicates that bits can be selected independently; say, even bits determine the $\cos\omega_c t$ sign and odd bits the $\sin\omega_c t$ sign. Here a "+" sign results from a logical 1 and a "−" sign from a logical 0.

It can now be recognized that $s(t)$ is the summation of two BPSK signals, both operating at the same transmission rate but with quadrature (90° offset) carriers. These two BPSK signals are independent. If the serial input data rate is R, then each of the BPSK signals is phase shifted at a rate of $R/2$ because only half of the bits appear in each. The two components are referred to as I and Q or in-phase and quadrature channels. The question is: What error rate can be expected at the receiver?

Recall that the ideal receiver multiplies the signal input by a reference signal phase locked to the input carrier and then filters the resulting output baseband pulse waveform. The QPSK receiver can be visualized as consisting of two channels: one that multiplies the input by $\cos\omega_c t$ and one that multiplies the input by $\sin\omega_c t$, with each of the two multiplier outputs filtered and decision made on each. The detection of bits in one channel will be independent of the other channel (no crosstalk) if the other channel adds zero voltage to the channel integrator output. To show that this is so for QPSK, select an arbitrary pair of signs in $s(t)$, for example

$$s(t)=\sqrt{P}\left[a\cos\omega_c t+b\sin\omega_c t\right],\qquad a=\pm1,\quad b=\pm1$$

The receiver integrator output for the in-phase channel is

$$y(T)=\frac{1}{T}\int_0^T s(t)\cos\omega_c t\,dt$$

$$=\sqrt{P}\left[\frac{1}{T}\int_0^T a\cos^2\omega_c t\,dt+\frac{1}{T}\int_0^T b\sin\omega_c t\cos\omega_c t\,dt\right]=\frac{a\sqrt{P}}{2}$$

This output value assumes that double-frequency components are suppressed. Note that $\sin \omega_c t$ and $\cos \omega_c t$ are orthogonal and that the second integral is equal to zero.

Independent signalling occurs in each channel, and each channel will have an error rate equal to that of BPSK. Note that while the total output power per channel is half that of a BPSK signal, the effective receiver noise bandwidth is halved because of the dividing of the input bit stream into I and Q components. Therefore, the same E_b/N_0 is maintained.

Offset QPSK

Offset QPSK (OK-QPSK) is a form of QPSK in which the digits in the quadrature channels have a relative delay in their transitions. If the serial input data have duration T, then the I and Q data will each have duration $2T$. The relative delay between channels is T. In conventional QPSK the transitions are coincident. The purpose of this delay is to restrict the carrier phase transitions from having 180° phase transitions. When filtered, the OK-QPSK will have less envelope fluctuation compared with QPSK.

In the unfiltered case, the introduction of a delay has no performance effect and OK-QPSK has the same error rate as does conventional QPSK.

Binary Frequency-Shift Keying (Coherent Detection)

Let the waveforms be one of two tones:

$$s_1(t) = \sqrt{2P} \, \sin \omega_1 t$$
$$s_2(t) = \sqrt{2P} \, \sin \omega_2 t$$

The correlation coefficient is then given by

$$\rho = \frac{1}{E_b} \int_0^T s_1(t) s_2(t) \, dt$$

$$= \frac{\sin(\omega_2 - \omega_1)T}{(\omega_2 - \omega_1)T} - \frac{\sin(\omega_2 + \omega_1)T}{(\omega_2 + \omega_1)T} \tag{2.3}$$

The carrier angular frequency is given by

$$\omega_c = \frac{\omega_2 + \omega_1}{2}$$

Equation (2.3) can be rewritten in terms of ω_c as

$$\rho = \frac{\sin(\omega_2 - \omega_1)T}{(\omega_2 - \omega_1)T} - \frac{\sin 2\omega_c T}{2\omega_c T} \tag{2.4}$$

Under the assumption that either (1) $2\omega_c T \gg 1$ or (2) $2\omega_c T = k\pi$, where k is an integer, (2.4) becomes

$$\rho = \frac{\sin(\omega_2 - \omega_1)T}{(\omega_2 - \omega_1)T} \qquad (2.5)$$

Note that the second assumption indicates that an integer number of quarter cycles of the carrier frequency is contained in one bit interval.

The FSK correlation coefficient given in (2.5) is plotted in Figure 2.6 [3] as a function of $(\omega_2 - \omega_1)T$. The minimum value of ρ occurs at $\rho = -\dfrac{2}{3\pi}$ for

$$h = \frac{(\omega_2 - \omega_1)T}{2\pi} = (f_2 - f_1)T = 0.715$$

where h is referred to as the modulation index.

The preceding results indicate that binary coherent FSK is not as efficient as BPSK as to error-rate performance. The best that FSK can offer is a correlation coefficient of $\rho = -\dfrac{2}{3\pi}$ compared with $\rho = -1$ for BPSK.

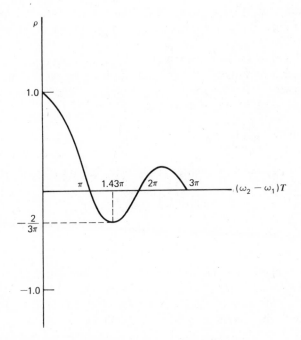

Figure 2.6 FSK correlation coefficient. (From Ref. 3; by permission of John Wiley & Sons, Inc.)

Fast Frequency-Shift Keying (FFSK)

The minimum value of the term $(\omega_2 - \omega_1)T$ for which orthogonal signalling can take place (see Figure 2.6) is

$$(\omega_2 - \omega_1)T = \pi$$

or

$$h = (f_2 - f_1)T = 0.5$$

With $h = 0.5$, the correlation coefficient $\rho = 0$. This would indicate that for this modulation index the best error rate performance will be 3 dB poorer in terms of E_b/N_0 than for BPSK. This is true if matched filtering is performed over a 1-bit interval. There is a modulation technique with $h = 0.5$ for which the same error-rate performance can be attained as for BPSK.* It is called fast frequency-shift keying (FFSK) and is also referred to as minimum-shift keying (MSK). These names arose as follows:

FFSK   The adjective "fast" is used because more bits per second can be transmitted in a given channel bandwidth compared with BPSK. This is elaborated upon in Section 4.5.

MSK   minimum-shift keying refers to the minimum modulation index ($h = 0.5$) for which orthogonal signalling occurs.

The modulation index h gives the separation between transmitted tones. For FFSK this separation is exactly half the bit rate. As will be seen, FFSK can be represented as a special type of offset QPSK where, rather than rectangular pulses, half-sinusoids are used.

Noncoherent Detection of FSK

Noncoherent FSK refers to the detection of the signalling tones

$$s_1(t) = \sqrt{2P} \, \sin(\omega_1 t + \theta)$$

$$s_2(t) = \sqrt{2P} \, \sin(\omega_2 t + \theta)$$

without knowledge of the carrier phase angle θ. The angle is assumed to be uniformly distributed over $(0, 2\pi)$. Detection is accomplished by passing the received signal through parallel demodulator paths. One path includes a bandpass filter with center frequency ω_1, followed by an envelope detector

---

*Actually, differentially encoded BPSK. This terminology is explained in Section 3.1.

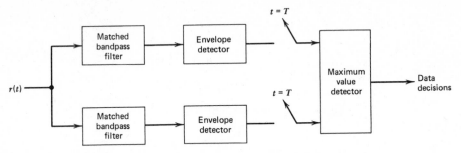

Figure 2.7 Noncoherent FSK receiver.

which in turn is followed by a filter matched to the expected envelope waveform. The second path is an equivalent processor for a signal at ω_2. The matched filter outputs are compared at $t=T$ and the largest output is used to decide which tone was sent. It can be shown [3] that an equivalent circuit is as shown in Figure 2.7, where a bandpass matched filter performs combined bandpass selection and matched filtering. Note that the phase angle is lost in envelope detection.

The envelope detectors are nonlinear and consequently the output noise is non-Gaussian. For the case of orthogonal signalling ($\rho=0$) it can be shown [3] that the error rate is given by

$$P_{\text{NCFSK}} = \tfrac{1}{2}e^{-E_b/2N_0}$$

Differential Detection of PSK

A differential PSK (DPSK) receiver avoids the need for provision of a phase-locked carrier as is the case for a coherent BPSK demodulator. The DPSK signal is an encoded BPSK waveform. A BPSK transmitter sends one of two signals:

$$s_1(t)=\sqrt{2P}\,\cos\theta_1(t), \qquad 0\leq t\leq T$$

$$s_2(t)=\sqrt{2P}\,\cos\theta_2(t), \qquad 0\leq t\leq T$$

where $\theta_1(t)=\omega_c t+\theta$ and $\theta_2(t)=\omega_c t+\pi+\theta$ and θ is an arbitrary phase. However, information is encoded as the difference in phase between consecutive transmissions:

Logical 0

$$q_1(t)\equiv\{s_1(t),s_1(t)\}, \qquad 0\leq t\leq 2T$$

$$q_2(t)\equiv\{s_2(t),s_2(t)\}, \qquad 0\leq t\leq 2T$$

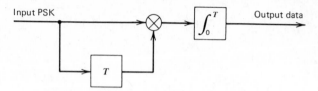

Figure 2.8   Differential detection of PSK.

Logical 1

$$q_3(t) = \{s_1(t), s_2(t)\}, \qquad 0 \le t \le 2T$$

$$q_4(t) = \{s_2(t), s_1(t)\} \qquad 0 \le t \le 2T$$

where $\{s_i(t), s_j(t)\}$ denotes $s_i(t)$ followed by $s_j(t)$. Note that information resides in the transmitted signal $\{q_i(t), i=1,4\}$, which is $2T$ in duration. Each symbol then has energy $2E_b$. The correlation coefficient for any pair of symbols $\{q_i(t), q_j(t); i \ne j\}$ is

$$\rho_{ij} = \frac{1}{2E_b} \int_0^{2T} q_i(t) q_j(t)\, dt$$

Because $s_1(t) = -s_2(t)$ it can be shown that $\rho_{ij} = 0$ and therefore the signal set is orthogonal with an arbitrary phase θ. It can be shown [3] that decoding of DPSK corresponds to noncoherent envelope detection (see noncoherent FSK) of orthogonal signals except that the energy per bit is twice as great. The DPSK error rate is therefore

$$P_{\text{DPSK}} = \tfrac{1}{2} e^{-E_b/N_0}$$

The bit information can be recovered by comparing the last half of the received symbol with the first half. If the symbols are the same, then a logical 0 was sent; whereas if the they are different, a logical 1 was sent. Figure 2.8 depicts the DPSK receiver where the comparison is effected by multiplying the received symbol by its T-second delayed replica.

It might appear at first glance that the DPSK receiver would have the same error rate performance as a BPSK demodulator. This is not the case because comparison (multiplication) is made between two noisy signals. However, as E_b/N_0 becomes greater, the DPSK error rate asymptotically approaches that of coherent BPSK.

Binary Modulation Error-Rate Curves

The probability of error is plotted for various modulation techniques in Figure 2.9. The curves apply to binary and four-phase coherent PSK (CPSK), fast frequency-shift keying (FFSK) and differentially encoded PSK (DECPSK),

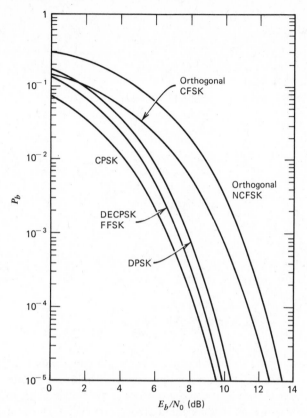

Figure 2.9 Ideal binary error-rate curves.

differentially detected PSK (DPSK), coherent and noncoherent detection of orthogonal FSK (CFSK and NCFSK, respectively).

DECPSK refers to coherent detection of differentially encoded data. In PSK systems data may be inverted at the demodulator output because of a phase ambiguity appearing in the carrier regenerated from the received signal. Differential encoding and decoding is used to circumvent this problem by representing the information as the difference between adjacent transmitted carrier phases. A phase ambiguity will shift both phases equally without distorting the difference between them.

Differential encoding is inherent in FFSK. Differential encoding will be discussed further in the modulation generation explanation (Section 3.1).

2.5 *M*-ARY PHASE-SHIFT KEYING

M-ary phase shift keying (MPSK) maps blocks of input bits into M distinct phases of a carrier. For n input bits per block there are $2^n = M$ combinations, for which a one-to-one correspondence with the carrier phase positions is made. For example, if bits at the modulator input are taken three at a time, $2^3 = 8$ phase positions result, giving eight-phase PSK.

At the receiver errors in detection are made when the detected phase of the received waveform (consisting of signal and noise) differs from predetermined limits about the actual phase. The limits are $\pm \pi/M$, as shown in Figure 2.10. It can be shown [4] that the probability of error of the block of bits (called the symbol-error probability) is given by

$$P_s = \frac{M-1}{M} - \frac{1}{2}\,\mathrm{erf}\left[\left(\sin\frac{\pi}{M}\right)\sqrt{\frac{E_s}{N_0}}\,\right]$$

$$- \frac{1}{\sqrt{\pi}}\int_0^{(\sin\pi/M)\sqrt{E_s/N_0}} e^{-y^2}\,\mathrm{erf}\left(y\cot\frac{\pi}{M}\right)dy$$

where E_s is the symbol energy, $E_s = (\log_2 M)E_b$ and $\mathrm{erf}(\cdot) = 1 - \mathrm{erfc}(\cdot)$.

A closed-form expression for P_s which applies to QPSK is given by

$$P_s = \mathrm{erfc}\left[\sqrt{\frac{E_s}{2N_0}}\,\right] - \frac{1}{4}\,\mathrm{erfc}^2\left[\sqrt{\frac{E_s}{2N_0}}\,\right] \tag{2.6}$$

This expression is identical to that obtained by considering QPSK as the summation of two BPSK signals in quadrature. Equation (2.6) can be obtained by considering the ways in which a symbol (a pair of bits) can be in error. It should be noted that P_s is the symbol-error probability. The probability of bit error is identical to that of BPSK. For $P_s < 10^{-3}$ a useful approximation valid for $M > 2$ is given by

$$P_s \simeq \mathrm{erfc}\left[\left(\sin\frac{\pi}{M}\right)\sqrt{\frac{E_s}{N_0}}\,\right]$$

The latter formula shows the poorer performance that results if the number of

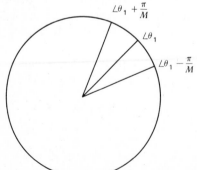

Figure 2.10 Decision regions for coherent detection of MPSK.

phases is increased. The bit-error probability, assuming encoding of the modulator input bits by a Gray code [5], is approximately given by

$$P_b \simeq \frac{P_s}{\log_2 M}$$

Figure 2.11 depicts the symbol-error probability for $2-$, $4-$, $8-$ and 16-phase PSK. Note that P_s is plotted as a function of $E_b/N_0 = (E_s/N_0)/\log_2 M$.

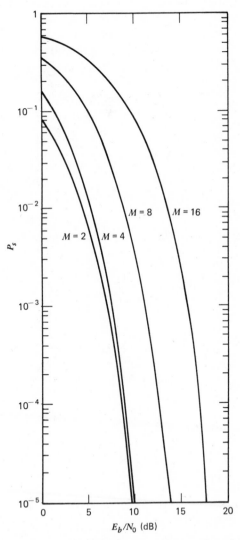

Figure 2.11 Symbol error probability for coherent detection of MPSK.

The preceding discussion has implicitly assumed that the receiver provides a perfect carrier reference. In practice, the carrier reference is usually produced by a nonlinear operation in the demodulator. For example, in QPSK the incoming signal is quadrupled to remove the phase modulation and the result divided by 4 to provide a reference at the carrier frequency. This action provides a carrier with a phase uncertainty or ambiguity of an integer multiple of $\pi/2$. Differential encoding has been described as one technique to circumvent this problem. The received signal can then be detected as in the coherent detection of PSK (CPSK), but a differential decoder must be provided. Alternatively, the signal can be differentially detected (DPSK) where phases of adjacent carrier segments are compared to determine the information symbols. This operation automatically provides the data without the need for a differential decoder.

The derivation of the symbol-error probabilities for coherent and differential detection of MPSK are complex. Only the final results are given below [4].

Coherent Detection of Differentially Encoded MPSK

$$\bar{P}_{s,1} = 2P_s \left[1 - \frac{1}{2}P_s - \frac{1}{2} \frac{\displaystyle\sum_{j=1}^{M-1} P_{e_k}^2}{P_s} \right]$$

where

P_s = symbol-error probability for coherent detection of m-ary CPSK

$$P_{e_k} = \frac{1}{\pi} \int_0^\infty \left\{ \exp\left[-\left(u - \sqrt{\frac{E_s}{N_0}} \right)^2 \right] \int_{u\tan[(2k-1)\pi/M]}^{u\tan[(2k+1)\pi/M]} \exp(-v^2)\,dv \right\} du$$

Differential Detection of Differentially Encoded MPSK

$$\bar{P}_{s,2} = 1 - \int_{-\pi/M}^{\pi/M} P_\phi(\phi)\,d\phi$$

where

$$P_\phi(\phi) = P_\phi\left(\frac{\pi}{2}\right) + \frac{E_s}{4N_0} \exp\left(\frac{-E_s}{N_0}\right) \left\{ I_1\left(\frac{E_s}{N_0}\cos\phi\right) + L_1\left(\frac{E_s}{N_0}\cos\phi\right) \right.$$

$$\left. + \cos\phi \left[I_0\left(\frac{E_s}{N_0}\cos\phi\right) + L_0\left(\frac{E_s}{N_0}\sin\phi\right) \right] \right\}$$

and

$$P_\phi\left(\frac{\pi}{2}\right) = \frac{(1+E_s/N_0)\exp(-E_s/N_0)}{2\pi}$$

and $I_n(a)$ and $L_n(b)$ are modified Bessel and Struve functions, respectively.

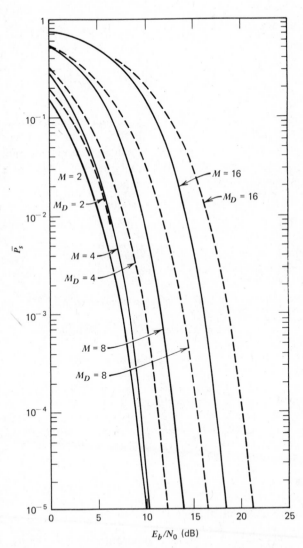

Figure 2.12 Symbol error probability for coherent and differentially coherent detection of differentially encoded MPSK. M_D refers to differential detection.

For large E_s/N_0, $\bar{P}_{s,2}$ is approximately given by

$$\bar{P}_{s,2} \simeq \mathrm{erfc}[W] + \frac{2W\exp(-W^2)}{\sqrt{\pi}\,(8E_s/N_0+1)}$$

where

$$W = \sqrt{2\frac{E_s}{N_0}}\,\sin\left(\frac{\pi}{2M}\right)$$

Another approximation also useful only for large E_s/N_0 is [6]

$$\bar{P}_{s,2} \simeq \mathrm{erfc}\left[\sqrt{\frac{E_s}{N_0}}\,\sin\left(\frac{\pi}{\sqrt{2}\,M}\right)\right]$$

Figure 2.12 shows the symbol-error probability for coherent and differential detection of differentially encoded MPSK. It can be shown [6, 7] that in the case of multiphase modulation for which the angular deviations are small (more than eight phases), the degradation introduced by phase comparison detection is essentially 3 dB over coherent detection.

REFERENCES

1 H. Van Trees, *Detection, Estimation and Modulation Theory*, Part 1, Wiley, New York, 1968.

2 J. M. Wozencraft and I. M. Jacobs, *Principles of Communication Engineering*, Wiley, New York, 1965, p. 83.

3 R. M. Gagliardi, *Introduction to Communications Engineering*, Wiley-Interscience, New York, 1978.

4 W. C. Lindsey and M. K. Simon, *Telecommunications Systems Engineering*, Prentice-Hall, Englewood Cliffs, N.J., 1973.

5 R. W. Lucky, J. Salz and E. J. Weldon, Jr., *Principles of Data Communications*, McGraw-Hill, New York, 1968.

6 E. Arthurs and H. Dym, "On the Optimum Detection of Digital Signals in the Presence of White Gaussian Noise—A Geometric Interpretation and a Study of Three Basic Data Transmission Systems", *IRE Trans. Commun. Syst.* Vol. CS-10, Dec. 1962, pp. 336–372.

7 C. R. Cahn, "Combined Digital Phase and Amplitude Modulation Communication Systems", *IRE Trans. Commun. Syst.*, Vol. CS-8, Sept. 1960, pp. 150–155.

PROBLEMS

1 Explain why QPSK has the same ideal error rate as BPSK.

2 A QPSK demodulator produces quadrature data outputs. Why is the overall error rate the same as that for either of the quadrature data paths? Assume that the data are random and equally likely to be ± 1 at the input to the QPSK modulator.

3 An integrate-and-dump circuit is used as a detector in various demodulators. The circuit integrates the received baseband data for a prescribed
   period of time, at which point the integrator output is sampled. Once the
   output has been sampled, the integrator is returned to a zero state, so that
   energy from one bit does not affect subsequent bit decisions. Suggest a
   circuit that would perform the function.

4 The probability of error for BPSK with Gaussian noise is given by

$$P_b = \frac{1}{2} \operatorname{erfc}\left[\sqrt{\frac{E_b}{N_0}}\right]$$

   This expression assumes that the detector is ideal, that is, if the demodulated data are greater than or equal to 0, the detector decides $+1$; and if
   the data are less than 0, the detector decides -1. Let a be an offset in the
   decision threshold of the detector. Determine P_b. If $a=0.1$ and $E_b/N_0=8.4$
   dB, find the degradation in error rate.

5 Show that the rules for accomplishing differential encoding are

   BPSK

$$C_n = C_{n-1} \oplus D_n$$

   where C_i represents the encoded data, D_i represents the input data and $\oplus$
   denotes the logical EXOR operation.

   QPSK

$$C_n = C_{n-1}\bar{I}_n\bar{Q}_n + \bar{C}_{n-1}I_nQ_n + D_{n-1}I_n\bar{Q}_n + \bar{D}_{n-1}\bar{I}_nQ_n$$

$$D_n = C_{n-1}\bar{I}_nQ_n + \bar{C}_{n-1}I_n\bar{Q}_n + D_{n-1}\bar{I}_n\bar{Q}_n + \bar{D}_{n-1}I_nQ_n$$

   where C_i and D_i are the quadrature differential encoder outputs and I_i and
   Q_i are the quadrature inputs to the encoder.
      What are the corresponding decoding rules?

6 If the channel bit error rate is given by P_e, show that the error rate at the
   output of the differential decoder for both BPSK and QPSK is given by

$$P_b = 2P_e(1-P_e)$$

   For small P_e, what is the effect of differential decoding? What is the E_b/N_0
   degradation if $P_e = 10^{-4}$ for BPSK and QPSK?

7 Show that the probability of symbol error for QPSK is given by

$$P_s = \operatorname{erfc}\left[\sqrt{\frac{E_s}{2N_0}}\right] - \frac{1}{4}\operatorname{erfc}^2\left[\sqrt{\frac{E_s}{2N_0}}\right]$$

   where E_s is the symbol energy.

8 A binary DPSK receiver multiplies the received signal by delayed version of itself. Determine the degradation if the delay is not exactly 1 bit. Suppose that the delay is $1.1T$. At what value should a phase shifter be set to compensate for this delay offset?

9 Explain why FSK with modulation index $0.5 \leq h \leq 1$ would be of primary interest in many communication systems.

CHAPTER 3

Generation and Detection of Modulated Signals

Chapter 2 provided ideal error-rate performance results for various modulation techniques. In this chapter conceptual techniques for the generation and detection of modulated signals are provided. Within this context detection refers to the recovery of the digital information from the received signal. Where a local reference is indicated, this signal is assumed to be ideal, that is, in phase lock and without jitter. Circuits that can be used for the regeneration of carrier and clock references are discussed in Chapter 5.

3.1 GENERATION OF WAVEFORMS

The digital modulator receives an input data stream and modifies the amplitude, frequency or phase of a sinusoidal carrier either individually or in combination. The modulated signal can be written as

$$s(t) = A(t)\cos[\omega_c t + \phi(t)] \tag{3.1}$$

Equation (3.1) can be expanded to give

$$s(t) = [A(t)\cos\phi(t)]\cos\omega_c t$$

$$- [A(t)\sin\phi(t)]\sin\omega_c t \tag{3.2}$$

Equation (3.2) indicates that the information is conveyed by the terms in brackets, which multiply carriers in phase quadrature. This expansion then provides an approach to the synthesis of various modulation techniques. The digital information is transformed into appropriate baseband waveforms which multiply quadrature carriers provided by a reference sinusoidal generator. The forms of $A(t)$ and $\phi(t)$ determine and characterize the output modulation.

In describing modulations by baseband waveforms it is convenient to use the concept of the phasor. The signal of (3.1) can be expressed as the real part of a complex signal as follows:

$$s(t)=\text{Re}\left[A(t)e^{j[\omega_c t+\phi(t)]}\right] \tag{3.3}$$

where $A(t)$ is assumed real. This can be verified by using the Euler formula

$$e^{jv}=\cos v+j\sin v$$

The complex signal can be written as

$$s_c(t)=\left[A(t)e^{j\phi(t)}\right]e^{j\omega_c t} \tag{3.4}$$

The signal phasor is the component of (3.4) given by

$$p_c(t)=A(t)e^{j\phi(t)} \tag{3.5}$$

The amplitude of the phasor is the amplitude of the signal, and the angle of the phasor with respect to the real axis is the phase of the signal [see (3.1)]. Equation (3.5) can be expanded to give

$$p_c(t)=A(t)\cos\phi(t)+jA(t)\sin\phi(t) \tag{3.6}$$

Comparison of (3.2) and (3.6) reveals that the real part of (3.6) is the baseband signal multiplying the reference $\cos\omega_c t$, and the imaginary part of (3.6) is the baseband signal multiplying the reference $(-\sin\omega_c t)$. In Figure 3.1 the phasor is represented by a vector in the complex plane and the real (x-axis) and imaginary (y-axis) components describe the modulation. Phasor diagrams are provided for some of the modulation techniques to be discussed.

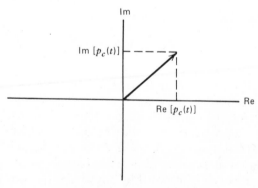

Figure 3.1 Phasor diagram.

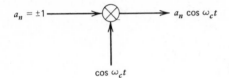

$a_n = \pm 1$ $\longrightarrow$ $\bigotimes$ $\longrightarrow$ $a_n \cos \omega_c t$

$\cos \omega_c t$

Figure 3.2 BPSK modulator.

Binary PSK (BPSK)

In a binary PSK modulator, input data bits cause discrete changes in the phase of the carrier. A data 1 causes no change (0°) whereas a data (-1) causes a 180° phase change. The generation of BPSK can be inferred from the following

    Data: (1)   output modulation: $\cos(\omega_c t + 0°) = +\cos \omega_c t$

    Data: (-1) output modulation: $\cos(\omega_c t + 180°) = -\cos \omega_c t$

This shows that the sign of the data bit determines the carrier phase; that is, over each bit interval multiply the carrier by the signed constant (± 1) representing the data. Figure 3.2 depicts a BPSK modulator.

If a phasor diagram is drawn representing the amplitude and phase of the carrier, two states are possible during a bit interval as shown in Figure 3.3. From bit interval to bit interval the phasor instantaneously changes from one state to the other as indicated by the arrows in the diagram. Note that there is no quadrature carrier term.

Quadriphase PSK (QPSK)

A QPSK modulator takes two input data bits at a time and produces a carrier whose phase is one of four values. These input bits can be consecutive bits for a serial input bit stream or corresponding bits for a parallel input. By using 2 bits at a time, the modulator achieves a signalling rate reduction of one-half. One relationship between data and carrier phase is as shown in Table 3.1.

The modulated carrier consists of the summation of quadrature sinusoids. The sign of the sinusoids corresponds to the data. When $a_n = +1$ the sign of $\cos \omega_c t$ is $+$, and it is $-$ when $a_n = -1$. The same holds for the relationship

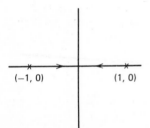

(−1, 0)   (1, 0)

Figure 3.3 BPSK phasor diagram.

Table 3.1   Data/Modulation Relationship for QPSK

Data		Carrier
a_n	b_n	
1	1	$\cos(\omega_c t - 45°) = \dfrac{1}{\sqrt{2}}(\cos \omega_c t + \sin \omega_c t)$
-1	1	$\cos(\omega_c t - 135°) = \dfrac{1}{\sqrt{2}}(-\cos \omega_c t + \sin \omega_c t)$
-1	-1	$\cos(\omega_c t - 225°) = \dfrac{1}{\sqrt{2}}(-\cos \omega_c t - \sin \omega_c t)$
1	-1	$\cos(\omega_c t - 315°) = \dfrac{1}{\sqrt{2}}(\cos \omega_c t - \sin \omega_c t)$

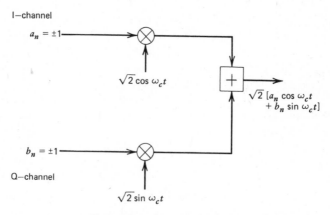

Figure 3.4   QPSK modulator.

between b_n and $\sin \omega_c t$. The QPSK modulator can be considered as two BPSK modulators whose outputs are added together. A QPSK modulator is shown in Figure 3.4. Note that the two quadrature channels have been labelled I and Q. The I-channel is the in-phase channel and the Q-channel is the quadrature channel as a result of the use of a carrier with a 90° phase offset.

Four carrier phase states (45°, 135°, 225°, 315°), each separated from its neighbour by 90°, are possible. The allowable transitions from symbol to symbol* are shown in the phasor diagram of Figure 3.5a. The vertices of the square represent the phase-state position during a symbol period, and the set of possible transitions are indicated by arrows.

---

*A symbol refers to the modulation state and represents a specific combination of input bits. The symbol time refers to the interval during which the input bits remain constant. In binary systems the bit time and the symbol time are the same.

Offset QPSK (OK-QPSK)

In the QPSK modulator described it has been implicitly assumed that the data transitions occur at the same time in each of the quadrature I and Q data channels. In other words, in an interval of time T, a_n and b_n are constant, giving a certain constant carrier phase. An instantaneous change in both values can occur only at the start of the next T-second interval. Figure 3.5a shows these instantaneous 180° phase jumps as the phasor passes through the origin. It will be shown later how these phase jumps influence the filtered output of a QPSK modulator. For now it suffices to say that large discontinuous phase jumps can lead to wide bandwidth occupancy. The question arises as to whether bandwidth conservation can be attained by limiting the phase-step changes.

Reduction of the amplitude of the phase step is possible by delaying the Q-channel digits by one-half symbol relative to the I-channel data. The top two

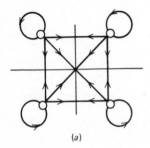

(a)

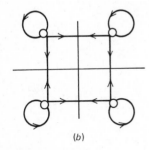

(b)

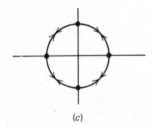

(c)

Figure 3.5 Phasor diagrams. (a) QPSK; (b) offset QPSK; (c) FFSK.

waveforms of Figure 3.6 show the timing for QPSK I- and Q-channel data. The third waveform in Figure 3.6 shows the Q data stream displaced by one-half symbol ($T/2$). For any transition in, say, the I-channel, the data in the Q-channel remain constant. This means that the carrier phase changes at most by 90° and that transitions

$$(1,1) \leftrightarrow (-1,-1)$$
$$(-1,1) \leftrightarrow (1,-1)$$

are not possible as in QPSK. Hence the size of the phase step has been reduced.

An offset QPSK (OK-QPSK) modulator is identical to a conventional QPSK modulator but for the time displacement between I- and Q-channel digits. The phasor/transition diagram for OK-QPSK is shown in Figure 3.5b.

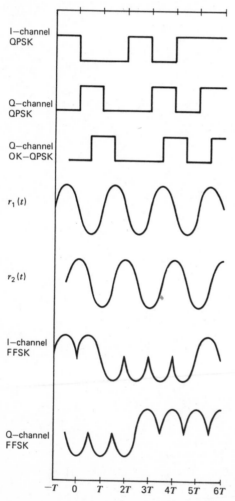

Figure 3.6 Quadrature modulator input signals.

Fast Frequency-Shift Keying (FFSK) and Minimum-Shift Keying (MSK)

Fast frequency-shift keying is a special case of continuous-phase frequency-shift keying (CPFSK) with modulation index $h=0.5$. It is an orthogonal form of FSK. Let the CPFSK signal be given by

$$s_{\text{CPFSK}}(t)=\cos\left[\omega_c t+hd_k\frac{\pi t}{T}+\psi_k\right], \qquad kT\leq t\leq(k+1)T$$

where ω_c = carrier frequency, rad/sec
   d_k = input data (±1) transmitted at rate $R=T^{-1}$
   ψ_k = constant phase valid for $kT\leq t\leq(k+1)T$

The frequency shift is reflected in the second term of the argument of the cosine term. For FFSK with $h=0.5$ the signal is given by

$$s(t)=\cos\left(\omega_c t+\frac{d_k\pi t}{2T}+\psi_k\right), \qquad kT\leq t\leq(k+1)T \qquad (3.7)$$

and the frequency shift is $\pm(4T)^{-1}=\pm R/4$ Hz, which is a positive or negative frequency shift by one-fourth of the data rate. The excess phase function of $s(t)$ is given by

$$\theta(t)=\left(\frac{d_k\pi}{2T}\right)t+\psi_k$$

The constant ψ_k represents the ordinate intercept of the phase $\theta(t)$. Over each interval T, $\theta(t)$ increases or decreases by $\pi/2$, depending upon d_k. The phase continuity requirement at a data transition results in the following relation:

$$\psi_k+\frac{d_k\pi kT}{2T}=\psi_{k-1}+\frac{d_{k-1}\pi kT}{2T} \qquad (3.8)$$

Rearranging terms in (3.8) leads to

$$\psi_k=\psi_{k-1}+(d_{k-1}-d_k)\frac{\pi k}{2}$$

For coherent detection an initial value of ψ_k, say ψ_0, can be set to zero. Consequently,

$$\psi_k=\begin{cases} 0, & d_k=d_{k-1} \\ \pi\bmod 2\pi, & d_k\neq d_{k-1} \end{cases}$$

The transition between states for all data inputs is shown in the phasor diagram of Figure 3.5c and in the phase trellis of Figure 3.7. In the phasor

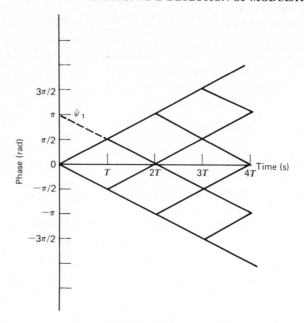

Figure 3.7   FFSK phase trellis.

diagram the phasor rotates clockwise or counterclockwise by $\pm\pi/2$, depending upon the bit transmitted. In the phase trellis the phase increases linearly over each signalling interval, and consequently the corresponding frequency shift is a constant ($\Delta f = d\phi/dt = $ constant). The dashed line in Figure 3.7 represents an extrapolation to indicate the phase ordinate intercept ψ_1.

Because $\psi_k = 0$, π modulo-2π, it can be shown that (3.7) can be expanded to give

$$s(t) = \left[(\cos\psi_k) \frac{\cos\pi t}{2T} \right] \cos\omega_c t$$

$$- \left[(d_k \cos\psi_k) \frac{\sin\pi t}{2T} \right] \sin\omega_c t \qquad (3.9)$$

The terms in (3.9) in the parentheses are related to the data input and are equal to ± 1. The $\cos\pi t/2T$ and $\sin\pi t/2T$ terms represent sinusoidal weighting factors. It can be shown (see Ref. 1 and Problem 6) that as a result of the phase continuity requirement, $\cos\psi_k$ can change value only at the zero crossings of $\cos\pi t/2T$, and $d_k\cos\psi_k$ can change value only at the zero crossings of $\sin\pi t/2T$. Therefore, the terms multiplying the quadrature carriers in (3.9) are each constant over successive $2T$ intervals. Because a 90° phase shift exists between $\cos\pi t/2T$ and $\sin\pi t/2T$, the symbols multiplying $\cos\omega_c t$ and $\sin\omega_c t$ have a relative displacement of T. The generation of the FFSK signal will now be explained in light of the preceding discussion.

FFSK is most easily thought of in terms of an offset-QPSK synthesis. In QPSK systems, the I- and Q-channel data consists of rectangular non-return -to-zero (NRZ) pulses over each symbol interval T. If half-sinusoid shaping is used in place of rectangular pulses, a different modulation results.

The modulator of Figure 3.8 can be used. Note that in Figure 3.8 the circuit contained within the dashed lines is a QPSK modulator provided that the inputs are NRZ. An OK-QPSK modulator results with appropriate relative time displacement of the I- and Q-channel input data.

The serial input data is divided into two data streams, with the I-channel formed from every even bit and the Q-channel from every odd bit. The rate in each channel is half the serial input data rate ($R/2$). The two data channels are derived as in offset QPSK (see Figure 3.6). The I data multiply a sinusoidal reference ($\omega_R = \pi/2T$) which has a period of twice the I-channel signalling interval. With the sinusoidal zero crossings coinciding with the data transitions as shown in the fourth waveform in Figure 3.6 the result is a half-sinusoidal pulse over an interval of $2T$. The half-sinusoidal pulse is then multiplied by a carrier sinusoid. A similar operation occurs in the Q-channel and the two channels are added.

The output of the modulator of Figure 3.8 can be written as

$$s(t) = \sum_n a_n p[t-(2n-1)T]\cos \omega_R t \cos \omega_c t$$

$$+ \sum_n b_n p(t-2nT)\sin \omega_R t \sin \omega_c t \qquad (3.10)$$

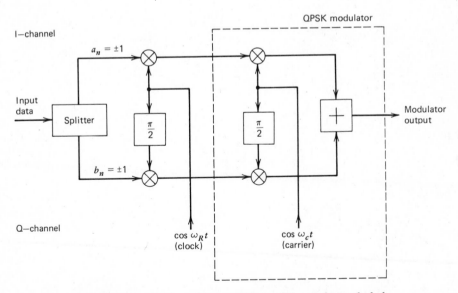

Figure 3.8 FFSK modulator. Differential encoder excluded.

where a_n and b_n represent information and

$$p(t)=\begin{cases} 1, & 0\le t\le 2T, \\ 0, & \text{elsewhere} \end{cases}$$

The arguments appearing in $p[\cdot]$ in (3.10) indicate the relative time displacement of the two data channels.

By using standard trigonometric identities, (3.10) can be written as

$$s(t)=\tfrac{1}{2}\sum a_n p\big[t-(2n-1)T\big](\cos 2\pi f_2 t+\cos 2\pi f_1 t)$$
$$+\tfrac{1}{2}\sum b_n p(t-2nT)(\cos 2\pi f_1 t-\cos 2\pi f_2 t) \qquad (3.11)$$

where $f_2=(\omega_c+\omega_R)/2\pi$
$\qquad\;\; f_1=(\omega_c-\omega_R)/2\pi$

Thus f_2 and f_1 represent the upper and lower transmitted frequencies, respectively.

Over each interval T, $p(t-2nT)$ and $p[t-(2n-1)T]$ are rectangular of unity amplitude and a_n and b_n can be equal to either ± 1. Inspection of (3.11) reveals that depending upon the combination of (a_n,b_n) one of two tones is produced by sideband cancellation as depicted in Figure 3.9. Table 3.2 shows the modulator output for each combination of (a_n,b_n).

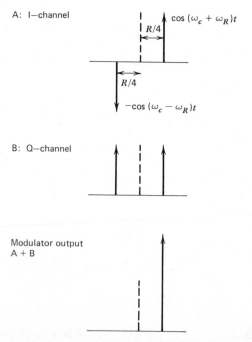

A: I—channel

$R/4$

$\cos(\omega_c+\omega_R)t$

$R/4$

$-\cos(\omega_c-\omega_R)t$

B: Q—channel

Modulator output
A + B

Figure 3.9 Sideband cancellation technique.

Table 3.2 Data/Modulation Relationship

a_n, b_n	Modulator Output
(1, 1)	$\cos 2\pi f_1 t$
(1, −1)	$\cos 2\pi f_2 t$
(−1, 1)	$-\cos 2\pi f_2 t$
(−1, −1)	$-\cos 2\pi f_1 t$

The output of this modulator is one of two frequencies and so is a form of FSK. However, which frequency is sent is not determined by one bit but by the combination (a_n, b_n). Inspection of Table 3.2 indicates that a tone at f_1 is sent when $a_n = b_n$ and a tone at f_2 is transmitted when $a_n \neq b_n$.

MSK is a generic term that refers to a modulation having the properties of an FSK signal as described in refs. 2 and 3. FFSK developed independently by de Buda [4] is essentially the same type of signal but with a one-to-one relationship between input data and transmitted frequency. FFSK is sometimes considered as one form of MSK by some authors.

One form (Type 1) of an MSK modulator has already been described in relation to Figure 3.8, wherein the data (a_n, b_n) represent the even and odd bits of the modulator input. A second form (Type 2) bears a closer resemblance to OK-QPSK. OK-QPSK is described by (3.10) with $\cos \omega_R t$ and $\sin \omega_R t$ removed. Note that the basic pulse shape is given by $p(t)$, which is a rectangular positive pulse. In MSK-type 1 the basic pulse shape is modified by the weighting factors of $\cos \omega_R t$ and $\sin \omega_R t$ to produce half-sinusoidal pulses. As depicted in Figure 3.6, the weighting is not a constant positive half-sinusoid but alternating positive and negative half-sinusoids. MSK-Type 2 differs from Type 1 in that the weighting is always a positive half-sinusoid.

The transformation of the circuit of Figure 3.8 into a FFSK modulator is accomplished by encoding the serial input data. The transformation is accomplished by a differential encoder having the rule

$$c_{n+1} = c_n \oplus d_n \qquad (3.12)$$

where c_n is the previous encoder output, d_n the present encoder input and $\oplus$ the logical Exclusive-OR (EXOR) function. Simply stated, the encoder function is to compare the input bit with the last transmitted bit. If they are the same, the next encoder output will be -1. If they are of opposite polarity, the encoder output will be $+1$. It is left as an exercise to the reader to verify that differential encoding results in a one-to-one relationship between modulator input data and transmitted frequency.

Whereas differential encoding is required to produce the one-to-one relationship between input data polarity and transmitted frequency, it not necessary otherwise. Where the data a_n and b_n are independent, the resulting modulation is still a form of CPFSK but without the one-to-one data/frequency relationship. This form of CPFSK is minimum-shift keying (MSK). The difference between FFSK and MSK is in the use of the differential encoding. In practice the encoder is required in both cases if resolution of the regenerated carrier ambiguity is needed in the demodulator. In this case FFSK and MSK are used synonymously.

3.2 COHERENT DETECTION OF SIGNALS

Figure 3.10 depicts a generalized quadrature modulator which can be applied to the modulation techniques previously discussed. The difference in the modulator for each technique will be reflected in the form of the baseband I and Q pulses produced by the baseband generator and the relative timing between them. In the case of BPSK only the upper half of the four-phase (4-ϕ) modulator is used.

Figure 3.11 depicts a generalized quadrature demodulator. It consists of a coherent demodulation section where the incoming signal is split into two paths which are each multiplied by locally regenerated quadrature carrier references. The carrier reference is derived from the incoming signal by a nonlinear operation which removes or suppresses the phase or frequency variation effect of the modulation. The I and Q baseband signals produced by the coherent demodulation are then individually processed through ideal matched filters. In the case of QPSK and OK-QPSK the matched filter is an integrate-and-dump, whereas for a fast FSK signal the matched filter integrate-and-dump accounts for the half-sinusoidal weighting. Timing for the integrate-and-dump and data recombining is provided by the clock recovery circuit, which can regenerate a local clock from the received modulated waveform or from the baseband outputs of the quadrature demodulators (described in Chapter 5).

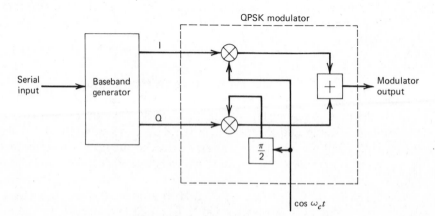

Figure 3.10 Generalized quadrature modulator.

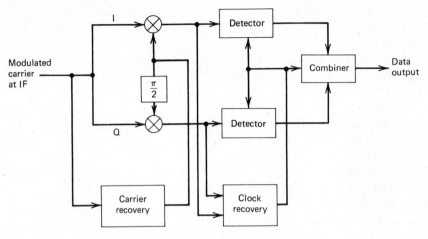

Figure 3.11 Generalized quadrature demodulator.

If differential encoding is employed to deal with recovered carrier ambiguity, then the error rate is degraded. The bit-error probability for DECPSK can be shown (Chapter 2, Problem 6) to be given by

$$P_{\text{DECPSK}} = 2P_e(1 - P_e)$$

where P_e is the channel bit-error rate.

3.3 DIFFERENTIAL DETECTION OF PSK

The coherent detection of PSK requires the use of locally generated carrier references. This may not be desirable because of the additional amount of circuitry required. An alternative approach is differential detection as depicted in Figure 2.8. The incoming signal is multiplied by a 1-bit delayed version of itself as follows:

$$m(t) = 2(a_n \cos \omega_c t)\left[a_{n-1}\cos \omega_c(t - T)\right]$$

$$= a_n a_{n-1} \cos \omega_c T$$

where double-frequency terms are ignored. If $\omega_c T = 2n\pi$ $(n = 1, 2, 3, \ldots)$, then

$$m(t) = \begin{cases} 1, & \text{if } a_n = a_{n-1} \\ -1, & \text{if } a_n \neq a_{n-1} \end{cases}$$

If the information is contained in the difference between adjacent bits, then $m(t)$ gives the information. Use of appropriate clock timing and detection results in the recovered bit stream. Note that if a frequency error exists, a reduction in the value of $m(t)$ will result because ω_c is now replaced by $\omega_c + \Delta\omega_c$.

In comparing DPSK and CPSK it should be recognized that DPSK will not perform as well. This is because the reference signal is in fact the incoming signal (delayed) and is perturbed by the same level of noise.

In addition to negating the requirement for a carrier recovery circuit, DPSK can provide valid data virtually immediately. Modems designed for burst applications (e.g., TDMA) require fast signal acquisition for efficiency reasons. While such modems may use PSK and coherent detection, the design of such modems is very exacting to attain lockup in the order of 30 symbols. DPSK can be used as an alternative if the E_b/N_0 penalty can be tolerated. It is interesting to speculate whether the E_b/N_0 advantage that CPSK offers is not significantly eroded in practical implementations, thus making DPSK attractive.

3.4 DETECTION OF FFSK

FFSK can be treated in different ways and consequently the data can be recovered by (a) coherent detection, (b) noncoherent frequency detection and (c) differential detection.

This flexibility can be very attractive, depending upon the specific application. For instance, if the spectral property of FFSK is of prime importance while the link provides adequate E_b/N_0, then the FFSK can be detected by noncoherent or differential techniques.

Coherent Detection

FFSK has been explained as a form of OK-QPSK and similarly can be demodulated in quadrature channels. In each of these channels the baseband waveforms are sinusoidal antipodal signals, and if they are detected by appropriate matched filters the same error rate as PSK results.

At this time it is important to explain the impact of differential encoding. It has been shown that differential encoding can be considered as part of the quadrature modulator. The encoding establishes a relationship between data bit and transmitted frequency (data $1 = f_2$, data $0 = f_1$). The differential encoding is inherent in FFSK.

If the quadrature demodulator structure of Figure 3.11 is used, the data output will have to be differentially decoded. Inspection of (3.12) shows that the information is retrieved by the decoder rule

$$d_n = c_n \oplus c_{n+1} \tag{3.13}$$

Equation (3.13) states that the decoder compares adjacent data to derive the information. Therefore, the error rate for FFSK is the same as for DECPSK. If differential encoding is not included and phase ambiguity is not a problem, as for example with delta-coded voice, then an E_b/N_0 advantage results, as in the case of CPSK over DECPSK. The modulation used in this case is non-differentially encoded MSK.

Although FFSK can be considered in the quadrature channel form, a variation on this approach described in Refs. 5 and 6 based on trellis decoding provides additional insight.

Trellis Decoding of FFSK*

If the initial phase constant is assumed to be equal to zero in (3.7), the two possible FFSK signals can be written as

$$\bar{s}_1(t) = \sqrt{2P} \cos\left(\omega_c t + \frac{\pi t}{2T}\right) \qquad (3.14)$$

$$\bar{s}_2(t) = \sqrt{2P} \cos\left(\omega_c t - \frac{\pi t}{2T}\right) \qquad (3.15)$$

where $P = E_b/T$. Recall from the discussion of binary FSK in Section 2.4 that the correlation coefficient is given by

$$\rho = \frac{\sin(\omega_2 - \omega_1)T}{(\omega_2 - \omega_1)T} - \frac{\sin 2\omega_c T}{2\omega_c T}$$

FFSK is an orthogonal modulation, and therefore both terms in the expression for ρ must be equal to zero. The second term is identically equal to zero for

$$2\omega_c T = k\pi, \qquad k = 1, 2, 3, \dots$$

It follows that $T = k/4fc$ and that the bit interval contains an integer number of quarter cycles of the carrier. Consequently, the carrier frequency can be written as

$$f_c = \left(N + \frac{K}{4}\right)R, \qquad K = 1, 2, 3, 4$$

where $R = 1/T$. Examination of (3.14) and (3.15) shows for the four values of K that at the end of the signalling interval the transition to the next waveform will occur either at a peak (negative or positive) or at a zero crossing of the carrier waveform (see Problem 7). Figure 3.12 depicts a waveform with $N = 4$ and $K = 4$. The transmission frequencies are f_1 and f_2. In the diagram $-f_1$

---

*This explanation was provided by L. Palmer, COMSAT Laboratories, Clarksburg, Maryland.

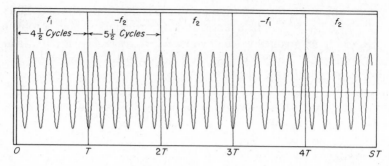

Figure 3.12 FFSK waveform showing transitions at positive and negative peaks.

refers to inversion of the cosine function. Now consider the case where

$$s_1(t) = \sqrt{2P}\cos\left(\omega_c t + \frac{\pi}{2T}t\right)$$

and

$$s_2(t) = \sqrt{2P}\cos\left(\omega_c t' - \frac{\pi}{2T}t'\right)$$

where $s_2(t)$ follows $s_1(t)$. Assume that the following case applies:

$$\omega_c = 2\pi f_c$$

$$= 2\pi\left(N + \tfrac{3}{4}\right)R$$

Then

$$s_1(t) = \sqrt{2P}\cos\left(\omega_c t + \frac{\pi}{2T}t\right)$$

$$= \sqrt{2P}\cos\left[2\pi R(N+1)t\right]$$

$$s_2(t) = \sqrt{2P}\cos\left[2\pi R\left(N+\tfrac{1}{2}\right)t\right]$$

Over an interval of duration T, $s_1(t)$ starts at a peak ($t=0$) and ends on a peak ($t=T$) while $s_2(t)$ starts on a peak and ends on a negative peak. For the resulting signal these peaks must match at the transition points for phase continuity to occur.

Define two states: 0 and π. State 0 denotes a positive peak and π a negative peak. Furthermore, when the input bit is a 0 the tone S_H is sent, and when the bit is a 1 the tone S_L is sent.* S_H denotes the high-frequency tone and S_L the low-frequency tone.

---

*Note that this convention is the inverse of that used previously. However, the explanation remains valid for both cases.

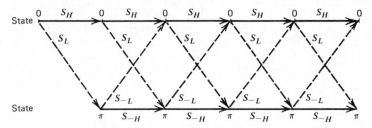

Figure 3.13 Trellis state diagram.

Evidently, for each input data bit the output signal will consist of S_H, $-S_H$, S_L or $-S_L$. If S_H is initially sent, then the next tone that is sent is determined by the modulation structure. This structure is shown in Figure 3.13.

In this diagram nodes represent bit transition points. A 0 denotes a positive peak, and π a negative peak, of the modulated signal at each node. Solid lines indicate that a 0 is sent, and a dashed line indicates that a 1 is sent. The convention $S_{-H} = -S_H$ and $S_{-L} = -S_L$ is used in the diagram. Note that the structure does not allow, for example, S_H to be followed immediately by S_{-H} or S_{-L} to be followed by S_{-H}. The trellis nature of the waveform indicates that an inherent coding exists.

During the kth interval one of four possible signals can be transmitted as follows:

$$S_H = \sqrt{2P} \cos\left(\omega_c t + \frac{\pi}{2T} t \right)$$

$$S_{-H} = -S_H$$

$$S_L = \sqrt{2P} \cos\left(\omega_c t - \frac{\pi}{2T} t \right)$$

and

$$S_{-L} = -S_L$$

The received signal is given by

$$r(t) = s(t) + n(t)$$

where $s(t) = S_H$, S_{-H}, S_L or S_{-L} and $n(t)$ is additive white Gaussian noise (AWGN). The demodulator first performs a correlation as shown in Figure 3.14. Table 3.3 gives the expected values of H_K and L_K for the four possible waveforms S_H, S_{-H}, S_L, S_{-L}. The probability density functions of the outputs

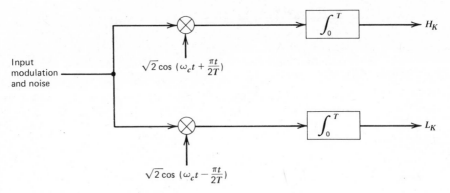

Figure 3.14   Demodulation of FFSK.

for two hypotheses are

$$p[H_K|S_H] = \frac{1}{\sqrt{\pi N_0}} \cdot \exp\left[-\frac{\left(H_K - \sqrt{E_b} \right)^2}{N_0} \right]$$

$$p[L_K|S_{-L}] = \frac{1}{\sqrt{\pi N_0}} \cdot \exp\left[-\frac{\left(L_K + \sqrt{E_b} \right)^2}{N_0} \right]$$

where $P[a|b]$ indicates the probability that a is detected if b was sent.

The correlator outputs are Gaussian random variables at the end of each bit interval with means given by the Table 3.3 values and variance $N_0/2$, where N_0 is the single-sided input noise spectral density.

The approach taken in demodulating a sequence of received waveforms is to maximize the probability that the demodulated data is the right sequence given a series of observations at the correlator outputs. Figure 3.13 indicates that with an assumed initial state of 0, if the state sequence is $0, 0, \pi, 0$, then the bit sequence was $0, 0, 1, 1$ and the transmitted tones were S_H, S_{-H}, S_L, S_{-L}. Maximum likelihood decoding maximizes the sum of the correlator outputs that lead to a node. These values are proportional to the joint probability,

Table 3.3   FFSK Correlator Outputs

Input Signal Component	$E\{H_K|\ \}$	$E\{L_K|\ \}$
S_H	$\sqrt{E_b}$	0
S_{-H}	$-\sqrt{E_b}$	0
S_L	0	$\sqrt{E_b}$
S_{-L}	0	$-\sqrt{E_b}$

$p[\bar{x}, \bar{H}, \bar{L}]$, so an accumulation of these metrics gives a sum proportional to the probability that the particular sequence was sent. Note that $\bar{x}$, $\bar{H}$ and $\bar{L}$ denote state and I and Q correlator output sequences, respectively.

The procedure is best explained by an example. Assume that the transmitted sequence is $0, 1, 1, 0, 1, 0, 0$. This gives the state diagram of Figure 3.15a. In Figure 3.15b an output indicated with $\sqrt{E_b}$ indicates that the incoming signal matches the correlator and produces a mean $\sqrt{E_b}$ output to which a noise sample N_{HK} is added. Where $\sqrt{E_b}$ does not appear, only noise is the output.

The receiver looks at all possible paths leading to a state and adds the metrics to that point (where the metrics are the correlator outputs). Consider the 0-state node at the end of the first interval ($K=1$) shown in Figure 3.16a. In the diagram,

$$M_0 = \text{correlator output } H_K = H_1$$

$$M_1 = \text{correlator output } L_K = L_1$$

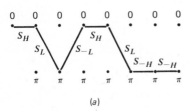

(a)

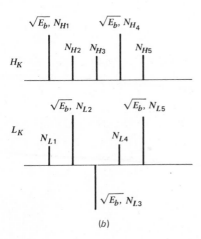

(b)

Figure 3.15 Example for $0, 1, 1, 0, 1, 0, 0$ sequence. (a) Trellis path; (b) correlator outputs.

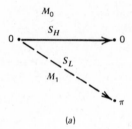

(a)

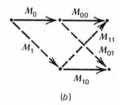

(b)

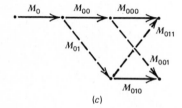

(c)   **Figure 3.16** Example: metric calculations.

Therefore,

$$M_0 = \sqrt{E_b} + N_{H1}$$

$$M_1 = N_{L1}$$

At this point the outputs M_0 and M_1 could be compared and a decision made. However, it is better to wait with a decision until the next state is viewed.

Now consider $k = 2$, that is, the first two states. The subscripts indicate what bits are transmitted to move through the trellis; for example, M_{01} indicates that bits 0 and 1 were transmitted (Figure 3.16b). To determine the metric at the second 0-state node corresponding to the transmission of 0,0, add the appropriately signed correlator outputs; that is, if 0,0 was sent, then the tones must have been S_H, S_H and so H_K correlator outputs are taken and added:

$$M_{00} = \sqrt{E_b} + N_{H1} + N_{H2}$$

Assume that the bits sent were 1, 1. Then the tones sent must have been S_L and $S_{-L} (= -S_L)$, so add the L_1 output to the negative of the L_2 output:

$$M_{11} = N_{L1} - \sqrt{E_b} - N_{L2}$$

Assume that the bits sent were 0, 1. Then the tones sent must have been S_H and S_L, so add H_1 and L_2:

$$M_{01} = \sqrt{E_b} + N_{H1} + \sqrt{E_b} + N_{L2}$$

Assume that the bits sent were 1, 0. Then the tones sent must have been S_L and S_{-H} so add L_1 and $-H_2$:

$$M_{10} = N_{L1} - N_{H2}$$

Which of the outputs is likeliest to be largest? Probably M_{01} because the mean is $2\sqrt{E_b}$ (the mean of the noise terms is 0). Note that in fact this was the first two-digit sequence sent.

It was stated previously that a delay in the decision is required. A decision on any one bit should only be made after 2 bits have been received. At each data transition two signal (tone) states are possible: a positive peak (0 state) or a negative peak (π state). It is convenient to use the following terminology. A surviving path into a node is that which represents the largest sum of correlator outputs to that node. For example, M_{00} and M_{11} are the metrics to state 0 at interval $k=2$. If M_{00} is the survivor, then

$$M_{00} > M_{11}$$

Now

$$M_{00} = \sqrt{E_b} + N_{H1} + N_{H2}$$

and

$$M_{11} = N_{L1} - \sqrt{E_b} - N_{L2}$$

Therefore,

$$\sqrt{E_b} + N_{H1} + N_{H2} > N_{L1} - \sqrt{E_b} - N_{L2}$$

and

$$2\sqrt{E_b} + N_{H1} + N_{L2} > N_{L1} - N_{H2}$$

which indicates that $M_{01} > M_{10}$. In other words, if $M_{00} > M_{11}$, then $M_{01} > M_{10}$.

This shows that the surviving paths (each path is 2 bits long) at a data transition both indicate that the same first bit was transmitted: that is, M_{00} is the largest metric into the 0 state and M_{01} is the largest metric into the π state. Both indicate that the first bit sent was a 0. The first bit of the 2-bit sequence can then be recovered from either of the surviving paths. This means that accumulating metrics any further does not provide additional benefits to the detection of the first bit.

This procedure can now be extended to the next bit. The state diagram (Figure 3.16c) assumes that the first bit is correctly detected. Here

$$M_{000} = M_{00} + H_3 = \sqrt{E_b} + N_{H1} + N_{H2} + N_{H3}$$

$$M_{011} = M_{01} - L_3 = 2\sqrt{E_b} + N_{H1} + N_{L2} - \left(-\sqrt{E_b}\right) - N_{L3}$$

$$M_{001} = M_{00} + L_3 = \sqrt{E_b} + N_{H1} + N_{H2} - \sqrt{E_b} + N_{L3}$$

$$M_{010} = M_{01} - H_3 = 2\sqrt{E_b} + N_{H1} + N_{L2} - N_{H3}$$

All terms contain the constant $\sqrt{E_b} + N_{H1}$, which can be subtracted from all metrics without influencing the result. This indicates that metrics spanning 2-bit intervals only are required. Decoding is accomplished by calculating metrics over 2 bits to determine the first bit and by using the second and third bits to determine the second bit, and so on. After 2 bits the mean correlator output giving a maximum of $2\sqrt{E_b}$ is twice as much as would have been the output after 1 bit:

$$M_{01} = 2\sqrt{E_b} + N_{H1} + N_{L2}$$

$$\text{mean value} = E(M_{01}) = 2\sqrt{E_b}$$

The variance of the noise term is given by

$$\sigma_2^2 = E\left[\left(N_{H1} + N_{H2}\right)^2\right]$$

$$= E\left[N_{H1}^2 + 2N_{H1}N_{H2} + N_{H2}^2\right]$$

$$= 2\sigma_1^2$$

where
$$E[N_{H1}^2] = E[N_{H2}^2] = \sigma_1^2$$
$$E(N_{H1} \cdot N_{H2}) = 0$$

The SNR after a 2-bit observation interval is given by

$$\text{SNR}_2 = \frac{\left(2\sqrt{E_b}\right)^2 R}{2\sigma_1^2} = 2\frac{E_b R}{\sigma_1^2} = 2\text{SNR}_1$$

which indicates a doubling of SNR compared with a 1-bit observation interval. Note that it has been assumed that the demodulator starts in the 0 state, which implies knowledge of the carrier phase.

Noncoherent Detection of FFSK

The FFSK signal is a frequency shift waveform. FSK can be noncoherently demodulated by means of a limiter and frequency discriminator. The frequency discriminator has a linear output voltage versus input frequency characteristic over a certain signal bandwidth. The limiter removes any amplitude variation

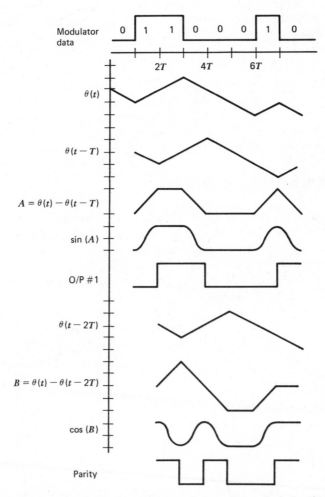

Figure 3.17 Differential MSK detection waveforms. (From Ref. 7: (© 1979 IEEE; reprinted from "Differential Detection of MSK with Non-redundant Error Correction", by T. Masamura, S. Samejima, Y. Morihiro and H. Fukada, from *IEEE Trans. Commun.*, June 1979.)

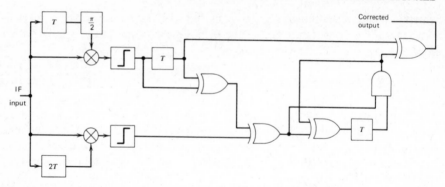

Figure 3.18 Differential MSK detector and decoder. (From Ref. 7: (© 1979 IEEE; reprinted from "Differential Detection of MSK with Non-redundant Error Correction", by T. Masamura, S. Samejima, Y. Morihiro and H. Fukada, from *IEEE Trans. Commun.*, June 1979.)

in the FSK signal. The output of the discriminator then is a binary data sequence corresponding to the input tone sequence.

This feature of FFSK, the ability to be demodulated by a relatively simple discriminator type of circuit, is attractive in situations where bandwidth conservation is of importance. A penalty of 3.6 dB is paid in E_b/N_0, however.

3.5 DIFFERENTIAL DETECTION OF MSK

Masamura et al. [7] describe a technique for differential detection of MSK in conjunction with a novel single-error-correction circuit. The outputs of a conventional differential detector circuit and one that compares alternating signalling intervals to form a parity check are used to improve error-rate performance. The waveforms used in the operation are shown in Figure 3.17 and the detection circuit is given in Figure 3.18.

The authors show that the resulting performance lies between that for DECPSK and DPSK giving a degradation relative to DECPSK of 0.1 dB at a 10-dB signal-to-noise ratio. The error-correction circuit provides significant improvements with a signal degraded by bandlimiting. For example, with $BT^* = 0.6$ and $P_b = 10^{-4}$, the E_b/N_0 values required before and after the error-correction circuit are 15 and 12.5 dB, respectively. The error-rate improvement for this BT is a factor of 10. The authors indicate that the technique can be applied to PSK as well.

REFERENCES

1   S. A. Gronemeyer and A. L. McBride, "MSK and Offset QPSK Modulation", *IEEE Trans. Commun.*, Vol. COM-24, Aug. 1976, pp. 809–820.

*BT is the filter bandwidth–bit interval product.

2   M. L. Doelz and E. T. Heald, Minimum-Shift Data Communication System, U.S. Patent No. 2 977 417, Mar. 1961.

3   W. M. Hutchinson and R. W. Middlestead, Data Demodulator Apparatus, U.S. Patent No. 3 743 755, July 1973.

4   R. de Buda, "Coherent Demodulation of Frequency Shift Keying with Low Deviation Ratio", *IEEE Trans. Commun.*, Vol. COM-20, June 1972, pp. 429–435.

5   G. D. Forney, Jr., "The Viterbi Algorithm", *Proc. IEEE*, Vol. 61, Mar. 1973, pp. 268–278.

6   M. G. Pelchat, R. C. Davis and M. B. Luntz, "Coherent Demodulation of Continuous Phase Binary FSK Signals", *1971 Int. Telem. Conf.*, Washington, D.C., pp. 181–190.

7   T. Masamura, S. Samejima, Y. Morihiro and H. Fukada, "Differential Detection of MSK with Non-redundant Error Correction", *IEEE Trans. Commun.*, Vol. COM-27, June 1979, pp. 912–918.

PROBLEMS

1   Conventional BPSK modulation entails the transmission of either $0°$ or $180°$ phase values. Consider the situation where the permissible phases in the signal are $60°$ and $120°$. Show that the resulting signal consists of two components, one of which is a BPSK waveform. What is the other signal and what is the power level relative to the BPSK signal? Can you explain why such a waveform would be used?

2   A maximal-length sequence (*m*-sequence) which approximates a random sequence is often used to scramble data before modulating. This is accomplished by modulo-2 adding the sequence to the data. Explain why this would be required.

3   If the data are scrambled in the transmitter, a de-scrambler will be required in the receiver. Explain how the data may be recovered. Describe modes of transmission where the de-scrambler (*a*) must be synchronized to the incoming data and (*b*) can operate without initialization. What types of errors, if any, are introduced in the latter case?

4   In a QPSK modulator NRZ I and Q data modulate quadrature carriers. The NRZ data do not contain a DC component. In practical circuits an offset could exist. What effect does such an offset have on the modulator output signal? Assume a relative offset of a volts. Find the modulation output power as a function of a.

5   Show that the detection of DPSK does not require a differential decoder.

6   An FFSK signal is given by

$$s(t)=\cos\left(\omega_c t+\frac{d_i\pi t}{2T}+\theta_i\right), \qquad iT\le t\le(i+1)T$$

where $f_c=\omega_c/2\pi$ is the centre frequency, the input data $d_i=\pm1$, T^{-1} is the transmission rate and θ_i is a phase term that is constant over the signal interval. Show that $s(t)$ can be expanded using a standard trigonometric expression to give

$$s(t) = \left(\cos \theta_i \cos \frac{\pi t}{2T} \right) \cos \omega_c t$$

$$- d_i \left(\cos \theta_i \sin \frac{\pi t}{2T} \right) \sin \omega_c t$$

With $\theta_i = 0$, π modulo 2π show that $\cos \theta_i$ can change only at the zero crossing of $\cos(\pi t/2T)$, and $d_i \cos \theta_i$ can change only at the zero crossings of $\sin(\pi t/2T)$. What is the symbol weighting on each of the quadrature carriers, and what is the symbol duration in terms of T?

7 Show that FFSK transitions can occur at peaks or zero crossings of the modulated waveform.

8 Show that for MSK-Type 2 the transformation between MSK and FFSK is given by:

(a) MSK to FFSK encoding: differentially decode the MSK data and polarity reverse every second bit of the result.

(b) FFSK to MSK encoding: polarity-reverse every second FFSK data bit and differentially encode the result.

9 Assume that the sequence of transmitted FFSK frequencies is $f_+, f_-, f_+,$ f_+, f_-, where f_+ and f_- denote the upper and lower transmitted frequencies, respectively. Assuming that the initial carrier phase is $0°$, draw the baseband quadrature waveforms.

CHAPTER 4

————————————————————————————

Performance Degradations

The comparative evaluation of modulation techniques on the basis of performance in additive white Gaussian noise provides a reference that is subject to deterioration as a result of imperfections in the subsystems comprising the satellite communications link as well as general system constraints and limitations. Usually, minimization of band occupancy is desired so as to maximize the number of modulated carriers occupying a given frequency band, and this influences the selection of bandlimiting filters. Power limitations require operating satellite and possibly Earth terminal amplifiers in nonlinear regions near saturation with attendant amplitude and phase distortions. Frequency translation and carrier recovery circuits introduce phase errors. Sensitivity to co-channel interference may be an important criterion in some systems. In this chapter these topics are reviewed and a comparison is presented of the performance of certain digital modulation techniques over satellite channels.

4.1 BANDLIMITATION

In general, the preceding discussions have assumed that the channel has infinite bandwidth. The introduction of a filter between transmitter and receiver causes waveform distortion and bit spreading. The waveform distortion is seen as fluctuations in the envelope of the modulated carrier, while bit spreading refers to a data pulse not being confined to the signalling interval T but lasting longer to interfere with adjacent bits causing intersymbol interference (ISI).

Consider the baseband system [1] shown in Figure 4.1. The encoder produces rectangular baseband pulses which are filtered by a linear time-invariant filter $H_b(\omega)$. To the output of the filter is added noise. The resulting signal is passed to a detector which assumes that the signal is not filtered; that is, the detector is an integrate-and-dump matched filter.

The transmitted pulses are given by $\pm\sqrt{E_b/T}\, u(t)$, where

$$u(t)=\begin{cases} 1, & 0\leq t\leq T \\ 0, & \text{elsewhere} \end{cases}$$

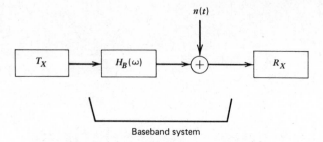

Figure 4.1 Bandlimiting filter system.

The pulse $u(t)$ and its $(\sin x/x)^2$ power spectrum are shown in Figure 4.2. The output of the filter is

$$x(t)=\sqrt{\frac{E_b}{T}}\ \sum_{i=0}^{\infty} a_i s(t+iT)$$

where $a_i=\pm 1$; a_0 occurs during $(0,T)$ and a_i is a previous bit. The modified pulse $s(t)$ is a filtered form of $u(t)$.

The Fourier transform of $s(t)$ is given by

$$S(\omega)=U(\omega)H_B(\omega)$$

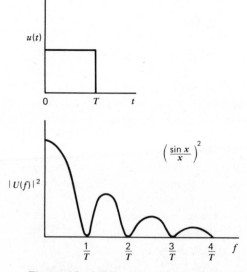

Figure 4.2 NRZ pulse and spectrum.

The receiver matched filter output is

$$y(T) = \int_0^T x(t)\, dt$$

$$= \sqrt{E}\, b_0 + \sqrt{E}\, b_I + \int_0^T n(t)\, dt \tag{4.1}$$

where $b_0 = (T)^{-1/2} \int_0^T a_0 s(t)\, dt$

$$b_I = (T)^{-1/2} \sum_{i=1}^{\infty} a_i \int_0^T s(t+iT)\, dt$$

The terms b_0 and b_I in (4.1) represent the information bit and the effect of intersymbol interference (ISI), respectively. The third term represents Gaussian noise with zero mean and variance given by

$$\sigma_n^2 = \frac{1}{2\pi} \int_{-\infty}^{\infty} S_d(\omega) |U(\omega)|^2\, d\omega$$

where $S_d(\omega)$ is the input noise spectral density.

Following the derivation given in Section 2.3, it can be shown that for a particular sequence of data $\{\bar{a}_i\}$ the probability of error is given by

$$P[\text{error}|\bar{a}_i] = \frac{1}{2} \text{erfc}\left[\sqrt{ \frac{E_b}{2N_0} \left(\frac{b_0 + b_I}{\sigma_n / N_0} \right) } \right]$$

Degradation to the error rate is caused by data patterns which effectively reduce the signal-to-noise ratio through b_I. Hartmann [2] has calculated the

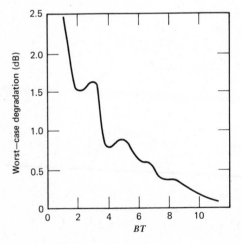

Figure 4.3 Degradation due to rectangular filter. (From Ref. 1; by permission of John Wiley and Sons, Inc.)

worst-case degradation with a rectangular filter frequency characteristic given by

$$H_b(\omega) = \begin{cases} 1, & |\omega| \leq 2\pi B \\ 0, & \text{elsewhere} \end{cases}$$

The degradation relative to antipodal signalling is plotted in Figure 4.3. It is seen that as the filter bandwidth is reduced the degradation increases. While the noise variance decreases, the ISI increases. This is because the receiver filter is no longer matched to the input signal.

4.2 TRANSMISSION RATES AND PULSE SHAPING

The rate at which pulses can be transmitted through a linear channel that filters the transmitted signal is a major concern. Consider the ideal low-pass filter with constant amplitude and linear phase characteristics in the passband as shown in Figure 4.4.

It can be shown that the impulse response of this filter has a $\sin x / x$ characteristic. Note that this is an ideal situation which cannot be realized for a number of reasons: (1) the impulse response is noncausal (i.e., the pulse would have to exist prior to $t = 0$) and (2) the sharp filter cutoff can only be attained with infinite delay and no phase distortion. However, this case is of interest because of the insight it gives to pulse transmission rates.

The zero crossings of the impulse response occur for $t = \pm nT$, where n is an integer and is not equal to zero. This means that if *impulses* are transmitted, the filter output consists of one impulse response waveform for each input impulse. Each impulse response waveform main-lobe peak will occur at the zero crossings of all other impulse responses. The impulse response peaks occur

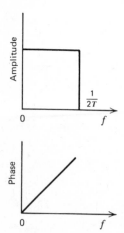

Figure 4.4   Ideal low-pass filter characteristic.

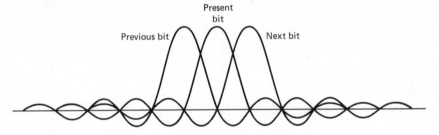

Figure 4.5 Signalling with $\sin x/x$ pulses.

every $t=\pm nT$. In other words, if the ideal low-pass filter cutoff frequency is $1/2T$, the channel will process data with a rate $R=1/T$. This rate is called the Nyquist rate. At the Nyquist rate there is no interference between symbols under ideal conditions.

In practice such a channel is not used both for the filter realization reasons given above and because of ISI considerations: if the timing from pulse to pulse is not exact, then at the sampling instant (impulse response peak) at the receiver the impulse responses for the other inputs will be nonzero. For this ideal low-pass filter the ISI from other pulses will add to change the amplitude of the pulse considered and cause an error in its detection at the receiver (see Figure 4.5).

In practice [1, 3, 4] filters are used which approximate the ideal low-pass filter except that the amplitude characteristic of the spectrum falls off more gradually. In such a filter a linear phase characteristic can be attained. The interference caused by other pulses is reduced.

If the filter amplitude characteristic is real and has odd symmetry about the cutoff frequency, then the zero crossings of the resulting impulse response will include at least those present in the $\sin x/x$ response. Figure 4.6 shows a representative filter* which has a sinusoidal roll-off and whose incremental frequency response relative to the cutoff frequency is given by

$$X(u)=\begin{cases}\dfrac{1}{2}\left(1-\sin\dfrac{\pi u}{2\omega_x}\right), & |u|<\omega_x \\[2mm] 0, & u>\omega_x \\[2mm] 1, & -\omega_1<u<-\omega_x\end{cases}$$

where $\omega_1=2\pi f_1$

$u=\omega-\omega_1$

$\alpha=\omega_x/\omega_1$

and α is the roll-off factor.

*There are an infinite number of filters satisfying the Nyquist symmetry criterion.

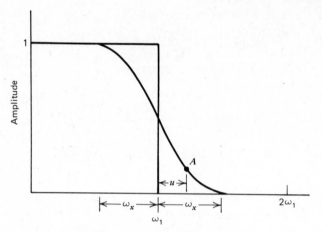

Figure 4.6 Nyquist filter.

Note that for $\alpha = 1$, namely $\omega_x = \omega_1$ (100% roll-off),

$$X(u) = \frac{1}{2}\left[1 - \sin\frac{\pi(\omega - \omega_1)}{2\omega_1}\right]$$

$$= \frac{1}{2}\left(1 + \cos\frac{\pi\omega}{2\omega_1}\right)$$

$$= \cos^2\frac{\pi\omega}{4\omega_1}$$

and the response is a cosine-squared or full-cosine roll-off spectrum. The

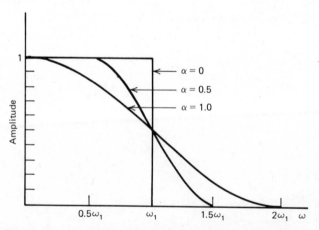

Figure 4.7 Sinusoidal roll-off filter for impulse transmission.

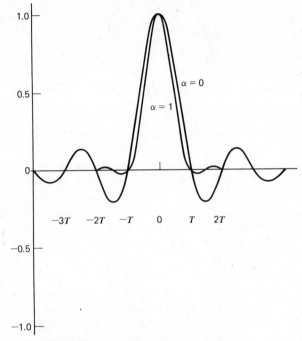

Figure 4.8 Normalized impulse response.

impulse response corresponding to the sinusoidal roll-off spectrum is given by

$$h(t) = \frac{\omega_1}{\pi} \frac{\sin \omega_1 t'}{\omega_1 t'} \frac{\cos \omega_x t'}{1 - (2\omega_x t/\pi)^2}$$

where t' denotes that in fact a realizable filter will have a transmission delay. Figure 4.7 gives the spectral amplitude response as a function of the spectral

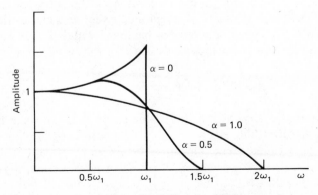

Figure 4.9 Sinusoidal roll-off filter for NRZ transmission.

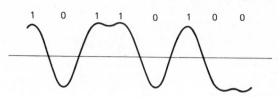

Figure 4.10 Baseband waveform with full-cosine filtering.

roll-off factor α. For this filter Figure 4.8 shows the normalized impulse response.

In data transmission NRZ pulses are transmitted rather than impulses. The NRZ pulse is a rectangular pulse of duration equal to the sampling interval. The NRZ spectrum has a $\sin^2 x / x^2$ characteristic, not a flat one as with impulses. To attain an overall Nyquist response, the channel filter must first perform an inverse transformation (i.e., multiply the incoming spectrum by $x/\sin x$) and then apply the required roll-off filtering. The overall characteristic is then modified as shown in Figure 4.9. Figure 4.10 shows a typical baseband pulse train shaped by a full-cosine filter.

In the Nyquist characteristic the upper frequency limit varies from ω_1 to $2\omega_1$, depending upon the sinusoidal roll-off factor α. The ideal low-pass filter is described by $\alpha = 0$. As α approaches 1 the cutoff is more gradual. The time waveforms have less oscillation in the tails and are therefore less sensitive to timing jitter. With $\alpha = 1$ a full-cosine characteristic results and the bandwidth is $2\omega_1$.

In general, a tradeoff exists. The transmission rate is normally given by

$$R = mB, \qquad 1 \leq m \leq 2$$

where B is the total occupied bandwidth. Values of m approaching 2 require greater care in system design. Often a 50% roll-off characteristic is selected. For this roll-off the low-pass occupied bandwidth will be 0.75 (i.e., $B = \frac{3}{4} R$).

In linear carrier modulation systems the occupied bandwidth is twice the low-pass bandwidth, owing to the translation of the spectrum to the radio frequency. The filtering can be accomplished either at a carrier frequency or the rectangular pulses can be filtered at baseband with low-pass filters. Linear frequency translation assures the equivalence of the two filtering approaches.

Factors Affecting Pulse-Shape Selection

The ISI present in a waveform is determined by the composite transmitter, channel and receiver filter characteristic. The apportioning of parts of the total characteristic between the elemental filters is an important system design problem [3, 5]. Several systems in which such analyses are considered are reviewed in Section 4.5.

The division of the composite filter characteristic is determined by two considerations:

1 Minimizing the effects of noise at the receiver.
2 Limiting the transmitter output spectrum to avoid interference into adjacent channels.

Note that while the system may contain many filters it is the overall pulse shaping seen at the receiver that determines the ISI at the receiver detector.

It has been shown in Chapter 2 that the optimum receiver consists of a filter matched to the received signal followed by a sampler that samples the receiver filter output at time $t = T$, where T is the sampling interval. For the bandlimited channel there is the additional requirement that no ISI be present at the receiver filter output at the sampling instant.

Let the transmitted signal have frequency response $X(\omega)$ and the receiver filter have response $H(\omega)$. Let $F(\omega)$ be the overall Nyquist filter response (the channel bandwidth is accounted for by the transmitter filter). The filter characteristics are related by

$$F(\omega) = X(\omega)H(\omega)$$

In the linear channel the theoretically best division of filter shapes between transmitter and receiver is one where the Nyquist filter response is equally shared. This means that each filter response is the square root of the overall characteristic if impulses are assumed to be present at the transmitter input. This is consistent with the following two criteria:

1 ISI $= 0$ at the sampling instant.
2 The receiver filter should be matched to the received signal.

Usually, the modulator uses NRZ pulses rather than impulses and compensation is required. Perfect matching demands that this compensation occur at the transmitter. For the single linear channel, the transmitter and receiver filter characteristics will not be the same. However, the Nyquist roll-off will be equally shared. This creates difficulties, because of the requirement to have separate designs and consequently different production filters. Compromises in performance are made so as to have identical filters at the transmitter and receiver or else to allocate all filtering to one site. Fortunately, performance degradation is only a few tenths of a decibel [5].

In the nonlinear channel, pulse superposition no longer applies. Whereas Nyquist's criterion holds in the linear channel, it is no longer applicable to the nonlinear case. Nyquist shaping is meant to reduce ISI at the sampling instant. However, at all other times the signal waveform amplitude is dependent upon pulse patterns and their oscillations. A saturating amplifier output is dependent upon the data sequence, and therefore the assumption of the message comprising a sequence of identical overlapping pulses as required by matched filter theory is no longer valid. If a portion of the overall filtering is done at the transmitter, the signals received at the transmit Earth station and satellite nonlinearities will contain ISI, which will eventually create problems at the receiver.

For the Nyquist filters, as the roll-off factor α tends to 0, the corresponding pulses will contain more and more oscillations in their tails, thus increasing the envelope fluctuations of the transmitted signals. The increase in envelope fluctuations causes increased distortion at the nonlinearity output. If filters are selected on the basis of linear channel theory, it is intuitive that increasing the spectral cutoff rate will result in increased performance degradation.

The selection of the filters in the satellite channel will then be based upon optimum performance in consideration of channel constraints such as adjacent channel interference, satellite operating point (degree of distortion introduced by the nonlinearities), noise and the practicalities and economics of the filter structures. While much has been written for the linear channel, solutions for the nonlinear channel are based largely upon computer simulations (see Section 4.5 for examples).

4.3   TRAVELLING-WAVE-TUBE (TWT) NONLINEARITY

The high-frequency, large-output TWT amplifier used in satellite communications exhibits two nonlinear distortion effects:

1 A nonlinear output–input power characteristic (amplitude modulation to amplitude modulation or AM/AM conversion).
2 A nonlinear output phase–input power characteristic (amplitude modulation to phase modulation or AM/PM conversion).

The single carrier power transfer curve and phase shift of an INTELSAT IV TWT are shown in Figure 4.11. Note that a different characteristic applies to multicarrier operation. For low input levels the output power is essentially a linear function of the input power. As the input drive increases, the output power increases nonlinearly until a point is reached at which any additional input level increase results in a *decreasing* output power. This point of maximum output power is referred to as saturation. The operating point of a TWT is usually given in terms of input or output backoff from saturation, that is, the input or output power in decibels relative to the saturation condition. Backoff is given as a positive quantity. For example, 3 dB output backoff means that only half of the possible output power is used.

The saturation region tends to be rather broad; that is, maximum output power occurs for a range of input drive. One definition of saturation is the

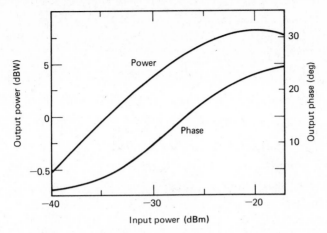

Figure 4.11 INTELSAT IV single-carrier TWT curves.

"7/11" rule of thumb. This states that saturation is the operating point from which an 11-dB decrease in input power results in a 7-dB decrease in output power.

In general, a transponder will carry a single wideband modulated carrier or a number of narrowband carriers separated in frequency over the transponder bandwidth. Because the TWT is a nonlinear device, it will cause the following effects once the signals approach the amplifier saturation region:

1 In the case of several carriers, it will introduce intermodulation interference. Signals will be suppressed and there will be disproportionate power sharing.

2 In the case of a single carrier, any envelope fluctuation will be translated into an unwanted phase modulation.

To maximize the available power of the TWT it is desirable to have a large drive level. An optimization is required to maximize output power and minimize the degradation due to distortion. In a multicarrier configuration the determination of the number of permissible carriers and their relative frequency locations in the band are important. In the case of a single carrier, the optimum drive level is required.

Significant effort, for example Refs. 6 to 9, has been made in attempting to develop analytic expressions that characterize the TWT. One technique [8, 9] entails determining approximations for the envelope nonlinearities in a quadrature model for the TWT (see Figure 4.12).

For an input signal

$$s(t) = A(t)\cos[\omega_c t + \theta(t)]$$

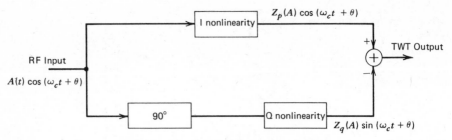

Figure 4.12   Quadrature model of TWT.

where $A(t)$ is the signal envelope and $\theta(t)$ is a phase function, the TWT output signal is given by

$$z(t) = Z_p(A)\cos\left[(\omega_c t + \theta(t)\right]$$

$$- Z_q(A)\sin\left[\omega_c t + \theta(t)\right]$$

$$= U(A)\cos\left[\omega_c t + \theta(t) + g(A)\right] \tag{4.2}$$

where

$$U(A) = \sqrt{\left[Z_p(A)\right]^2 + \left[Z_q(A)\right]^2}$$

$$g(A) = \tan^{-1}\left[\frac{Z_q(A)}{Z_p(A)}\right]$$

Here $U(A)$ and $g(A)$ are the AM/AM distortion and AM/PM conversion, respectively.

The quadrature model can be used to assess the link performance of a modulation technique by means of a computer simulation. The quadrature components $Z_p(A)$ and $Z_q(A)$ are obtained by converting the measured power and phase characteristics of an actual TWT amplifier into quadrature components using the relationship given in (4.2). Samples of the quadrature curves which provide sufficient resolution can be stored and used to determine quadrature output values for a range of input values by means of interpolation. Alternatively, approximations to these quadrature curves can be used on a "best-fit" basis. For the INTELSAT IV TWT (Hughes Corporation 261H tube) these two envelope nonlinearities are given by a least-squares fit [9]

$$Z_p(A) = C_1 A e^{-C_2 A^2} I_0\left[C_2 A^2\right]$$

$$Z_q(A) = S_1 A e^{-S_2 A^2} I_1\left[S_2 A^2\right]$$

where $I_n(\alpha)$ is the modified Bessel function of order n and argument α, A refers to $A(t)$ and the constant coefficients are given by

$$C_1 = 1.61245, \qquad C_2 = 0.053557$$

$$S_1 = 1.71850, \qquad S_2 = 0.242218$$

Polynomial approximations can be used as well, but more coefficients are required to be determined as compared with the four indicated by these expressions.

As an indication of the type of performance measured with a TWT, Figure 4.13 [10] depicts the P_b measured at an Earth station in the TELESAT TDMA link. In this link a 61.248-Mbps QPSK signal was transmitted to a satellite and received from the satellite at a distant station. The transmitting station varied its transmit power (satellite input backoff) while the P_b was measured at the distant station. It is seen that for decreasing satellite input backoff (increasing transmit power) the P_b decreases. This is the thermal noise region where the E_b/N_0 is increasing. Close to saturation, degradations due to the nonlinearity dominate. It is seen that an optimum operating point exists at which the minimum P_b is attainable. This operating point of 2-dB input backoff is a typical value.

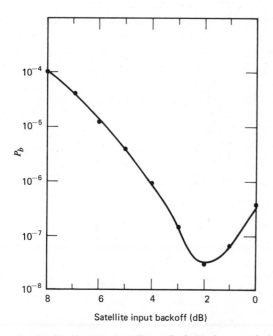

Figure 4.13 Input back-off effect on P_b. (From Ref. 10; by permission of the World Telecommunications Forum, *Telecom' 79*, International Telecommunications Union, Geneva, 1979.)

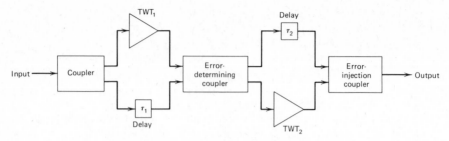

Figure 4.14   Feed-forward TWT compensation.

The TWT parameters can be compensated so that the overall characteristic approaches that of a linear amplifier. Two techniques that can be used are feed-forward compensation [11] and predistortion [12]. Feed-forward compensation refers to a cancellation technique (see Figure 4.14) whereby the signal is divided between two parallel paths, one comprising a TWT(TWT$_1$) and the other a low-level delay line. The delay line is matched to the transit time through TWT$_1$. An error signal determined by comparing the outputs of TWT$_1$ and the delay line is amplified in a subsequent TWT(TWT$_2$). The output of TWT$_1$ is delayed by an amount equal to the transit time of TWT$_2$. The amplified error signal and the output of TWT$_1$ are compared in a coupler whose output is the required compensated signal.

The feed-forward technique does not incorporate a closed loop and therefore is unconditionally stable. With properly matched elements a power-enhancement advantage of 2 to 3 dB is possible [11, 13]. Cost is a factor with this technique because of the requirement for matched elements and a second TWT.

Predistortion entails using a predistorting nonlinearity which in conjunction with the TWT generates an approximately piecewise-linear combined characteristic having negligible AM/PM. An advantage of this method is that predistortion can be attained by baseband or IF design techniques because it is the signal envelope that is being altered. Predistortion does not require a compensating TWT and therefore is a less costly approach.

4.4   MEASUREMENT OF E_b/N_0

The parameter E_b/N_0 is the parameter used as the figure of merit at a specified P_b in a digital communications system. It is calculated as follows:

$$\frac{E_b}{N_0} = \frac{C}{N} \times \frac{B}{R}$$

$$= 10\log_{10}\left(\frac{C}{N}\right) - 10\log_{10}\left(\frac{R}{B}\right) \qquad \text{dB}$$

where C/N = carrier-to-noise ratio in demodulator bandwidth
    R = overall serial transmission data rate
    B = demodulator noise bandwidth

Measurement of C/N after the demodulator filter is not always convenient. It can be measured at IF at the demodulator input if a filter of known noise bandwidth (B_N) is used to measure the noise power. The noise spectral density is then given by

$$N_0 = \frac{N}{B_N}$$

Although the 3-dB filter bandwidth is an approximation, it is more accurate to use the noise bandwidth.

B_N is related to the actual filter characteristic by

$$B_N = \frac{1}{|H(f_c)|^2} \int_0^\infty |H(f)|^2 \, df$$

As shown in Figure 4.15, the noise bandwidth is the bandwidth of a rectangular filter having the same square amplitude as that of the filter characteristic at centre frequency and the same area under the amplitude curve. By integrating the measured characteristic of the filter and normalizing it, one can calculate the noise bandwidth.

The determination of E_b/N_0 in a satellite link requires that the measurement be made at the satellite operating point. This is because the noise

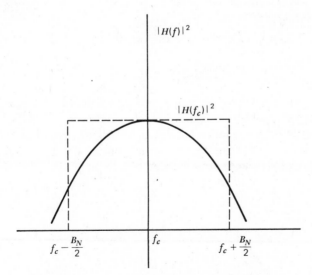

Figure 4.15 Equivalent noise bandwidth filter.

contribution of the TWT is a function of drive level. E_b/N_0 can be measured by means of a filter prior to the demodulator input which is narrow compared with the transponder. The procedure is as follows:

1   Measure the received $C+N$ at the demodulator input by transmitting an unmodulated carrier at band centre at the operating point. The calibrating filter is assumed to be centred at the received IF.

2   Offset the carrier so that the received signal falls outside the filter bandwidth. The filter bandlimits the noise added by the system.

3   Measure the filter noise output power. Subtracting this value from the power measured in (2) gives the actual carrier power and C/N.

4   Measurement of the noise bandwidth of the narrowband filter leads to the calculation of E_b/N_0.

4.5   COMPARISON OF QPSK, OK-QPSK AND FFSK

The three modulation techniques QPSK, OK-QPSK and FFSK have received significant attention because of their properties, which are well suited to the satellite link. These characteristics include power and bandwidth efficiency as well as a constant envelope.

In the comparison of modulation techniques, their sensitivity to a number of parameters is of interest. These include

1   Thermal noise
2   Bandlimitation and delay distortion
3   Adjacent channel interference
4   Co-channel interference
5   Phase and amplitude nonlinearities present in amplifiers such as TWTs

In this section these parameters are of principal concern. Relative merits in terms of implementation and cost tradeoffs are not included but are relevant. Although other modulation techniques such as BPSK are being used in certain satellite systems, these are not considered here within the context of bandwidth and power efficient modulation techniques. It should be noted that although many new modulation techniques are being reported, the use of a scheme in an operational environment is usually preceded by extensive analytical and experimental investigation. The inertia to introduce new processing techniques may be significant.

Performance in the Presence of Thermal Noise

For the case where any one of the modulations is transmitted unfiltered through a channel which adds noise whose spectral density is constant with frequency (additive white Gaussian noise, AWGN), the probability that a bit is in error is the same as that for antipodal signalling. This was shown to be

$$P_b = \frac{1}{2}\mathrm{erfc}\left[\sqrt{\frac{E_b}{N_0}}\right]$$

and is marked as the coherent PSK or CPSK curve in Figure 2.9. This result assumes that the demodulator performs a matched filter detection (rectangular impulse response for QPSK and OK-QPSK and a half-sinusoid for FFSK) with perfect carrier and clock recovery. In practice, regeneration of the carrier at the receiver is accomplished by a nonlinear operation. In the case of QPSK, the nonlinearity is a ($\times 4$) device and for FFSK it is a doubler. In either case all or a significant portion of the modulation is removed, leaving carrier-related tones. Division by 4 or 2 results in a recovered carrier with a phase ambiguity. If this carrier is used for demodulation, the decoded bits could be all correct or all inverted, depending upon the phase of the recovered carrier. By encoding the data as differences between adjacent symbols, the effect of the ambiguity is removed. This is called differential coding. Differential coding was reviewed in Section 3.1 but is explained here in detail with BPSK used as an example.

Suppose that the sequence of input bits to the differential encoder is 1011. The differential encoder is assumed to first transmit a 1. This bit is compared with the first input bit. If they are the same, a 0 is transmitted; if they are different, a 1 is transmitted. The rule of encoding is: the next transmitted bit is the EXOR of the previous transmitted bit and the input bit. At the receiver, the demodulator output bits are differentially decoded by comparing adjacent bits. If they are the same, the source bit was a 0; if they are different, the source bit was a 1.

The following shows the decoder output with and without an ambiguity.

Transmitter

Input bits   1 0 1 1

Transmitted bits   1 0 0 1 0

Receiver—No Ambiguity

Demodulator output   1 0 0 1 0

Decoder output   1 0 1 1

Receiver—Ambiguity (Bits Inverted)

Demodulator output   0 1 1 0 1

Decoder output   1 0 1 1

Note that by encoding the information as differences between bits, the possibility of error due to inversion of all bits is removed. The resolution of the ambiguity occurs with a penalty. If one demodulator output bit is in error, the decoder output will tend to have double errors because 2 bits are compared for each output bit, as indicated by the following:

3 adjacent demodulator bits: a   b   c

(b incorrect)

decoded bits   d   e

(both wrong)

A similar explanation can be applied to the modulation techniques under discussion. The error rate for all three modulations is given by the following if differential encoding is used:

$$P_{\text{DECPSK}} = 2P_e(1 - P_e)$$

where DECPSK refers to differentially encoded coherent PSK and P_e is the channel bit-error rate. Note that for $P_e \ll 1$, the differentially encoded error rate is twice the uncoded value. P_{DECPSK} is shown in Figure 2.9 and exhibits a 0.5-dB degradation relative to P_{CPSK} for $P_e = 10^{-4}$. It should be noted that differential encoding is not always necessary. If the original source of the data bits is a data generator, the bits must be received without inversion. However, if the original source was voice that was digitized by means of a delta encoder, then ambiguity resolution is not required. This feature can be exploited especially at low E_b/N_0 to make efficient use of the available power in the satellite link. In this case the relevant error-rate curve is that for P_{CPSK}.

Spectra

PSK Systems

It can be shown that the power spectral density relative to centre frequency for PSK systems is given by

$$\mathcal{G}(f) = \frac{2P}{R_S} \frac{\sin^2(2\pi f/R_s)}{(2\pi f/R_s)^2} \tag{4.3}$$

where P = modulated signal power
R_s = symbol rate

This is the well-known $(\sin x/x)^2$ characteristic. The assumption here is that rectangular pulses are used as inputs to modulate the quadrature carriers and that filtering is not used. The expression of (4.3) applies to BPSK, QPSK, $8 - \phi$ PSK and so on. If the input bit rate is R then $R_s = R$ for BPSK, $R_s = R/2$ for QPSK, $R_s = R/3$ for $8 - \phi$ PSK and so on. Consequently, QPSK occupies one-half the bandwidth of BPSK and $8 - \phi$ occupies one-third the bandwidth

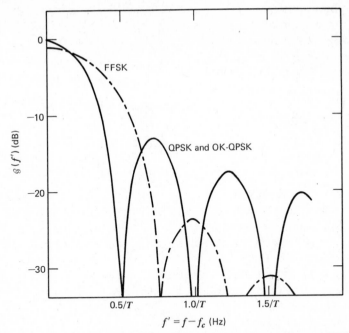

Figure 4.16 Power spectra of unfiltered QPSK, OK-QPSK and FFSK.

of BPSK. When it is said that QPSK occupies one-half the bandwidth of BPSK, it is meant that the spectral shape is exactly the same, but is compressed by a factor of one-half and that spectral nulls are at one-half the frequency separation from the carrier as for BPSK.

With QPSK, the main lobe contains 92.5% of the power and spectral nulls are at $\pm(\frac{1}{2} \times \text{bit rate})$ from the carrier centre frequency. The first side-lobe peak is 13.3 dB below the main lobe peak and the spectral roll-off is 6 dB/octave. The QPSK spectrum is given in Figure 4.16. Offset QPSK has the same spectrum as QPSK. A relative delay in the I and Q channels of QPSK does not affect the ideal spectrum.

FFSK Systems

FFSK is a phase-coherent binary frequency-shift-keyed modulation with a modulation index of $h=0.5$. Bennett and Davey [3] have shown that the spectral density for continuous phase FSK systems is given by

$$\mathcal{G}(\omega) = \frac{2A^2 \sin^2\left[(\omega-\omega_1)/2\right] T \sin^2\left[(\omega-\omega_2)/2\right] T}{T\left[1 - 2\cos(\omega-\alpha)T\cos\beta T + \cos^2\beta T\right]} \left(\frac{1}{\omega-\omega_1} - \frac{1}{\omega-\omega_2}\right)^2$$

$$+ \frac{2A^2 \sin^2\left[(\omega+\omega_1)/2\right] T \sin^2\left[(\omega+\omega_2)/2\right] T}{T\left[1 - 2\cos(\omega+\alpha)T\cos\beta T + \cos^2\beta T\right]} \left(\frac{1}{\omega+\omega_1} - \frac{1}{\omega+\omega_2}\right)^2$$

where T=bit duration
$\quad A$=signal amplitude
$\quad \omega_1, \omega_2$=signalling angular frequencies
$\quad \alpha=(\omega_2+\omega_1)/2$
$\quad \beta=(\omega_2-\omega_1)/2$

For the special case of FFSK the power spectral density with respect to the centre frequency is given by

$$\mathcal{G}(f)=\frac{8P_c}{\pi^2 R}\cdot\frac{1+\cos4\pi f/R}{\left[1-(16/R^2)f^2\right]^2}$$

where $R=T^{-1}$ is the bit rate.

With FFSK the main lobe contains 99.5% of the power and spectral nulls are at $\pm(0.75\times$bit rate) from the carrier. The first side lobe in 23 dB below the main lobe and the spectral roll-off is 12 dB/octave. The FFSK spectrum is given in Figure 4.16. Note that the FFSK main lobe is wider than that of QPSK, but decays at a faster rate. Figure 4.17 depicts the relative out-of-band power as a function of the bandwidth B normalized to the binary data rate for the two modulation techniques.

The out-of-band power is given by

$$G_{\text{ob}}=1-\frac{\displaystyle\int_{-B}^{B}\mathcal{G}(f)\,df}{\displaystyle\int_{-\infty}^{\infty}\mathcal{G}(f)\,df}$$

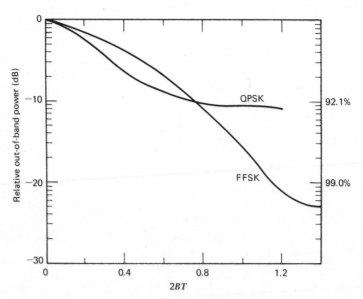

Figure 4.17 Relative out-of-band power.

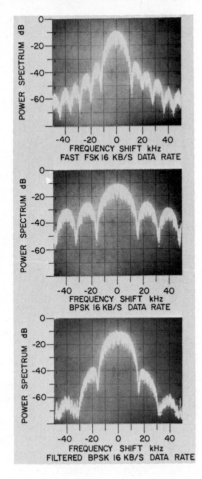

Figure 4.18 PSK and FFSK spectra. (Courtesy J. L. Pearce.)

where $\mathcal{G}(f)$ is the spectrum relative to centre frequency. Note that the equal power bandwidth occurs at an offset from the carrier frequency of $\sim 0.76/T$. Closer to the carrier QPSK has more relative power, while beyond this crossover point FFSK contains more power.

Figure 4.18 depicts photographs of FFSK and BPSK modulator outputs. Note that the filtered BPSK photograph refers to a BPSK signal at the output of a 16-kHz Butterworth filter with a 3-dB bandwidth equal to the data rate. It should be recognized that a QPSK spectrum will be identical to the BPSK result except for a scale change.

Adjacent Channel Interference

Adjacent channel interference (ACI) refers to the degradation in performance in one channel arising from spillover from an adjacent channel. The proximity with which channels can be located (the separation of their centre frequencies),

Table 4.1   Centre   Frequency   Separation   for
E_b/N_0 **Loss of 1 dB Maximum**

Modulation	Relative Interference Level (dB)	
	0	+10
QPSK	4.5R	13.5R
OK-QPSK	5.0R	14.0R
FFSK	1.5R	2.5R

Source: Ref. 14.

assuming the same type of modulation and data rate, is determined by the spectral roll-off and the width and shape of the main spectral lobe. It has been seen that the main lobe of FFSK is wider than for QPSK, but the sideband roll-off is sharper. The power crossover point occurs for a bandwidth of $0.76/T$.

White [14] has performed an analysis of the effects of worst-case crosstalk between adjacent channels. This is indicated by the portion of the output of an integrate-and-dump matched filter which is due to the interfering signal. Two mutually interfering waveforms of the same modulation type were used in the analysis and transmit filtering was not included.

Table 4.1 shows the closeness with which modulations can be spaced for no more than a 1-dB E_b/N_0 degradation. The spacing is given as a function of the relative power of the interfering channel to the main channel. These results were based upon a linear AWGN channel with ideal demodulation.

Pearce [15] has reported on experimental measurements using 16-kbps FFSK modems. These measurements indicate that for a channel separation of 25 kHz ($1.6R$) and an error rate of $P_b = 1 \times 10^{-4}$, the adjacent channel power should be no greater than 13.5 dB relative to the signal for a performance (E_b/N_0) degradation of 1 dB. These results are more optimistic than White's, but it should be noted that White's was a worst-case analysis.

For the linear wideband situation it can be concluded that FFSK is superior to QPSK modulations in terms of adjacent channel interference. This means that channels can be spaced closer and less control of transmit power is required. In a multiuser environment with significant uplink power variations, FFSK appears to be very attractive.

Co-channel Interference

Co-channel interference refers to the degradation caused by an interfering waveform appearing within the signal bandwidth. This interference could be, among others, an unmodulated carrier or a low-power interferer of the same modulation type.

Pearce [15] reports that FFSK is no more sensitive to a varying amplitude interference than to a second FFSK signal. It was found that for a $P_b = 10^{-5}$ and no more than a 1 dB, E_b/N_0 degradation the interfering FFSK co-channel signal should be at least -15 dB in relative power.

Spilker [16] and Wachs and Weinreich [17] give E_b/N_0 degradation as a function of signal-to-interference ratio (S/I). With $P_b = 10^{-5}$, approximately a 2-dB E_b/N_0 degradation occurs for QPSK with $S/I \simeq 15$ dB.

Phase Noise

Regeneration of the carrier at the receiver implies that the local carrier reference is imperfect and is a noisy phase reference. That the carrier is not exactly in phase with the received signal means that the I- and Q-channel inputs to the quadrature bit detectors are (1) reduced in level and (2) not mutually independent. The interference that the Q-channel component causes to the I-channel is different for QPSK, OK-QPSK, and FFSK.

In the case of QPSK, the alignment of bits is the same in the two channels and a constant interference results over the I-channel bit interval. With OK-QPSK, during a bit duration in the I-channel a data transition may occur in the Q-channel so that the average interference is reduced. FFSK is similar to QPSK but benefits from a sinusoidal weighting reduction of the interference. The effects of a noisy phase reference on FFSK detection have been investigated in ref. 18. Figure 4.19 gives the E_b/N_0 degradation for the modulation techniques as a function of the SNR of the recovered carrier. The error mechanism is explained in detail in Section 5.1.

Note that in the comparison shown in Figure 4.19, the degradation was determined for a specified recovered carrier SNR. These results do not reflect the ease with which each demodulator attains the SNR value. It should be noted that carrier recovery in FFSK is accomplished by a doubling ($\times 2$) operation, while a times four ($\times 4$) operation is used in QPSK.

Bandwidth Limitations and Phase Distortion

A filter is characterized by an amplitude characteristic and a phase characteristic. Both of these may lead to degradations in performance. The amplitude characteristic alters the modulation by changing the pulse shape. The phase characteristic describes the relative transmission times through the filter of the pulse frequency components. The net effect is pulse smearing wherein successive pulses are affected and pulse amplitude fluctuations results. Intersymbol interference (ISI) refers to the pulse-smearing effect. The envelope fluctuations induced in the modulated signal become important when the modulation is further passed through a nonlinear amplifier such as a TWT, which has an AM/PM transfer characteristic that can cause additional distortions. If the filter at IF is not symmetric about the centre frequency, then for quadrature modulation signals mutual interference between the I and Q channels also occurs.

Jones [19] investigated filter distortion and ISI on PSK signals by means of a computer. One of the cases considered was the transmission of QPSK through a Chebyshev filter having 0.1-dB ripple in the passband. Jones

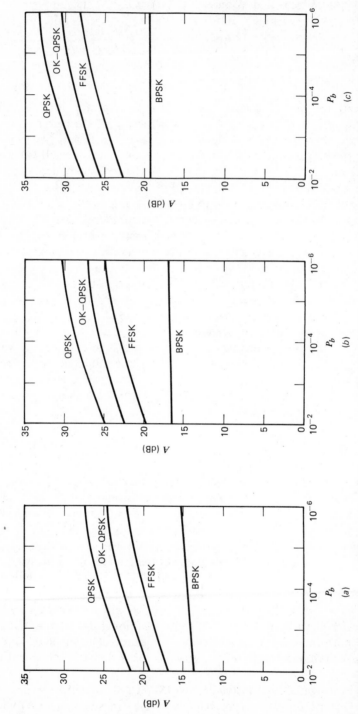

Figure 4.19 Phase reference SNR versus P_b. (a) 0.2-dB loss; (b) 0.1-dB loss; (c) 0.05-dB loss. $A \stackrel{\triangle}{=} 10\log_{10}\alpha$; α, phase reference SNR. (From Ref. 18: © 1978 IEEE; reprinted from "Effect of Noisy Phase References on Coherent Detection of FFSK Signals", by R. Matyas, from *IEEE Trans. Commun.*, June 1978).

determined the degradation from theoretical E_b/N_0 for $P_b = 10^{-6}$ as a function of the signalling rate to 3-dB RF filter bandwidth. Detection was accomplished with an integrate-and-dump filter as well as with a two-pole Butterworth filter having $BT = 1$ ($T =$ symbol duration). Results are given in Figure 4.20. Figure 4.20 indicates that as the transmission symbol rate approaches $0.7 \times$ (filter 3-dB bandwidth), the degradation becomes severe. For example, consider 60-Mbps QPSK transmission with a rate of 30 Msymbols/s. The transmission bandwidth would then be $30/0.7 = 42.9$ MHz. It should be noted that as the filter becomes very narrow, neither the integrate-and-dump nor the Butterworth filters are well matched to the received waveform and degradation can be expected. Jones's results also indicate that the integrate-and-dump detector can be effectively replaced by a simpler filter in band-limited situations.

The effect of phase distortion [19] is shown in Figure 4.21, where the degradation in E_b/N_0 for QPSK is shown for a pure parabolic (no amplitude) distortion. In this graph degradation is plotted versus the parabolic shape determined by the phase at the first null in the QPSK spectrum. Detection is with an integrate-and-dump circuit.

Pearce [15] reports measurements of the effect of an IF bandlimiting filter on the performance of FFSK. It was found that significant E_b/N_0 degradation (>2 dB) occurs once the filter bandwidth is less than $0.9 \times$ (data rate). Included in this degradation is 0.75 dB for modem implementation.

The preceding examples give an indication of the effects of bandlimitation on the performance of QPSK and FFSK. It should be noted that some results are based upon computer evaluation while others represent empirical data, and therefore some care should be taken in making absolute conclusions.

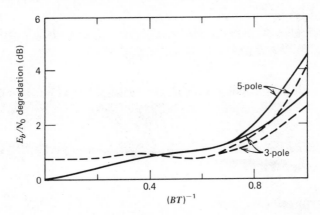

Figure 4.20 Effect of bandlimitation on QPSK. P_b, 10^{-6}; B, 3-dB RF bandwidth; T, symbol duration. Solid line, integrate-and-dump; dashed line, two-pole Butterworth data filter ($BT = 1$). (From Ref. 19: © 1971 IEEE; reprinted from "Filter Distortion and Intersymbol Interference Effects on PSK Signals", by J. Jay Jones, from *IEEE Trans. Commun.*, Apr. 1971.)

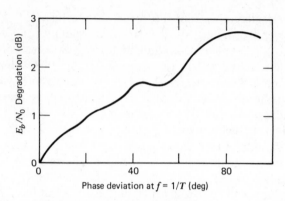

Figure 4.21 QPSK degradation versus parabolic phase distortion P_b, 10^{-6}; B, 3-dB RF bandwidth; T, symbol duration. From Ref. 19: © 1971 IEEE; reprinted from "Filter Distortion and Intersymbol Interference Effects on PSK Signals", by J. Jay Jones, from *IEEE Trans. Commun.*, Apr. 1971.)

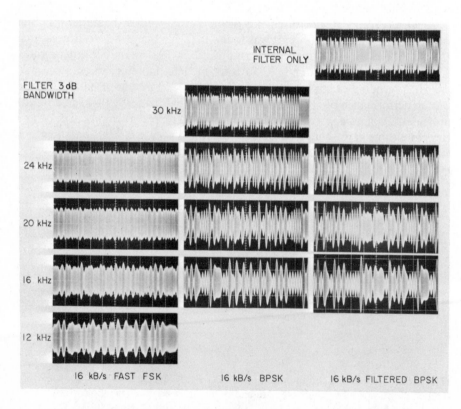

Figure 4.22 Filtered BPSK and FFSK waveforms. (Courtesy J. L. Pearce.)

It should also be realized that better performance may be possible at reduced bandwidth where filters are properly designed to minimize ISI. What these results indicate is that there is a threshold in performance beyond which the modulation is not suitable.

Figure 4.22 depicts the envelope fluctuations of FFSK and BPSK as a function of filter bandwidth. These results are based on measurements of 16-kbps modems. The filter used to determine the envelope fluctuations was a Chebyshev type. The BPSK modem included an internal filter, and the third column of photographs indicates the effect of the cascaded filters. The internal filter had a Butterworth characteristic with a 3-dB bandwidth of 16 kHz. Comparable results would be obtained with a 32-kbps QPSK modulator.

Combined Effects

Previous sections have described individually the effects of various system parameters on the modulation techniques. In any operational system all parameters must be accounted for collectively. System performance evaluation is usually based upon computer simulations and/or hardware implementation and measurement.

Of primary concern is the satellite configuration wherein the modulator output is filtered and the result transmitted by the Earth station to the satellite, where it is amplified by a TWT nonlinearity. The signal is filtered at the receiver and data are recovered by the demodulator. The transmit terminal amplifier effect may also be of interest.

A significant amount of information has been generated during the past number of years comparing QPSK, OK-QPSK and FFSK over satellite links. In attempting to derive general conclusions from this information, care must be exercised because the basic assumptions used in the various evaluations are not always the same. In reviewing this material it is important to consider

1 Whether the systems were equivalent. For example, were the TWTs and the filter the same?
2 How bandwidth was defined.
3 Whether carrier and clock recovery degradations were included.

Effects of Bandlimitation and Hard Limiting

The various modulation techniques are attractive because they exhibit bandwidth efficiency with relatively little envelope fluctuation. These together mean that operation with bandlimited channels having power amplifiers driven to saturation is feasible.

Rhodes [20] used computer simulations to determine the power spectra of the modulation techniques before and after a hard limiter. In addition, he evaluated the effect of filtering the modulated signal prior to the limiter.

Rhodes found that both OK-QPSK and FFSK retain most of the bandlimitation introduced prior to the limiter. In the case of QPSK, side lobes were restored to the unfiltered levels. This result was found for $BT=1$ (B=baseband bandwidth and T=symbol duration). With $BT=0.5$, the postlimiter results were similar to the $BT=1$ case. Although the FFSK result was slightly better than that of OK-QPSK, Rhodes's results indicated little difference between the two.

Miscellaneous Evaluations

Mathwich et al. [21] presented results derived from experimental data on the effects of tandem band and amplitude limiting on MSK. In the experimental configuration the MSK modulator output was passed through a bandlimiting channel which also added noise. The demodulator recovered data by means of an integrate-sample-and-dump circuit. To show the effects of filtering and limiting, only hardwired references were provided from the modulator.

Bit rates of 5, 5.5 and 6 Mbps were used. Two types of transmit and receive filters were used: (1) Gaussian to the 6-dB point and (2) a three-pole Butterworth. The Gaussian filters had 3-dB RF bandwidths of 3.56, 5.2 and 6.8 MHz. One Butterworth filter was used which had a 3-dB bandwidth of 5.5 MHz. Channel limiting was provided by a wideband Hewlett-Packard Corporation 461A amplifier driven 10 dB into saturation. The maximum static AM/PM coefficient was 1.1 degrees/dB, dropping to 0 degrees/dB for ± 3-dB level variation about the worst-case drive point. Parametric data were obtained by using combinations of the data rates and channel filters.

The MSK spectrum was measured after filtering and limiting. The results based on a three-pole Butterworth prelimiter filter and $BT=1.1$ compared well with Rhodes's computer simulations with a four-pole Butterworth filter giving $BT=1.0$ and an ideal limiter. The suppression of the first side lobe at the limiter output was about 7 dB compared with the unfiltered MSK side lobe. It was also found that while narrow filters could be used, the limiter expanded the spectrum such that the 99% power bandwidth occupied nearly the same channel bandwidth as for unfiltered MSK (i.e., $\sim 1.15R$), although there was a reduction in the side lobes.

The E_b/N_0 degradation at $P_b=10^{-5}$ was determined by using combinations of the data rates and transmit–receive filters. The conclusion reached was that allowing a degradation of no more than 0.5 dB requires that the channel bandwidth be greater than $0.65R$. It should be reiterated that carrier and clock recovery degradations were excluded.

Gronemeyer and McBride [22] compared MSK and OK-QPSK over nonlinear and severely bandlimited channels. These comparisons were made by means of digital computer simulation to estimate bit error-rate performance on double-hop satellite links. Although the receivers were not optimum for the received pulse waveforms, the results were presented as providing relative performance measures.

The simulations were performed at baseband. The modulator output was filtered, noise was added and the result was hard-limited. Noise was added to the limiter output. The result was filtered in the receiver and then passed through a matched filter for bit detection. The matched filter was matched to the rectangular or sinusoidal transmitter pulse shape.

The transmit and receive system filters were chosen to be identical, with channel bandwidth defined as the noise bandwidth of the cascaded filter response. These filters were seven-pole, 0.1-dB-ripple Chebyshev low-pass equivalent filters. The condition of downlink limited operation was of primary interest. The effect of imperfect reference signals was not included. The hard-limiter model did not include AM/PM effects.

The results indicated that a crossover in relative E_b/N_0 performance occurred. For channel bandwidths greater than $1.1R$, MSK was superior to OK-QPSK. For $B<1.1R$, the converse was true. With the channel bandwidth $B=R$ and $P_b=10^{-7}$, the E_b/N_0 degradation was found to be about 1 dB, degrading sharply for more reduced bandwidths. The authors noted that reception was not optimum and that improved performance could probably be attained by means of phase equalization.

Taylor et al. [23] performed a computer simulation study to evaluate the performance of high-data-rate satellite communications using several digital modulation techniques. The single-carrier wideband case was analysed.

The channel model consisted of a cascade of a transmit filter, a nonlinearity and a receive filter. The simulation was performed at baseband. The nonlinearity was represented by a quadrature model which represented a curve-fitting approximation of the Hughes Corporation 261-H TWT. This TWT has been used in several satellites, such as INTELSAT IV. The transmit and receive filters used were four-pole, 0.5-dB-ripple Chebyshev. Only downlink noise was considered. Perfect carrier references were assumed at the receiver. Symbol timing was derived from the data. Integrate-and-dump matched filters were used.

It was found that for a $P_b=10^{-4}$, OK-QPSK and FFSK performed better than QPSK for $1<BT\leq2$ ($T=$symbol duration, $B=3$-dB bandwidth). In this range the E_b/N_0 degradation was ~1 dB. The difference between OK-QPSK and FFSK as far as E_b/N_0 degradation is concerned was not significant, while QPSK was 0.5–1.0 dB worse. For $BT\leq1$ all modulations performed poorly, but the relative order of performance was the same. The findings applied for 12- or 1-dB input backoff from TWT saturation.

Fielding et al. [24] compared 600 and 800 Mbps OK-QPSK* and MSK in laboratory measurements and modeled the system by computer simulation. The two-hop satellite configuration consisted of a modulator, a wideband IF amplifier, an up-converter, a TWT, a down-converter, a 650-MHz-wide IF amplifier, an up-converter, a TWT, and a down-converter whose output was

---

*Although the authors refer to QPSK in their paper, the text actually describes an OK-QPSK system.

corrupted by additive noise and the demodulator. The TWTs had 1-GHz bandwidths and 7.5-degree/dB AM/PM conversion. Bit synchronization was performed in the receiver. Matched filters at these data rates were difficult to produce, so two-pole Butterworth detection filters were used instead. The 3-dB lowpass bandwidth of the detection filters was approximately $R/4$.

Measurements and computer simulation both indicated that MSK required 1 to 2 dB less E_b/N_0 at $P_b = 10^{-5}$ than did OK-QPSK. Note that the minimum channel bandwidth was introduced by the second IF amplifier. The corresponding $BT = 1.08$ and 0.81 for 600- and 800-Mbps operation. Table 4.2 gives the E_b/N_0 required for the complete link for MSK and OK-QPSK to attain $P_b = 10^{-5}$. This is to be compared with a theoretical $E_b/N_0 = 9.9$ dB required for $P_b = 10^{-5}$.

The impact on overall link performance of the 3-dB low-pass detection filter was assessed in terms of the E_b/N_0 required for $P_b = 10^{-5}$. This is shown in Table 4.3. These results show that MSK is less sensitive than OK-QPSK to the detection filter bandwidth.

Murakami et al [25] compared QPSK, OK-QPSK and MSK over wide- and narrowband channels. The analysis was based upon computer simulation. The narrowband model was representative of an INTELSAT channel, while the wideband channel applied to a Japanese domestic system.

A modulator output was filtered and passed through a high-power amplifier (HPA) and combined with similar outputs of upper and lower adjacent channels. The composite signal was passed through a filter which suppressed the adjacent channels. The result was passed through a TWT. Noise was added at the TWT output. The noisy signal was passed through the receiver filter and the data recovered. In the narrowband model the channel spacing was $1.3R$ and in the wideband case it was $2.3R$. The adjacent channel suppression filter had a sinusoidal roll-off with a suppression of 6 dB at the channel-spacing separation point. The roll-off factor α was 0.15. Four combinations of transmitter and receiver filters were used. These included a Nyquist shaping filter together with an aperture equalizer and a sharp-cutoff filter (sinusoidal roll-off factor = 0.1).

In the simulation the filters were optimized with the Nyquist shaping filter roll-off factor and the sharp-cutoff filter cutoff frequency as the two parameters. The optimization was based upon determining the P_b at the demodulator output. Carrier and clock references were assumed connected directly from the modulator. It was found that the best combination of filters for the nonlinear

Table 4.2 End-to-End Performance: E_b/N_0 Required for $P_b = 10^{-5}$ (dB)

	MSK	OK-QPSK
600 Mbps	11.6	13.4
800 Mbps	13.8	15.1

Source: Ref. 24.

Table 4.3 E_b/N_0 Degradation versus Detection Filter Bandwidth ($P_b =$ 10^{-5})

Filter 3-dB Bandwidth	600 Mbps		800 Mbps	
(MHz)	MSK	OK-QPSK	MSK	OK-QPSK
130	2.1	—	5.2	—
145	1.9	—	—	—
160	2.3	—	4.1	6.2
175	2.0	5.1	—	5.5
190	1.9	3.8	4.4	6.9
205	—	4.0	—	—
220	—	—	4.6	—

Source: Ref. 24: © 1977 IEEE; reprinted from "Performance Characterization of a High Data Rate MSK and QPSK Channel", by R. M. Fielding, H. L. Berger and D. L. Lochhead, from *Conf. Rec. 1977 Int. Conf. Commun.*, 12–15 June 1977, Chicago, (77CH1292-2 CSCB).

channel consisted of a sharp-cutoff filter at the transmitter and a Nyquist shaping filter with equalizer at the receiver for the TWT backoff points considered (0 to 12 dB input back-off).

Which of the three modulation techniques was best was determined primarily by whether the channel was wideband or narrowband and to a lesser degree by the HPA/TWT drive levels, as shown in Table 4.4. In Table 4.4 large backoff refers to a 14-dB HPA input backoff and a 4-dB TWT input backoff, while saturation refers to saturation of both amplifiers. These results indicate that for large backoff, MSK is best. MSK and OK-QPSK are best for the wideband model. For the narrowband saturated application, QPSK is best.

The carrier phase rms jitter was found to be 0.4, 1.0 and 1.3 degrees for QPSK, MSK and OK-QPSK, respectively, in the wideband model. In the narrowband case, the values were 0.9, 5.0 and 5.0 degrees, respectively. These results would indicate that QPSK suffers the least phase jitter. A difficulty in generalizing these conclusions arises from the fact that carrier recovery was based upon a specific circuit using a ($\times 4$) multiplier, followed by a bandpass

Table 4.4 E_b/N_0 Degradation (dB) for $P_b = 10^{-4}$

	Wideband		Narrowband	
	Large Backoff	Saturation	Large Backoff	Saturation
QPSK	0.7	1.1	1.2	2.8
OK-QPSK	0.3	0.3	1.5	3.5
MSK	0.2	0.4	1.1	4.0

Source: Ref. 25.

filter whose output was divided by 4. As is well known (see de Buda [26]), carrier recovery for MSK is accomplished with a ($\times 2$) circuit.

Lundquist [27] focused on techniques for use in INTELSAT-type transponders where 60 Mbps would be transmitted through 36-MHz channels with 40-MHz channel spacing. Computer simulation was used to show that QPSK would be superior to OK-QPSK and MSK would be worse than OK-QPSK. Simulations were confirmed by measurements. Of specific concern was the effect of pulse shaping on performance. The model consisted of a wanted channel with upper and lower adjacent channel interference. Each signal path included a transmitter nonlinearity as well as the satellite nonlinearity.

It was concluded that with proper Nyquist shaping (composite transmitter/receiver response) a QPSK channel could be designed to operate to within 1 to 2 dB of the E_b/N_0 required for $P_b = 10^{-4}$ in a linear channel. This analysis included degradations due to filters, nonlinear amplifiers and interchannel interference only. Optimum performance was attained with 50% Nyquist shaping at the transmitter and 40 to 50% roll-off at the receiver.

A similar analysis was performed for OK-QPSK. While the transmitter amplifier was saturated for OK-QPSK, the QPSK results assumed no backoff. However, it was claimed that no difference in qualitative results were noted. The TWT was operated at the same drive in both cases. OK-QPSK was found to be more sensitive to the pulse shaping (Nyquist roll-off) than was QPSK. The optimum roll-off was determined to be 50% with a 25%/75% division of the characteristic between transmitter and receiver, respectively. For MSK the best receiver filter was a 100% cosine roll-off.

It was found that adjacent channel interference considerations precluded the use of MSK. The three-channel model was used in comparing QPSK and OK-QPSK. The transmitter amplifier outputs were not filtered. Amplifier drive levels were adjusted to improve linearity. For this case OK-QPSK was worse by 1.3 dB for $P_b = 10^{-4}$ than QPSK for 1 to 9-dB HPA output backoff. The adjacent channel interference reduction provided by OK-QPSK was offset by worse in-band performance.

In single-channel operation with 50% cosine roll-off filters, it was found that OK-QPSK was worse than QPSK by 0.5 dB with a linear channel and 1 dB in the nonlinear case.

Harris [28] considered various modulation techniques for use in the European Space Agency (ESA) 120-Mbps TDMA system with channel spacing of $0.7R$. Citing work by Castellano [29], Harris showed that for a model including half of the overall cosine-shaping done at each of the transmitter and receiver and a saturated TWTA in between, OK-QPSK was better than QPSK for more gradual roll-offs, whereas the reverse was true for sharper roll-off factors. The crossover in performance occurred for a 60% cosine roll-off factor. Although while OK-QPSK provided a slight improvement in uplink adjacent channel interference characteristics, QPSK was chosen as being preferred for the narrowband system.

Finally, Chakraborty et al. [30] addressed the problem of selecting among QPSK, OK-QPSK and MSK for bandlimited nonlinear channels. Included in the evaluation were filter optimization, power/bandwidth tradeoffs, operating point selection and the sensitivity of synchronization. A computer simulation program was used.

For a typical INTELSAT V channel, $P_b = 10^{-4}$, no adjacent channel interference and phase/timing jitter considerations, a transmitter amplifier input backoff of 14 dB and a 4-dB TWTA input backoff, the equivalent power loss was measured as a function of receive BT product (T = symbol duration). A low-pass transmit Nyquist filter with a roll-off factor of 0.45 and a five-section elliptic function receive filter were used. QPSK exhibited the least loss (1 dB) for $BT = 1.1$. Both OK-QPSK and MSK were consistently worse for $0.9 \le BT \le 1.4$, with losses at $BT = 1.1$ in E_b/N_0 of 2 and 3 dB, respectively.

With an adjacent channel interference constraint and a tight transmit $BT = 1$ bandpass filter, the values of E_b/N_0 required for a $P_b = 10^{-4}$ were 16, 16.8 and 17.0 dB for QPSK, MSK and OK-QPSK, respectively. For wider channels ($BT = 1.5$) OK-QPSK was best followed by MSK and QPSK with $E_b/N_0 = 15.8$, 17.3 and >21 dB, respectively. For this case each demodulator used the optimum BT receiver filter. In any event, with and without an adjacent channel interference constraint, QPSK was best for the narrow channel.

The optimum transmit filter was determined to have a 0.45 Nyquist roll-off and a maximally flat five-pole elliptic receive filter with $BT = 1.1$ for mild nonlinearities, while the reciprocal arrangement was best for near saturation. With carrier recovery and clock recovery degradations included, the degradation from theoretical E_b/N_0 for $P_b = 10^{-4}$ was 2 dB with mild nonlinearity and 3 dB at saturation of the transmitter and satellite amplifiers.

4.6 SUMMARY

Several papers have been reviewed for the specific purpose of showing that which of the modulations QPSK, OK-QPSK or MSK is best cannot be simply stated. The selection of a particular optimum modulation technique depends upon the specific system configuration. The system parameters of importance are (1) the BT product that describes the bandwidth efficiency, (2) adjacent channel constraints, (3) co-channel interference, (4) tolerable total E_b/N_0 degradation, (5) the design of the synchronization circuitry and (6) the specific characteristics of the filters used.

In a number of situations MSK has been shown to provide the best performance. Several authors have concluded that MSK and OK-QPSK are preferred for wider-band applications where $BT > 1$, whereas QPSK is best for the very narrowband situation.

Murakami, Lundquist, Harris and Chakraborty [25, 27, 28, 30] all provide the same explanation for the better QPSK performance over a narrowband satellite link such as that employed in INTELSAT or the ESA links. The poorer performance of OK-QPSK and MSK is explained as follows. Although the overall envelope fluctuations are reduced for these two techniques, at the critical receiver sampling points the amplitude variations are in fact greater than for QPSK and so will result in greater sensitivity through the nonlinearity, thus requiring greater care in the design of an optimum receiver.

In both OK-QPSK and MSK a transition (through zero) can occur in one channel while sampling occurs in the quadrature channel. With filtering followed by limiting, crosstalk occurs between the I and Q channels.

The nonlinearity restricts the dynamic range of the input signal and severely distorts the waveform about the sampling point. The limiter itself does not introduce distortion at the sampling point. Postlimiter filtering distributes the waveform distortion across the sampling point.

Evidently, a major parameter is the filtering used in the system. An optimum filter for QPSK may not necessarily be the best for MSK. It is conceivable that phase equalization of the filter could be more important for MSK than QPSK because of the large power present at the band edges.

REFERENCES

1 R. Gagliardi, *Introduction to Communications Engineering*, Wiley, New York, 1978.

2 H. Hartmann, "Degradation of SNR Due to Filtering", *IEEE Trans. Aerosp. Electron. Syst.*, Vol. AES, Jan. 1969, pp. 22–32.

3 W. R. Bennett and J. R. Davey, *Data Transmission*, McGraw-Hill, Toronto, 1965.

4 A. B. Carlson, *Communications Systems*, McGraw-Hill, Toronto, 1968.

5 R. A. Harris and R. A. Gough, "Factors Influencing the Choice of Pulse Shaping Filters for the ECS System", in *Proc. ECS–Swiss PTT Colloquium*, Technical Centre, Berne, 24–25 June 1976, pp. 101–120.

6 R. G. Lyons, "The Effect of a Bandpass Nonlinearity on Signal Detectability", *IEEE Trans. Commun. Technol.*, Vol. COM-21, Jan. 1973, pp. 51–60.

7 P. C. Jain and N. M. Blackman, "Detection of a PSK Signal Transmitted through a Hard-Limited Channel", *IEEE Trans. Inf. Theory*, Vol. IT-19, Sept. 1973, pp. 623–630.

8 M. J. Eric, "Intermodulation Analysis of Nonlinear Devices for Multiple Carrier Inputs", Commun. Res. Centre (CRC) Rep. No. 1234, Dept. of Commun., Canada, Nov. 1972.

9 P. Hetrakul and D. P. Taylor, "The Effects of Transponder Nonlinearity on Binary CPSK Signal Transmission", *IEEE Trans. Commun.*, Vol. COM-24, May 1976, pp. 546–553.

10 R. M. Lester, "The Planning and Application of Digital Communication Techniques in the Canadian Domestic Satellite System", in *Proc. World Telecommun. Forum Tech. Symp.*, Geneva, 6–8 Oct. 1975, pp. 2.4.4.1–2.4.4.7.

11 D. Chakraborty, "Considerations of a 8-Phase CPSK-TDMA Signal Transmission through Bandlimited Satellite Channels", *IEEE Trans. Commun.*, Vol. COM-25, Oct. 1977, pp. 1233–1237.

12 R. E. Markle, "The AR6A Single-Sideband Long Haul Radio System", in *Proc. Int. Conf. Commun.*, June 1977, pp. 40.1.78–40.1.82.

13 CCIR, *Recommendations and Reports of the CCIR*, XIVth Plenary Assembly, Vol. IV, Kyoto, 1978, pp. 134–135.

14 B. E. White, "A Worst-Case Crosstalk Comparison among Several Modulation Schemes", *IEEE Trans. Commun.*, Vol. COM-25, Sept. 1977, pp. 1032-1037.

15 J. L. Pearce, "Measured Performance Comparison of Fast FSK and BPSK", *Can. Elect. Eng. J.*, Vol. 2, No. 4, 1977, pp. 33–37.

16   J. J. Spilker, Jr., *Digital Communications by Satellite*, Prentice-Hall, Englewood Cliffs, N.J., 1977.

17   M. R. Wachs and D. E. Weinreich, "A Laboratory Study of the Effects of CW Interference on Digital Transmission over Nonlinear Satellite Channels", in *Proc. ICDSC-3*, Kyoto, 1974, pp. 65–72.

18   R. Matyas, "Effect of Noisy Phase References on Coherent Detection of FFSK Signals", *IEEE Trans. Commun.*, Vol. COM-26, June 1978, pp. 807–815.

19   J. J. Jones, "Filter Distortion and Interference Effects on PSK Signals", *IEEE Trans. Commun.* Vol. COM-19, Apr. 1971, pp. 120–132.

20   S. A. Rhodes, "Effects of Hardlimiting on Bandlimited Transmissions with Conventional and Offset QPSK Modulation", in *Proc. Natl. Telecommun. Conf.*, Houston, Tex., 1972, pp. 20F1–20F7.

21   H. R. Mathwich, J. F. Balcewicz and M. Hecht, "The Effect of Tandem Band and Amplitude Limiting on the E_b/N_0 Performance of Minimum (Frequency) Shift Keying (MSK)", *IEEE Trans. Commun.*, Vol. COM-22, Oct. 1974, pp. 1525–1540.

22   S. A. Gronemeyer and A. L. McBride, "MSK and Offset QPSK Modulation", *IEEE Trans. Commun.*, Vol. COM-24, Aug. 1976, pp. 809–820.

23   D. P. Taylor, H. C. Chan and S. S. Haykin, "A Simulation Study of Digital Modulation Methods for Wide-band Satellite Communications", *IEEE Trans. Commun.*, Vol. COM-24, Dec. 1976, pp. 1351–1354.

24   R. M. Fielding, H. L. Berger and D. L. Lochhead, "Performance Characterization of a High Data Rate MSK and QPSK Channel", in *Proc. Int. Conf. Commun.*, Chicago, 1977, pp. 3.2.42–3.2.46.

25   S. Murakami, Y. Furuya, Y. Matsuo and M. Sugiyama, "Modulation Scheme Comparative Study for Nonlinear Satellite Channels", in *Proc. Int. Conf. Commun.*, Toronto, 1978, pp. 19.2.1–19.2.5.

26   R. de Buda, "Coherent Demodulation of FSK with Low Deviation Ratio", *IEEE Trans. Commun.*, Vol. COM-20, June 1972, pp. 429–435.

27   L. Lundquist, "Modulation Techniques for Band and Power Limited Satellite Channels", in *Proc. ICDSC-4*, Montreal, Oct. 1978, pp. 94–100.

28   R. A. Harris, "Transmission Analysis and Design for the ECS System", in *Proc. ICDSC-4*, Montreal, Oct. 1978, 81–93.

29   F. Castellano, "Relative Performance of Conventional QPSK and Staggered QPSK Modulations in a Nonlinear Channel", *Eur. Space Agency J.*, No. 1, 1978.

30   D. Chakraborty, T. Noguchi, S. J. Campanella and C. Wolejsza, "Digital Modem Design for Non-linear Satellite Channels", in *Proc. ICDSC-4*, Montreal, Oct. 1978, pp. 123–130.

PROBLEMS

1   Show that the autocorrelation function of PSK NRZ input data is given by

$$\mathcal{R}(\tau) = \begin{cases} 1 + \dfrac{|\tau|}{2T}, & -2T \leq \tau \leq 2T \\ 0, & |\tau| > 2T \end{cases}$$

where T is the symbol duration. Also show that the power spectrum relative to centre frequency is given by

$$\mathcal{G}(f) = 2PT\left[\frac{\sin(2\pi fT)}{2\pi fT}\right]^2$$

where P is the power of the modulated signal. How do these expressions differ for BPSK, QPSK, offset QPSK, and eight-phase PSK?

2 Show that the autocorrelation function of an FFSK baseband pulse is given by

$$\mathcal{R}(\tau) = \begin{cases} \left(1 + \dfrac{|\tau|}{2T}\right)\cos\left(\dfrac{\pi|\tau|}{2T}\right) + \dfrac{1}{\pi}\sin\left(\dfrac{\pi|\tau|}{2T}\right), & -2T \le \tau \le 2T \\ 0, & |\tau| > 2T \end{cases}$$

where T is the signalling interval. Also show that the power spectrum relative to the centre frequency is given by

$$\mathcal{G}(f) = \frac{8PT(1 + \cos 4\pi fT)}{\pi^2(1 - 16T^2f^2)^2}$$

where P is the modulation signal power.

3 The Butterworth low-pass filter has a square magnitude characteristic given by

$$|H(j\omega)|^2 = \frac{1}{1 + \omega^{2n}}$$

where n is the order of the filter. Find the 3-dB bandwidth for $n = 1, 2, 10$. Find the ratio of the noise bandwidth to the 3-dB bandwidth for $n = 1, 2$.

4 The Gaussian low-pass filter has a magnitude characteristic given by

$$|H(j\omega)| = e^{-0.347\omega^2}$$

Find the 3-dB and equivalent noise bandwidths.

5 An E_b/N_0 calibration is to be performed over a satellite link using the procedure given in Section 4.4. The QPSK symbol rate is 30 Msymbols/s. With a carrier transmitted in the band of the calibrating filter, the filter output power is 0 dBm. The output power is -30 dBm, with the carrier suitably offset from centre frequency so that its contribution is negligible. The calibration filter has a 3-dB IF bandwidth of 2.15 MHz and is a second-order Butterworth. Find the available E_b/N_0 and the maximum attainable P_b with a 1.5-dB implementation margin.

6 For the MSK modulator of Figure 3.8, find the output spectrum for input sequences of 101010... and 1111....

7 A rectangular pulse of unity amplitude is detected at the output of an *RC* filter. Assuming a pulse duration of 1 μs and *RC*=0.5 μs determine the degradation compared with matched filtering detection. Assume that the filter is initially at a zero state.

8 Assume that the signal at the input to the TWT whose transfer characteristics are given in Figure 4.11 is

$$s(t) = 4\cos\frac{\pi t}{2T}\cos\omega_c t, \qquad -T \le t \le T$$

where $s(t)$ is in millivolts. Determine and draw the baseband output prior to the receiver detector assuming perfect carrier recovery. Compare the result with an undistorted signal. If there is a quadrature component, draw it as well.

CHAPTER 5

Carrier and Clock Recovery

Previous discussions on the demodulation process assumed that the receiver provided the required ideal carrier and clock timing signals. By "ideal" what is meant is that the references are perfect, unaffected by channel distortions. In this section approaches that can be used to recover these reference waveforms are reviewed. Emphasis is placed on techniques applicable to PSK and FFSK signals, as these are the modulations of current interest in satellite communications.

Carrier recovery for coherent demodulation can be classified according to application: continuous or burst transmission. Where transmission is continuous, the demodulator signal acquisition time need not be rapid. However, in time-division multiple-access (TDMA) systems a succession of short-duration bursts (typically of the order of tens of microseconds) emanating from a number of different stations is presented to the demodulator. Each burst has its own independent carrier phase and consequently rapid-acquisition modems are mandatory. In the latter case, bursts are usually structured so that each starts with a preamble containing bits that accentuate carrier and clock line spectra and so assist lock-up during the training interval.

Demodulator reference circuit design entails making compromises to satisfy a number of system constraints. Although minimizing the degradation in error rate relative to a theoretical curve is the ultimate objective, penalties accrue as a result of design compromises (if necessary) for rapid acquisition, carrier frequency offset due to Earth station and satellite translation oscillators, bandwidth, power and data-rate limitations.

In multiburst systems it should be noted that demodulator recovery circuits can be used which operate serially or in parallel. In serial recovery the same recovery circuit operates on each successive burst. This requires rapid phase acquisition at the beginning of each burst. In parallel recovery, a circuit is dedicated to each burst and retains phasing information from burst to burst arriving from the same source. The latter approach is complex and will not be discussed further.

124

It should be noted that carrier recovery is necessary for coherent demodulation of PSK and FFSK. However, if differential detection is used, only clock recovery is required.

Carrier and clock recovery are aspects of synchronization that relate directly to modem operation. These areas represent part of the overall problem of system synchronization. Within an operating system framework, synchronization [1] may be required to isolate a segment of communications from a specific source, and word synchronization may be required to demultiplex word groups. These concepts will be explored further in the discussion of time-division multiple-access techniques (Chapter 8).

5.1 EFFECT OF NOISY PHASE REFERENCES

Before considering circuits that derive carrier reference signals, it is informative to consider the effect of a nonideal reference supplied by the receiver upon the probability of error. A noisy carrier phase reference degrades the error-rate performance of each of the coherent modulation techniques discussed. Consider first a BPSK signal received in additive Gaussian noise

$$r(t) = \sqrt{2} \sum_n a_n p(t-nT) \cos \omega_c t + n(t) \tag{5.1}$$

where $\sqrt{2}$ is the signal amplitude, $p(t)$ is a rectangular pulse having unity amplitude for $0 \leq t \leq T$ and zero amplitude elsewhere and a_n refers to the digit transmitted in the nth interval. The receiver provides a reference

$$R(t) = \sqrt{2} \cos(\omega_c t + \theta) \tag{5.2}$$

where θ is the phase error. Multiplication of (5.1) by (5.2) gives (with double-frequency terms ignored)

$$r(t)R(t) = \sum_n a_n p(t-nT) \cos \theta + n(t)R(t)$$

For the case of a rectangular pulse shape the matched filter is an integrate-and-dump and its output corresponding to the nth bit is

$$y_n(T) = \frac{1}{T} \int_0^T r(t)R(t) dt$$

$$= a_n \cos \theta + \frac{1}{T} \int_0^T n(t)R(t) dt \tag{5.3}$$

The noise component of (5.3) can be shown to be Gaussian with zero mean and variance σ^2. The probability of error conditioned on θ can now be calculated as in Section 2.3 to give

$$P_b(\theta) = Q\left[\sqrt{\frac{2E_b}{N_0}} \cos \theta\right] \tag{5.4}$$

where

$$Q(x) \triangleq \frac{1}{\sqrt{2\pi}} \int_x^\infty e^{-u^2/2} du$$

The average error rate is then computed by integrating $P_b(\theta)$,

$$P_{BPSK} = \int_{-\pi}^{\pi} P_b(\theta) p(\theta)\, d\theta$$

where $p(\theta)$ is the probability density function (pdf) of θ.

One well-known pdf is that derived for a first-order phase-locked loop with an unmodulated carrier input [2]

$$p(\theta) = \frac{\exp(\alpha \cos \theta)}{2\pi I_0(\alpha)}, \qquad -\pi \leq \theta \leq \pi$$

where α is the phase reference signal-to-noise ratio and $I_0(\alpha)$ refers to the modified Bessel function of order zero and argument α. For large α

$$p(\theta) \simeq \frac{\exp[\alpha(\cos \theta - 1)]}{(2\pi/\alpha)^{1/2}}$$

In the case of QPSK, demodulation is accomplished in quadrature channels using references

$$R_1(t) = \sqrt{2} \cos(\omega_c t + \theta)$$

$$R_2(t) = \sqrt{2} \sin(\omega_c t + \theta)$$

and parallel integrate-and-dump circuits. If $\theta = 0$, then orthogonal detection occurs. However, if $\theta \neq 0$, mutual interference between channels occurs. An analysis similar to that for BPSK shows that the conditional error rate is given by

$$P_q(\theta) = \frac{1}{2}\left\{Q\left[\sqrt{\frac{2E_b}{N_0}}(\cos \theta + \sin \theta)\right] + Q\left[\sqrt{\frac{2E_b}{N_0}}(\cos \theta - \sin \theta)\right]\right\}$$

With offset QPSK [3] one channel is offset with respect to the other channel by T, where T is the serial input bit duration (see Figure 5.1). As a

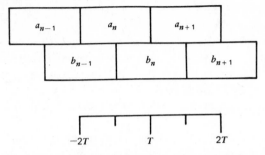

Figure 5.1 Alignment of data in quadrature channels.

result, a transition in one channel occurs at the midpoint of the signalling interval of the other channel. If at the transition the data changes polarity, then the interference from the channel in the first half of the interval is cancelled by that in the second half. In this case the conditional error rate is the same as for BPSK. If no transition occurs, the interference remains constant and the conditional error rate is the same as that for QPSK. Because the probability of a transition is one-half, the conditional error rate for offset QPSK is

$$P_{0q}(\theta) = \tfrac{1}{2}\left[P_b(\theta) + P_q(\theta) \right]$$

Thus offset QPSK is not as sensitive to phase as QPSK.

FFSK is similar to offset QPSK except that the pulse shape is a half-sinusoid rather than rectangular. An analysis similar to that described above gives for the conditional error rate [4]

$$P_f(\theta) = \tfrac{1}{4}\left[P_{f1}(\theta) + P_{f2}(\theta) \right]$$

where

$$P_{f1}(\theta) = Q\left[\sqrt{\frac{2E_b}{N_0}} \left(\cos\theta + \frac{2}{\pi}\sin\theta \right) \right] + Q\left[\sqrt{\frac{2E_b}{N_0}} \left(\cos\theta - \frac{2}{\pi}\sin\theta \right) \right]$$

$$P_{f2}(\theta) = 2Q\left[\sqrt{\frac{2E_b}{N_0}} \cos\theta \right]$$

Note that in all expressions for the conditional error rate it has been assumed that clock timing is perfect. A comparison of the performance of the four modulation techniques appears in Figure 4.19. This comparison shows that for the same phase reference SNR the order of increasing sensitivity to phase is BPSK, FFSK, offset QPSK and QPSK.

5.2   CARRIER RECOVERY

Both PSK and FFSK are double-sideband suppressed carrier techniques whose signal spectra are symmetric with respect to the carrier frequency. Because a discrete carrier component is absent, carrier regeneration is required. This regeneration cannot be based upon a linear addition of received signal components but rather must depend upon processing the signal through a nonlinearity.

Carrier recovery can be achieved in a number of ways, including the Mth power method, the Costas loop and the decision-directed feedback circuit. With $M=2$, the Mth power method is known as a squaring loop. It can be shown that the Costas loop has equivalent theoretical performance to the squaring loop. The decision-directed feedback loop provides improved performance by using the modulation power. These techniques will be reviewed for PSK. FFSK will be treated separately.

Squaring Loop

Figure 5.2 depicts a circuit [5, 6] which uses a square-law nonlinearity to regenerate a BPSK carrier. This circuit is called a squaring loop. The received signal $r(t)$ is filtered at an intermediate frequency (IF) to remove unwanted interference and the result is squared. The effect of the squaring operation is to remove the modulation. For a BPSK signal (ignoring filtering effects) the doubler output is

$$r^2(t) = \left\{ \sqrt{2P}\, d(t)\sin\left[\omega_c t + \theta(t)\right] + \sqrt{2}\, N(t)\sin\left[\omega_c t + \theta_N(t)\right] \right\}^2$$

$$= Pd^2(t)\left[1 - \cos\left[2\omega_c t + 2\theta(t)\right]\right] + N^2(t)\left[1 - \cos\left[2\omega_c t + 2\theta_N(t)\right]\right]$$

$$+ 2\sqrt{P}\, d(t)N(t)\left\{\cos\left[\theta(t) - \theta_N(t)\right] - \cos\left[2\omega_c t + \theta(t) + \theta_N(t)\right]\right\}$$

$$(5.5)$$

where P = signal power
$\quad d(t)$ = binary information = ± 1
$\quad \theta(t)$ = signal phase
$\quad N(t)$ = noise amplitude
$\quad \theta_N(t)$ = noise phase

The DC terms can be ignored. Now $d^2(t) = 1$ for BPSK and (5.5) becomes

$$r^2(t) = -P\cos\left[2\omega_c t + 2\theta(t)\right] - N^2(t)\cos\left[2\omega_c t + 2\theta_N(t)\right]$$

$$- 2\sqrt{P}\, d(t)N(t)\cos\left[2\omega_c t + \theta(t) + \theta_N(t)\right]$$

$$(5.6)$$

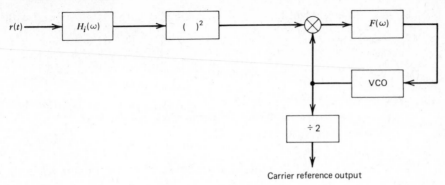

Figure 5.2 Squaring loop.

The first term in (5.6) is essentially a discrete component at twice the carrier frequency. The second term is the noise component and the third term depends upon both the noise and the modulation. The phase-locked loop (PLL) [7] serves as a narrowband filter and tracks the double-frequency component. Frequency division by two of the PLL output yields the required carrier reference. Division by 2 results in a phase ambiguity (zero or π) in the carrier phase which can be accommodated by differential encoding of the source data.

The PLL voltage-controlled oscillator generates a reference signal given by

$$R(t) = \sqrt{2}\,K_1 \sin\left[2\omega_c t + 2\hat{\theta}(t)\right]$$

where K_1 is a constant and $2\hat{\theta}(t)$ is the loop-phase estimate. Multiplication of the first term of (5.6) by $R(t)$ and ignoring the term at $4\omega_c$, which is removed by the loop filter, yields

$$-PR(t)\cos\left[2\omega_c t + 2\theta(t)\right] \simeq \frac{PK_1}{\sqrt{2}} \sin 2\left[\theta(t) - \hat{\theta}(t)\right]$$

$$= \frac{PK_1}{\sqrt{2}} \sin 2\phi(t)$$

where $\phi(t) = \theta(t) - \hat{\theta}(t)$. For small ϕ the input control to the VCO is given approximately by $\sqrt{2}\,PK_1\phi(t)$. Loop operation will be degraded by the noise effects.

The narrower the bandwidth of the PLL, the greater will be the noise it rejects. However, a more important consideration may be the received carrier frequency uncertainty and the required acquisition time. An adaptive technique can be used where (1) for initial acquisition the bandwidth is large while for tracking purposes the bandwidth is reduced or (2) the VCO can be swept during the acquisition period [7].

Layland [8] has shown that the SNR at the PLL input is maximized with

$$H_i(s) = k \left[\frac{G(s)}{G(s) + N_0/2} \right]^{1/2}$$

where $H_i(s)$ = transfer function of input pre-squaring filter
$\quad\quad G(s)$ = power spectral density of the modulated signal
$\quad\quad k$ = normalizing constant

For large SNR an ideal rectangular bandpass filter is optimum, while for small SNR

$$H_i(s) = k \left[\frac{2G(s)}{N_0} \right]^{1/2}$$

and the filter is matched to the modulation spectrum.

Lindsey and Simon [6] show that the variance of the phase jitter of the recovered reference at the desired frequency, with a PLL tracking circuit is given by

$$\sigma_\phi^2 = \frac{\sigma_{2\phi}^2}{4} = (\rho S_L)^{-1}$$

where $\sigma_{2\phi}^2$ = variance of the phase jitter of the reference at $2\omega_c$
$\quad\quad S_L$ = squaring loss of the PLL
$\quad\quad \rho$ = equivalent SNR in the loop bandwidth of a second-
$\quad\quad\quad$ order loop = $2P/N_0 W_L$

The squaring loss is given by

$$S_L = \left[1 + \frac{2}{PN_0} \int_{-\infty}^{\infty} \mathcal{R}^2(\tau)\, d\tau \right]^{-1}$$

where

$$\mathcal{R}(\tau) = \frac{N_0}{2} \int_{-\infty}^{\infty} |H_L(j2\pi f)|^2 e^{j2\pi f\tau}\, df$$

and $H_L(j2\pi f)$ is the low-pass equivalent of $H_i(j2\pi f)$. For example,

$$S_L = \frac{1}{1 + (2\rho\gamma)^{-1}} \quad\quad \text{for an } RC \text{ filter function}$$

$$= \frac{1}{1 + (\rho\gamma)^{-1}} \quad\quad \text{for an ideal bandpass filter}$$

$$= \frac{1}{1 + 2(3\rho\gamma)^{-1}} \quad\quad \text{for a filter matched to NRZ BPSK modulation}$$

where γ is the ratio of the two-sided noise bandwidth of the PLL to that of $H_L(\omega)$: $2W_L/W_i$. Note that while the matched filter provides the maximum PLL input SNR, it results in a larger squaring loss than for an RC-type filter.

For the situation where the PLL is replaced by a narrowband filter $G(s)$, Oberst and Schilling [9] have shown that the SNR at the output of $G(s)$ at twice the carrier frequency $(2\omega_c)$ is given for the general vth order nonlinearity by the following asymptotic relationships:

Large SNR$_i$

$$\text{SNR}_o \simeq \frac{(B_1/B_2)\text{SNR}_i}{2\left[1+(v/2)^2\right]}$$

Small SNR$_i$

$$\text{SNR}_o \simeq \frac{(B_1/B_2)(v/2)^2(\text{SNR}_i)^2}{4\sum\limits_{l=1}^{\infty}\left[f(2l)(v/2-l+1)_{l^2}/(l-1)!(l+1)!\right]}$$

where

$$(a)_n = a(a+1)\cdots(a+n-1)$$

For the special case of a doubler,

$$\text{SNR}_o = \frac{\frac{1}{2}(B_1/B_2)(\text{SNR}_i)^2}{f(2)+2(\text{SNR}_i)}$$

where SNR$_i$ is the input SNR measured after $H_i(s)$, and B_1 and B_2 are the noise bandwidths of $H_i(s)$ and $G(s)$, respectively, and are given by

$$B_1 = (2\pi)^{-1}\int_0^\infty \frac{|H_i(\omega)|^2}{|H_i(\omega_c)|^2}\,d\omega$$

$$B_2 = (2\pi)^{-1}\int_0^\infty \frac{|G(\omega)|^2}{|G(2\omega_c)|^2}\,d\omega$$

The function $f(u)$ is given by

$$f(u) = \frac{S_{N,u}(0)}{N_0(N_0 B_1)^{u-1}}$$

where $S_{N,u}(0)$ is the convolution of the low-pass equivalent spectrum of the noise at the output of $H_i(s)$ with itself $(u-1)$ times evaluated at $\omega=0$. The

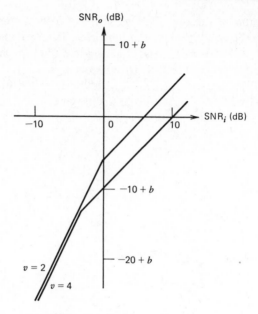

Figure 5.3 SNR_0 versus SNR_i for doubler and ($\times 4$) for single-tuned IF filter. $b = 10\log_{10}(B_1/B_2)$. (From Ref. 9: © 1971 IEEE; reprinted from "The SNR of a Frequency Doubler", by J. F. Oberst and D. L. Schilling, from *IEEE Trans. Commun.*, Feb. 1971.)

low-pass noise spectrum is given by

$$S_N(\omega) = 2S_n(\omega - \omega_0)\mathcal{U}(\omega - \omega_0)$$

$$= 2S_n(\omega + \omega_0)\mathcal{U}(\omega + \omega_0)$$

where $\mathcal{U}(\omega)$ is the unit step function. Oberst and Schilling show that for a single-tuned $H_i(S)$ the function $f(u) = u^{-1}$. Asymptotic limits for SNR_o for this filter are shown in Figure 5.3.

Costas and Decision Feedback Loops

The error-control signal to the VCO can be derived by means of baseband processing in situations where IF multiplication is not practical. In the Costas [10] loop (see Figure 5.4) the received signal is multiplied by quadrature outputs of a VCO. The in-phase signal, which is a function of $\cos \phi(t)$, is used to derive the data, and the quadrature channel produces a term proportional to $\sin \phi(t)$. Multiplication of the two channel outputs produces an error signal that is a function of $\sin 2\phi(t)$, as in the squaring loop. In fact, the two loops provide the same performance provided that (1) the Costas quadrature arm filters provide the same noise filtering as that of the low-pass equivalent of the

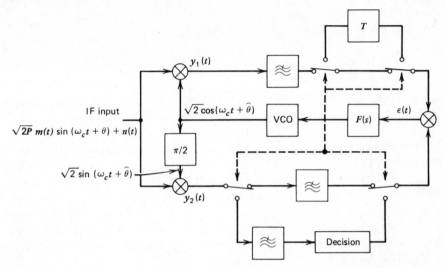

Figure 5.4 Costas and decision-directed feedback loops. (Switches shown in Costas loop position.)

squaring loop IF filter, (2) the VCO prefilters are identical and (3) the Costas VCO gain measured in rad/s-V must be one-half that of the squaring loop's VCO because of operation at ω_c rather than $2\omega_c$ [6].

The Costas loop arm filters can be replaced by integrate-and-dump (ID) filters, resulting in improved performance in noise. However, this implementation requires the prior derivation and provision of a clock signal to the filters, which may not be desirable. Alternatively, the ID filters may be approximated by low-pass filters having a bandwidth $\simeq 2/T$, where T is the data duration.

The decision feedback loop (see the alternate switch positions in Figure 5.4) is a modification of the Costas loop wherein the quadrature channel output is multiplied by an estimate of the data (determined by an ID circuit) rather than by the in-phase channel output directly. The product of the received signal and the VCO reference is a function of $a_i(t)\sin\phi(t)$, where $a_i(t)$ is the data sequence. Multiplication of this signal (delayed by 1 bit) by the data estimate $\hat{a}_i(t)$ produces a VCO input control signal [6] which is given approximately by

$$\varepsilon(t) \simeq E\big[a_i(t)\hat{a}_i(t)\big]\sin\phi(t)$$

$$= \big\{(+1)P\big[a_i(t)=\hat{a}_i(t)|\phi(t)\big]$$

$$+(-1)P\big[a_i(t)\neq\hat{a}_i(t)|\phi(t)\big]\big\}\sin\phi(t)$$

$$= \big\{1-2P_b\big[\phi(t)\big]\big\}\sin\phi(t)$$

where $E[\cdot]=$ expected value

$P_b[\phi(t)]=$ data error probability conditioned on $\phi(t)$

$$= \frac{1}{2}\,\mathrm{erfc}\left[\sqrt{\frac{E_b}{N_0}}\,\cos\phi\right]$$

and ϕ is constant over several bit intervals. It should be noted that the decision feedback loop exhibits a 180° phase ambiguity, as in the squaring and Costas loops.

The decision feedback loop provides improved performance relative to the squaring loop. For example, it tends to be insensitive to the ratio $\gamma = 2W_L/W_i$, while the squaring loop degrades with increasing prefilter-to-loop filter bandwidth ratio. With zero input frequency offset, the decision-directed loop has a phase error variance of $\sigma^2 \simeq 0.05$ rad^2 for $E_b/N_0 = 1$ dB. The reader is referred to Lindsey and Simon [6, pp. 68–69] for further details. More elaborate loops, such as the data-aided and hybrid carrier tracking loops, are considered in Refs. 11 to 13.

5.3   TECHNIQUES APPLICABLE TO RAPID QPSK CARRIER RECOVERY

Although the techniques discussed so far can be extended to QPSK, only those circuits of interest for rapid carrier recovery will be reviewed. The interested reader is referred to Lindsey and Simon [6] for a discussion of N-phase Costas and decision-directed loops. QPSK modulation is the technique that has received significant attention for TDMA applications. In these situations conventional PLL recovery circuits suffer from the "hang-up" phenomenon and require excessive acquisition times [14].

In carrier recovery loops employing a PLL, an error voltage is produced which is a sinusoidal function of the loop phase error (actually other characteristics are possible—the sinusoidal relationship will be used for the purposes of demonstration). The loop control voltage results from a multiplication of the input signal by a reference in a mixer that is a phase detector. The error voltage is given by

$$e(t)=K_1\sin\phi(t)$$

where K_1 is the mixer gain and $\phi(t)=\theta(t)-\hat{\theta}(t)$ is the phase error between the signal phase $\theta(t)$ and the loop phase estimate $\hat{\theta}(t)$. The loop reference signal is the output of a VCO. The VCO instantaneous frequency is given by $\omega_c + K_c e(t)$, where ω_c is the nominal centre frequency and K_c is a constant. The VCO output phase is then given by

$$\hat{\theta}(t)=\omega_c t + \int_0^t K_c e(t)\,dt$$

$$=\omega_c t + K\int_0^t \sin\phi(t)\,dt$$

where $K=K_1K_c$. The loop phase error and its time derivative are

$$\phi(t)=\theta(t)-\hat{\theta}(t)$$

$$=\omega_it-\omega_ct-K\int_0^t\sin\phi(t)\,dt$$

$$\frac{d\phi(t)}{dt}=(\omega_i-\omega_c)-K\sin\phi(t)$$

where $\theta(t)=\omega_it$. If $\omega_i=\omega_c$ and letting $K=1$, the time derivative of the loop phase error is given by

$$\dot{\phi}=-\sin\phi$$

and is drawn in Figure 5.5a. Note the occurrence of two nulls per cycle. With phase errors close to 0 rad the slope of the characteristic is negative, whereas for phase errors close to $-\pi$ the slope is positive. For small u, $\sin u\simeq u$ and $\sin(u-\pi)\simeq -u$ and

$$\dot{\phi}\simeq\begin{cases} -u, & -\psi<|\phi|<\psi \\ u, & -\pi-\psi<|\phi|<-\pi+\psi \end{cases} \tag{5.7}$$

where ψ indicates an arbitrarily small nominal phase value limit.

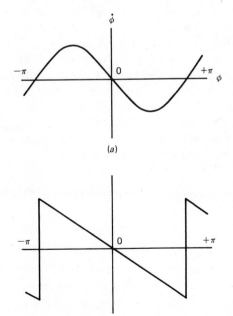

(a)

(b)

Figure 5.5 Phase detector characteristics. (a) Sinusoidal; (b) sawtooth.

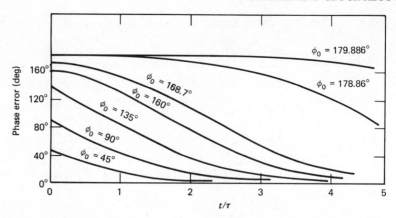

Figure 5.6 Transient phase error in first-order PLL with sinusoidal phase detector. Initial phase error, ϕ_0; zero frequency error between PLL VCO and signal; τ, closed-loop time constant. (From Ref. 15; reprinted by courtesy of the European Space Agency from thesis ESA TM-169.)

An examination of (5.7) indicates that the zero null is stable whereas the $-\pi$ null is not. With a phase error near the zero null the PLL tends to remain locked, whereas for a phase error near $-\pi$ the PLL tends to drift away from the null. The rate of phase correction in the PLL depends upon the phase detector output voltage. If the PLL state initially results in a $-\pi$ null, the error control voltage is small. The loop state is unstable at this null. However, the rate of divergence will be slow, as will be the rate of convergence to a stable null. Figure 5.6 [15] depicts the convergence time for specified initial phase errors. The convergence time is given in terms of τ/T, where τ is the closed-loop time constant of a first-order PLL [the loop filter $H(s)=1$].

A modified phase detector characteristic is shown in Figure 5.5b. The sawtooth characteristic provides a large restoring voltage at the $\pm\pi$ nulls. Although the sawtooth characteristic would appear to combat the problem of low correction voltages at the unstable nulls, it has been found [15] that additive channel noise and certain data patterns degrade the characteristic by essentially rounding the discontinuities, and the hang-up phenomenon persists.

As a result of these findings it has been generally concluded that deterministic filters should be used in place of PLLs as the post-nonlinearity filter where rapid acquisition is mandatory. As to the type of nonlinearity, two types are usually considered: the times four multiplier ($\times4$) and the remodulator.

Carrier Recovery: ($\times4$) and Remodulator

In M-ary PSK systems the set of phase states is a multiple of $2\pi/M$ and passing the signal through an M-power nonlinearity will provide a component at M times the carrier frequency. This can be seen by an expansion. Let the noise-free received signal be given by

$$r_i(t) = \cos(\omega_c t + \theta_i), \qquad 0 \le t \le T$$

where $\theta_i = 2\pi k/M$; $k = 0, 1, \ldots, M-1$. Passing $r_i(t)$ through the nonlinearity yields

$$r_i^M(t) = \cos^M(\omega_c t + \theta_i) = \cos^{2p}(\Theta)$$

$$= \frac{1}{2^{2p}}\binom{2p}{p} + \frac{1}{2^{2p-1}}\left[\cos 2p\Theta + \binom{2p}{1}\cos(2p-2)\Theta\right.$$

$$\left. + \cdots + \binom{2p}{p-1}\cos 2\Theta\right]$$

where $p = M/2$ and $\Theta = \omega_c t + \theta_i$. Thus the nonlinearity output contains a component at even harmonics of ω_c and the component at $M\omega_c$ has zero phase modulo-2π. Recovery of a carrier component is then accomplished by filtering and dividing by M. The resulting recovered carrier has an M-fold phase ambiguity.

With a fourth-order nonlinearity as in Figure 5.7, the $\times 4$ multiplier circuit entails passing the received signal through the nonlinearity, filtering the output in a narrowband filter, limiting the filter output to remove amplitude fluctuations and finally frequency dividing the result to yield a reference at the carrier frequency. It should be noted that only in the case of a constant-envelope infinite bandwidth input signal can a pure carrier reference be recovered. In practice, the signal is bandlimited and the output of the nonlinearity will contain a modulation-dependent interference called pattern noise.

In the remodulator [15–17] (or more correctly remodulator/demodulator) circuit of Figure 5.8 the received signal is demodulated by means of quadrature versions of the reference signal, and the result of each demodulation is passed through a low-pass filter to remove unwanted components of the mixing operation. The mixer outputs are each limited in the remodulator section, and the limiter outputs multiply quadrature versions of the input signal. For correct operation the input signal must be delayed to account for baseband filter delays in the quadrature demodulator paths.

Summation of the quadrature multiplier outputs (with appropriate sign) yields a signal containing a discrete carrier component. Interference components are suppressed by a narrowband filter and a limiter serves to remove amplitude fluctuations.

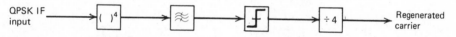

Figure 5.7 Carrier recovery using ($\times 4$) multiplier.

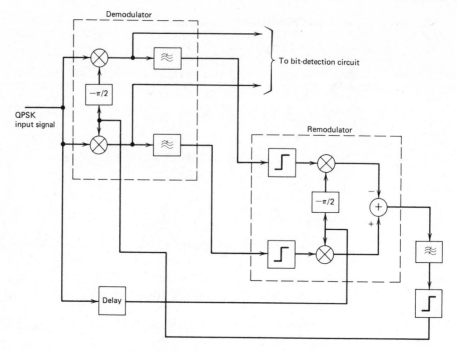

Figure 5.8   Demodulator/remodulator.

A decision-aided version of the remodulator results if the limiters are replaced by data detectors. These detectors form decisions on the received signals by an integrate-and-dump (for rectangular pulse shape) operation. However, ID circuits require correct timing, which implies prior clock recovery.

Gardner [15, 18] has conducted a detailed link computer simulation with the two recovery loops. The link consisted of a QPSK modulator, a travelling-wave-tube amplifier and a QPSK demodulator with Gaussian noise added to the signal at the demodulator input. It was found that the phase jitter for both circuits was dominated by additive noise for low E_b/N_0 and limited by pattern noise for large E_b/N_0. No significant difference in output phase jitter was found between a single-tuned and a more complex double-tuned carrier selection filter after the nonlinearity. The single-tuned bandpass resonant circuit transfer function is given by

$$H(s) = \frac{s\omega_0/Q}{s^2 + s\omega_0 Q^{-1} + \omega_0^2}$$

where ω_0 is the resonant frequency and Q is the quality factor. The 3-dB bandwidth is given by $B_{3\text{dB}} = \omega_0 Q^{-1}$. The simpler filter has a faster transient response, so the filter transition time from an initial state at the start of a burst to a steady state suitable for demodulation will be shorter. Both loops require filtering to remove interference that may occur due to adjacent channels in multicarrier systems. The $\times 4$ multiplier by necessity of design requires an IF filter, whereas the remodulator can operate with easier-to-implement baseband filters.

The $(\times 4)$ multiplier was found to be data pattern sensitive and in fact could not be used for a 0, $-\pi/2$, $-\pi$, and $-3\pi/2$ continuously repetitive sequence. The signal appears to be an unmodulated carrier offset from centre frequency by $\frac{1}{4T}$, where T is the symbol duration.

Figure 5.9 compares the phase jitter performance of the two recovery loops with a single tuned recovery circuit filter having $B_N T = 0.07$ with

$$B_N = \frac{\pi B_{3\text{dB}}}{2}$$

where B_N is the noise bandwidth of the filter. It was also found that the remodulator could operate to lower E_b/N_0 values than could the $(\times 4)$ circuit. For the particular system simulated, the remodulator operated to $E_b/N_0 \simeq 2$ to 3 dB, whereas the $(\times 4)$ circuit could operate satisfactorily only to $E_b/N_0 \simeq 6$ to 7 dB. In addition, to attain an error rate of 10^{-4} the $(\times 4)$ circuit required 0.1 to 0.3 dB more E_b/N_0.

It should be noted that the $(\times 4)$ multiplier operation can be improved by "quenching" the circuit filter. Quenching refers to the operation of initializing the filter to a zero state by rapid discharge at the start of each new received burst signal. The purpose of quenching is to reduce the initial filter output

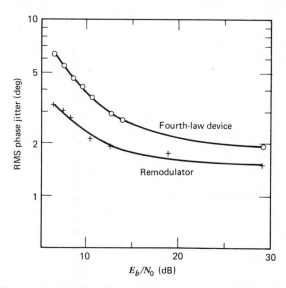

Figure 5.9   Reference carrier phase jitter. Single-tuned carrier filter. $B_N T = 0.07$. (From Ref. 15; reprinted by courtesy of the European Space Agency from thesis ESA TM-169.)

transient by ensuring that the initial state is not a random state determined by the previous burst signal. Although quenching is applicable to the (×4) circuit, it cannot be used in a remodulator. For the remodulator to operate, a local reference must be available. If the quadrature arm filters are initially set to zero output, no carrier reference is available. It should be noted that a difficulty with this approach is establishing the correct time to initiate quenching.

The preceding discussion tends to imply that the remodulator offers improved performance without any problems. Unfortunately, the simulations by Gardner indicated the occurrence of the hang-up phenomenon where the initial phase error between remodulator and received carrier phases was $\pi/4$. Both pattern noise and additive channel noise could cause the remodulator to dwell at this state for long periods of time, preventing convergence to an equilibrium state (an integer multiple of 90°).

A solution to this problem [15, 17] is to operate the remodulator initially in an open-loop mode, where the feedback path is disabled and the remodulator is forced into an equilibrium state by a preamble of unmodulated carrier. Certain advantages result from this approach. The fact that an unmodulated carrier is used at the beginning of each burst means that the carrier derived by bypassing the remodulator does not contain a phase ambiguity. Furthermore,

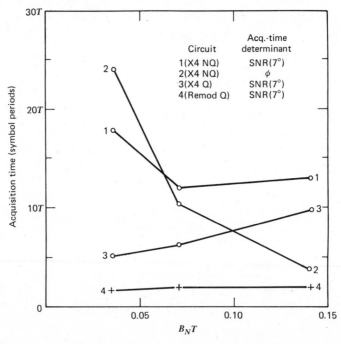

Figure 5.10 Carrier acquisition times for (×4) and remodulator circuits. Q, quenched; NQ, not quenched. (From Ref. 15; reprinted by courtesy of the European Space Agency from thesis ESA TM-169.)

the nonlinearity loss due to the remodulator action is not present, implying more rapid acquisition during the preamble period. This approach is not without its own problem: clock recovery cannot occur until the preamble of continuous carrier stops to be followed by a modulated signal.

As a final comparison, acquisition times for a number of ($\times 4$) and remodulator configurations are shown in Figure 5.10 [15]. Some care is required in the use of the term "acquisition time". For the purposes of the simulation Gardner determined two parameters to be significant: the acquisition phase transient and the SNR acquisition time. The phase acquisition time was defined to be the time for the phase transient to decay to less than the steady-state rms phase jitter observed in the simulation results. The SNR acquisition time was defined somewhat arbitrarily as the time corresponding to recovery of rms phase jitter to less than 7°. For each recovery circuit performance was determined by the greater of the phase and SNR acquisition times. The various curves demonstrate the superior performance of the remodulator. Note that acquisition in all cases is less than 30 symbol intervals.

A significant number of papers relating to PSK recovery techniques have appeared in the literature. In addition to those already cited, the reader is referred to Refs. 19 to 24.

5.4 AUTOMATIC FREQUENCY CONTROL

Automatic frequency control (AFC) is necessary to compensate for carrier offsets introduced by the various conversion oscillators and Doppler shifts appearing in the system. In present 4/6-GHz satellite systems the major contributor is the satellite's translation oscillator, which can introduce uncertainties of typically ± 40 kHz. Doppler shift contributions are negligible for synchronous satellites. Offsets become more critical for lower data rates. For example, the modulation bandwidth of 32 kbps BPSK is comparable to the system offset, and without frequency compensation, performance can be expected to be unacceptable because of the wideband receiver filters required and the resulting effect on reference recovery circuits. In differential BPSK systems the error-rate performance is a cosine function of the product $2\pi f_c T$, where f_c is the carrier frequency and T is the bit duration. Evidently, DPSK systems are very sensitive to carrier frequency offset.

One technique appropriate for low-rate modulation signals involves the transmission of a separate pilot tone through the transponder. This pilot tone can be used by the receiver for spectrum alignment. The AFC circuit compares the pilot frequency with an accurate local reference and uses the difference to centre the received signal (see Figure 5.11). This technique can be employed in systems where many narrowband signals pass through the same satellite transponder. The AFC circuit in the receiver will compensate for the satellite and receiving Earth station translation oscillators. Provided that all transmitting stations have relatively stable up-converters, which can be tracked by the receiver, correct demodulation is possible.

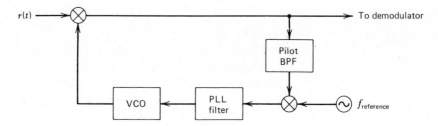

Figure 5.11   Pilot tone recovery AFC.

A second technique uses the received signal directly and operates in conjunction with the carrier recovery circuit (see Figure 5.12). The signal is processed by a nonlinearity and the output is compared with a local stable reference. The detector output controls a VCO which centres the spectrum.

Figure 5.13 [15] depicts a circuit that can be used to generate a frequency error voltage. In this detector the phase shift through the carrier filter is measured by comparing input and output waveforms. The group delay through the filter is given by

$$\tau = \frac{d\theta}{d\omega}$$

where θ is phase and ω is angular frequency (θ is assumed set to zero at centre frequency). If the group delay is relatively constant throughout the filter passband, the phase shift is directly proportional to frequency. The frequency error signal can then be made proportional to $\sin\theta$. The phase shifter is used to ensure a quadrature relationship between signals at the phase detector inputs.

Figure 5.14 [15, 25] depicts a circuit that operates on the demodulated baseband waveforms. If the IF input signal is given by (noiseless case)

$$r(t) = x(t)\cos(\omega_c t + \theta_i) + y(t)\sin(\omega_c t + \theta_i)$$

and the reference carrier by

$$R(t) = \cos(\omega_c t + \theta_0)$$

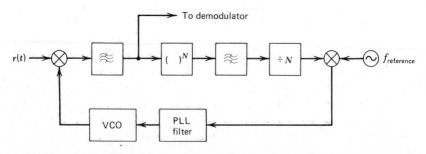

Figure 5.12   Nonlinearity recovery AFC.

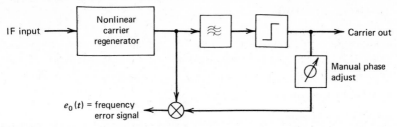

Figure 5.13 Filter input/output phase comparison derivation of frequency error control signal.

then it can be shown that the average value of the resulting error voltage is given by

$$\overline{e_0(t)} = \text{average} \{[-x\sin\phi + y\cos\phi] \cdot \text{SGN}[x\cos\phi + y\sin\phi]$$

$$-[x\cos\phi + y\sin\phi] \cdot \text{SGN}[-x\sin\phi + y\cos\phi]\}$$

where $\phi = \theta_i - \theta_0$ and $\text{SGN}(\cdot)$ denotes the signum function (hard limiter). For rectangular pulses the resulting voltage/phase transfer characteristic is periodic modulo-$\pi/2$. The voltage is given approximately by

$$\overline{e_0(t)} \simeq -K\phi, \qquad -\pi/4 < \phi < \pi/4$$

where K is a constant.

For additional papers dealing with AFC techniques the reader is referred to Yokoyama et al. [24] and Asahara et al. [26].

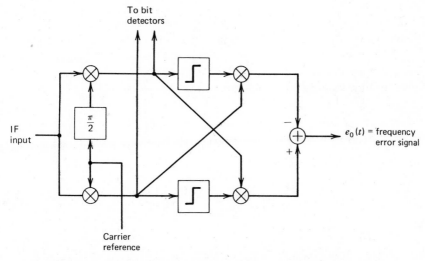

Figure 5.14 Baseband derivation of frequency error control signal.

5.5   CLOCK RECOVERY

As was the case with carrier recovery for PSK signals, clock recovery requires a nonlinear operation to be performed on the received signal. This is because the signal does not contain discrete spectral lines at the clock frequency. Clock recovery (or symbol timing recovery, STR) entails two operations:

1   The extraction of the timing information from the phase or bit transitions.
2   Using this information to synchronize a local reference.

In some applications the second operation consists of a narrow filter that isolates the clock component. Alternatively, phase-locked-loop type circuits can be used (either analog or digital), but these are subject to the same difficulties as described in the carrier acquisition discussion (Section 5.3).

Clock recovery can occur subsequent to or coincident with carrier recovery. In the former case, recovery circuits operate on the demodulated (not necessarily detected) baseband waveforms, whereas in the latter situation, circuits operate directly on the modulated carrier signal. That clock recovery can occur simultaneously with carrier recovery is significant because in systems such as TDMA the time devoted to carrier/clock recovery is an overhead that should be kept at a minimum to maximize the information throughput. In such systems STR is accomplished by a nonlinear/filter operation. These techniques are usually suited to good E_b/N_0 operating conditions. For continuous transmission systems or in situations where minimizing the combined carrier/clock recovery time is secondary to desired operation at low E_b/N_0 conditions, techniques utilizing the synchronization of a local reference can be employed.

Performance Criteria

Distinguishing the performance of synchronizers requires a criterion of comparison. One approach is to compare the rms error $(\overline{\varepsilon^2})^{1/2}$ or the mean magnitude error $\overline{|\varepsilon|}$. Alternatively, the degradation in error rate for a given E_b/N_0 can be determined. For baseband data in Gaussian noise it can be shown that the conditional probability of error is given by [27]

$$P(\text{error}|\varepsilon) = \frac{1}{2}\int_0^{1/2}\left\{1 - \frac{1}{2}\text{erf}\left[\sqrt{\frac{E_b}{N_0}}\left(\frac{\mathcal{R}(\varepsilon)+\mathcal{R}(T-\varepsilon)}{\mathcal{R}(0)}\right)\right]\right.$$

$$\left. - \frac{1}{2}\text{erf}\left[\sqrt{\frac{E_b}{N_0}}\left(\frac{\mathcal{R}(\varepsilon)-\mathcal{R}(T-\varepsilon)}{\mathcal{R}(0)}\right)\right]\right\}p(\varepsilon)\,d\varepsilon \qquad (5.8)$$

where $p(\varepsilon)$ is the probability density function of the timing error ε and $\mathcal{R}(\varepsilon)$ is the autocorrelation function of the positive symbol pulse and is given by

$$\mathcal{R}(\varepsilon)=E[s(t)s(t+\varepsilon)] \qquad (5.9)$$

where $E[\cdot]$ denotes expectation. Note that the degradation is a function of the input E_b/N_0, the signal shape reflected by $\mathcal{R}(\varepsilon)$ and the synchronizer through $p(\varepsilon)$. The basic baseband data pulse $s(t)$ is confined to the interval $0 \leq t \leq T$.

Optimum STR

The optimum synchronization technique for a signal received with additive Gaussian noise and based upon a maximum *a posteriori* estimation of the symbol location has been described by Wintz and Luecke [27]. Although the resulting synchronizer is complex and impractical to implement, it does lead to suboptimum synchronizers which perform nearly as well.

Figure 5.15 portrays a transmitted antipodal signal $s(t)$ and the noisy antipodal received signal $r(t)$. The basic signalling pulse is given by $p(t)$, $p(t)$ ≥ 0 for $0 \leq t \leq T$. The observation interval is KT, where K is referred to as the memory (K is an integer) and T is the bit duration. τ is the parameter that indicates the occurrence of the first zero crossing of the signal relative to an arbitrary time reference $t=0$. Wintz and Luecke showed that the optimum value of τ is determined by a correlation and weighting operation as follows.

An initial value of $\tau = \tau_0$ is selected. The portion of the observation interval for $0 \leq t \leq \tau_0$ is correlated with the last τ_0 second portion of the positive signalling pulse $p(t)$. The natural logarithm of the hyperbolic cosine (ln cosh)

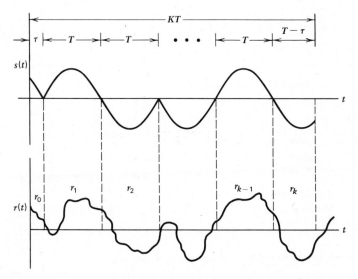

Figure 5.15   Transmitted signal and received noisy signal.

of the correlator output is stored in an accumulator. The correlation and ln cosh operations are repeated for the portion of the record between $\tau_0 \leq t \leq T + \tau_0$ and the positive pulse and the result is added to the accumulator. This procedure is repeated for the balance of the signal record. For the complete observation interval the accumulator contains

$$\bar{S}_{A_0} = \sum_{j=0}^{k} \ln \cosh \left[r_j s_j(\tau_0) \right]$$

This calculation must now be repeated for all selections of τ in $-T/2 \leq \tau \leq T/2$. The accumulator sum $\bar{S}_{A_i}$ having the largest value indicates the optimal choice of $\tau = \tau_i$.

It can be readily seen that the approach is a computationally time-consuming one. Wintz and Luecke have obtained performance results with a finite but large number of values of τ and evaluated the performance of this synchronizer for equal-energy raised cosine, half-sine and rectangular pulse shapes using $K = 1, 2, \ldots, 8$. The pulse shapes are given by:

Half-Sine

$$p(t) = \begin{cases} \sqrt{\dfrac{2E_b}{T}} \, \sin \dfrac{\pi t}{T}, & 0 \leq t \leq T \\ 0, & \text{elsewhere} \end{cases}$$

Raised Cosine

$$p(t) = \begin{cases} \sqrt{\dfrac{2E_b}{3T}} \left(1 - \cos \dfrac{2\pi t}{T} \right), & 0 \leq t \leq T \\ 0, & \text{elsewhere} \end{cases}$$

Rectangular

$$p(t) = \begin{cases} \sqrt{\dfrac{E_b}{T}}, & 0 \leq t \leq T \\ 0, & \text{elsewhere} \end{cases}$$

For the observation intervals used in the digital computer evaluation and a raised cosine pulse it was found that $\overline{|\varepsilon|}/T$ varies inversely with $\sqrt{E_b/N_0}$ and directly with $\sqrt{K}$. For $0.5 \leq E_b/N_0 \leq 5$ and $K = 8$, $\overline{|\varepsilon|}/T$ is approximately the same for the three pulse shapes. However, for $E_b/N_0 > 5$ the order of pulse shapes for decreasing $\overline{|\varepsilon|}/T$ is half-sine, raised cosine and square (see Ref. 27, Fig. 5b).

Suboptimum Clock Recovery: Nonlinearity Followed by a Filter

The MAP synchronizer performs a correlation or matched filter operation together with a nonlinear weighting by ln cosh. The ln cosh nonlinearity is approximately square-law for low-input signals and has approximately a magnitude function for large inputs x:

$$\ln \cosh x \simeq \begin{cases} \dfrac{|x|}{2}, & |x| \gg 1 \\[2mm] \dfrac{x^2}{2}, & |x| \ll 1 \end{cases}$$

A suboptimum configuration for a baseband input is given in Figure 5.16, wherein a low-pass filter preceding the nonlinearity approximates the matched filter operation of the optimum synchronizer and a post-nonlinearity bandpass filter tuned to the clock frequency performs an averaging operation. The bandpass filter output is approximately a sine wave whose positive-going zero crossings estimate the data transition times. The bandpass clock filter bandwidth determines the synchronizer memory. An output limiter can be used to provide a rectangular output and a manual phase shifter may be required for the clock to be used elsewhere in the demodulator.

For the three pulse shapes discussed, Wintz and Luecke found that a single-pole RC filter with time constant $RC \simeq T$ was a good approximation to a matched filter. The memory of the bandpass clock filter is defined as

$$K \triangleq \frac{(\Sigma C_i)^2}{\Sigma C_i^2}$$

where C_i is the relative filter weighting applicable to the ith preceding time interval. (Note that a bandpass filter has exponential weighting rather than the optimum circuit's equal averaging.) For a square-law nonlinearity the expected magnitude of the timing error as a fraction of the symbol interval for a raised-cosine pulse is [28]

$$\frac{\overline{|\varepsilon|}}{T} \simeq \frac{1}{3\sqrt{KE_b/N_0}} \qquad (5.10)$$

for $E_b/N_0 > 5$ and $K \geq 18$. The result corresponding to a square pulse and a hard-limiter nonlinearity based upon the results in Ref. 27, Fig. 10b is

$$\frac{\overline{|\varepsilon|}}{T} \simeq \frac{1}{1.18\sqrt{KE_b/N_0}} \qquad (5.11)$$

Figure 5.16 Suboptimum synchronizer.

for $E_b/N_0 > 12$. A comparison of (5.10) and (5.11) shows that the square-pulse -based recovery circuit requires a factor of 6.5 increase in E_b/N_0 over the raised cosine recovery circuit to attain the same jitter performance.

While the best nonlinearity has a $\ln\cosh$ characteristic, it was found that square-law and absolute value characteristics provided nearly-optimum results for rounded pulses, whereas a hard-limiter was best for square pulses. Overall, rounded pulses were found to provide better error-rate performance. This can be seen by recognizing the dependence of the conditional error rate in (5.8) upon the pulse autocorrelation function given by (5.9). The autocorrelation function of a square pulse is triangular, whereas that of the rounded pulses is flatter for ε near 0. As ε increases, $\mathcal{R}(\varepsilon)$ decreases more rapidly for the square pulse than for the rounded pulses. Thus the error rate for a system using square pulses is more sensitive to timing jitter.

It should be noted that the suboptimum configuration can be applied to the situation where the input is an IF signal, provided that the input filter is a bandpass filter. For a bandlimited system a filter matched to the received signal maximizes the SNR at the sampling instant. However, a large degree of jitter in the zero crossings of the baseband data modulating the carrier may exist. It follows that the best channel filter for data detection is not necessarily the best filter for clock recovery. For example, Gardner [15] found that for the case of QPSK signals the amplitude of the derived clock signal based upon a Nyquist prefilter and a square-law detector is proportional to the roll-off factor α of the filter.

Gardner [15, 29] compared the performance of several clock recovery circuits by means of a digital satellite communications link simulation program. The link consisted of a QPSK modulator whose output was bandlimited, a travelling-wave-tube amplifier (TWTA) and downlink additive noise. Channel filtering consisted of a 50% roll-off Nyquist characteristic evenly divided between the transmitter and the receiver. Two STR nonlinearity prefilters were used: (1) the receiver portion of the Nyquist response and (2) a flat filter having an equivalent IF bandwidth of 1.6 times the symbol rate. The STR circuits were of the rectifier type:

1   Square law
2   IF absolute value
3   Baseband absolute value

The baseband absolute value nonlinearity actually consists of two nonlinearities, one for each QPSK demodulated channel. The nonlinearity outputs are summed and the result passed through the clock filter.

Figure 5.17 shows the effect of a constant offset in the clock on the performance of coherent demodulation of QPSK. In this figure the E_b/N_0 required to maintain a constant bit-error rate of $P_b = 10^{-4}$ is plotted as a function of the clock error represented as a fraction of the symbol period. Note that the ideal $E_b/N_0 = 8.4$ dB for coherent demodulation. The graph shows

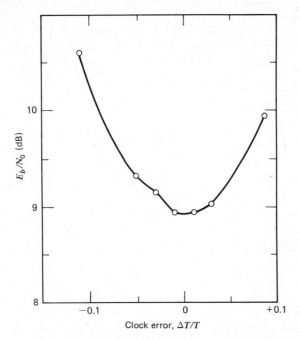

Figure 5.17 Effect of constant clock offset on E_b/N_0 required for $P_b = 10^{-4}$. Perfect carrier recovery and no clock jitter. (From Ref. 15; reprinted by courtesy of the European Space Agency from thesis ESA TM-169.)

that the degradation is negligible for 2 to 3% timing offset but increases to 1 dB for a 7% offset (perfect carrier recovery is assumed).

Figure 5.18 represents one set of simulation results for square-law, absolute value (IF and baseband) nonlinear detectors. The graph depicts rms clock error versus E_b/N_0. The clock filter was a single tuned filter having a noise-bandwidth symbol interval product $B_N T = 0.035$. The results indicate that the square-law nonlinearity yields the poorest performance (greatest rms jitter), while the baseband absolute value rectifier provides the best performance in the steady state. It was found that this was true for any channel filter simulated.

Gardner found that the square-law (IF and baseband) and IF envelope rectifiers in general are insensitive to carrier phase and can be used independent of carrier recovery. However, baseband rectifiers other than square-law are carrier-phase dependent. It was also determined that the flat prefilter resulted in less rms clock jitter than the more bandlimiting Nyquist filter. Nyquist prefilters having greater roll-off factors (wider bandwidths) provided less clock jitter. Furthermore, less clock jitter resulted with narrower clock filters.

Franks and Bubrouski [30] have developed criteria for eliminating pattern jitter for pulse amplitude modulation for a linear channel with a baseband synchronizer employing a square-law detector as follows:

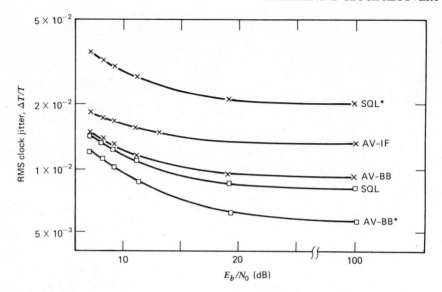

Figure 5.18 Jitter of reference clock. T, symbol duration; SQL, square-law; AV, absolute value; BB, baseband; IF, intermediate frequency; *, with carrier jitter. (From Ref. 15; reprinted by courtesy of the European Space Agency from thesis ESA TM-169.)

1 The baseband spectrum of the data pulses prior to the rectifier (nonlinearity) must be symmetrical above $f_0 = 0.5/T$ and bandlimited to $0.25/T \leq f \leq 0.75/T$.

2 The postrectifier filter must be symmetrical about $f = 1/T$.

Lyon [31] investigated clock recovery for quadrature amplitude modulation (also known as QASK). The clock recovery circuit employed parallel prefilters $G(f)$ and nonlinearities (square-law and absolute value) in each of two demodulated quadrature I- and Q-channels. The outputs of the parallel nonlinearities were added and the sum passed through a narrow clock filter $H(f)$. The following constraints on $G(f)$ and $H(f)$ were found to be necessary in addition to those given in Ref. 30:

1 The filters must have Hermitian symmetry; that is,

$$H\left(f + \frac{1}{T}\right) = H^*\left(f - \frac{1}{T}\right)$$

$$G\left(f + \frac{1}{2T}\right) = G^*\left(f - \frac{1}{2T}\right)$$

2 Both filters must be tuned to have zero-phase shift at their centre frequencies.

The preceding discussion has pertained to suboptimum synchronizers derived from the optimum synchronizer. Other techniques will now be reviewed.

Delay-Line Detector

Figure 5.19 depicts a delay-line detector that can operate on IF or baseband signals. In this circuit the received signal is multiplied by its delayed replica. That this process provides clock information may not be obvious. For purposes of demonstration an expression for the clock output will be derived [15] assuming that the received signal is a noise-free QPSK waveform. The input signal is given by $r(t)$ and the output is given by

$$z(t)=r(t)r(t-\tau)$$

$$=\left[x(t)\cos\omega_c t+y(t)\sin\omega_c t\right]$$

$$\cdot\left[x(t-\tau)\cos\omega_c(t-\tau)+y(t-\tau)\sin\omega_c(t-\tau)\right]$$

where τ is the detector delay. If high-frequency components and multiplier constants are neglected, the output becomes

$$z(t)=\left[x(t)x(t-\tau)+y(t)y(t-\tau)\right]\cos\omega_c\tau$$

$$-\left[x(t)y(t-\tau)+y(t)x(t-\tau)\right]\sin\omega_c\tau$$

The expected value of $z(t)$ is given by

$$E\left[z(t)\right]=E\left[x(t)x(t-\tau)+y(t)y(t-\tau)\right]\cos\omega_c\tau$$

$$-E\left[x(t)y(t-\tau)+y(t)x(t-\tau)\right]\sin\omega_c\tau \qquad (5.12)$$

Now $x(t)$ and $y(t)$ can be represented as an infinite summation of delayed basic pulse shapes

$$x(t)=\sum_{i=-\infty}^{\infty}a_i p(t-iT)$$

$$y(t)=\sum_{i=-\infty}^{\infty}b_i p(t-iT)$$

where a_i, $b_i=\pm1$ and $p(t)$ is the pulse shape generated for the interval

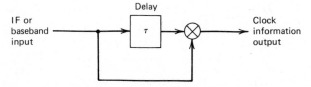

Figure 5.19 Delay-line detector.

$0 \leq t \leq T$. The quadrature data symbols a_i and b_i are assumed statistically independent. By recognizing that $a_i^2 = b_i^2 = +1$, the expectations in (5.12) become

$$E[x(t)x(t-\tau)] = E\left[\sum_{i=-\infty}^{\infty} a_i p(t-iT) \sum_{i=-\infty}^{\infty} a_i p(t-iT-\tau)\right]$$

$$= \sum_{i=-\infty}^{\infty} p(t-iT)p(t-iT-\tau)$$

$$= E[y(t)y(t-\tau)]$$

and

$$E[x(t)y(t-\tau)] = E\left[\sum_{i=-\infty}^{\infty} a_i p(t-iT) \sum_{i=-\infty}^{\infty} b_i p(t-iT-\tau)\right]$$

$$= 0$$

$$= E[y(t)x(t-\tau)]$$

Therefore, $E[z(t)]$ reduces to

$$E[z(t)] = 2 \sum_{i=-\infty}^{\infty} p(t-iT)p(t-iT-\tau)\cos\omega_c\tau \tag{5.13}$$

Note that (5.13) is maximized when $\omega_c\tau = 2n\pi$ (n an integer). The periodicity of (5.13) can be verified by replacing t by $t+T$. Therefore, the average output of the delay detector is periodic in T and can be represented by a series expansion (assuming that $\cos\omega_c\tau = 1$),

$$v_0(t) = \sum_{n=-\infty}^{\infty} c_n e^{j2\pi nt/T}$$

where

$$c_n = \frac{1}{T}\int_{-T/2}^{T/2} v_0(t) e^{-j2\pi nt/T}\, dt$$

The complex amplitude of the clock component is then

$$c_1 = \frac{2}{T}\int_{-T/2}^{T/2} \sum_{i=-\infty}^{\infty} p(t-iT)p(t-iT-\tau) e^{-j2\pi t/T}\, dt$$

$$= \frac{2}{T} \sum_{i=-\infty}^{\infty} \int_{-T/2}^{T/2} p(t-iT)p(t-iT-\tau) e^{-j2\pi t/T}\, dt$$

The expression for c_1 is an infinite sum of adjacent finite integrals with the

Table 5.1 Maximization of Delay Detector Output

Pulse Shape	Maximizing Delay, τ_{max}
Rectangular pulse $$p(t)=\begin{cases} 1, & -\dfrac{T}{2}\leq t\leq\dfrac{T}{2} \\ 0, & \text{elsewhere} \end{cases}$$	$T/2$
Raised cosine pulse $$p(t)=\begin{cases} \frac{1}{2}\left(1+\cos\dfrac{\pi t}{T}\right), & \lvert t\rvert\leq T \\ 0, & \text{elsewhere} \end{cases}$$	$0.2T$ (approx.)
Sinc pulse $$p(t)=\dfrac{\sin \pi t/T}{\pi t/T}$$	Not applicable
Nyquist pulse For pulse shape, see Section 4.2	0

Source: Ref. 15

same integrand and may be rewritten as

$$c_1=\frac{2}{T}\int_{-\infty}^{\infty} p(t-iT)p(t-iT-\tau)e^{-j2\pi t/T}dt$$

If the delay-line detector output is passed through a narrowband filter, then the fundamental component can be isolated. The clock signal produced in this manner is a function of three parameters:

1 $\omega_c\tau$
2 The baseband pulse shape
3 The value of τ

Given $\omega_c\tau=2n\pi$ there exists an optimum delay for a given pulse shape to maximize the clock detector output. Table 5.1 gives the delay required for rectangular, raised cosine, $(\sin x)/x$ and Nyquist pulses.

Note that $\sin x/x$ pulses cannot be processed because the output is zero regardless of the value of τ. In the case of Nyquist pulses the delay detector reduces to a doubler, i.e., $\tau=0$.

It should be noted that although the delay-line detector has been described in terms of an IF input, it is applicable to baseband inputs as well.

In-Phase/Midphase Bit Synchronizer

The in-phase/midphase bit synchronizer, which is also known as a data transition tracking loop (DTTL) [6, 32], is a closed-loop synchronizer. Two channels are used (as shown in Figure 5.20) with the in-phase channel

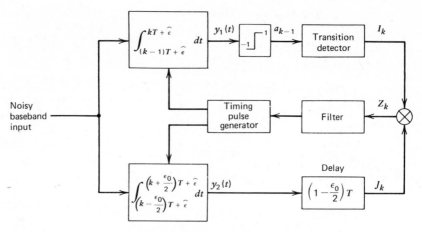

Figure 5.20   In-phase/midphase synchronizer.

indicating the occurrence of a transition in the data and the midphase channel providing a measure of the offset between the incoming data transitions and a local clock.

The in-phase channel includes an integrate-and-dump (ID)* whose output is sampled every T seconds (T=duration of NRZ pulse). A hard limiter decides on the polarity of the pulse. Decisions on adjacent pulses are compared in a transition detector. The detector output is determined by the rule

$$I_k = \frac{a_{k-1} - a_k}{2}$$

so that $I_k = 0$ when no transition occurs ($a_{k-1} = a_k$) and $I_k = \pm 1$ (depending upon a_{k-1}, a_k) when a transition does occur.

The midphase integrate-and-dump operates on a staggered time basis relative to the in-phase ID. Its output is proportional to the difference between the loop estimate and actual transition times. The integrate-and-dump is sampled every T seconds. In the basic DTTL the quadrature integration window is T. However, decreasing the interval to $\varepsilon_0 T (0 \leq \varepsilon_0 \leq 1)$ can provide improved performance. The ID output must be delayed by $(1 - \varepsilon_0/2)T$ for proper operation. With $\varepsilon_0 = 1$ the required delay is $T/2$. The output signals of the two channels are multiplied together to give $Z_k = I_k J_k$, which is filtered, and the filter output in turn controls the instantaneous frequency of a generator, which provides the required timing. Figure 5.21 depicts loop signals for the case of perfect timing and a T second midphase window. It is left as an exercise to the reader (Problem 13) to derive similar waveforms for imperfect timing ($Z_k \neq 0$) and reduced windows.

An advantage of the DTTL is its ability to "flywheel" through a long sequence of ones and zeros at its input. In such a situation transitions are

---

*The ID is the matched filter for the input pulse waveform, here taken as NRZ.

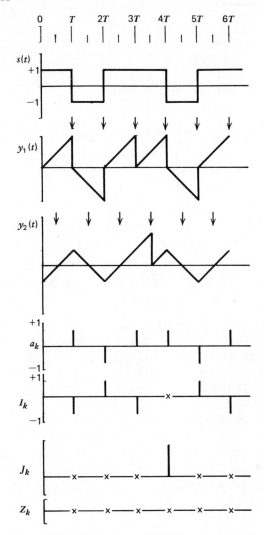

Figure 5.21   Derivation of DTTL control signals.

infrequent and the loop maintains its last adjustment until new timing information is provided by the data. Note that this operation assumes that noise does not introduce any errors. Furthermore, the closed-loop bandwidth must account for the inherent jitter in the received signal transitions. This bandwidth is also a function of the midphase window interval.

Absolute Value Early–Late Gate Synchronizer

The absolute value early–late gate synchronizer utilizes two integrate-and-dump circuits (see Figure 5.22). Each integrates over a pulse interval T, with one

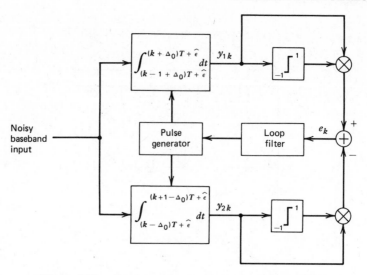

Figure 5.22   Absolute value early–late gate synchronizer.

starting $\Delta_0 T$ early relative to the transition time estimate and the other starting $\Delta_0 T$ late. The integrator outputs pass through hard limiters. The limiter outputs are multiplied by their inputs to remove the sign of the ID outputs. The error signal e_k is then a measure of the difference in output magnitudes of the two channels. The instantaneous frequency of the local reference is then advanced or retarded in proportion to this difference. If a transition has not occurred, the channel outputs are identical and $e_k = 0$. Otherwise, e_k is linearly proportional to the offset λT between actual and estimated transition times.

Simon [33] has shown that for large E_b/N_0 the mean-square timing error for the absolute value early–late gate synchronizer (AVGS) and DTTL synchronizer with full integration in the quadrature ID is given asymptotically as follows:

AVGS

$$\sigma_\lambda^2 = \frac{W_L T}{8(E_b/N_0)\,\mathrm{erf}\left[\dfrac{\sqrt{E_b/N_0}}{2}\right]}$$

DTTL

$$\sigma_\lambda^2 = \frac{W_L T}{4(E_b/N_0)\,\mathrm{erf}\left[\dfrac{\sqrt{E_b/N_0}}{2}\right]}$$

where W_L is the two-sided loop bandwidth. Thus in the linear region of operation the AVGS offers a 3-dB advantage over the DTTL.

Binary Quantized Digital Phase-Locked Loop

The binary quantized digital phase-locked loop [34] (BQDPL) is shown in Figure 5.23. The BQDPL compares transitions in the signal input with those of a local clock derived from a stable reference oscillator. The comparison is a coarse quantization (binary) which indicates whether the local clock transitions are leading or lagging the received signal transitions. This binary information is processed by a sequential filter that time-"averages" the lead/lag information. The loop filter provides advance or retard commands to a circuit, which adds or deletes a cycle from the reference clock. The adjusted reference is divided by ($N \pm k$), where N and k are integers, to give a local clock with the appropriate phase step changes.

The loop is characterized by its nonlinear behaviour. Clock phase adjustments are made in discrete steps at discrete points in time. The filter accumulates the lag/lead pulses and indicates an advance/retard correction only when the appropriate threshold is reached. The time at which a threshold is reached is a function of the transition statistics of the incoming signal. The existence of a threshold indicates that a minimum number of symbol intervals are needed before a correction can be made.

Figure 5.24 depicts one way of deriving the lag/lead information. For each zero crossing in the input waveform a narrow positive pulse is generated. The local clock is assumed to have an offset error ε. (Note that the data duration is T while the data clock completes a full cycle in T.) Two gating signals, denoted as A and B, are derived from the positive and negative portions of the clock, respectively. Transition pulses are gated by these signals to produce the required lag/lead pulses. Note that in Figure 5.24 the transition pulses will be gated only by signal B, thus indicating that the clock needs a correction in one direction.

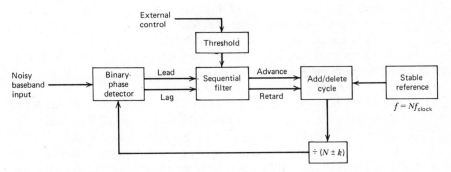

Figure 5.23 Binary quantized digital phase locked loop (BQDPL).

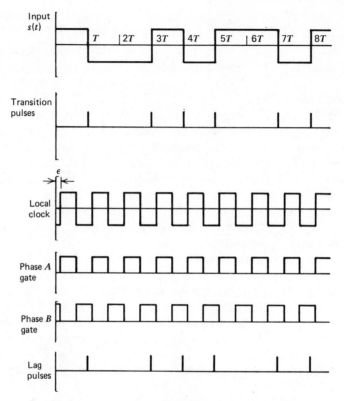

Figure 5.24   BQDPL lag/lead pulse derivation.

Two types of filters that can be used are the N-before-M and the random walk filter (RWF). In the N-before-M filter (see Figure 5.25a) the lag and lead inputs are accumulated in registers of length N and their sum is accumulated in a register of length M. With an initial all-zero state the filter operates as follows. If either lag/lead register reaches its maximum limit prior to or coincident with the sum register reaching its limit, a corresponding advance/retard signal is produced. The three counters are reinitialized and the measurement repeats. However, if the lag/lead registers do not reach their maxima, no phase correction is made and the counters are reset.

In the RWF an up/down counter reacts to the lag/lead pulses by counting up or down for each input pulse. When either an upper or lower threshold is reached, an appropriate phase step command results.

The sequential filter determines the rate of phase correction of the local clock or equivalently the bandwidth of the BQDPL. The use of dual thresholds provides a convenient mechanism for acquiring and tracking the signal transitions: a low absolute value threshold (causing more frequent phase adjustments) can be used for acquisition and a large absolute value threshold (with a

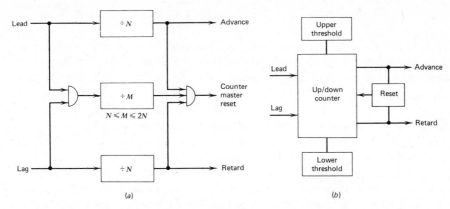

Figure 5.25 BQDPL sequential filter types. (*a*) *N*-before-*M*; (*b*) random walk.

longer time between corrections) can be used for tracking purposes. Furthermore, if the tracking bandwidth is made small enough, the loop will "flywheel" through long sequences of ones and zeros. Figure 5.23 shows that the threshold can be adjusted by an external control. This control can be based upon received signal strength or the detection of a word at the start of a message burst [35].

The basic BQDPL has a maximum correction rate (assuming no noise) of

$$R_c = \frac{T_i \Delta}{360 N}$$

where T_i = number of transitions occurring per second in a periodic input
    signal
Δ = phase-step correction in degrees made by the BQDPL for each
    advance/retard command
N = RWF threshold

The local master oscillator is not coherent with the received signal, and therefore perfect synchronization by discrete phase-step changes is not possible. The minimum phase error value δ is given by $|\delta| \le \Delta/2$.

Note that more elaborate circuits may be used rather than a simple binary lead/lag pulse where a quantized measurement is made of the phase offset between signal and estimated transitions and a finer correction phase step is used. Furthermore, a null zone about zero phase error could be used so that an error would have to exceed a prescribed limit ($\pm \varepsilon_L$) before corrections would be made. The basic BQDPL is analogous to a first-order loop, and these alternative approaches form the bases for higher-order loops.

5.6   COMPENSATION FOR INFREQUENT ZERO CROSSINGS

The amplitude of the derived clock at the output of a suboptimum synchronizer clock filter depends upon randomness in the data. This randomness is required to produce sufficient zero crossings. If insufficient transitions occur, the regenerated clock may slip a cycle, leading to loss of bit count integrity and skewing of the data. In cases where the data at the source is expected to have long sequences of ones or zeros, compensation is required. (Note that these long sequences may also be undesirable in terms of the resulting modulator transmitted spectrum. Components may exist that contravene regulatory limitations on spectral density.) One solution to this problem is to scramble or randomize the data by adding modulo 2 a pseudo-random sequence prior to the modulator (the same operation is required in the receiver to recover the original data).

An alternative approach or one that could be used in conjunction with scrambling has been described by Yokoyama et al. [24]. The circuit is shown in Figure 5.26. The first stage is a half-symbol delay-line detector followed by a bandpass clock filter. The filter output is divided into two paths. In one the signal is hard-limited to remove the amplitude fluctuations. The limiter output is phase shifted as required for use in the demodulator. The second filter output path detects the signal envelope. If the amplitude falls below a predetermined threshold, a clock signal is recirculated through a one-symbol delay to provide a stable "flywheel" effect during the infrequent transition period. The limiter output path is disabled during this interval.

A third approach applicable to the closed-loop circuits discussed previously is based upon the inherent stability of the system clock oscillators and uses a synchronizer to maintain clock integrity during periods of infrequent transitions. Two situations may exist. In one, the duration of the communications is known and the stability of a locally available reference clock is such that timing accuracy can be maintained to less than a threshold value provided that the initial phase has been adjusted at the beginning of the burst. Another case is one where a digital phase-locked loop (DPLL) tracks the clock information in the form of data zero crossings. The DPLL uses these zero crossings to

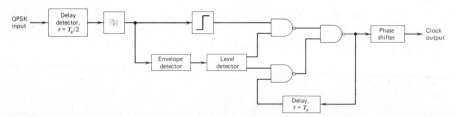

Figure 5.26   Symbol timing recovery circuit with infrequent transition compensator. (From Ref. 24: © 1975 IEEE; reprinted from "The Design of a PSK Modem for the Telesat TDMA System", by S. Yokoyama, K. Kato, T. Noguchi, N. Kusami and S. Otani, from *Proc. Int. Conf. Commun.,* June 1975.)

adjust a local reference clock (either an advance or a retard). If the number of zero crossings is temporarily reduced, the inherent stability of the reference clock and the DPLL adjustment rate (loop bandwidth) are such that clock integrity can be maintained.

5.7   CARRIER AND CLOCK RECOVERY FOR FFSK

FFSK signalling entails the transmission of tones at frequencies f_1 and f_2. These tones are separated in frequency by half the bit rate. The FFSK power spectrum contains no discrete components.

However, use can be made of the fact that FSK, having a modulation index $h = 1$, contains discrete components at the signalling frequencies. Transformation of the FFSK into a unity modulation index FSK is attained by passing the FFSK through a doubler whose output contains spectral lines at $2f_1$ and $2f_2$. Carrier frequency (f_c) and clock frequency (f_{CP}) are established by the following relationship:

$$f_c = \frac{2f_1 + 2f_2}{4} = \frac{f_1 + f_2}{2}$$

$$f_{CP} = 2f_2 - 2f_1$$

Figure 5.27 given by de Buda [36] demonstrates an implementation. The frequency doubler output is passed through narrowband filters at $2f_1$ and $2f_2$

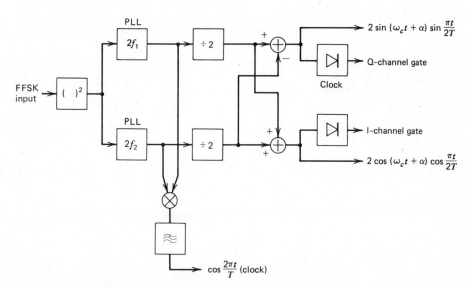

Figure 5.27   FFSK carrier and clock recovery. (From Ref. 36: © 1972 IEEE; reprinted from "Coherent Demodulation of Frequency-Shift Keying with Low Deviation Ratio", by R. de Buda, from *IEEE Trans. Commun.*, June 1972.)

(in this representation the filters are phase-locked loops). The clock is obtained by multiplying the PLL outputs and low-pass filtering the result.

The quadrature carrier references could be obtained by passing the multiplier output through a high-pass filter and dividing the output by 4. In such an implementation the phase is ambiguous by a multiple of $\pi/2$. However, FFSK is a binary modulation technique and the $\pi/2$ ambiguity can be resolved by using the circuit shown in Figure 5.27. In this representation the PLL outputs are each divided by 2. The divider outputs are added and subtracted as shown. Because the sign of the divider outputs is not known, the summer output signs are unknown. In any case, the outputs will be $\pm\cos\omega_c t\cos\omega_R t$ and $\pm\sin\omega_c t\sin\omega_R t$, where $\omega_R=\pi/2T$. These signals are the required I and Q reference signals when they are gated between the nulls of their respective envelopes.

The I and Q reference signals multiply the received signal in quadrature channels. The resulting baseband waveforms are optimally processed in parallel integrate-and-dump circuits which are controlled by the recovered clock. The ID outputs are then differentially decoded by combining their outputs in a logical EXOR gate.

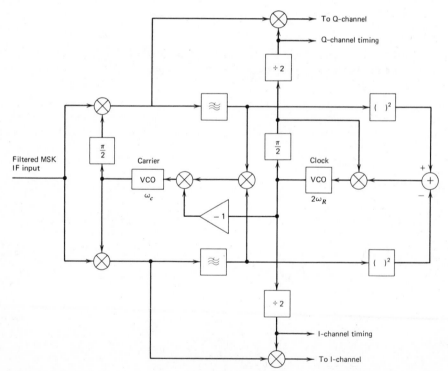

Figure 5.28 MSK I-Q recovery circuit. [From Ref. 37: © 1978 IEEE; reprinted from "Carrier Phase and Bit Sync Regeneration for the Coherent Demodulation of MSK," by R. W. D. Booth, from *NTC '78 conf. Rec.*, 3–6 Dec. 1978, Birmingham, Ala. (78CH1354-0 CSCB).]

In situations where doubling the input IF signal is not desirable, an I–Q version of the FFSK carrier and clock recovery loop, similar to the Costas loop used for PSK, can be used. Figure 5.28, taken from Ref. 37, demonstrates the alternative implementation. A demodulator/remodulator recovery technique can be used with MSK and is described in Ref. 38.

REFERENCES

1 W. Schrempp and T. Sekimoto, "Unique Word Detection in Digital Burst Communications", *IEEE Trans. Commun. Technol.*, Vol. COM. 16, Aug. 1968, pp. 597–605.

2 A. J. Viterbi, *Principles of Coherent Communication*, McGraw-Hill, New York, 1966, p. 111.

3 S. A. Rhodes, "Effect of Noisy Phase References on Coherent Detection of Offset-QPSK Signals", *IEEE Trans. Commun.*, Vol. COM. 22, Aug. 1974, pp. 1046–1055.

4 R. Matyas, "Effect of Noisy Phase References on Coherent Detection of FFSK Signals", *IEEE Trans. Commun.*, Vol. COM-26, June 1978, pp. 807–815.

5 R. M. Gagliardi, *Introduction to Communications Engineering*, Wiley, New York, 1978.

6 W. C. Lindsey and M. K. Simon, *Telecommunications Systems Engineering*, Prentice-Hall, Englewood Cliffs, N.J., 1973.

7 F. M. Gardner, *Phaselock Techniques*, Wiley, New York, 1966.

8 J. W. Layland, "An Optimum Squaring Loop Filter", *IEEE Trans. Commun.*, Vol. COM-18, Oct. 1970, pp. 695–697.

9 J. F. Oberst and D. L. Schilling, "The SNR of a Frequency Doubler", *IEEE Trans. Commun. Technol.*, Vol. COM-19, Feb. 1971, pp. 97–99.

10 J. P. Costas, "Synchronous Communications", *Proc. IRE*, Vol. 44, Dec. 1956, pp. 1713–1718.

11 W. C. Lindsey and M. K. Simon, "Data-Aided Carrier Tracking Loops", *IEEE Trans. Commun. Technol.*, Vol. COM-19, Apr. 1971, pp. 157–169.

12 W. C. Lindsey, "Hybrid Carrier and Modulation Tracking Loops", *IEEE Trans. Commun. Technol.*, Vol. COM-20, Feb. 1972, pp. 53–55.

13 R. Matyas and P. J. McLane, "Decision-Aided Tracking Loops for Channels with Phase Jitter and Intersymbol Interference", *IEEE Trans. Commun.*, Vol. COM-22, Aug. 1974, pp. 1014–1023.

14 F. M. Gardner, "Hangup in Phase-Lock Loops", *IEEE Trans. Commun.*, Vol. COM-25, Oct. 1977, pp. 1210–1214.

15 F. M. Gardner, "Carrier and Clock Synchronization for TDMA Digital Communications", Eur. Space Agency Rep. ESA TM-169 (ESTEC), Dec. 1976.

16 C. L. Weber, "Design of a Demod/Remod Receiver Operating On Burst QPSK Data", in *Proc. AIAA/CASI 6th Commun. Satellite Conf.*, Montreal, 5–8 Apr. 1976, pp. 479–511.

17 H. Yamamoto, K. Hirade and Y. Watanabe, "Carrier Synchronizer for Coherent Detection of High Speed Four-Phase-Shift-Keyed Signals", *IEEE Trans. Commun.*, Vol. COM-20, Aug. 1972, pp. 803–807.

18 F. M. Gardner, "Comparison of QPSK Carrier Generator Circuits for TDMA Applications", in *Proc. Int. Conf. Commun.*, Minneapolis, Minn., 1974, pp. 43B.1–43B.5.

19 C. J. Wolejsza, Jr., A. M. Walker and A. M. Werth, "PSK Modems for Satellite Communications", in *Proc. ICDSC-1*, London, 1969, pp. 127–143.

20 K. Nozaka, T. Muratani, M. Ogi and T. Shoji, "Carrier Synchronization Techniques of PSK-Modem for TDMA Systems", in *Proc. ICDSC-1*, London, 1969, pp. 154–165.

21 S. Yokoyama and T. Noguchi, "Theoretical and Experimental Considerations of the Carrier and the Bit Timing Recovery in the Burst Mode Operation", in *Proc. ICDSC-1*, London, 1969, pp. 106–115.

22   K. Nosaka, A. Ogawa and T. Muratani, "PSK Demodulator with Delay Line for the PCM-TDMA System", *IEEE Trans. Commun. Technol.*, Vol. COM-18, Aug. 1970, pp. 427–434.

23   A. M. Walker, "High Data Rate PSK Modems for Satellite Communications", *Telecommun. Mag.*, July 1976, pp. 27–31.

24   S. Yokoyama, K. Kato, T. Noguchi, N. Kusama and S. Otani, "The Design of a PSK Modem for the TELESAT TDMA System", in *Proc. Int. Conf. Commun.*, San Francisco, June 1975, pp. 44.11–44.14.

25   J. Y. Huang, "An Investigation of a Differentially Coherent Detector for Reception of QPSK Signals", in *Proc. Natl. Telecommun. Conf.*, Birmingham, Ala., Dec. 1978, pp. 27.1.1–27.1.5.

26   M. Asahara, N. Toyonaga, S. Sasaki and T. Miyo, "Analysis of Carrier Recovery Adopting a Narrow Band Passive-Filter with AFC Loop", in *Proc. ICDSC-3*, Kyoto, Nov, 1975, pp. 99–104.

27   P. A. Wintz and E. J. Luecke, "Performance of Optimum and Suboptimum Synchronizers", *IEEE Trans. Commun. Technol.*, Vol. COM-17, June 1969, pp. 380–389.

28   J. J. Spilker, Jr., *Digital Communications by Satellite*, Prentice-Hall, Englewood Cliffs, N.J., 1977, p. 433.

29   F. M. Gardner, "Clock Recovery for QPSK-TDMA Receivers", in *Proc. Int. Conf. Commun.*, San Francisco, June 1975, pp. 28.11–28.15.

30   L. E. Franks and J. P. Bubrouski, "Statistical Properties of Timing Jitter in a PAM Timing Recovery Scheme", *IEEE Trans. Commun.*, Vol. COM-22, July 1974, pp. 913–920.

31   D. L. Lyon, "Envelope-Derived Timing Recovery in QAM and SQAM Systems", *IEEE Trans. Commun.*, Vol. COM-23, Nov. 1975, pp. 1327–1331.

32   M. K. Simon, "An Analysis of the Steady-State Phase-Noise Performance of a Digital-Data-Transition Tracking Loop", in *Proc. Int. Conf. Commun.*, 1969, pp. 20.9–20.15.

33   M. K. Simon, "Nonlinear Analysis of an Absolute Value Type of an Early–Late Gate Bit Synchronizer", *IEEE Trans. Commun. Technol.*, Vol. COM-18, Oct. 1970, pp. 589–596.

34   J. R. Cessna and D. M. Levy, "Phase Noise and Transient Times for a Binary Quantized Digital Phase-Locked Loop in White Gaussian Noise", *IEEE Trans. Commun.*, Vol. COM-20, Apr. 1972, pp. 94–104.

35   R. Matyas, C. Jagger, R. Robitaille and R. de Buda, "Digital Implementation of a FFSK Modem", in *Proc. Natl. Telecommun. Conf.*, Birmingham, Ala., Dec. 1978, pp. 30.2.1–30.2.5.

36   R. de Buda, "Coherent Demodulation of Frequency-Shift Keying with Low Deviation Ratio", *IEEE Trans. Commun.*, Vol. COM-20, June, 1972, pp. 429–435.

37   R. W. D. Booth, "Carrier Phase and Bit Sync Regeneration for the Coherent Demodulation of MSK", in *Proc. Natl. Telecommun. Conf.*, Birmingham, Ala., Dec. 1978, pp. 6.1.1–6.1.5.

38   Y. Morihiro, S. Nakajima and N. Furuya, "A 100 Mbit/s Prototype MSK Modem for Satellite Communications", *IEEE Trans. Commun.*, Vol. COM-27, Oct. 1979, pp. 1512–1518.

PROBLEMS

1   In a binary coherent BPSK receiver the incoming signal is multiplied by a generated carrier to produce the demodulated baseband data. If the carrier has a phase error $\hat{\theta}$, find an expression for the probability of error. If the phase error is a constant 45°, find P_b if $E_b/N_0 = 8.4$ dB. How does this compare with the ideal P_b?

2   The relative phase offset between a received BPSK signal and a local carrier reference is 45°. Assume that the carrier recovery loop phase error

follows an exponential law

$$\phi = \phi_0 e^{-0.5t/T}$$

where ϕ_0 = initial recovery loop phase offset = 45°
T = symbol interval

After what period (in symbols) will the error rate be 10^{-4}? Assume that $E_b/N_0 = 8.4$ dB.

3　The output of the phase detector of a PLL is given by $\sin[\phi(t)]$, where $\phi(t)$ is the phase-error signal. Why is a sine function used? Can the phase detector output be $\cos[\phi(t)]$? Explain.

4　A typical expression for the phase probability density function of a high-gain second-order PLL is given by

$$p(\theta) = \frac{e^{\alpha \cos \theta}}{2\pi I_0(\alpha)}$$

What is the effect of large α?

5　If an FFSK signal is passed through a doubler, a form of FSK results which has discrete components (called Sunde lines). Determine the modulation index of the resulting signal and the proportion of the signal power contained in the discrete components.

6　Derive the expressions for the conditional error probability for QPSK and offset QPSK for detection with a noisy carrier phase reference. Assume perfect clock timing.

7　Implicit in the discussions of ideal BPSK and QPSK modulation has been the use of non-return-to-zero (NRZ) input data. A number of formats are possible, as shown in Appendix B. Explain what effect the use of a return-to-zero format would have on the output spectrum of a BPSK modulator. Why would such a format be used?

8　Show that the probability of error conditioned on the clock error ε for NRZ data is given by

$$P(\text{error}|\ \varepsilon) = \tfrac{1}{4} \text{erfc}\left(\sqrt{\frac{E_b}{N_0}}\right) + \tfrac{1}{4} \text{erfc}\left[\sqrt{\frac{E_b}{N_0}}\ (1 - 2|\varepsilon|)\right]$$

9　Why is a 111111 pattern used for clock recovery in differentially encoded PSK?

10　Show that the magnitude of the clock component at the output of a delay-line detector for rectangular, $\sin x/x$ and raised cosine pulse shapes is as given in Table 5.1.

11 Show the equivalence in clock recovery at IF and baseband for delay-line detection.

12 Show that square-law IF and baseband detectors are equivalent, whereas absolute value IF and baseband rectifiers are not.

13 For a data-transition tracking loop, derive circuit waveforms for the following conditions:

(a) Synchronization offset $= T/8$

(b) Midphase window $= T/2$

Assume an input NRZ sequence of 1011101.

14 Suppose that an n-bit-long sequence is to be detected out of a random sequence. Show that the probability of miss P_m and the probability of false detection P_{fd} are given by

$$P_m = \sum_{i=\varepsilon+1}^{n} \binom{n}{i} p^i q^{n-i}$$

$$P_{fd} = \frac{1}{2^n} \sum_{i=0}^{\varepsilon} \binom{n}{i}$$

where ε is the number of errors allowed and $p = 1 - q$ is the bit-error rate. What happens to P_m and P_{fd} as ε increases?

CHAPTER **6**

IIIIIIIIIIIIIIIIIIIIIIIIIIIIIIIIIIIIIIIIIIIIIIIIII

Multi-*h* Phase-Coded
Modulation

Of particular interest in satellite communications are modulation techniques that are relatively constant envelope and power and bandwidth efficient. These requirements have been necessitated to date by limitations that are both technological and regulatory.

Modulation formats that satisfy these requirements use phase or frequency as the data-dependent parameter. A significant amount of investigation into PSK has resulted in its use in a number of satellite systems. FSK modulation has been of lesser interest because of the E_b/N_0 penalty associated with the more conventional formats.

In this chapter FSK modulation techniques are examined which provide performance as good as or better than PSK systems. These techniques rely upon the use of inherent coding in the modulation structure and require more elaborate receiver configurations. Although these techniques may not represent the ultimate modulation scheme, they do indicate an area of significant potential improvement.

6.1 CONTINUOUS-PHASE FREQUENCY-SHIFT KEYING

FFSK is a special case of continuous-phase frequency-shift keying (CPFSK) with the modulation index $h=0.5$. The more general case with $h \neq 0.5$ is of interest because error-rate performance superior to that of FFSK is possible. Recall that FFSK is the FSK modulation having the minimum correlation coefficient for which orthogonal transmission exists. With a 1-bit observation interval, coherent orthogonal detection was shown to result in an error-rate performance 3 dB worse than CPSK. However, with a 2-bit observation interval, performance equal to that of differentially encoded PSK could be attained. It is reasonable to wonder whether improved performance can result with longer observation intervals and different h.

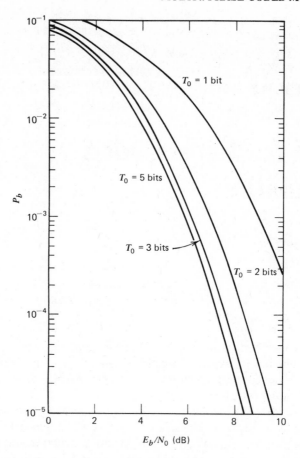

Figure 6.1 CPFSK performance—upper bounds as function of observation interval (T_0). $h=0.715$. (From Ref. 2: © 1974 IEEE; reprinted from Coherent and Noncoherent Detection of CPFSK", by W. P. Osborne and M. B. Luntz, from *IEEE Trans. Commun.*, Aug. 1974.)

An observation interval longer than 2T is unnecessary in the case of FFSK [1]. This can be explained by considering the phase trellis. Starting at an initial phase position, the two paths corresponding to the data sequences 01 and 10 merge after 2T at the same phase position. Any information subsequent to the merging of the two paths cannot resolve information previous to the merge. Therefore, a 2T observation interval is sufficient. For different values of h, improved results can result with longer observation intervals. Examples include CPFSK with $h=\frac{2}{3}$ or $\frac{5}{7}$ where the signal passes through a finite set of phase states. De Buda has shown that with $h=\frac{2}{3}$ and an observation interval of 4T, the performance improvement is 0.8 dB; that is, for a specified error rate, the required E_b/N_0 is less by 0.8 dB than for FFSK.

Osborne and Luntz [2] addressed the problem of basing the detection of one bit on observing n bits. This decision is based upon observing the bit to be detected as well as the $(n-1)$ preceding bits. The optimum receiver was shown to correlate the received signal with the set of m ($=2^n - 1$) possible transmitted waveforms for the cases where the bit to be detected is first assumed to be (1) a one and then (2) a zero. The sequence providing the largest correlation sum over n bits is used to determine the required bit.

In Figure 2.6 the minimum correlation coefficient for binary FSK is shown to occur for $h=0.715$. Osborne and Luntz have determined the error-rate performance of this particular form of CPFSK. Figure 6.1 depicts the performance for observation intervals (T_0) ranging from $1 \leq T_0 \leq 5T$. It should be noted that the curves represent upper bounds on the error-rate performance. The authors concluded that an observation interval of $3T$ is usually sufficient in CPFSK detection. The performance improvement relative to CPSK is approximately 1 db.

6.2   MULTI-*h* PHASE-CODED MODULATION

In the preceding discussion binary data were considered together with a single modulation index, thus indicating two-tone FSK transmission. Extensions to CPFSK can be envisaged where a number of modulation indices as well as different pulse values are used. In such a system the signal is given by

$$s_i(t) = \sqrt{\frac{2E_s}{T}} \cos\left[\omega_c t + \frac{a_i \pi h_i [t - (i-1)T]}{T} \right.$$

$$\left. + \pi h_i \sum_{j=1}^{i-1} a_j + \theta_0 \right], \qquad (i-1)T \leq t \leq iT \quad (6.1)$$

where E_s = symbol energy per interval T
$\quad \omega_c$ = carrier angular frequency
$\quad a_i$ = digital information $[\pm 1, \pm 3, \ldots, \pm(M-1)]$
$\quad M$ = even constant and a power of 2
$\quad h_i$ = one of a set of modulation indices $(h_1, h_2, \ldots, h_K)$
$\quad \theta_0$ = initial phase

The summation term in (6.1) is the excess phase $\phi(iT)$, which indicates that symbols are not independent and that memory exists in the signal. The transmitted information resides in the phase function.

An example of a system having multiple digit levels but otherwise equivalent in structure to MSK is multiamplitude minimum-shift keying (MAMSK), which has been described by Weber et al. [3]. MAMSK is bandwidth compressive in that side lobes are sufficiently low in level that they can be removed by straightforward filtering to give an essentially bandlimited signal. The spectrum is identical to that of MSK except for a frequency scale factor which depends upon the number of bits represented by each symbol. This is analogous to the relationship between BPSK and M-ary PSK spectra. Note that for 4-bit MAMSK the required $E_b/N_0 \simeq 12.4$ dB for $P_b = 10^{-4}$.

Another technique of interest is that described by Anderson and Taylor [4]. A class of modulation called multi-h phase-coded modulation (MHPM) is based upon binary signals but uses several modulation indices in a prescribed manner to attain improved performance efficiencies relative to MSK. A similar technique has been described in Ref. 5. Throughout the remainder of this Chapter the structure, demodulation and performance of MHPM will be described in some detail. Investigation of this technique is of interest for the following reasons:

1 MHPM exhibits the constant envelope characteristic of MSK but offers E_b/N_0 performance improvements of as much as 4 dB and therefore is of interest in satellite communications applications.

2 The modulation is representative of the trend toward the integration of coding and modulation into one process.

3 It demonstrates how E_b/N_0 savings of 2 to 4 dB can be attained without the $1/r$ (r=rate of error-correcting coder) data-rate expansion that results from the use of separate codecs and modems.

4 It introduces the concepts of the vector-space description of signals and their maximum likelihood sequence estimation (MLSE).

Fundamental Concepts of MHPM

FFSK was described as a special form of CPFSK with $h=0.5$ and depicted pictorially by its phase trellis (see Figure 3.7), wherein phase is continuous at the data transitions with linear positive or negative increments of $\pi/2$ over the symbol interval. MHPM similarly has continuous phase at the data transitions and has piecewise-linear phase changes between transitions. However, the slope of the phase changes from interval to interval. A set of modulation indices $\{h_i; i=1,\dots,K\}$ is used in a cyclical rotation; that is, h_1 is available during the first bit interval, h_2 during the second and so on until K indices have been used. For the $(K+1)$st interval h_1 becomes available once more. For each h_i two tones, corresponding to a data 1 or 0, can be transmitted. The tones are displaced from the nominal carrier frequency by $\pm h_i/2T$.

If the indices $\{h_i\}$ are restricted to be a multiple of $1/q$, where q is an integer, all phase values at the transition times $t=nT$ are a multiple of $2\pi/q$. This defines a set of phase states for each $t=nT$. As has been shown for FFSK, the receiver can use this phase state information to demodulate the data provided that

1 The demodulator is synchronized to the incoming signal (i.e., interval timing in terms of which h_i applies is known).

2 The set of possible transitions from state to state (i.e., the trellis) is known.

An example of a phase trellis for the case $(h_1, h_2) = (\frac{1}{4}, \frac{2}{4})$ with $q=4$ is shown in Figure 6.2. Note that phase is modulo-2π and the phase diagram actually lies on a cylinder rather than on a plane. Figure 6.2 then represents the cylinder cut along its z-axis and laid flat.

The constraint length of the trellis $(K+1)$ is used to indicate the number of intervals over which any pair of phase paths do not merge. The set of indices cannot have a constraint length longer than $(K+1)$. This can be attained [4] by selecting K indices of which no two subsets have the same sum modulo 1. The set of indices can be represented as

$$C = \{h_i; i=1,\ldots,K\} = \left\{\frac{L_i}{q}; i=1,\ldots,K\right\}$$

Sets C meeting these conditions have $q \geq 2^K$.

The number of phases impacts upon the complexity of the receiver and should be limited. Furthermore, the indices should be ratios of small integers to reduce demodulator complexity.

It can be shown [6] that the trellis exhibits two fundamental properties:

1 The period of the trellis (T_p) depends upon the sum

$$\Gamma = \sum_{i=1}^{K} L_i$$

where $h_i = L_i/q$. $T_p = KT$ if Γ is even and $T_p = 2KT$ if Γ is odd.

2 The number of phase states is q if Γ is even and $2q$ if Γ is odd.

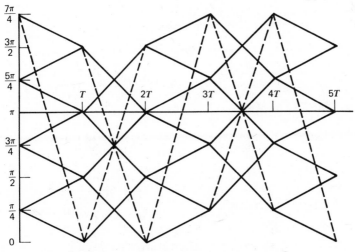

Figure 6.2 MHPM phase trellis. $(\frac{1}{4}, \frac{1}{2})$ code.

For a given code structure the MHPM signal can be demodulated in a manner similar to that described for trellis decoding of MSK (Section 3.4); that is, the received signal is multiplied in a bank of mixers by appropriately selected reference signals. The mixer outputs are integrated over each symbol interval to provide a set of metrics which are accumulated over a period of time, after which the most likely symbol path sequence is estimated.

The generation of local references is an important aspect of receiver design. However, for the purposes of this presentation ideal reference signals are assumed. The interested reader is referred to ref. 7 for an investigation of reference regeneration.

Generation of MHPM

In binary CPFSK one modulation index is used and the transmitted frequencies are $(f_c \pm h/2T)$, where f_c is the nominal centre frequency. MHPM can be conceived similarly except that the modulation index is one of K predetermined values and the indices are used consecutively and cyclically, one in each interval. The transmitted frequency in each symbol interval is now $(f_c \pm h_i/2T)$, where $i = 1, 2, \ldots, K$. This generation procedure can be represented as in Figure 6.3. For each h_i two sideband tones are generated. The tone selected by the switch depends upon the binary data. For example, arbitrarily select the upper tone for a data "1" and the lower tone for a data "0". Note that continuity in the MHPM output is required at the data transition point.

The MHPM modulated signal has a quadrature baseband representation. This consists of portions of sinusoids rather than the half-sinusoids present in

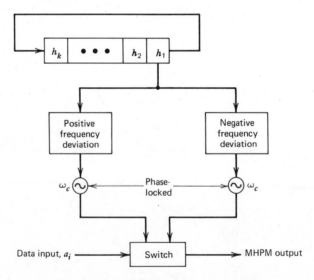

Figure 6.3   MHPM generator concept.

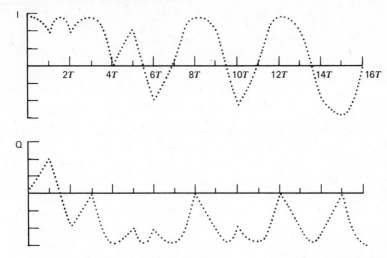

Figure 6.4 Typical MHPM quadrature channel baseband signals.

FFSK. The form of the quadrature signals is found by expanding the signal $s(t)$ as follows,

$$s(t) = A \cos\left(\omega_c t + \frac{a_i \pi h_i t}{T} + \phi_i \right), \qquad iT \leq t \leq (i+1)T$$

$$= I(t) \cos \omega_c t - Q(t) \sin \omega_c t$$

where A = signal amplitude

$\qquad \phi_i$ = phase at the beginning of the ith interval

$$I(t) = A \cos\left(\frac{a_i \pi h_i t}{T} + \phi_i \right)$$

$$Q(t) = A \sin\left(\frac{a_i \pi h_i t}{T} + \phi_i \right)$$

An example of representative I and Q waveforms is given in Figure 6.4.

Demodulation of MHPM

Recall from the discussion of Section 2.4 that MSK represents an orthogonal form of FSK for which the modulation index is $h = 0.5$. Two signals $s_1(t)$ and $s_2(t)$ are said to be orthogonal when their integrated product is 0, namely

$$m(t) = \int_0^T s_1(t) s_2(t) \, dt = 0$$

This leads to the demodulator concept given previously in Figure 3.14, wherein the received signal is correlated with local replicas

$$R_1(t) = \cos\left(\omega_c t + \frac{\pi t}{2T}\right)$$

$$R_2(t) = \cos\left(\omega_c t - \frac{\pi t}{2T}\right)$$

In such a receiver (neglecting additive noise) the output of the upper tone correlator is 0 when the lower tone is transmitted, and vice versa.

A similar approach can be used in *m*-ary CPFSK (see Schonhoff et al. [8]). A difficulty in using this approach with MHPM is that the transmitted signals are not necessarily orthogonal. In order to determine an appropriate receiver structure the concept of the vector-space representation of signals will now be introduced.

6.3   VECTOR REPRESENTATION OF SIGNALS

Implicit in previous discussions where phasor diagrams and quadrature baseband representations of signals have been described is the concept of vectors. Vector representation proves to be a useful means for expressing waveforms. These waveforms can be conceived as a linear combination of N basic elemental waveforms as follows

$$s_i(t) = \sum_{j=1}^{N} s_{ij}\psi_j(t), \qquad i = 0, 1, \dots, M-1 \tag{6.2}$$

where $s_i(t)$ is one of M possible transmitted waveforms
     s_{ij} is a coefficient
     $\psi_j(t)$ is one of the basic waveforms of duration T
The transmitter synthesis operation may be viewed as filtering or mixing operations (Figures 6.5 and 6.6). In the former case, for each desired output waveform a generator provides a set of predetermined impulses to a bank of filters each having an impulse response $\psi_j(t)$. Alternatively, as shown in Figure 6.6, a generator produces a set of coefficients that multiply the outputs of function generators in a bank of parallel multipliers.

The selection of the basic waveforms usually depends upon channel considerations and the practical aspect of implementing the transmitter. An often used elemental waveform is a sinusoidal pulse. For example, QPSK can be represented by

$$s_i(t) = s_{i1}\sqrt{\frac{2}{T}}\cos\omega_c t + s_{i2}\sqrt{\frac{2}{T}}\sin\omega_c t, \qquad 0 \le t \le T \tag{6.3}$$

where $s_{i1}, s_{i2} = \pm\frac{1}{2}$ for phases a multiple of $\pi/4$.

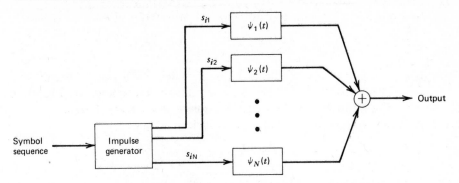

Figure 6.5 Generation of waveform by summation of filter outputs.

More generally, the functions $\psi_j(t)$ are selected as orthonormal, that is

$$\int_0^T \psi_i(t)\psi_j(t)dt = \begin{cases} 1, & i=j \\ 0, & i\neq j \end{cases}$$

The term "orthonormal" arises from the fact that the basic functions $\psi_j(t)$ are mutually orthogonal and normalized to unit energy.

Note that (6.2) can be represented as

$$\begin{bmatrix} s_0(t) \\ s_1(t) \\ \vdots \\ s_{M-1}(t) \end{bmatrix} = \begin{bmatrix} s_{01} & s_{02} & \cdots & s_{0N} \\ s_{11} & s_{12} & \cdots & \\ \vdots & & \ddots & \vdots \\ s_{M-1,1} & & & s_{M-1,N} \end{bmatrix} \begin{bmatrix} \psi_1(t) \\ \psi_2(t) \\ \vdots \\ \psi_N(t) \end{bmatrix}$$

which states that once a set of orthogonal functions has been selected, the

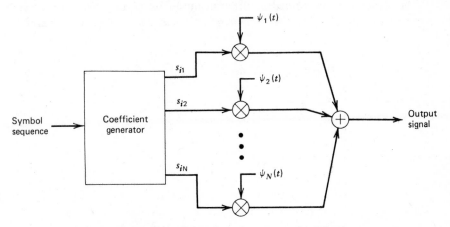

Figure 6.6 Generation of waveform by summation of mixer outputs.

various transmitter waveforms can be described in terms of vector coefficients

$$\mathbf{s}_i = (s_{i1}, s_{i2}, \ldots, s_{iN}), \qquad i = 0, 1, \ldots, M-1$$

In such a format the elemental waveforms $\psi_j(t)$ describe a basis of vectors in a geometrical signal space. The vectors $\boldsymbol{\psi}_j$ are mutually perpendicular. The signal vectors $\mathbf{s}_i$ are formed by a linear combination of the basis vectors.

Returning to the QPSK example, the vector coefficients are

$$\mathbf{s}_0 = \left(\tfrac{1}{2}, \tfrac{1}{2}\right)$$

$$\mathbf{s}_1 = \left(-\tfrac{1}{2}, \tfrac{1}{2}\right)$$

$$\mathbf{s}_2 = \left(-\tfrac{1}{2}, -\tfrac{1}{2}\right)$$

$$\mathbf{s}_3 = \left(\tfrac{1}{2}, -\tfrac{1}{2}\right)$$

and the basis vectors can be visualized as vectors along the X- and Y-axes of a Cartesian coordinate system as shown in Figure 6.7.

At the receiver, the set of coefficients can be recovered by using the knowledge that the basic waveforms are orthogonal. For example, if the signal $s_i(t)$ is multiplied by one of the basic waveforms and integrated, the result is

$$\int_0^T s_i(t)\psi_k(t)\,dt = \int_0^T \left[\sum_{j=1}^N s_{ij}\psi_j(t)\right]\psi_k(t)\,dt$$

$$= \sum_{j=1}^N s_{ij} \int_0^T \psi_j(t)\psi_k(t)\,dt$$

$$= s_{ik}$$

The signal $s_i(t)$ is formed at the transmitter by adding the appropriately weighted functions $\psi_j(t)$. The weight or coefficient for each basic waveform

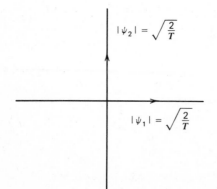

Figure 6.7 Basis vectors for QPSK transmission.

can be recovered at the receiver by multiplying $s_i(t)$ by the basic waveform and integrating. If this multiplication is performed for each waveform $\psi_j(t)$, then the vector of coefficients that pertain to a given $s_i(t)$ can be determined. This operation is shown in Figure 6.8.

As a result of the orthonormality condition, the energy in each signal E and the correlation coefficient between any two signals ρ_{ij} are given by

$$E = \int_0^T s_i^2(t)\,dt = \sum_{k=1}^N s_{ik}^2$$

$$\rho_{ij} = \frac{1}{E}\int_0^T s_i(t)s_j(t)\,dt = \frac{1}{E}\sum_{k=1}^N s_{ik}s_{jk}$$

respectively. With the signal expressed as an orthonormal expansion or equivalently as a vector, it can be recognized that E is the square magnitude of the vector and ρ_{ij} is proportional to the dot or inner product of two signal vectors. In signal space E is the distance from the origin to the tip of the vector, while ρ_{ij} indicates the angle between vectors. Furthermore, the distance squared (d^2) between two vectors is related to ρ_{ij} as follows:

$$d^2 = |\mathbf{s}_i - \mathbf{s}_j|^2 = \left| \sum_{k=1}^N s_{ik}\psi_k - \sum_{k=1}^N s_{jk}\psi_k \right|^2$$

$$= \sum_{K=1}^N s_{ik}^2 + \sum_{k=1}^N s_{jk}^2 - 2\sum_{k=1}^N s_{ik}s_{jk}$$

$$= 2E - 2E\rho_{ij}$$

$$= 2E(1 - \rho_{ij}) \tag{6.4}$$

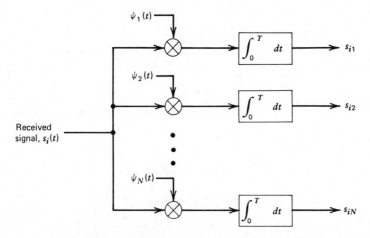

Figure 6.8 Recovery of signal vector at receiver.

   Equation (6.4) therefore shows that if the constant term is neglected (as it may because it is common to all correlations), the likeliest transmitted signal may be determined by a straightforward dot product of the received vector with the set or matrix of reference vectors.

Determination of MHPM Vector Coefficients

Now that the vector representation of signals has been reviewed, the appropriate format for MHPM waveforms will be derived. The received MHPM signal (ignoring noise) can be written as

$$r(t)=\sqrt{\frac{2E}{T}}\cos\left(\omega_c t+\frac{a_i\pi h_i t}{T}+\phi_i\right) \tag{6.5}$$

where E is the energy and ϕ_i, $iT\leq t\leq(i+1)T$, is the phase at the beginning of the ith interval. Expansion of (6.5) yields

$$r(t)=\sqrt{\frac{2E}{T}}\left[\cos\phi_i\cos\left(\omega_c t+\frac{a_i\pi h_i t}{T}\right)\right.$$

$$\left.-\sin\phi_i\sin\left(\omega_c t+\frac{a_i\pi h_i t}{T}\right)\right]$$

$$=\sqrt{\frac{2E}{T}}\left\{\cos\phi_i\left[\cos\omega_c t\cos\frac{a_i\pi h_i t}{T}-\sin\omega_c t\sin\frac{a_i\pi h_i t}{T}\right]\right.$$

$$\left.-\sin\phi_i\left[\sin\omega_c t\cos\frac{a_i\pi h_i t}{T}+\cos\omega_c t\sin\frac{a_i\pi h_i t}{T}\right]\right\}$$

$$=\sqrt{\frac{2E}{T}}\left\{\cos\phi_i\left[\cos\omega_c t\cos\frac{\pi h_i t}{T}-a_i\sin\omega_c t\sin\frac{\pi h_i t}{T}\right]\right.$$

$$\left.-\sin\phi_i\left[\sin\omega_c t\cos\frac{\pi h_i t}{T}+a_i\cos\omega_c t\sin\frac{\pi h_i t}{T}\right]\right\} \tag{6.6}$$

In (6.6) the relationships

$$\cos(ax)=\cos x$$

$$\sin(ax)=a\sin x$$

where $a=\pm1$, have been used.

Standard trigonometric identities for $\cos u \cos v$, $\cos u \sin v$, $\sin u \sin v$ and $\sin u \cos v$ can be used and (6.6) becomes

$$r(t) = \sqrt{\frac{2E}{T}} \, (\cos \phi_i) \left(\frac{1+a_i}{2} \right) \cos \left(\omega_c t + \frac{\pi h_i t}{T} \right)$$

$$- \sqrt{\frac{2E}{T}} \, (\sin \phi_i) \left(\frac{1+a_i}{2} \right) \sin \left(\omega_c t + \frac{\pi h_i t}{T} \right)$$

$$+ \sqrt{\frac{2E}{T}} \, (\cos \phi_i) \left(\frac{1-a_i}{2} \right) \cos \left(\omega_c t - \frac{\pi h_i t}{T} \right)$$

$$- \sqrt{\frac{2E}{T}} \, (\sin \phi_i) \left(\frac{1-a_i}{2} \right) \sin \left(\omega_c t - \frac{\pi h_i t}{T} \right) \qquad (6.7)$$

Although each pair of terms in (6.7) with $+\pi h_i t / T$ and $-\pi h_i t / T$ is orthogonal, terms in one pair are not orthogonal with terms in the second pair. It would therefore be convenient to transform (6.7) into an orthonormal expansion. This can be attained by means of the Gram–Schmidt orthogonalization procedure, which is described next.

Gram–Schmidt Procedure

The Gram–Schmidt procedure [9] provides a systematic approach for finding an orthogonal representation for signals. The procedure determines a new orthonormal bases by subtracting the nonorthogonal components from the original bases as follows:

1 Given a set of nonorthogonal $\{s_i(t)\}$ let

$$\psi_1(t) = \frac{s_0(t)}{\sqrt{E_0}}$$

where

$$E_0 \triangleq \int_0^T s_0^2(t) \, dt$$

Thus $\psi_1(t)$ is the normalized version of $s_0(t)$ and

$$s_0(t) = \sqrt{E_0} \, \psi_1(t)$$

$$= s_{01} \psi_1(t)$$

The vector $\mathbf{s}_0$ is depicted in Figure 6.9a.

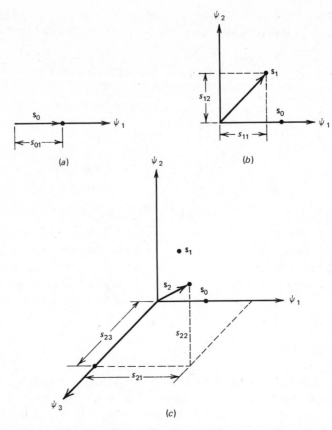

Figure 6.9 Gram–Schmidt orthogonalization. (*a*) Vector s_0; (*b*) vector s_1; (*c*) vector s_2.

2 The auxiliary function $f_1(t)$ is defined as

$$f_1(t) \stackrel{\triangle}{=} s_1(t) - s_{11}\psi_1(t)$$

where

$$s_{11} = \int_0^T s_1(t)\psi_1(t)\,dt$$

Provided that $f_1(t)$ is nonzero, an orthonormal function can be obtained from $f_1(t)$ by letting

$$\psi_2(t) = \frac{f_1(t)}{\sqrt{E_1}}$$

where

$$E_1 = \int_0^T f_1^2(t)\,dt$$

It follows that

$$s_1(t) = s_{11}\psi_1(t) + f_1(t)$$

$$= s_{11}\psi_1(t) + \sqrt{E_1}\,\psi_2(t)$$

and $s_{12} = \sqrt{E_1}$. The function $\psi_2(t)$ has unit energy. That $\psi_1(t)$ and $\psi_2(t)$ are orthogonal can be shown as follows:

$$\sqrt{E_1}\int_0^T \psi_1(t)\psi_2(t)\,dt = \int_0^T f_1(t)\psi_1(t)\,dt$$

$$= \int_0^T \left[s_1(t) - s_{11}\psi_1(t)\right]\psi_1(t)\,dt$$

$$= \int_0^T s_1(t)\psi_1(t)\,dt - s_{11}\int_0^T \psi_1^2(t)\,dt$$

$$= s_{11} - s_{11}$$

$$= 0$$

The signal vector $\mathbf{s}_1$ is shown in Figure 6.9b. $\mathbf{s}_2$ is found in a similar manner and is shown in Figure 6.9c. The procedure is continued until all M signals comprising the set $\{s_i(t)\}$ have been used in the generation of the orthonormal set $\{\psi_j(t)\}$.

The orthonormal set thus derived is not unique. For example, a different set can be generated by changing the order of selection of the M waveforms: for example, $s_5(t)$, $s_1(t)$, $s_0(t),\ldots$ instead of $s_0(t)$, $s_1(t)$, $s_2(t),\ldots$. Regardless of the orthonormal set used, once the set has been determined, each signal is described by the set of orthonormal coefficients. Although a different orthonormal set may be used, the geometrical relationship between the signal vectors $\{\mathbf{s}_i\}$ remains the same. As this geometry determines the receiver performance, each orthonormal representation will yield the same result.

6.4   ORTHOGONAL REPRESENTATION OF MHPM

The Gram–Schmidt orthogonalization procedure is now applied to the MHPM signal representation. By inspection of (6.7), two of the orthonormal functions can be assigned as

$$\psi_1(t) = \sqrt{\frac{2}{T}} \cos\left(\omega_c t + \frac{\pi h_i t}{T}\right) \tag{6.8}$$

$$\psi_2(t) = \sqrt{\frac{2}{T}} \sin\left(\omega_c t + \frac{\pi h_i t}{T}\right) \tag{6.9}$$

The relationship can be verified by integration:

$$\int_0^T \psi_1(t)\psi_2(t)\,dt = \frac{2}{T}\int_0^T \cos\left(\omega_c t + \frac{\pi h_i t}{T}\right)\sin\left(\omega_c t + \frac{\pi h_i t}{T}\right)dt$$

$$= \frac{1}{T}\int_0^T \sin\left[2\left(\omega_c t + \frac{\pi h_i t}{T}\right)\right]dt$$

$$= \frac{-1}{2T(\omega_c + \pi h_i / T)}\left\{\cos\left[2\left(\omega_c + \frac{\pi h_i}{T}\right)t\right]\Big|_0^T\right\}$$

$$= \frac{-1}{2T(\omega_c + \pi h_i / T)}\left[\cos(2\omega_c T + 2\pi h_i) - 1\right] \tag{6.10}$$

In most systems of practical interest the carrier frequency is much greater than the bit (or symbol) rate. This means that

$$f_c \gg \frac{1}{T}$$

or

$$\omega_c T \gg 2\pi$$

Therefore, (6.10) is approximately equal to zero.
 The norm of $\psi_1(t)$ is determined by

$$N_1 = \int_0^T \psi_1^2(t)\,dt$$

$$= \frac{2}{T}\int_0^T \cos^2\left(\omega_c t + \frac{\pi h_i t}{T}\right)dt$$

$$= \frac{1}{T}\int_0^T\left[1 + \cos 2\left(\omega_c t + \frac{\pi h_i t}{T}\right)\right]dt$$

$$= 1 + \frac{\sin 2(\omega_c + \pi h_i / T)t}{2T(\omega_c + \pi h_i / T)}\Big|_0^T$$

$$\approx 1 \quad \text{if } \omega_c T \gg 2\pi$$

The remaining orthonormal functions are now determined as follows. With $s_3(t)$ given by

$$s_3(t) = \sqrt{\frac{2}{T}} \cos\left(\omega_c t - \frac{\pi h_i t}{T}\right)$$

$$s_{31} = \int_0^T s_3(t)\psi_1(t)\,dt$$

$$= \frac{2}{T}\int_0^T \cos\left(\omega_c t - \frac{\pi h_i t}{T}\right)\cos\left(\omega_c t + \frac{\pi h_i t}{T}\right)dt$$

$$= \frac{1}{T}\int_0^T \left(\cos 2\omega_c t + \cos\frac{2\pi h_i t}{T}\right)dt$$

$$= \frac{1}{T}\left(\left.\frac{\sin 2\omega_c t}{2\omega_c}\right|_0^T + \left.\frac{T\sin(2\pi h_i t/T)}{2\pi h_i}\right|_0^T\right)$$

$$\simeq \frac{\sin 2\pi h_i}{2\pi h_i} = C_{1,i}$$

Similarly,

$$s_{32} = \int_0^T s_3(t)\psi_2(t)\,dt$$

$$= \frac{2}{T}\int_0^T \cos\left(\omega_c t - \frac{\pi h_i t}{T}\right)\sin\left(\omega_c t + \frac{\pi h_i t}{T}\right)dt$$

$$= \frac{1}{T}\int_0^T \left(\sin 2\omega_c t + \sin\frac{2\pi h_i t}{T}\right)dt$$

$$= \frac{1}{T}\left(\left.\frac{-\cos 2\omega_c t}{2\omega_c}\right|_0^T - \left.\frac{T\cos 2\pi h_i t/T}{2\pi h_i}\right|_0^T\right)$$

$$\simeq 1 - \frac{\cos 2\pi h_i}{2\pi h_i} = C_{2,i}$$

Therefore,

$$f_3(t) = s_3(t) - s_{31}\psi_1(t) - s_{32}\psi_2(t)$$

$\psi_3(t)$ is now determined by normalizing $f_3(t)$:

$$\psi_3(t) = \frac{f_3(t)}{\sqrt{E_3}}$$

where

$$E_3 = \int_0^T f_3^2(t)\,dt$$

$$= \int_0^T \left[\sqrt{\frac{2}{T}} \cos\left(\omega_c t - \frac{\pi h_i t}{T} \right) - C_{1,i} \sqrt{\frac{2}{T}} \cos\left(\omega_c t + \frac{\pi h_i t}{T} \right) \right.$$

$$\left. - C_{2,i} \sqrt{\frac{2}{T}} \sin\left(\omega_c t + \frac{\pi h_i t}{T} \right) \right]^2 dt$$

It can be shown that E_3 reduces to

$$E_3 = 1 + C_{1,i}^2 + C_{2,i}^2 - 2C_{1,i}^2 - 2C_{2,i}^2$$

$$= 1 - C_{1,i}^2 - C_{2,i}^2$$

Therefore,

$$\psi_3(t) = \frac{s_3(t) - C_{1,i}\psi_1(t) - C_{2,i}\psi_2(t)}{\sqrt{D_i}} \tag{6.11}$$

where

$$D_i = 1 - C_{1,i}^2 - C_{2,i}^2$$

Finally,

$$f_4(t) = s_4(t) - s_{41}\psi_1(t) - s_{42}\psi_2(t) - s_{43}\psi_3(t)$$

The coefficients s_{4i} are determined as follows:

$$s_{41} = \int_0^T s_4(t)\psi_1(t)\,dt$$

$$= \int_0^T \sqrt{\frac{2}{T}} \sin\left(\omega_c t - \frac{\pi h_i t}{T} \right) \sqrt{\frac{2}{T}} \cos\left(\omega_c t + \frac{\pi h_i t}{T} \right) dt$$

$$= \frac{1}{T}\left(\left. \frac{-\cos 2\omega_c t}{2\omega_c} \right|_0^T + \left. \frac{T\cos 2\pi h_i t/T}{2\pi h_i} \right|_0^T \right)$$

$$\simeq \frac{\cos 2\pi h_i - 1}{2\pi h_i}$$

$$= -C_{2,i}$$

$$s_{42} = \int_0^T s_4(t)\psi_2(t)\,dt$$

$$= \int_0^T \sqrt{\frac{2}{T}}\,\sin\!\left(\omega_c t - \frac{\pi h_i t}{T}\right)\sqrt{\frac{2}{T}}\,\sin\!\left(\omega_c t + \frac{\pi h_i t}{T}\right)dt$$

$$= \frac{1}{T}\left(\left.\frac{T\sin(2\pi h_i t/T)}{2\pi h_i}\right|_0^T - \left.\frac{\sin 2\omega_c t}{2\omega_c}\right|_0^T\right)$$

$$\simeq \frac{\sin 2\pi h_i}{2\pi h_i}$$

$$= C_{1,i}$$

$$s_{43} = \int_0^T \sqrt{\frac{2}{T}}\,\sin\!\left(\omega_c t - \frac{\pi h_i t}{T}\right)\sqrt{\frac{2}{T}}\,\cos\!\left(\omega_c t - \frac{\pi h_i t}{T}\right)dt$$

$$= 0$$

$\psi_4(t)$ is now found by normalizing $f_4(t)$:

$$\psi_4(t) = \frac{f_4(t)}{\sqrt{E_4}} \tag{6.12}$$

where

$$E_4 = \int_0^T f_4^2(t)\,dt$$

$$= \int_0^T \left[\sqrt{\frac{2}{T}}\,\sin\!\left(\omega_c t - \frac{\pi h_i t}{T}\right) - C_{3,i}\sqrt{\frac{2}{T}}\,\cos\!\left(\omega_c t + \frac{\pi h_i t}{T}\right)\right.$$

$$\left. - C_{4,i}\sqrt{\frac{2}{T}}\,\sin\!\left(\omega_c t + \frac{\pi h_i t}{T}\right)\right]^2 dt$$

It can be shown that E_4 reduces to

$$E_4 = 1 + C_{3,i}^2 + C_{4,i}^2 - 2C_{3,i}^2 - 2C_{4,i}^2$$

$$= 1 - C_{1,i}^2 - C_{2,i}^2$$

Therefore,

$$\psi_4(t) = \frac{s_4(t) + C_{2,i}\psi_1(t) - C_{1,i}\psi_2(t)}{\sqrt{D_i}} \tag{6.13}$$

Equations (6.8), (6.9), (6.11) and (6.13) represent the required orthonormal set for each signalling interval. In turn, the set of quadrature waveforms at $(\omega_c \pm \pi h_i / T)$ which appear in (6.7) can be written as

$$s_1(t) = \sqrt{\frac{2}{T}} \cos\left(\omega_c t + \frac{\pi h_i t}{T}\right) = \psi_1(t)$$

$$s_2(t) = \sqrt{\frac{2}{T}} \sin\left(\omega_c t + \frac{\pi h_i t}{T}\right) = \psi_2(t)$$

$$s_3(t) = C_{1,i}\psi_1(t) + C_{2,i}\psi_2(t) + \sqrt{D_i}\,\psi_3(t)$$

$$s_4(t) = -C_{2,i}\psi_1(t) + C_{1,i}\psi_2(t) + \sqrt{D_i}\,\psi_4(t)$$

These expressions for $s_i(t)$ can now be substituted into (6.7) to give

$$r(t) = \sqrt{E}\,(\cos\phi_i)\left(\frac{1+a_i}{2}\right)\psi_1(t)$$

$$- \sqrt{E}\,(\sin\phi_i)\left(\frac{1+a_i}{2}\right)\psi_2(t)$$

$$+ \sqrt{E}\,(\cos\phi_i)\left(\frac{1-a_i}{2}\right)\left[C_{1,i}\psi_1(t) + C_{2,i}\psi_2(t) + \sqrt{D_i}\,\psi_3(t)\right]$$

$$- \sqrt{E}\,(\sin\phi_i)\left(\frac{1-a_i}{2}\right)\left[-C_{2,i}\psi_1(t) + C_{1,i}\psi_2(t) + \sqrt{D_i}\,\psi_4(t)\right] \quad (6.14)$$

Rearranging terms in (6.14) gives

$$r(t) = \sqrt{E}\left[A_{1,i}\psi_1(t) + A_{2,i}\psi_2(t) + A_{3,i}\psi_3(t) + A_{4,i}\psi_4(t)\right] \quad (6.15)$$

where

$$A_{1,i} = \cos\phi_i\left(\frac{1+a_i}{2}\right) + C_{1,i}\cos\phi_i\left(\frac{1-a_i}{2}\right)$$

$$+ C_{2,i}\sin\phi_i\left(\frac{1-a_i}{2}\right)$$

$$A_{2,i} = -\sin\phi_i\left(\frac{1+a_i}{2}\right) + C_{2,i}\cos\phi_i\left(\frac{1-a_i}{2}\right)$$

$$- C_{1,i}\sin\phi_i\left(\frac{1-a_i}{2}\right)$$

$$A_{3,i} = \cos\phi_i\left(\frac{1-a_i}{2}\right)\sqrt{D_i}$$

$$A_{4,i} = -\sin\phi_i\left(\frac{1-a_i}{2}\right)\sqrt{D_i}$$

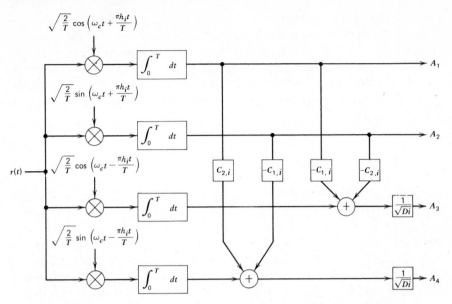

Figure 6.10 Multi-h demodulation: determination of basis coefficients.

The coefficients reflect the data $\{a_i = \pm 1\}$, the excess phase $\{\phi_i\}$ and the modulation index applicable to each signalling interval. Therefore, the coefficients A_{1i}, A_{2i}, A_{3i} and A_{4i} fully describe the signal vector over each signalling interval.

Equation (6.15) states that recovery of the vector $(A_{1i}, A_{2i}, A_{3i}, A_{4i})$ is accomplished by passing the received signal through a bank of multipliers followed by integrators. The kth multiplier has as its other input $\psi_k(t)$. Because the vectors are orthogonal, the kth path integrator output will yield A_{ki}.

A form of demodulator can be visualized as follows. Consider a receiver having four multiplier/integrator paths. The reference inputs to the multipliers are taken as $\cos(\omega_c t \pm \pi h_i t/T)$ and $\sin(\omega_c t \pm \pi h_i t/T)$. It has been shown previously that these references are not all orthogonal. However, the coefficients $\{A_{ji}; j=1, 2, 3, 4\}$ may be recovered by using

$$\psi_1(t) = \sqrt{\frac{2}{T}} \cos\left(\omega_c t + \frac{\pi h_i t}{T}\right) \tag{6.16}$$

$$\psi_2(t) = \sqrt{\frac{2}{T}} \sin\left(\omega_c t + \frac{\pi h_i t}{T}\right) \tag{6.17}$$

$$\psi_3(t) = \frac{1}{\sqrt{D_i}}\left[\sqrt{\frac{2}{T}} \cos\left(\omega_c t - \frac{\pi h_i t}{T}\right) - C_{1,i}\psi_1(t) - C_{2,i}\psi_2(t)\right] \tag{6.18}$$

$$\psi_4(t) = \frac{1}{\sqrt{D_i}}\left[\sqrt{\frac{2}{T}} \sin\left(\omega_c t - \frac{\pi h_i t}{T}\right) + C_{2,i}\psi_1(t) - C_{1,i}\psi_2(t)\right] \tag{6.19}$$

The demodulator based upon the use of orthonormal references is given in Figure 6.10. The first multiplier uses $\psi_1(t)$ as reference. The resulting integrator output is A_{1i}. Similarly, the second multiplier/integrator output is A_{2i}. A_{3i} is derived by combining multiplier/integrator paths 1, 2 and 3 with appropriate weightings. Note that because $\psi_3(t)$ is not specifically used as a multiplier input reference, an equivalent operation need be used. Equation (6.18) indicates the required operation. Similarly, (6.19) provides $A_{4,i}$.

6.5 VITERBI DECODING OF MHPM

With a relatively simple signal a maximum likelihood determination of a complete MHPM symbol sequence follows a similar procedure to that described in Section 3.4 for trellis decoding of FFSK. However, for more powerful phase-coded signals where the number of indices K is typically ≤ 4 and the integer common denominator of the modulation index $q \leq 16$, an efficient procedure called the Viterbi algorithm (VA) is used. The VA is usually associated with the decoding of convolutional codes, but it can be applied to an assortment of situations, as described by Forney [10]. Although a detailed explanation of the algorithm is provided in Section 12.4, a brief description of the technique and its use in the decoding of MHPM will be provided here.

The algorithm is best explained by a simple example (refer to Figure 6.11). Suppose that a car is located at checkpoint A_1 and another at checkpoint A_2. Each car's objective is to reach either of checkpoints C_1 and C_2 and in so doing to cover the minimum distance. In driving to checkpoints C_1 or C_2, each car must pass through checkpoints B_1 and B_2. Which car will have the shortest route?

An examination of Figure 6.11 shows that there are eight possible paths:

$$A_1B_1C_1 \qquad A_2B_1C_1$$
$$A_1B_1C_2 \qquad A_2B_1C_2$$
$$A_1B_2C_1 \qquad A_2B_2C_1$$
$$A_1B_2C_2 \qquad A_2B_2C_2$$

The laborious approach is to calculate all eight possible distances. An alternative is to recognize the following. Each car can pass through B_1 or B_2. The

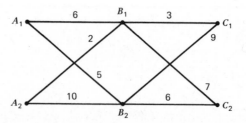

Figure 6.11   Viterbi decoding: car race example.

shortest distance to B_1 is $A_2B_1 = 2$ and the shortest distance to B_2 is $A_1B_2 = 5$. Is it necessary to consider the remaining paths leading to B_1 and B_2?

The cars at A_1 and A_2 can both drive through B_1. Once they have arrived at B_1, they will both have identical choices for getting to either C_1 or C_2. Therefore, the car travelling the minimum distance through B_1 to C_1 or C_2 must be the one travelling the least distance to B_1. This reasoning leads to the following simpler procedure:

1 For each checkpoint B_1 and B_2 determine the minimum distance from each of A_1 and A_2.
2 Record the minimum path and path length to each one.
3 Repeat for checkpoints C_1 and C_2.

The minimum path is then determined by the minimum overall distance $(A_2B_1C_1 = 5)$.

Application of the VA to the decoding of a MHPM signal now follows. The simple checkpoint diagram is replaced by the phase state diagram. Distances are now determined for transitions from each phase state at $t = iT$ to phase states at $t = (i+1)T$. The distances or metrics are determined by the dot product of the received signal vector during the ith interval with each of the locally available reference signal vectors. For each phase-to-phase transition there is a corresponding "0" or "1" modulator data input. Over an appropriate number of intervals (called the decision depth) the path histories (the "0" or "1" symbols) together with the path metrics are accumulated in accordance with the Viterbi algorithm. Once the decision depth has been reached the path having the minimum accumulated metric is used to resolve the first transmitted symbol. After the next symbol interval the decision depth is immediately reached for the second transmitted symbol. This procedure is continued for as many symbols as are transmitted. Of course, the received signal will be contaminated by noise and other channel distortions and errors in decoding will result.

It should be noted that the reference signals will change from interval to interval as the applicable modulation indices are rotated. The preceding discussion has assumed that the appropriate starting index is known to the receiver. An investigation of the techniques for establishing synchronization and generating the reference signals is beyond the scope of this text.

6.6 PERFORMANCE

Of primary concern in evaluating the relevance of a technique such as MHPM to satellite communications is the performance as a function of bandwidth and signal-to-noise ratio (SNR) constraints. As will be shown below, tradeoffs are possible in terms of having (a) bandwidth and power-efficient codes and (b) bandwidth or power-efficient techniques. The error-rate performance in additive white Gaussian noise (AWGN) will be reviewed.

Error Rate-High SNR

If $\mathbf{s}_1$ and $\mathbf{s}_2$ are two possible transmitted signal vectors, then the probability of erroneously decoding one (say $\mathbf{s}_2$) given that the other was transmitted ($\mathbf{s}_1$) can be shown (see Problem 2) to be

$$P(\mathbf{s}_2|\mathbf{s}_1) = Q\left[\tilde{d}(1,2)\sqrt{\frac{E}{2N_0}}\right] \qquad (6.20)$$

where

$$\tilde{d}(1,2) = \frac{d(1,2)}{\sqrt{E}}$$

with $d(1,2) = d$ as in (6.4) and $Q(x)$ is the function applicable to Gaussian distributions.

Anderson and Taylor [4] have shown that if two phase paths are separated over L symbol intervals, the total distance between paths is the sum of the

Table 6.1 Worst-Case Free Distance for Best Multi-*h* Code Occupying *q* Phases with 2, 3 and 4 h_i

Number of Phases, q	$d^2_{\text{free,min}}$ for:		
	$K=2$	$K=3$	$K=4$
4	5.57		
5	6.14		
6	6.90		
7	6.65		
8	7.10	7.58	
9	6.92	5.40	
10	7.25	7.63	
11	7.14	7.46	
12	7.36	7.63	
13	7.28	8.34	
14		8.23	
15		7.76	
16		8.68	9.29
17		8.46	8.02
18			8.12
19			8.83
20			9.55
21			8.54
22			9.78

Source: Courtesy J. B. Anderson and D. P. Taylor.

distances for each of the L intervals

$$d^2 = \sum_{k=1}^{L} d_k^2(i, j)$$

where d_k is the distance in the kth interval. For large SNR the probability of codeword error is dominated by the distance between the closest codewords in the codeword set, d_{min}. As (6.20) shows, distance directly affects the error probability: increased distances result in lower required SNR for a given error rate. Table 6.1 demonstrates the reduction in required SNR (or coding gain) for all MHPM signals having two, three and four indices per code for the number of phases shown. The d_{min}^2 values should be compared with $d_{min}^2(\text{BPSK}) = 4$. Table 6.1 indicates that coding gains of 2 to 4 dB over BPSK are possible without the bandwidth expansion associated with separate coding and modulation schemes.

In poorer SNR situations ($E_b/N_0 \leq 12$ dB) an estimate of error probability can be bounded by a maximum value by applying the union bound. The union bound is of sufficient importance generally that a discussion of the concept is now appropriate.

Probability of Error: The Union Bound and the Low-E_b/N_0 Case

In certain communication systems, a group of information digits (a symbol) can be transmitted in the form of a number of signal waveforms. Consider the situation where the set of possible equiprobable waveforms is $\{s_i(t), i = 1, N\}$. The probability of symbol error is then the conditional error probability for each symbol averaged over all N symbols:

$$P_s = \frac{1}{N} \sum_{i=1}^{N} (P_s|i) \tag{6.21}$$

where $(P_s|i)$ is the probability of error for the ith symbol given that the ith symbol was transmitted. Although (6.21) is an exact expression, it proves difficult to evaluate in specific situations. Limits on the error probability are often used because of their simpler derivation. One such expression which provides an upper limit or bound on the symbol error probability is the union bound.

If $y_1, \ldots, y_N$ are N random values, the probability that one, say y_i, is less than all $N-1$ remaining values is given by

$$P_{yi} = P[(y_i < y_1|i) + (y_i < y_2|i) + \cdots + (y_i < y_N|i)]$$

$$\leq \sum_{\substack{j=1 \\ j \neq i}}^{N} P[(y_i < y_j|i)] \tag{6.22}$$

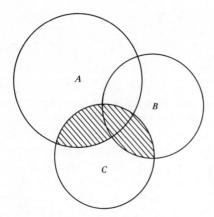

Figure 6.12 Venn diagram showing $P[A+B+C]\leq P[A]+P[B]+P[C]$. Removal of hatched area results in equality.

where $P[(y_i<y_j|i)]$ denotes the probability that y_i is less than y_j given y_i. The bound expressed in (6.22) can be explained by means of a Venn diagram (see Figure 6.12), where the probability of the union of events A, B and C must account for the occurrence of overlapping events; that is, the events A, B and C are not necessarily mutually exclusive and therefore

$$P[A+B+C]\leq P(A)+P(B)+P(C)$$

If the y_i are taken as the outputs of the maximum likelihood receiver, then the probability of symbol error given that $s_i(t)$ was transmitted is the probability that y_i is less than all other receiver outputs. Substituting (6.22) into (6.21) yields

$$P_s\leq\frac{1}{N}\sum_{i=1}^{N}\sum_{\substack{j=1\\j\neq i}}^{N}P[(y_i<y_j|i)]$$

It should be noted that $P[(y_i<y_j)|i]$ represents the probability of error for a two-symbol system wherein $s_i(t)$ and $s_j(t)$ are transmitted. As such, a binary system is represented and the determination of the conditional error probability is usually readily calculated.

The expression for the symbol error probability can also be represented in terms of distances between signal vectors. The signal is now represented in terms of an orthonormal expansion and the receiver assumes the form of Figure 6.8. The receiver outputs are the coefficients of the vector basis. The probability of error is then the probability that the received vector, given that the ith vector was transmitted, is closer to one of the remaining $N-1$ vectors. The proximity of vectors is the Euclidean (geometric) distance between the vectors. This is equivalent to determining the correlation coefficient of the corresponding waveforms.

The corresponding conditional error probability is

$$P[\mathbf{s}_i, \mathbf{s}_j] = Q\left[\frac{|\mathbf{s}_i - \mathbf{s}_j|}{2 N_0}\right]$$

It should be noted that significant effort has been placed into the derivation of upper and lower bounds which attempt to locate the error rate in tighter ranges. These bounds have been applied to cases where distortions such as intersymbol interference degrade the received signal. The interested reader is referred to Lugannani [11] and McLane [12].

The probability of incorrectly decoding, given a transmitted codeword, can be overbounded by applying the union bound. Figure 6.13 provides an indication of performance improvement for MHPM in the low-SNR situation

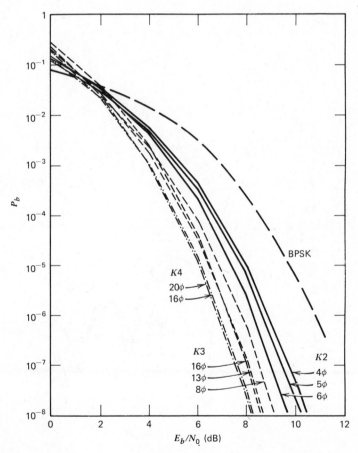

Figure 6.13 Overbound to probability of error event versus E_b/N_0, given correct decoding until present. Codes with 2, 3 and 4 h_i and phases as marked. (Courtesy J. B. Anderson and D. P. Taylor.)

assuming that calculations are made for the worst-case assumed transmitted codeword and a long observation interval. Note the improving performance with the number of h_i and the characteristic crossover or performance reversal relative to uncoded BPSK at very low E_b/N_0.

6.7 MHPM SPECTRA

It can be shown that the power spectral density of an MHPM signal [6] is given by

$$\mathcal{G}(f) = \lim_{\lambda \to \infty} \frac{2}{\lambda} \mathcal{G}_\lambda(f), \qquad f > 0$$

where

$$\mathcal{G}_\lambda(f) = E\left\{ \left| \int_0^\lambda e^{-j2\pi ft} \sqrt{\frac{2E}{T}} \right. \right.$$

$$\left. \left. \cdot \exp\left\{ j\left[\left(\omega_c t + \int_0^t \psi(\tau)\,d\tau + \theta \right) \right] \right\} dt \right|^2 \right\}$$

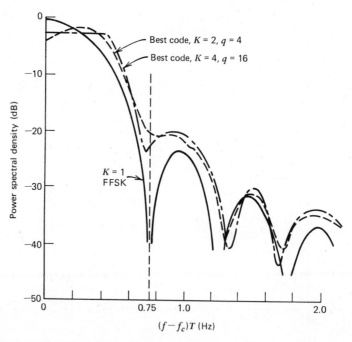

Figure 6.14 One-sided power spectral densities for FFSK, best 4-phase code of constraint two and best 16-phase code of constraint four. (Courtesy J. B. Anderson and D. P. Taylor.)

and $\psi(\tau)$ is the modulating signal and θ is an initial phase. The MHPM signal is given by

$$s(t) = \sqrt{\frac{2E}{T}} \cos\left[\omega_c t + \int_0^t \psi(\tau)d\tau + \theta\right]$$

Figure 6.14 is a plot of the spectra for codes having 2 and 4 modulation indices and 4 and 16 phases. While the MHPM signals have similar spectral occupancy to FFSK, SNR advantages of 1.7 and 3.8 dB are provided.

6.8 SUMMARY

A detailed explanation of MHPM signals has been provided to show what may well be an evolutionary process in combined coding and modulation systems. This examination has introduced a number of important concepts that will be of assistance in approaching other signal space codes. Although MHPM has been selected for the purpose of demonstration, it should be recognized that extensions (such as m-ary rather than binary signalling) and other signal-space-based techniques will be of interest in the future.

The reader is referred to Refs. 13 and 14 for additional information.

REFERENCES

1 R. de Buda, "About Optimal Properties of Fast Frequency-Shift Keying," *IEEE Trans. Commun.*, Vol. COM-22, Oct. 1974, pp. 1726–1727.

2 W. P. Osborne and M. B. Luntz, "Coherent and Noncoherent Detection of CPFSK," *IEEE Trans. Commun.*, Vol. COM-22, Aug. 1974, pp. 1023–1036.

3 W. J. Weber, P. H. Stanton and J. T. Sumida, "A Bandwidth Compressive Modulation System Using Multiple Amplitude Minimum Shift Keying (MAMSK)," *IEEE Trans. Commun.*, Vol. COM-26, May 1978, pp. 543–551.

4 J. B. Anderson and D. P. Taylor, "A Bandwidth-Efficient Class of Signal-Space Codes," *IEEE Trans. Inf. Theory*, Vol. IT-24, Nov. 1978, pp. 703–712.

5 H. Miyakawa, H. Harashima and Y. Tanaka, "A New Digital Modulation Scheme—Multimode Binary CPFSK," in *Proc. ICDSC-3*, Kyoto, Nov. 1975, pp. 105–112.

6 A. T. Lereim, "Spectral Properties of Multi-*h* Phase Codes," CRL Intern. Rep. CRL-57, McMaster Univ., Hamilton, Ontario, July 1978.

7 D. P. Taylor and B. Mazur, "Research on Self-Synchronization of Phase Codes," CRL Intern. Rep. CRL-63, McMaster Univ., Hamilton, Ontario, Apr. 1979.

8 T. A. Schonhoff, H. E. Nichols and H. M. Gibbons, "Use of the MLSE Algorithm to Demodulate CPFSK," in *Proc. Int. Conf. Commun.*, Toronto, June 1978, pp. 25.4.1–25.4.5.

9 J. M. Wozencraft and I. M. Jacobs, *Principles of Communication Engineering*, Wiley, New York, 1967, pp. 266–273.

10. G. D. Forney, "The Viterbi Algorithm," *Proc. IEEE*, Vol. 61, Mar. 1973, pp. 268–278.

11 R. Lugannani, "Intersymbol Interference and Probability of Error in Digital Systems," *IEEE Trans. Inf. Theory*, Vol. IT-15, Nov. 1969, pp. 682-688.

12. P. J. McLane, "An Upper Bound for Error Pattern Probabilities for Digital Transmission by PAM," *IEEE Trans. Commun.*, Vol. COM-21, Mar. 1973, pp. 230–233.

13 G. Ungerboeck, "Channel Coding with Multilevel Phase Signals," *IEEE Trans. Inf. Theory*, to appear.

14 T. Aulin and C-E. Sundberg, "Bounds on the Performance of Binary CPFSK Type of Signalling with Input Data Symbol Pulse Shaping," in *Proc. Natl. Telecommun. Conf.*, Birmingham, Ala., Dec. 1978, pp. 6.5.1–6.5.5.

PROBLEMS

1 Draw the phase trellis for a $(3/4, 2/4)$ multi-*h* phase code modulation.

2 Given two signal vectors s_1 and s_2, show that the probability of decoding s_2 given that s_1 was transmitted is

$$P(s_2|s_1) = Q\left[\tilde{d}(1,2)\frac{E}{2N_0}\right]$$

3 Three signal waveforms $\{s_i(t), i=1,2,3\}$ span a time interval $3T$. Over each interval T each waveform has a constant value:

$$s_0(t) = 2p(t) - 2p(t-T) + 2p(t-2T)$$

$$s_1(t) = -p(t) + 3p(t-T) + p(t-2T)$$

$$s_2(t) = p(t) - 2p(t-T)$$

where

$$p(t) = \begin{cases} 1, & 0 \le t \le T \\ 0, & \text{elsewhere} \end{cases}$$

Find an othogonal vector representation for the signals.

4 For MHPM signals show that

$$d^2(1,2) = \sum_{n=1}^{L} d_n^2(1,2)$$

where

$$d_n^2(1,2) = \begin{cases} 2\left(1 - \dfrac{\sin \Delta\phi_{n+1} - \sin \Delta\phi_n}{\Delta\phi_{n+1} - \Delta\phi_n}\right), & \Delta\phi_{n+1} \ne \Delta\phi_n \\ 2(1 - \cos \Delta\phi_n), & \text{otherwise} \end{cases}$$

$\Delta\phi_n$ is the phase difference between s_1 and s_2 at the beginning of the nth interval and the two phase paths do not merge for L intervals.

5 If an MHPM signal has K indices $\{h_i; i=1,2,...,K\}$, show that it has constraint length $K+1$ if and only if all 2^K subset sums are different modulo 1.

6 For an MHPM signal to have constraint length $K+1$, show that $q \geq 2^K$ where the h_i are multiples of $1/q$.

PART 2

Multiple Access Techniques

||||||||||||||||||||||||||||||||||||||||||

CHAPTER 7

Frequency-Division Multiple Access

Multiple access is the shared use of the capacity of a satellite channel; this usually means sharing the bandwidth and power. The most common form of multiple access is FDMA. This chapter presents FDMA principles, methods and descriptions of systems. A design is presented for a single-channel-per-carrier system.

7.1 FDMA CONCEPTS AND DEFINITIONS

In frequency-division multiplexing (FDM), signals occupying nonoverlapping frequency bands are combined (added) and a specific signal can be recovered by filtering. **Frequency-division multiple access** (FDMA) is FDM applied to satellite repeaters. Each uplink RF carrier occupies its own frequency band b and is assigned a specific location within the repeater bandwidth B. **Guard bands** between accesses allow for imperfect filters and oscillators. The special case of a single access of a wideband repeater is included in this chapter. Receiving Earth stations select a desired carrier by RF and IF filtering. No clocking control exists between accesses. Other than remaining "on frequency", there is no coordination between accessing stations in elementary forms of FDMA.

FDMA is extensively used with frequency modulation (FM) and demodulation, as in television distribution and broadcasting. Two TV signals on separate carriers through the same RF channel are also feasible in a 36-MHz or more bandwidth. **Multi-channel-per-carrier** (MCPC) signals consist of single-sideband suppressed-carrier (SSB/SC) signals which are FDM'd before or at the Earth station. Historically, the elementary signals have been voice signals, but data are now also included. Then the composite signal frequency modulates a carrier for transmission to the satellite. This is usually called FDM/FM/FDMA, with some authors calling this MCPC in contradistinction with SCPC. Typical systems have 24 or 60 voice channels or multiplexed data per access. **Single-channel-per carrier** (SCPC) telephony and data are also provided using FM [1], for example in Algeria, Indonesia, in the Alaska Bush Network [2] and in the Canadian pilot projects with HERMES and ANIK-B [3]. Such systems are labelled FM/SCPC and the information signal is analog. Such analog modulation and multiple-access systems are not treated here, but it is well to remember that a repeater might be supporting such accesses. There are important examples of mixes of analog and digital systems in the same repeater; for example, FM TV services can be accompanied by high-speed data, including burst types (see Chapters 8 and 10), and by low-speed data or voice signals.

A variety of digital services can be supported by satellite communications: multichannel telephony, where the voice channels are digitized and multiplexed in the Earth station; digital TV of several types; high-speed facsimile or data; single-channel-per-carrier telephony or data; low-speed data and packet-switched services. Of course, the transmission rates are different, the performances can be selected for each system and multiple-access considerations are significant. A system designed for one or more of the foregoing services or of analog services is said to give a **mix of services**. Systems designed for the same type, capacity and performance of each access are called **uniform networks**. An overview of FDMA is provided in a paper by Dicks and Brown [4].

7.2   FDMA METHODS

The major items of a typical SCPC station are illustrated in Figure 7.1. A **channel unit** (CHU) is a combination of transmit and receive hardware for a single data or voice circuit. The **voice processor** is usually PCM type (PCM encoder) or delta modulation type (delta-mod, and many variations). The use of the terms "codec", "encoder" and "modulation" to refer to baseband source processors is confusing but present throughout the literature. One or both synthesizers (programmable sources of required frequencies) might be replaced by fixed oscillators in economical systems. Typical modems are discussed in previous chapters and principles of codecs are treated in later chapters. A power supply can serve several channel units.

FDMA Technologies

Some of the technologies that affect FDMA design are reviewed briefly here. Clearly, digital modems are very significant and the previous chapters treat the many types and problems that exist. Error-control techniques are presented in later chapters. FDMA requires stable sources for IF systems, for up- and down-conversions. In stations that have requirements to process several carriers, there are multiple sources required; often, this results in frequency synthesizers being used. For more flexibility and fast response, networks use

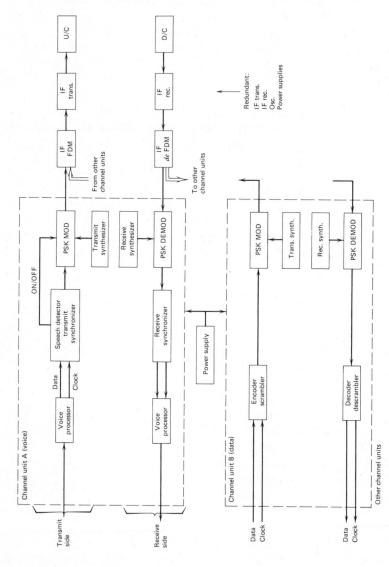

Figure 7.1 Block diagram of a typical SCPC station. A channel unit for voice is shown, including voice activation; and a data channel unit is illustrated with FEC codecs. Similar channel units can be added. IF equipment is usually redundant.

demand assignment (DA). (This is often called DAMA, for demand-assigned multiple access, and usually implies SCPC/FDMA.) DA signalling, network monitoring and control and many other techniques are employed in FDMA networks. Particularly significant to cost effectiveness and maintenance are digital implementations, notably MSI and LSI (medium- and large-scale integration), of modems, controllers, error control equipment and so on.

Some significant factors are discussed below. Nonlinear amplifiers, such as the station HPA or the repeater TWTA, produce several effects and have a strong influence on repeater capacity and multiple access.

Nonlinear Amplifiers

Only ideal amplifiers are linear for all input signals. A **linear function** is defined as

$$Z(ax+by)=aZ(x)+bZ(y) \tag{7.1}$$

where a and b are constants, x and y are variables and where a, b, x, y and $Z(\cdot)$ can be complex. In the models of subsystems discussed here, $Z(\cdot)$ is the power transfer function or the $Z_p(\cdot)$, $Z_q(\cdot)$ of the quadrature model. In a linear model, phase angles are not distorted (linear phase) and amplitudes are given by (7.1). *Real* amplifiers and devices do not have this property and are called **nonlinear**. There are many effects due to nonlinear properties:

1 Spectral spreading
2 Signal suppression
3 Intermodulation
4 Modulation transfer

There is always some **spectral spreading**; the wanted signal components appear in portions of the band beyond the input spectrum. Spectral spreading is more important when one wideband signal uses most of the TWTA power and has the possibility of interfering with adjacent repeater channels. For digital modulations, the mechanism that produces spectral spreading can be understood from intermodulation principles. For multiple accesses there is **signal suppression**, where the wanted carrier is affected by the level of the other carriers. The most significant effect for multiple access is intermodulation. **Modulation transfer** is the conversion by AM/PM and AM/AM of one modulation to another carrier. In the case of TDMA, audible tones may result in the baseband of FM carriers [5].

Intermodulation and Its Effects

In order to understand FDMA system design and performance, it is essential to consider **intermodulation** (IM), which is the presence of unwanted signal-dependent spectral components. **Intermodulation products** (IPs) are produced when two or more spectral components (frequencies or bands of frequencies) pass through a nonlinear device. Only active IM is considered here, such as that produced by HPAs in Earth stations and TWTAs in satellite repeaters.

The single-carrier *power* transfer function is what is usually measured, although multicarrier models are required for FDMA. An example appears in Chapter 4, where saturation is formally defined. For a typical TWTA, the following approximation for this power transfer function is valid near $X = -10$ dB:

$$Y(X) = -0.03X^2 + 0.07X \quad \text{dB} \tag{7.2}$$

where $X = P_i/P_s = -BO_i$ dB. The input X is normalized by the power required to saturate the TWTA. The output Y is also normalized by the output power at saturation, for a single carrier.

This section presents, by analysis and example, the fundamentals of IM effects and magnitudes of IPs. For illustration purposes and to keep the example simple, a quasi-linear or practically linear amplifier is considered first. Let the *voltage* transfer function be represented by

$$g(x) = c_1 x + c_2 x^2 + c_3 x^3 \tag{7.3}$$

where x is the input voltage and the coefficients c_i are determined by the particular device in its operating environment. Under the usual conditions (called **bandpass** for brevity) that carrier frequencies are high with respect to the bandwidth ($f \gg B$) and that a zonal filter is placed after the nonlinear device to remove out-of-band harmonics, only the odd powers are of interest (see Problem 1). It can be shown that a single access produces no IM and its output power is determined according to the power transfer curve. Also, double accesses cause signal suppression (but no IM after the zonal filter for this polynomial model) (see Problem 2). For three or more accesses, there is signal suppression and there are significant IPs.

Example 7.1   Three Accesses

The following analysis shows how IPs are produced and what their relative sizes are. Three narrowband bandpass signals are assumed at the input.

$$x = A_1 \cos \theta_1 + A_2 \cos \theta_2 + A_3 \cos \theta_3 \tag{7.4a}$$

where

$$\theta_i = \omega_i t + \phi_i(t), \qquad i = 1, 2, 3 \tag{7.4b}$$

and time dependence is suppressed. At this stage in the analysis, let there be no phase modulation ($\theta_i = \omega_i t$), so that only three carriers are present. Using (7.3), the output is

$$g(x) = c_1(A_1\cos\theta_1 + A_2\cos\theta_2 + A_3\cos\theta_3) + c_2x^2$$

$$+ c_3(A_1^3\cos^3\theta_1 + A_2^3\cos^3\theta_2 + A_3^3\cos^3\theta_3)\} \qquad \text{monomial}$$

$$\left.\begin{array}{l} + c_3(3A_1^2A_2\cos^2\theta_1\cos\theta_2 + 3A_1A_2^2\cos\theta_1\cos^2\theta_2 \\[4pt] + 3A_1^2A_3\cos^2\theta_1\cos\theta_3 + 3A_1A_3^2\cos\theta_1\cos^2\theta_3 \\[4pt] + 3A_2^2A_3\cos^2\theta_2\cos\theta_3 + 3A_2A_3^2\cos\theta_2\cos^2\theta_3) \end{array}\right\} \text{binomial}$$

$$+ c_3(6A_1A_2A_3\cos\theta_1\cos\theta_2\cos\theta_3)\} \qquad \text{trinomial} \qquad (7.5)$$

The linear, quadratic and cubic terms have been written separately. The terms labelled *monomial* involve one θ_i and have only one spectral component per input, since $3\omega_i t$ terms do not pass through the zonal filter; *binomial* terms include all combinations of two inputs with coefficients of 3 in each case; there is only one *trinomial* term involving all three inputs and it has coefficient 6. This form of the output does not indicate the spectral components, so the following identities are used to identify them.

$$\cos^3 A = \tfrac{1}{4}(\cos 3A + 3\cos A) \qquad (7.6a)$$

$$\cos^2 A\cos B = \tfrac{1}{4}[2\cos B + \cos(2A - B) + \cos(2A + B)] \qquad (7.6b)$$

$$\cos A\cos B\cos C = \tfrac{1}{4}[\cos(A + B - C) + \cos(A - B - C)$$
$$+ \cos(A - B + C) + \cos(A + B + C)] \qquad (7.6c)$$

Note that (7.6a) and (7.6b) can be derived from (7.6c), using $\cos X = \cos(-X)$. The terms of type $3A$, $(2A + B)$ and $(A + B + C)$ do not pass through the zonal filter. The $(2A - B)$ terms and the $(A + B - C)$ terms are called **dominant**, pass through the zonal filter and are IPs from the cubic term. The zonal filter output is

$$y = \left.\begin{array}{l} \left[c_1A_1 + \tfrac{3}{2}c_3(A_1A_2^2 + A_1A_3^2) + \tfrac{3}{4}c_3A_1^3\right]\cos\theta_1 \\[4pt] + \left[c_1A_2 + \tfrac{3}{2}c_3(A_1^2A_2 + A_2A_3^2) + \tfrac{3}{4}c_3A_2^3\right]\cos\theta_2 \\[4pt] + \left[c_1A_3 + \tfrac{3}{2}c_3(A_1^2A_3 + A_2^2A_3) + \tfrac{3}{4}c_3A_3^3\right]\cos\theta_3 \end{array}\right\} \text{wanted}$$

$$\left.\begin{array}{l} + \tfrac{3}{4}c_3\left[A_1^2A_2\cos(2\theta_1 - \theta_2) + A_1A_2^2\cos(2\theta_2 - \theta_1)\right. \\[4pt] + A_1^2A_3\cos(2\theta_1 - \theta_3) + A_1A_3^2\cos(2\theta_3 - \theta_1) \\[4pt] + A_2^2A_3\cos(2\theta_2 - \theta_3) + A_2A_3^2\cos(2\theta_3 - \theta_2)] \\[4pt] + \tfrac{3}{2}c_3A_1A_2A_3[\cos(\theta_1 + \theta_2 - \theta_3) \\[4pt] \qquad + \cos(-\theta_1 + \theta_2 + \theta_3) \\[4pt] \qquad + \cos(\theta_1 - \theta_2 + \theta_3)] \end{array}\right\} \begin{array}{l} \text{intermod} \\ \text{products} \quad (7.7) \\ \text{(IPs)} \end{array}$$

The wanted output signals are shown first. All combinations of dominant IPs are present. Table 7.1 contains an example of the IPs for a specific polynomial model and four narrowband input signals as in the model of (7.4). The notation $M(k_1, k_2, k_3)$ means the amplitude of the spectral term for frequency

$$f = k_1 f_1 + k_2 f_2 + k_3 f_3 \quad \text{Hz} \tag{7.8}$$

where $\omega_i = 2\pi f_i$ rad/s, under the bandpass constraint

$$k_1 + k_2 + k_3 = 1 \tag{7.9}$$

where k_i are negative, zero or positive integers. Using this simple model, suppression effects are now analysed.

Table 7.1 Intermod Products for a Quasi-linear Amplifier with Four Inputs

k_1	k_2	k_3	k_4		$M(k_1, k_2, k_3, k_4)$	
					Example 1	Example 2
1	0	0	0	4×1 terms	0.625	0.6250
0	1	0	0			0.6250
0	0	1	0			0.6250
0	0	0	1			0.4240
2	−1	0	0			−0.0750
2	0	−1	0			−0.0750
2	0	0	−1			−0.0375
−1	2	0	0			−0.0750
0	2	−1	0			−0.0750
0	2	0	−1	4×3 terms	−0.075	−0.0375
−1	0	2	0			−0.0750
0	−1	2	0			−0.0750
0	0	2	−1			−0.0375
−1	0	0	2			−0.0187
0	−1	0	2			−0.0187
0	0	−1	2			−0.0187
1	1	−1	0			−0.150
1	1	0	−1			−0.075
1	−1	1	0	$\dfrac{4 \times 3}{2}$ terms	−0.150	−0.150
1	−1	0	1			−0.075
−1	1	1	0			−0.150
−1	1	0	1			−0.075

Example 1:

$$A_1 = A_2 = A_3 = A_4 = 1$$

Example 2:

$$A_1 = A_2 = A_3 = 2A_4 = 1$$

Notes: (a) $c_1 = 1$, $c_3 = -1$; (b) select all combinations of k_i subject to (1) $|k_i| <$ order of polynomial; (2) signs to obtain positive frequencies in (7.8); (3) $\Sigma k_i = 1$ in (7.9)

Signal Suppression

Signal suppression, or sometimes called gain suppression, occurs because of a particular nonlinear effect; the output is not proportional to the input. We recall the concept of small-signal gain, g_{ss}, and model the effect as

$$c_1 x + c_3 x^3 = g(x) < g_{ss} x \qquad \text{for all } x \qquad (7.10a)$$

and

$$\frac{g(x_2)}{x_2} < \frac{g(x_1)}{x_1} \qquad \text{for } x_1 < x_2 \qquad (7.10b)$$

EXERCISE For any x_1 and x_2 in a region of interest, find the following expressions for c_1 and c_3:

$$c_1 = \frac{x_2^2 \dfrac{g(x_1)}{x_1} - x_1^2 \dfrac{g(x_2)}{x_2}}{x_2^2 - x_1^2} \qquad (7.11a)$$

$$c_3 = \frac{-\dfrac{g(x_1)}{x_1} + \dfrac{g(x_2)}{x_2}}{x_2^2 - x_1^2} \qquad (7.11b)$$

Also, using (7.10), verify that $0 < c_1 < g_{ss}$ and $c_3 < 0$.

In the following illustrations, P_i is the input power for saturation and P_1, P_2, P_3 are the three signal powers. Case 1 is for equal powers at 10-dB input backoff. Case 2 represents the situation when carrier 1 is off. Case 3 is for definitely unequal carriers. The input and output amplitudes are given in Table 7.2.

CASE 1   Let $P_1 = P_2 = P_3 = P_i - 10$ dB. Then the wanted signals in (7.7) have equal amplitude.

$$|y_1| = |y_2| = |y_3| = \left[c_1 - 3|c_3|(0.25)A^2 \right] \sqrt{\frac{2P_i}{10}} \qquad (7.12)$$

CASE 2   Let $A_1 = 0$ and calculate the amplitudes for the other outputs as

$$|y_2| = |y_3| = \left[c_1 - 3|c_3|(0.15)A^2 \right] \sqrt{\frac{2P_i}{10}} \qquad (7.13)$$

which shows the slight increase in amplitude.

CASE 3   For $P_1 = P_i - 3$ dB, $P_2 = P_i - 10$ dB and $P_3 = P_i - 20$ dB, this case models large, medium and small carriers.

Table 7.2 Illustrating Signal Suppression

Case	Input Condition	Wanted Outputs
1	$A_1 = A_2 = A_3 = \sqrt{\dfrac{2P_i}{10}}$	Each: $[c_1 - 3\lvert c_3\rvert(0.25)P_i]\sqrt{\dfrac{2P_i}{10}}$
2	$A_1 = 0;\ A_2 = A_3 = \sqrt{\dfrac{2P_i}{10}}$	$0 = \lvert y_1\rvert;\ \lvert y_2\rvert = \lvert y_3\rvert = [c_1 - 3\lvert c_3\rvert(0.15)P_i]\sqrt{\dfrac{2P_i}{10}}$
3	$A_1 = \sqrt{\dfrac{2P_i}{2}}$	$[c_1 - 3\lvert c_3\rvert(0.36)P_i]\sqrt{\dfrac{2P_i}{2}}$
	$A_2 = \sqrt{\dfrac{2P_i}{10}}$	$[c_1 - 3\lvert c_3\rvert(0.56)P_i]\sqrt{\dfrac{2P_i}{10}}$
	$A_3 = \sqrt{\dfrac{2P_i}{100}}$	$[c_1 - 3\lvert c_3\rvert(0.605)P_i]\sqrt{\dfrac{2P_i}{100}}$

For case 1, each access receives its equal share of the output power. Each case shows that the outputs are not linearly proportional to the inputs but the output levels have the general relation as the input levels. In case 3, $\lvert y_1\rvert$ is the strongest carrier and is affected least (-0.36); $\lvert y_3\rvert$ is the weakest carrier and is affected most (-0.605). Case 2 represents the event of a strong carrier being shut off: the remaining carriers increase. Hence when x_1 is present, it is said to suppress the others. Such suppression is very significant if there are few and strong carriers, but not so harmful for many equal carriers.

Suppression is usually regarded as harmful. However, Loo has studied the beneficial effects of a saturating carrier (such as for TV) on smaller SCPC

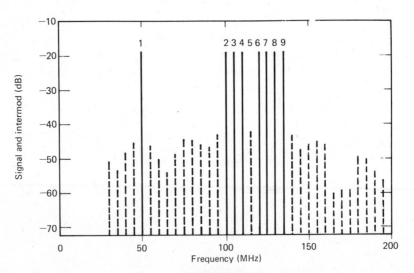

Figure 7.2 Calculated output spectrum for eight equal carriers with each at an input backoff of 25 dB. Solid line, signal level; dashed line, intermod level.

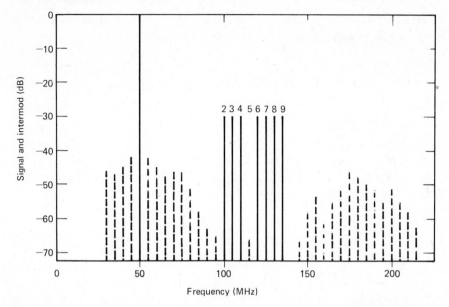

Figure 7.3 Calculated output spectrum for carrier 1 at saturation and carriers 2 to 4 and 6 to 9 each at an input backoff of 25 dB. Solid line, signal level; dashed line, intermod level.

carriers in the same repeater bandwidth [6]. For eight equal carriers at 25-dB input backoff, the frequency plan consists of 5-MHz spacing from channel numbers 2 to 9. Number 5 is missing in the input (to show the large IP in the output) and number 1 is placed 50 MHz below number 2. The calculated output spectrum is shown in Figure 7.2. The levels of IPs near the small carriers are large (min $C/I = 19$ dB). In Figure 7.3 carrier 1 is adjusted to saturate the TWTA. The calculated signal suppression is 11.5 dB and the IP in the vacant channel 5 is suppressed by 28.5 dB. In the SCPC band of interest, the C/I is enhanced by 17 dB. Observe also the slightly decreased C/I in carrier 1. These results agree within 2 dB with experimental measurements.

To this point, only unmodulated carriers are considered, as modelled by (7.4). When there is ideal phase modulation with constant envelopes, input and output spectra are "filled in" and smooth. For large numbers of accesses, the IM effects are therefore reasonably modelled as IM noise.

Other Models

The voltage transfer function of (7.3) might be an adequate model for a quasi-linear amplifier or for a TWTA which has been linearized, but in general, higher-order models are necessary to characterize HPAs and TWTAs operating near saturation. If higher-order polynomials are used, there are needs for trinomial coefficients (in this case of three accesses) and multinomial coefficients (in the m-access case). The binomial closed form

$$\binom{n}{m}=\binom{n}{n-m}=\frac{n!}{m!(n-m)!}=(n;m,n-m) \qquad (7.14)$$

has a simple extension. $(n; n_1, n_2, \ldots, n_m)$ is the number of ways of putting $n=(n_1+n_2+\cdots+n_m)$ different objects into m different boxes with n_k in the kth box, $k=1,2,\ldots,m$ [7]. In this analysis, n is the order of the polynomial term and m is the number of multiple accesses, and

$$(n; n_1, n_2, \ldots, n_m)=\frac{n!}{n_1!n_2!\cdots n_m!} \qquad (7.15)$$

Although these multinomial coefficients are easy to compute, as with binomial coefficients, there is convenience in having a table of coefficients. For the trinomial case, the table is in three dimensions. Most entries are from the binomial table, but there are significant trinomial terms. Figure 7.4 illustrates the table for $n=1$, 2, 3, 4, 5, 6 and 7. Problem 3 develops some useful recurrence relations. For example, consider

$$(6;3,2,1)=\frac{6!}{3!2!1!}=60$$

On Figure 7.4, move 3 units along n_1 axis, 2 units up n_2 axis and 1 unit along n_3 axis; read the trinomial coefficient 60.

Now consider

$$(7;3,2,2)=\frac{7!}{3!2!2!}=\frac{7\times6\times5\times4}{2\times2}=210$$

Also,

$$(7;3,2,2)=(6;3,2,1)+(6;3,1,2)+(6;2,2,2)$$

This entry of 210 is illustrated with a dashed triangle in the figure.

Other models for nonlinear amplifiers are much better in characterizing the amplitude and phase transfer properties of TWTAs and other amplifiers [8, 9]. IM analysis is then complicated by the functional form, but fewer terms are required in expansions. This is because the functional form is close to the shape of the transfer function of the nonlinear device. For m complex inputs,

$$x(t) = \sum_{i=1}^{m} A_i(t) \exp[j\omega_0 t + j\theta_i(t)] \tag{7.16}$$

and the complex transfer function

$$Z_p(\rho) + jZ_q(\rho) = g(\rho) \exp[jf(\rho)] \tag{7.17}$$

The output is given by

$$y(t) = \exp(j\omega_0 t)\} \text{ center carrier}$$

$$\cdot \sum_{\substack{-\infty \\ k_i}}^{\infty} \exp\left[j \sum_{i=1}^{m} k_i \theta_i(t)\right]\} \text{ phase modulation}$$

$$\cdot M(k_1, k_2, \ldots, k_m)\} \text{ complex coefficients} \tag{7.18}$$

where $-\infty < k_i < \infty$, k_i integer, $\Sigma_1^m k_i = 1$, and where

$$M(k_1, k_2, \ldots, k_m) = \int_0^{\infty} \gamma \prod_{i=1}^{m} J_{k_i}(\gamma A_i) \, d\gamma$$

$$\cdot \int_0^{\infty} \rho g(\rho) \exp[jf(\rho)] J_i(\gamma\rho) \, d\rho \tag{7.19}$$

The Bessel functions $J_n(\cdot)$ arise from the use of Fourier transforms in the analysis. See Ref. 9 for details. The condition on the sum of k_i arises from the bandpass characterization of the output of the zonal filter. This condition is the generalization of the removal of terms of type $3A$, $(2A+B)$, $(A+B+C)$ and those of type $2A$ and $(A-A)$ (DC term). Given gain and phase characteristics, it is possible to compute $M(k_1, k_2, \ldots, k_m)$ directly from the numerical data [10] or by numerical integration of (7.19). With proper precautions, the average spectral components are reported to compare to within 0.5 dB with measured spectra. For a specific model,

$$g(\rho) \exp[jf(\rho)] = \sum_{s=1}^{L} b_s J_1(\alpha s \rho) \tag{7.20}$$

where L and α are parameters of best fit and b_s, $s = 1, 2, \ldots, L$, are complex

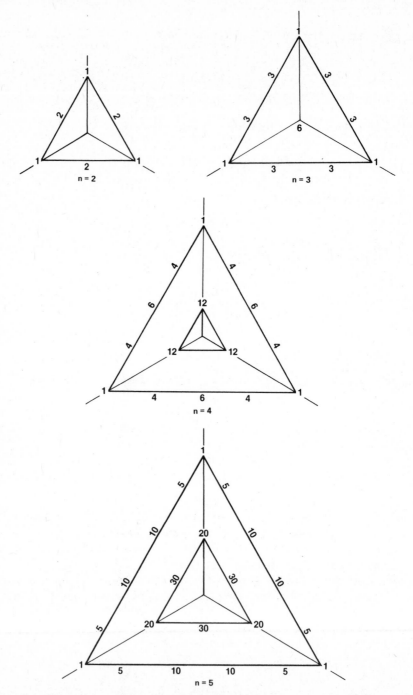

Figure 7.4 Trinomial coefficients in a three-dimensional figure. In any plane for two variables, the familiar binomial coefficients result. Trinomial coefficients for a given *n* are on the appropriate Figure.

212

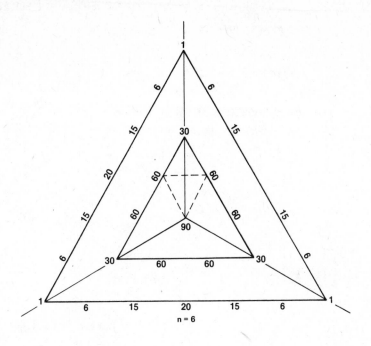

n = 6

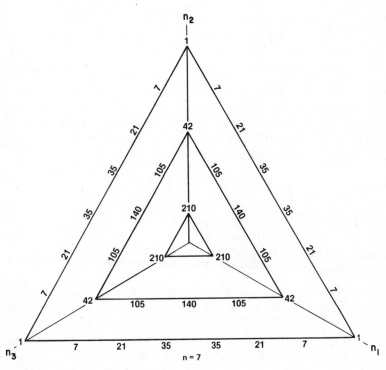

n = 7

213

coefficients. For this model, the complex M are given by

$$M(k_1, k_2, \ldots, k_m) = \sum_{s=1}^{L} b_s \prod_{i=1}^{m} J_{k_i}(\alpha s A_i) \qquad (7.21)$$

For the quadrature model, the real $\mathrm{Re}(\cdot)$ and imaginary $\mathrm{Im}(\cdot)$ parts of M can be calculated separately.

$$\mathrm{Re}M(k_1, k_2, \ldots, k_m) = \int_0^\infty \gamma \prod_{i=1}^{m} J_{k_i}(\gamma A_i)\, d\gamma$$

$$\cdot c_1 \int_0^\infty \rho^2 \exp(-c_2 \rho^2) I_0(c_2 \rho^2) J_i(\gamma\rho)\, d\rho \quad (7.22)$$

$$\mathrm{Im}\, M(k_1, k_2, \ldots, k_m) = \int_0^\infty \gamma \prod_{i=1}^{m} J_{k_i}(\gamma A_i)\, d\gamma$$

$$\cdot s_1 \int_0^\infty \rho^2 \exp(-s_2 \rho^2) I_1(s_2 \rho^2) J_i(\gamma\rho)\, d\rho \quad (7.23)$$

where c_1, c_2 and s_1, s_2 are parameters of the model; some typical values are given in Chapter 4. No closed forms similar to (7.21) are known to the authors.

C/I versus Input Backoff

Several authors [8–11] have produced curves of carrier-to-intermodulation ratio (C/I). This is the ratio of the wanted carrier power to the total power due to all IPs. As already noted, this is dependent on the input backoff, $X = -BO_i$ dB. A representative case for many equal SCPC accesses is shown in Figure 7.5. For use in an example, the approximation

$$Z(X) = 0.05 X^2 - 0.5 X \qquad \mathrm{dB} \qquad (7.24)$$

is valid near $X = -10$ dB.

Backoffs and Accesses for SCPC

For SCPC networks, the repeater bandwidth B is shared by multiple accesses through FDMA and the repeater output power is shared proportionally to input power. However, this sharing is not linearly proportional and controlled power sharing is a major aspect of all SCPC systems. For a uniform network discussed here, each accessing station has the same bandwidth and operates so as to take the same output power. To be able to use the foregoing IM analysis, the frequency assignments are assumed uniform across the band; extensions are discussed below.

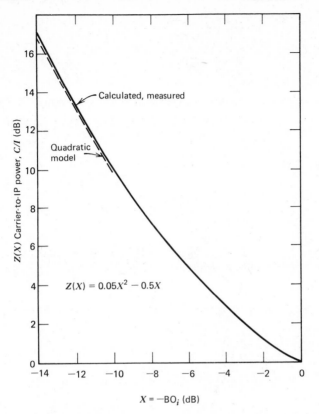

Figure 7.5 C/I for a uniform SCPC network having many accesses. $B = 50$ MHz.

The following relations reflect the nonlinearities:

$$\left.\frac{C}{N_0}\right)_u = \left.\frac{C}{N_0}\right)_s + X = 105 + X \qquad \text{dB-Hz} \qquad (7.25)$$

$$\left.\frac{C}{N_0}\right)_d = \left.\frac{C}{N_0}\right)_s + Y(X) = 90 - 0.03X^2 + 0.07X \qquad \text{dB-Hz} \qquad (7.26)$$

$$\frac{C}{I_0} = 10\log B + Z(X) = 77.0 + 0.05X^2 - 0.5X \qquad \text{dB-Hz} \qquad (7.27)$$

where $C/N_0)_s$ is at saturation of the TWTA, B is the reference bandwidth for C/I and C/I_0 is the equivalent intermodulation noise spectral density. The numerical examples on the right of (7.25), (7.26), and (7.27) are used below to illustrate procedures in finding the optimum X and the maximum number of accesses. The total C/N_0 is calculated in numerical ratio as

$$\left(\frac{C}{N_0}\right)_t^{-1} = \left(\frac{C}{N_0}\right)_u^{-1} + \left(\frac{C}{N_0}\right)_d^{-1} + \left(\frac{C}{I_0}\right)^{-1} \qquad 1/\text{Hz} \qquad (7.28)$$

not in dB. For the power-limited case in this uniform network,

$$\frac{C}{N_0}\bigg)_t = m\frac{C}{N_0}\bigg)_{ch} \tag{7.29}$$

where m = number of accesses supported by the links
$C/N_0)_{ch} = C/N_0$ required per channel, taking into account performance, rate, modem type—all margins.

Solving for m as a ratio, we obtain

$$m = \frac{(C/N_0)_{ch}^{-1}}{(C/N_0)_t^{-1}} \tag{7.30}$$

To maximize m for a fixed $C/N_0)_{ch}$, $C/N_0)_t$ should be maximized, or equivalently its inverse is minimized by selecting X. Because numeric ratios of powers

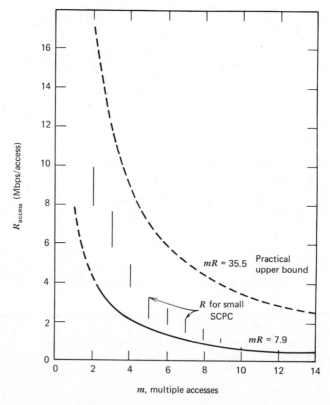

Figure 7.6 Capacities for digital SCPC, with Mbps per access versus number of accesses. See example case for link parameters, no coding.

of 10 are summed, there is no known algebraic minimization technique. Graphical or equivalent procedures are quite straightforward; a typical design is illustrated in Section 7.4.

Capacity versus Number of Accesses

For the optimum input backoff and the maximum m accesses, there is a simple relation for the capacity per access, R.

$$10\log m + \frac{E_b}{N_0} + 10\log R + \sum \text{margins} = \frac{C}{N_0}\bigg)_t \qquad \text{dB-Hz} \qquad (7.31a)$$

$$10\log m + 10\log R = \text{constant} \qquad \text{dB-Hz} \qquad (7.31b)$$

$$mR = \text{constant} \qquad (7.31c)$$

Thus R versus m is linear on log-log scales. On linear-linear scales, (7.31c) is a hyperbola and represents the asymptotic relation for large m (>16). For small m, the assumptions do not apply because C/I is smaller and controllable by assignment of channel centre frequencies. Figure 7.6 illustrates the asymptote for $mR = 7.9$ Mbps, the likely cases for small m, and the upper bound $mR = 35.5$ Mbps. The numbers 7.9 and 35.5 are for the example of Section 7.4. The upper bound for SCPC is obtained from the single-access case.

7.3 FDMA SYSTEMS

This section describes some of the digital FDMA systems that are now in operation or under development. The operating companies and their suppliers are best sources for details, but specific design and performance information is not available in each case. Some history, implementation notes and some performance figures are given.

Multichannel PSK/FDMA Systems

FDM/PSK/FDMA INTELSAT, TELESAT, Western Union, RCA Americom and American Satellite Corporation (ASC) operate a variety of FDM/PSK/FDMA systems for telephony [12, 13].*

---

*INTELSAT is the international organization, with over 100 signatories, operating the space segments and providing management for international telecommunications by satellite. TELESAT Canada has a charter to provide all domestic satellite communications services in Canada. Western Union, RCA Americom, American Satellite Corporation and COMSAT General are specialized common carriers authorized by the Federal Communications Commission (FCC) to provide satellite communications within the United States.

TDM/PSK/FDMA   PCM voice and data are TDM'd and transmitted in systems of specialized design by Western Union, RCA Americom, COMSAT General, ESA and SYMPHONIE [13–15].*

PCM/PSK Single Carrier   TELESAT is preparing for ANIK-C digital services which will begin in 1982 and operate at 90.148 Mbps using QPSK modulation. This all-digital satellite system will provide message services for Heavy-Route interconnection between major Canadian cities [16]. Field trials with ANIK-B are reported in Ref. 3.

Experimental Systems   Experiments at 180 Mbps [17], many experiments and field trials in digital TV [18, 19] and new high-speed modems indicate the general direction for wideband digital services.

Digital SCPC Systems

Digital SCPC systems use FDMA as the multiple-access mode and obtain their label SCPC from single-channel-per-carrier. Historically, the name applied to systems with one voice channel but now includes any multiple access in FDMA mode by similar capacity channels. Digital SCPC refers to the use of digital modulation techniques; BPSK and QPSK are most common.

SPADE   The best known SCPC system is the SPADE system [4, 20] operated by INTELSAT; the name SPADE is loosely derived from Single-channel Per carrier Access-on-Demand Equipment. It is also reported to come from sCPC/pSK/FDMa/dA eQUIPMENT. SPADE is characterized as PCM/PSK/SCPC/DA. For small users with fluctuating requirements, SPADE provides voice and data services and includes a demand-assignment subsystem (itself a TDMA system). Table 7.3 illustrates some of the larger connections to about 34 countries in the global networks. The success of SPADE has caused problems, since many connections are more than thin-route and require many channel units. The data services within SPADE are called DIGISAT; little activation gain similar to voice activation is available on these circuits. Demand assignment (DA) is used to establish connections between pairs of stations, by drawing from a pool of 800 one-way (half) circuits. Each carrier is voice-activated, so that with a speech activity factor of 0.4, only 320 accesses need to be supported by the link budgets. Channel bandwidths of 45 kHz, QPSK modems at 64 kbps, rate-$\frac{7}{8}$, forward-acting, error-correcting codecs and 56-kbps PCM voice processors are used.

In experiments with SPADE [21], 64-kbps BPSK modems using two adjacent SPADE channels have been used to support a 28-kbps, rate-$\frac{1}{2}$,

---

*ESA, The European Space Agency, is an organization of 22 European members and is responsible for research and development in many areas of satellite technologies. SYMPHONIE is the cooperative program between France and West Germany, and operates two experimental satellites by that name.

convolutional-encoded, threshold-decoded digital channel. A BER of 10^{-3} was achieved at a C/N of 4.6 dB. The significance of this is that data and delta-modulated voice can be provided to small Earth stations, with 5-m-diameter antennas, low power and moderate receiver front ends.

TELESAT THIN-ROUTE   This was the first domestic digital SCPC system and now operates with over 50 stations, many with several channel units. About 265 circuits (two-way) were reported in use in 1980. Figure 7.7 illustrates the THIN-ROUTE network, which operates with ANIK A or ANIK B using the 6/4-GHz repeaters. In contrast to SPADE, the TELESAT THIN-ROUTE system uses two hops to connect user stations through a central station, and preassigned channels are used. The system uses BPSK modems operating at 40 kbps. For a repeater bandwidth $B = 36$ MHz and equal channel spacings of $b = 60$ kHZ, up to 600 one-way circuits are possible in the frequency plan; the link budgets do not support this many. In the first version, companded delta-modulated (CDM) voice processors operated at 40 kbps [22]. This was reduced to 37.5 kbps in some stations in order to provide 2400-bps data services or multiplexed telex. Also, the system is being changed to QPSK modems.

ASC Networks   The American Satellite Corporation (ASC) provides a number of service types to many groups of customers. SCPC networks provide voice and data from Teletype speeds to 56 kbps. "Shared FX" is a multiple-access foreign exchange service for business telephony. Satellite Data Exchange (SDX) provides up to 1.544-Mbps digital links with a reported BER $= 10^{-7}$ with a high availability. The company uses several techniques to increase data throughput; one is accomplished by *satellite delay compensation units* [23]. These units are used in ARQ schemes to return an acknowledge (ACK) to the sending station immediately; then the actual satellite transmission is accomplished reliably. Private and secure data transmissions are offered through encryption systems. The flexibility and mix of services are the outstanding features of these digital SCPC networks [24].

RCA Americom System   RCA Americom first demonstrated a hybrid analog/digital SCPC system of up to 20 channels in 1977 [25]. Services offered include Teletype, facsimile, 1200-bps, 9600-bps and 56-kbps data, together with analog links. The data are distributed (broadcast) to a large number of newspapers and radio stations. A rate-$\frac{7}{8}$ codec and BPSK modems are used and a BER of 1×10^{-7} is reported. This network has effectively traded channel capacity against costs of receive-only stations while achieving different quality requirements for groups of users. Press services, national radio networks, regional networks and specialized networks are users.

There are a number of suppliers of channel units, subsystems and complete Earth stations for the systems just described.

Table 7.3 SPADE Service Billable Time

SPADE Service Billable Time —by Terminal Pair (min)

From \ To	Algeria	Angola	Argentina	Belgium	Brazil	Canada	Egypt	France	Germany (West)	Greece	Iran	Iraq	Italy	Jordan	Kuwait	Libya	Mexico
Algeria			251		1866	5896					539	3748		362	1568		166
Angola																	
Argentina	12			668	135	15	32			2152	221	60	405			38	107
Belgium			5567		940		2309				1765	2681		969	3104		7886
Brazil	893		17048	1486			422			28842	921	2464		1297	1239		753
Canada							384				1042	5226	12	3581	7121		
Egypt				44	666	75	1566										
France										1539	151	54				50	
Germany (West)																	
Greece			8673		15159						2205	4785		1547			3985
Iran	37		260	1351	764	2521				477		304		178			258
Iraq	1523		139	5347	2286	6177	23663			5913	37		1206		47	978	
Italy			5250		1940	6460					4034	6543		5465			
Ivory Coast	735		1014	25	5207	17220	459			1841					777		
Jordan	15			2439	386	266				4388	617	82	621		1293		11
Kuwait	1176		499	5370	2260	8140											277
Libya			41		36												
Mexico				5440						1005	911				279		
Netherlands			7371		2843	10381					101			1850	4582		11855
Nigeria				73	33	30		24		17647			28				
Nordic countries			19467		5119		15522				11643	21946		2009	10012	17006	13823
Paraguay				526	349	1134				378		36		85			
Peru																	
Romania	41		718		479	20	1596				1538	4221		3038	1965	5526	
Saudi Arabia	1637		58		195	642	83	52		26			828	819	21		
South Africa			13431		8369						1315		5276				
Spain			22											2435			109
Sudan	129			332	203	353	338	23		2070	44	35	8138		658	283	
Switzerland			5122				2464					2744		4724	540		250
United Kingdom			3110											9			
United States			6656										1319		8326		
USSR	119		497		394	81					824	900			1538		1045
Venezuela				24				11			298	17		499			
Yugoslavia			1465		739	190	128				1551	6081		646	4106		664
Total	6317	—	96703	23747	49777	50711	67781	99		66289	29757	63246	16514	25304	47176	23881	41189

MARISAT The MARISAT* maritime system has been operational since 1976 [26]. Voice/FM/SCPC and data/TDM/PSK/SCPC are combined in the shore-to-ship and ship-to-shore links, with a random-access channel for requests by ships. On a single, continuous carrier with BPSK modulation, the transmission rate is 1200 bps. The data are put in TDM format at the shore station, in frames of 0.29 s (348 bits) consisting of a 20-bit burst codeword (BCW), a 63-bit assignment channel and 22 slots of 12 bits each for telegraph, plus 1 bit. Frames, burst codewords and slots are discussed in Chapter 8. Centralized network control and in-band signalling are used. The shore station for each ocean area transmits and receives pilot carriers to assist in automatic frequency control (AFC).

COMSAT* has reported on a receive-only MARISAT terminal [27] which is a complete, high-quality communications system for the reception of facsi-

*MARISAT is a consortium, consisting of COMSAT General Corporation, ITT World Communications, RCA Global Communications and Western Union International.
*COMSAT, Communications Satellite Corporation, is a regulated U.S. common carrier for satellite communications. COMSAT is a signatory to INTELSAT, has partnership in MARISAT and SBS, and operates the COMSAT Laboratories.

Rotated label at far right of table: SPADE Service Billable Time

Netherlands	Nigeria	Nordic countries	Paraguay	Peru	Romania	Saudi Arabia	South Africa	Spain	Sudan	Switzerland	United Kingdom	United States	USSR	Venezuela	Yugoslavia	Total	Terminals To / From
					2980	2549			50				3154	79		24155	Algeria
																—	Angola
849	21	909	227		34	82	377	39		96		951	176	117	22	7761	Argentina
	1688		325			2337			68					907		33944	Belgium
2996	2124	7330	645		1185	1512	5773						741		1979	81338	Brazil
			1564		13210	1758			290				252		7907	45976	Canada
848		889			30			787	15						48	6762	Egypt
																—	France
									14							14	Germany
	1444		126			1478			1373							41505	Greece
120		6197			843	1347	289	11					37	33	1277	16344	Iran
		35280			2924	1765	7603		95	429		558	5397			101367	Iraq
	146				3434				1491							34763	Italy
5687	624	3866			84		569		11		437			23	317	38896	Ivory Coast
2192		3662				2925	88	1477	67	7683	960			127	1457	32686	Jordan
9642	56	13396	17		2698				1961	619		2711	857	793	10865	61608	Kuwait
																77	Libya
9068		18714			466								459		1092	37434	Mexico
	3860		352		5211	3874	4169		261				2573	1403		62076	Netherlands
55		377						1025	23	380		85				19780	Nigeria
	4358		499		9608	7862			961					119		144887	Nordic countries
433		654			115	241				1317	3076	258				8602	Paraguay
																—	Peru
	78								92					791		20160	Romania
23	224	1138						3314	208	80	453			29		42019	Saudi Arabia
6473		15987	414			2312				10476					4468	68880	South Africa
	243		15		1011				66							3901	Spain
237	15	272			163	6398		211		8512	12797	3882			152	45245	Sudan
			1339			3782	7555		1873					428		38825	Switzerland
			2180						12761					6427		33668	United Kingdom
			1489		5308		40		148		3256		125	23039	5728	55434	United States
361								54								5813	USSR
33			60		1902	14	215									3127	Venezuela
		530					1134		139				1933	707		21078	Yugoslavia
39017	17996	108731	11094	—	38620	44962	31411	11207	21967	29592	20979	10378	13771	35022	35312	1138125	Total

Total billable minutes → 2276250

mile, data or telex traffic via a geostationary satellite from one or more stations. The antenna subsystem on board ship uses a 1.2-m (4-ft) parabolic dish, which is mounted on a stable platform. After the receiver locks on to the satellite signal, the antenna step-tracks the satellite automatically.

Quick FAX To illustrate another type of digital service by satellite, we describe briefly the Quick FAX system for facsimile transmission by satellite [28, 29]. The first reference describes the principle of Relative Address Coding, which is an efficient line-by-line source encoding scheme to reduce redundancy. ARQ techniques and the High Level Data Link Control (HLDLC) protocol are used. To maximize throughput, a frame length of 2032 bits is used. The transmission rates are 4800 and 2400 bps. Field trials were conducted over one-hop and two-hop circuits, and regular service between Japan and the United States was started in March 1978.

MAROTS System MAROTS [30] is being planned for maritime operation by the European Space Agency (ESA) and PTTs in the CEPT organization, the European association for post and telecommunications services. Telephony and telex services are provided through shore-to-shore, shore-to-ship and

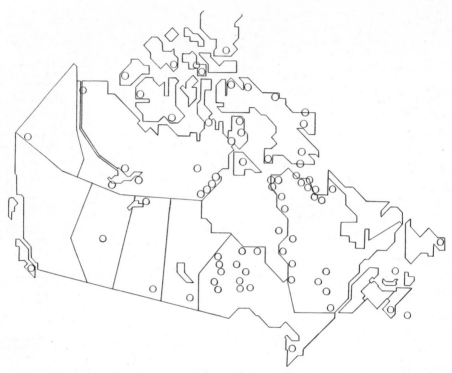

Figure 7.7 TELESAT THIN-ROUTE network, Summer 1980, showing radio programme distribution and message services.

ship-to-shore circuits. Narrowband FM with threshold extension and companding was chosen as the voice modulation method. For telegraphy and signalling, TDMA is the access method to the 50-kHz channels. For requests, random accesses are used. AFC (frequency) and power control are functions of the shore stations.

MARISAT and MAROTS systems and services are likely to be combined by a new international organization, INMARSAT,* which was established 16 July 1979 [31].

7.4 SCPC SYSTEM DESIGN

In this section some design techniques and system tradeoffs are presented. An optimum input backoff is found and C/N_0 and the allowed accesses are calculated, with and without coding in the channel. Voice activation is applied to increase the number of accesses for the same performance.

*INMARSAT is the INternational MARitime SATellite communications organization, providing a variety of services to ships and mobile users.

Example 7.2

For the specific example of "Backoffs and Accesses for SCPC" in Section 7.2, find X to minimize

$$\left(\frac{C}{N_0}\right)_t^{-1} = 10^{-10.5-X/10} + 10^{-9.0-Y/10} + 10^{-7.7-Z/10}$$

$$= 10^{-7.7}(10^{-2.8-X/10} + 10^{-1.3-Y/10} + 10^{-Z/10}) \qquad 1/\text{Hz}$$

$$(7.32)$$

Table 7.4 shows a few steps in the calculations and Figure 7.8 indicates the best value for $X = -10.9$ dB. This is an input backoff of 10.9 dB, for which the output backoff is 4.3 dB. The corresponding multiaccess $C/I = 11.4$ dB and the optimum $C/N_0)_{t,\text{opt}} = 83.4$ dB-Hz. We also observe that X_{opt} is well defined in Figure 7.8 and that X_{opt} is very sensitive to the parameters and to $Y(X)$ and $Z(X)$.

Example 7.3

A specific service on each channel is $R = 64$ kbps at 10^{-4} BER. Using BPSK modems with a 2.0-dB implementation margin and other margins totalling 4.0 dB, the required $C/N_0)_{\text{ch}}$ is

$$\frac{C}{N_0}\bigg)_{\text{ch}} = \frac{E_b}{N_0} + R + M_i + M_f + M_s \qquad \text{dB-Hz}$$

$$= 8.4 + 48.1 + 2.0 + 3.0 + 1.0$$

$$= 62.5 \text{ dB-Hz} \qquad (7.33)$$

Table 7.4 Finding X_{opt}

X	Y	Z	$\left(\dfrac{C}{N_0}\right)_u^{-1}$	$\left(\dfrac{C}{N_0}\right)_d^{-1}$	$\left(\dfrac{C}{I_0}\right)^{-1}$	$\left(\dfrac{C}{N_0}\right)_t^{-1}$
−13.0	−5.98	14.95	0.031623	0.198609	0.031989	0.262221
−12.0	−5.16	13.20	0.025119	0.164437	0.047863	0.237419
−11.5	−4.77	12.36	0.022387	0.150400	0.058043	0.230830
−11.0	−4.40	11.55	0.019953	0.138038	0.069984	0.227974
−10.9	−4.33	11.39	0.019498	0.135747	0.072602	0.227847
−10.8	−4.26	11.23	0.019055	0.133512	0.075301	0.227868
−10.5	−4.04	10.76	0.017782	0.127131	0.083898	0.228811
−10.0	−3.70	10.00	0.015849	0.117490	0.100000	0.233339

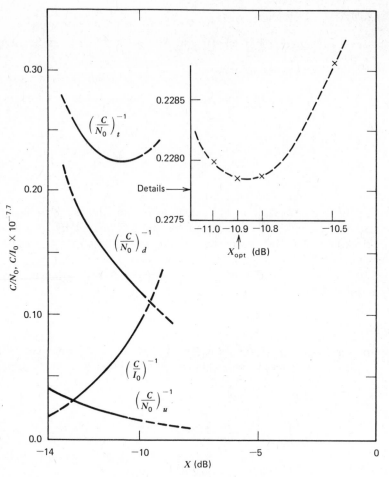

Figure 7.8 Graphs of $\left(\dfrac{C}{N_0}\right)_u^{-1}$, $\left(\dfrac{C}{N_0}\right)_d^{-1}$, $\left(\dfrac{C}{I_0}\right)^{-1}$ and $\left(\dfrac{C}{N_0}\right)_t^{-1}$ versus X, showing $X_{opt} = -10.9$ dB.

and the number of accesses is

$$\text{maximum } m = \frac{(C/N_0)_{ch}^{-1}}{\min_X (C/N_0)_t^{-1}} = \frac{10^{-6.25}}{10^{-8.34}} = 124 \text{ accesses} \qquad (7.34)$$

Use of Coding

The foregoing calculations assume power-limited operation; we check that there is no bandwidth limitation. For each access,

$$b = \frac{B}{m} = \frac{50\ 000}{124} = 403 \text{ kHz} \qquad (7.35)$$

is the available bandwidth, which is much more than needed. Only about $1.2R = 77$ kHZ bandwidth plus guard bands are required for 64-kbps BPSK, so bandwidth-expansion techniques can be applied to improve performance (lower BER) or to reduce $C/N_0)_{ch}$. For a rate $r = \frac{1}{2}$ code, the coding gain is assumed to be 2.2 dB and

$$\left. \frac{C}{N_0} \right)_{ch, coding} = 8.4 - 2.2 + 48.1 + 2.0 + 3.0 + 1.0$$

$$= 60.3 \text{ dB-Hz} \tag{7.36}$$

For this coding rate, 2 bits are transmitted for every 1-bit input to the codec. The information rate rR is the same, but the transmission rate R is doubled and the channel bandwidth is doubled.

Example 7.4

Using (7.34),

$$\text{maximum } m = 205 \text{ accesses, with coding} \tag{7.37}$$

The available bandwidth $b = 244$ kHz, still more than needed. In the bandwidth $B = 50$ MHz of this example, and for a product $bT = 1.2$, the available number of channels is $50,000/(1.2 \times 128) = 325$.

Use of Voice Activation

For the case of voice transmission, voice activation is applicable when large numbers of carriers are accessing the same repeater. **Voice activation** requires the detection of speech activity and refers to the control of the uplink by inhibiting carrier transmission during pauses in the speech. The resulting average transmit power is lower and the average satellite power is reduced for each channel.

Example 7.5

For a speech activity factor of 0.4, there can be $205/0.4 = 512$ channels, although only 205 accesses are supported by the $C/N_0)_t$ and only 325 are available in the repeater bandwidth B; 325 channels are assigned uniformly across the band, and each access has equal power in this example of a uniform network.

When voice activation is used on a specific channel, the IPs involving that channel vanish, because $A = 0$ and $J_n(0) = 1$ in the output voltage of (7.21). Hence the IM spectrum is affected in shape and level; for large numbers of accesses it is assumed that the shape is about the same. The C/I or $Z(X)$ is enhanced by the reciprocal of the activity factor:

$$Z(X)_{va} = Z(X) - 10\log(0.4)$$
$$= Z(X) + 4 \text{ dB} \tag{7.38}$$

It is necessary to iterate the foregoing design procedure if $Z(X)$ has an effect on the optimum X.

The design presented above has assumed the worst-case IM, which usually occurs for the center channel. There are also methods of estimating the average IM [8]. Staggered frequency plans help to reduce the IM when not all dominant IM terms fall on or near wanted carriers. Another variant on frequency plans is to make denser assignments toward the band edges and fewer assignments near the center. Amplitudes of the carriers can also be adjusted; a system with two carrier levels, alternately high and low across the band, has been studied [32]. With a uniform frequency plan, the amplitudes can be made progressively higher toward the band edges. For mixed services, with one large carrier and many small ones, the advantages of suppression of IM have been studied [6]. A simple design of digital SCPC is given by Gray [33]. All of these design approaches must be carefully analysed and verified, and operational factors should be considered [34].

REFERENCES

1   B. G. Evans and R. J. Kernot, "Satellite Communication System Employing Single-Channel/Carrier Frequency Modulation with Syllabic Companding", *Proc. IEE*, Vol. 122, July 1975.

2   S. P. Browne, "The Alaska Bush Communications Network", *Telecommunications*, Oct. 1979, pp. 61–64.

3   "Record of the ANIK-B User's Meeting", Dept. of Commun., Ottawa, Canada, June 1980, pp. 7–31.

4   J. L. Dicks and M. P. Brown, "Frequency-Division Multiple Access (FDMA) for Satellite Communication Systems", in *Proc. EASCON '74*, Washington, D.C., Oct. 1974, pp. 167–178. Also in ref. 34.

5   R. J. F. Fang, "Modulation Transfer from TDMA to FM Carriers in Memoryless Nonlinear Devices", in *Proc. Int. Conf. Commun.*, Chicago, June 1977, pp. 16.5.348–16.5.352.

6   C. Loo, "Calculation of the Suppression of Signals and Intermodulation Noise When Multiple Unequal Carriers Are Amplified by a TWT", *Can. Elec. Eng. J.*, Vol. 2, No. 4, 1977, pp. 29–32.

7   M. Abramowitz and I. A. Stegun, *Handbook of Mathematical Functions*, U.S. Dept. of Commerce, National Bureau of Standards, AMS 55, Washington, D.C., June 1964.

8   J. L. Pearce, "Intermodulation Performance of Solid-State UHF Class-C Power Amplifiers", *IEEE Trans. Commun.*, Vol. COM-25, Mar. 1977, pp. 304–310.

9   J. C. Fuenzalida, O. Shimbo and W. L. Cook, "Time-Domain Analysis of Intermodulation Effects Caused by Nonlinear Amplifiers", *COMSAT Tech. Rev.*, Vol. 3, Spring 1973, pp. 89–143.

10 C. Loo, "Calculations of Intermodulation Noise Due to Hard and Soft Limiting of Multiple Carriers", in *Proc. Int. Conf. Commun.*, St. Paul, Minn., June 1974, pp. 3A.1–3A.4.

11 G. Berretta and R. Gough, "Improvements in the Characterization of High Power Amplifiers in Multi-carrier Operation", *ESA Sci. Tech. Rev.*, Vol. 2, No. 2, 1976, pp. 103–113.

12 W. L. Pritchard, "Satellite Communication—An Overview of the Problems and Programs", *Proc. IEEE*, Vol. 65, Mar. 1977, pp. 294–307. Also in ref. 34.

13 G. Knouse, "Terrestrial/Land Mobile Satellite Consideration, NASA Plans and Critical Issues", in *Proc. Natl. Telecommun. Conf.*, Washington, D.C., Nov. 1979, pp. 30.1.1–30.1.6.

14 A. Lautier, A. Loevenbruck and J. Y. L'Honnen, "High Speed Data Transmissions via the SYMPHONIE Satellite", in *Proc. ICDSC-4*, Montreal, Oct. 1978, pp. 23–29.

15 D. Lombard, J. C. Bic and J. C. Imbeaux, "A Frequency Division Multiple Access Digital Data Transmission System Using Digital Speech Interpolation Equipment and a Viterbi Decoder", in *Proc. ICDSC-4*, Montreal, Oct. 1978, pp. 306–311.

16 F. H. Smart, "The ANIK C 90 Mb/s Digital Service", in *Proc. ICDSC-4*, Montreal, Oct. 1978, pp. 30–35.

17 C. D. Hughes, "Results of Measurements Made on an Experimental Link Simulating a 180 Mbit/s Satellite Channel", in *Proc. ICDSC-4*, Montreal, Oct. 1978, pp. 116–122.

18 T. Ishiguro et al., "NETEC System: Interframe Encoder for NTSC Color Television Signals", in *Proc. ICDSC-4*, Montreal, Oct. 1978, pp. 309–314.

19 A. Gatfield, "CODIT, A Digital TV System for Communication Satellites", in *Proc. INTELCOM '77*, Atlanta, Ga., Oct. 1977, paper 3.8.2.

20 A. M. Werth, "SPADE: A PCM FDMA Demand Assignment System for Satellite Communication", in *Proc. ICDSC-1*, London, 1969.

21 E. R. Cacciamani, "The SPADE System as Applied to Data Communications and Small Earth Station Operation", *COMSAT Tech. Rev.*, Vol. 1, No. 1, 1971.

22 P. Rossiter, "Systems Aspects of the Initial TELESAT THIN-ROUTE Satellite Communication System", in *Proc. Int. Conf. Commun.*, June 1973, pp. 8.6–8.9.

23 E. R. Cacciamani and K. Kim, "Circumventing the Problem of Propagation Delay on Satellite Data Channels", *Data Commun.*.

24 E. R. Cacciamani and A. S. Dohne, "A 1.3 Megabit Satellite Data Transmission Network Using SCPC", in *Proc. ICDSC-3*, Kyoto, Nov. 1975, pp. 337–343.

25 R. M. Lansey and M. R. Freeling, "RCA's Satellite Distribution System for Small-Dish Earth Terminals", IEEE, CH1352-4, 1978, pp. 362–367.

26 D. W. Lipke, D. W. Swearingen, J. F. Parker, E. E. Steinbrecher, T. O. Calvit and H. Dodel, "MARISAT—A Maritime Satellite Communications System", *COMSAT Tech. Rev.*, Vol. 7, Fall 1977.

27 R. Cooperman and J. Kaiser, "A Receive-Only MARISAT Terminal", IEEE, CH1352-4, 1978, pp. 356–361.

28 H. Teramura, Y. Yamazaki, T. Isago and S. Yamamoto, "Field Trial for the Digital Facsimile Service over the Satellite Circuit", in *Proc. ICDSC-4*, Montreal, Oct. 1978, pp. 312–318.

29 N. Hattori, "Digital Facsimile Communication over Satellite Links", *AIAA 8th Commun. Satellite Syst. Conf.*, Orlando, Fla., Apr. 1980, pp. 538–545.

30 PREPCOM/TECH/United Kingdom, INMARSAT Preparatory Committee, Technical Panel, May 1977.

31 E. J. Martin and R. D. Bourne, "INMARSAT: A New Venture in International Cooperation", *Satellite Commun.*, May 1980, pp. 22–23.

32 M. Horstein and D. T. LaFlame, "Intermodulation Spectra for Two SCPC Systems", *IEEE Trans. Commun.*, Vol. COM-25, Sept. 1977, pp. 990–994.

33 J. S. Gray, "Digital Signal Processing for Small Aperture Earth Terminals", IEEE, CH1352-4, 1978, pp. 990–994.

34 H. L. Van Trees (Ed.), *Satellite Communications*, IEEE Press Book, New York, 1979. Also distributed by Wiley, New York.

PROBLEMS

1 Show that even powers in polynomial expansions produce only DC and sums of pairs of frequencies.

2 Calculate the coefficients of all outputs for the polynomial model of (7.3) for an input of $A_1\cos\theta_1 + A_2\cos\theta_2$. Observe the effects of relative magnitudes of A_1 and A_2.

3 The trinomial coefficients are readily calculated by recursive relations. Verify the formula

$$(n; n_1, n_2, n_3) = (n-1; n_1, n_2, n_3-1)$$
$$+ (n-1; n_1, n_2-1, n_3)$$
$$+ (n-1; n_1-1, n_2, n_3)$$

where $n_1, n_2, n_3 \geq 1$.

CHAPTER 8

Time-Division
Multiple Access

This chapter begins with a formal definition of time-division multiple access (TDMA) and an historical perspective on TDMA developments. A typical block diagram and the functions of the blocks are outlined. TDMA technologies are presented, including extensive sections on synchronization and acquisition. Many of the TDMA systems are described, always with ample references. A hypothetical TDMA system is designed to illustrate procedures.

8.1 ADVENT OF TDMA

Time-division multiple access (TDMA) is the sharing of a satellite repeater by several Earth stations which transmit in bursts timed and interleaved so as not to overlap each other in the repeater. TDMA is a close counterpart to time-division multiplexing (TDM), so widely used in terrestrial communications. The distinguishing feature is that TDMA must achieve interleaving at a remote satellite that is in relative motion to all user stations. This definition of TDMA encompasses a variety of TDMA systems, which are discussed in this chapter; the definition excludes several concepts having time-domain transmissions in unstructured or nonregimented control, some of which are described in Chapter 10. TDMA can be applied with repeaters aboard synchronous or nonsynchronous satellites, with satellite switches (discussed below) and with on-board processing repeaters (some of which are identified in Chapter 15).

Some of the reasons for using TDMA are noted. For high-capacity, full-bandwidth TDMA systems, there is only one carrier present in the TWTA at any one instant in time, so there is no need for input backoff and the capacity is very high. Compared to FDMA, *increased capacity* is the main reason for development of TDMA. For INTELSAT IV repeaters operating with global beams and standard A Earth stations (30-m antenna, $G/T=41.7$ dB/K) in a network of 10 stations, FM/FDMA has typical capacities of 450 one-way voice channels, whereas PCM/PSK/TDMA provides 900 voice channels.

Another major reason for choosing TDMA is its *flexibility*. Flexibility is not only a significant benefit to large systems, but is often the key to the system viability in smaller systems. Nonuniform accesses pose no problem in TDMA, because time-slot assignments are easy to adjust. This applies to initial network configuration, assignments, reassignments and demand assignments. That all traffic is digital, integrated, buffered and multiplexed is a key feature of TDMA. The possibilities of channel coding are discussed in later chapters. Such flexibilities permit change and growth as networks evolve.

TDMA concepts were first published in 1968 [1], with patents and internal publications preceding, and a large number of papers appeared in the first International Conference on Digital Satellite Communications (ICDSC), held in 1969. INTELSAT sponsored the first round of TDMA research and development, with several of its signatories and contractors participating. By the time of the second ICDSC in 1972, domestic and regional systems were being studied and planned in the United States, Japan, Canada, France and West Germany. By 1975 and the time of ICDSC-3, the first commercially operational TDMA system was described [2] and a variety of technologies and applications were presented. ICDSC-4 in 1978 witnessed a maturing technology, not stagnant but cost-conscious. These and several other conferences amply illustrate the dynamic state of TDMA technologies. It was equally evident that cost impacts were a major factor in the apparently delayed introduction of TDMA; INTELSAT medium-capacity Earth stations found difficulty in justifying the costs of TDMA equipment and, in some cases, of new or additional uplink and downlink chains. In mid-1980, no commercial TDMA system was in operation, but planned were three in the United States, one in Europe and one in Canada. It is in such specialized applications that TDMA has its first major uses, where the satellite and terrestrial systems are regarded as a whole, where digital conversion is not an added penalty, and where connections to the existing facilities are minimal, well defined or nonexisting. Such is the brief, dynamic history of TDMA over its first 15 years.

Figure 8.1 illustrates a typical block diagram of TDMA equipment at an Earth station. TDMA technologies are in constant development by many institutions, so the terminology is neither common, consistent, nor static, but in this chapter it is hopefully representative and useful. Alternative but not always equivalent terms are given in parenthesis in the following paragraphs. Many terms are formally defined in later sections of this chapter.

In the transmit portion of the block diagram, there are **ports** [terrestrial interface modules (TIMs), data modules, group interface boards, etc.] for input of voice, data, facsimile or video services. For analog services, there must be conversion to digital form, before the port or in the port itself. The function of these ports is to provide electrical, mechanical and format interfacing with the sources of the digital services. Continuous, serial, synchronous (sometimes asynchronous) data are written into the associated **compression buffer** at the

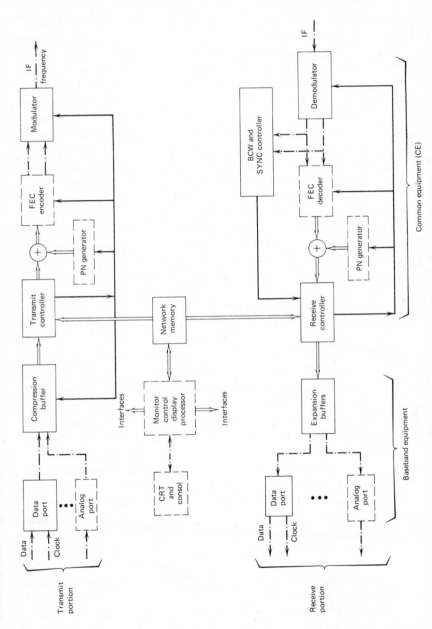

Figure 8.1 Block diagram of a TDMA system, showing typical blocks and signal flow. Dashed units are sometimes optional. ———, control; —·—·—, data; — — —, optional.

designated operating speed of the port. At high speed, the **transmit controller** (multiplexer) reads out the bursts of data, to form a multiplexed burst; the compression buffer is so named because it assembles the continuous, slow data bits for rapid output to the modulator during the burst for that station. The transmit controller also has control of the modulo-2 ($\oplus$) addition of a PN-sequence generator (scrambler) used in some systems. Note that one forward-error-correcting (FEC) coder can be inserted in the high-speed line or at each of the data ports as required. If a single unit is used selectively in the burst, the controller also has the required control. The controller provides for guard time, gates on the preamble generator and appends the postamble if needed. The composite burst is given to the burst modulator, which transforms the data stream into a modulated carrier. The controller gates the signal for transmission. The interface to the rest of the Earth station is usually at some intermediate frequency (IF) but can also be at radio frequency (RF).

In the receive portion, the **receive controller** (demultiplexer) provides a **gated period** (window), during which the demodulator processes the received signal; demodulation includes carrier recovery, symbol timing recovery and detection to recover the message symbols. The **synchronization** functions are detailed below; the major equipment include BCW (burst codeword) detectors and a sync* controller (burst synchronizer). The FEC decoder operates on the encoded portions of bursts and the PN-sequence is removed, all under control of the receive controller. The data contained in each received burst are demultiplexed and the high-speed sub-bursts are gated to the appropriate **expansion buffers**, where the data are read at low speed into the ports for output.

The controllers are now usually implemented in microprocessor or LSI (large-scale integration) form and are reminded to perform their functions at designated times by the **network memory**. In fixed configurations, the memory is programmed once; in systems having features of reconfiguration (e.g., monthly), variable assignments (e.g., manual or daily) or demand assignment (DA), the memory is usually in two parts: active and standby. The standby portion can be changed without interruption or interference to services.

A minicomputer or microprocessor is used to process operator commands, to display status, to process system requests (e.g., assignments, verifications) and can be used for gathering statistics and producing billing forms. Many of these functions are the same for TDMA and SCPC systems. The literature uses the terms **common equipment** (CE) or ground common equipment (GCE) to designate all equipment that is needed once per station and is not duplicated when more ports are added. Some TDMA systems have provision for **re-dundancy** of equipment, such as 1 : 1 for the HPAs and 1 : n (one spare for n operating units) for some subsystems. A monitor, alarm and control unit is used to detect failures and replace failed units by available units. The control to switch over is also provided.

---

*The short form "sync" is frequently used for the word "synchronization".

8.2 TDMA TECHNOLOGIES

Definitions

Timing Hierarchy

Figure 8.2 illustrates the essential features of a TDMA timing hierarchy, which is an ordered set of time durations or intervals. A **superframe** is a number of sequential frames organized to distribute system and network control or signalling information. There are other names and further extensions of this idea—levels of superframes, master frame, cycle. A frame time or simply **frame** is the time interval over which the signal format is established and then repeated indefinitely. A frame is subdivided into time **slots**, and a **burst time** consists of an integer number of slots. The literature uses frame, burst, slot and symbol terminologies somewhat interchangeably.* A **burst** is an accessing signal that occupies an assigned (number of) slot(s) in the frame. Because of the imperfect timing of bursts, a **guard time** is required between bursts to assure that no overlap occurs. A burst typically consists of a preamble, the message portion and in some coded systems a postamble.

A **preamble** is the initial portion of a burst and consists of a part for carrier recovery in coherent demodulating systems, a part for symbol-timing recovery (STR) [also called clock and bit-timing recovery (BTR)], a burst codeword (BCW) [frequently called a unique work (UW)] for burst synchronization, a station-identification code (SIC) and some housekeeping symbols. The housekeeping part may contain signalling bits, an order wire (OW) for voice or data, command and control signalling and error-monitoring symbols.

Example 8.1

For a binary BCW that is to be processed digitally, verify that the error rate for BCW is related to the error rate for the channel by these expressions:

$$P(\text{BCW error}) = \sum_{i=e+1}^{n} \binom{n}{i} p^i (1-p)^{n-i}$$

$$= 1 - \sum_{i=0}^{e} \binom{n}{i} p^i (1-p)^{n-i} \tag{8.1}$$

where n = number of bits in BCW
  e = number of bit errors allowed
  $\binom{n}{i}$ = binomial coefficient = $n!/(n-i)!i!$
  p = BER, bit error rate of the channel = $P(\text{bit error})$

This is the probability of missing the BCW and must clearly be very small for reliable performance of the station.

*It is common practice to use the signal name (symbol) in place of time duration of the signal (symbol time). This avoids the cumbersome use of expressions such as frame time, burst time, slot time and symbol time, but clarity sometimes requires these more exact forms.

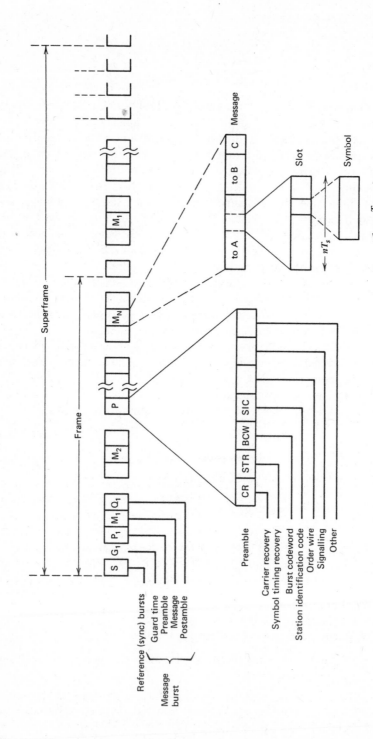

Figure 8.2 Timing hierarchy and typical frame format. Not all levels are always used.

234

The **message** (information, traffic) portion of a burst contains the desired data, time-multiplexed for all destinations of the burst, channel-coded and modulated onto the carrier. In some applications, a **postamble** is used for decoder quenching (initialization) for the next burst.

Proceeding down the hierarchy in Figure 8.2, a burst occupies a number of **slots** and is made up of **symbols**, which are selected from the **alphabet**, which is the collection of available symbols in the modulation type. The **symbol rate** or clock rate of a burst is the working rate of the modem and is the inverse of the symbol duration (symbol length) T_s. For convenient reference, a **word** is a structured set of continuous symbols. It is important to realize that these hierarchial terms are related to modems and accesses; however, we also need concepts that are invariant for different systems.

A bit, in the strict information-theoretic sense, is a measure or quantification of information content of a transmission. In digital communications, such absolute measure of information is usually neither possible nor desirable. (Who's information is it? Where is it to be measured?) The term "bit" has come to mean the apparent information in one or more symbols. Hence a binary symbol has 1 bit, a quaternary symbol (one of an alphabet of four) has 2 bits. In general, the number of bits per symbol are calculated as $\log_2 A$, where A is the **alphabet size**. For comparison of different systems of multiple access, modems and codecs, the **transmission rate** is the instantaneous bit rate of the signal. Only in binary systems are bit and symbol, transmission rate and symbol rate the same when viewed at the channel.

Efficiencies

There are several types of efficiencies which are considered in TDMA literature. **Frame efficiency** (eff) is the ratio of the portion available for messages to the total frame length. Equivalently, it measures the sum of all the guard times, preambles and postambles in a normalized form. It is useful in comparing frame formats for different synchronization schemes and modulations; it is used in selecting frame and buffer sizes. For a large class of systems it has the general form

$$\text{eff} = 1 - \frac{\left[S + \sum_{i=1}^{n} (G_i + P_i + Q_i)\right] T_s}{F} \tag{8.2}$$

where F = frame length, μs

S = number of symbols in synchronization bursts

G = guard time, μs

P = number of symbols in preamble

Q = number of symbols in postamble

T_s = symbol length, μs

$$= \frac{\log_2 A}{R} \tag{8.3}$$

A = number of symbols in the alphabet

R = transmission rate, Mbps

n = number of accesses

Of course, any other set of consistent units can be used. For efficient systems, G, P and Q should be small; G is controlled by synchronization, P is predominantly affected by the demodulation and signalling needs, and Q is entirely determined by codec considerations. It is also evident that F should be as large as practical, keeping in mind delay, buffer and demodulation constraints. For moderate n (30) and very large F, efficiency is very high and not sensitive to each of G, P and Q. It is noteworthy that spare slots do not contribute to frame inefficiency.

Transmission efficiency of a burst η is the ratio of useful message information to the total bits transmitted.

$$\eta = \frac{rM}{P+M+Q} \tag{8.4}$$

where $r =$ coding rate of the codec
$\quad$ $M =$ message symbols per burst
$\quad$ $P =$ number of symbols in preamble
$\quad$ $Q =$ number of symbols in postamble

By extension, the **transmission efficiency of a frame** is the weighted average of the burst efficiencies.

Another useful measure is **system efficiency**, defined as the ratio of the useful capacity (paying traffic) to the available capacity. Useful capacity is the amount of message traffic measured in bits per second. Available capacity is used in the information-theoretic sense of the channel capacity: for the Gaussian channel,

$$\mathcal{C} = B \log_2 \left(1 + \frac{E_s T_s}{N_0 B} \right) \tag{8.5}$$

where $B =$ channel bandwidth
$\quad$ $E_s =$ energy per symbol
$\quad$ $T_s =$ symbol duration
$\quad$ $N_0 =$ spectral density of white Gaussian noise

Clearly, system efficiency depends on traffic patterns, networking, multiple-access method, modulation and demodulation, coding and decoding. It is difficult to calculate and to compare dissimilar systems. However, there is a growing need for such comparisons.

TDMA Synchronization Methods

The discussions on perturbations and imperfections of geostationary orbits in Chapter 1 showed that there are very significant time-delay variations. In TDMA systems, several difficulties arise in synchronizing transmissions from

Table 8.1 Categories of classes of TDMA Synchronization

Category	Class
Packet satellite network	Random access + others (Chapter 10)
Open loop	Coarse sync
	Clock-controlled method
	Reference burst
Closed loop	Conventional
	Global-spot
	m-Sequence synchronization
Sync window	
Hybrid types	Ranging and extrapolation
	Centralized sync and ranging
	Open- and closed-loop selection

Earth stations and the satellite. Sync methods encompass approaches where each transmitting station determines sync, to where a station is provided with the necessary information.

A number of classes and subclasses of TDMA synchronization methods have been identified [3, 4]. There are now clearly four categories of synchronization methods—random-access, open-loop, closed-loop, and SS-TDMA window methods and another category of hybrid classes is described. In this section some classes are described. Table 8.1 summarizes the new classification; Section 8.4 gives details on some of these.

Random-Access Methods

Multiple access in the time domain can be accomplished through random access by each station. Although this is not true TDMA as just defined, this synchronization is included here for comparative purposes; several classes and systems are discussed in Chapter 10. In the basic form of random access no network timing is present; each station transmits bursts or packets as necessary at random and some bursts overlap. Random access is notably inefficient and variations have been devised for better efficiency.

Open-Loop Methods

An important category of sync methods has come to be known as open loop, and several classes are identified. **Open-loop** methods are characterized by the property that a station's transmitted burst is not received by that station—hence the synchronization loop is not closed and called open. The category merits attention because less equipment is required and because there are operational advantages.

Coarse Sync Without any active form of synchronization, it is possible to achieve accuracies from 5 μs to 1 ms through **coarse sync.** Based on approximate orbit parameters and free-running clocks, burst positioning has been done to 200-μs accuracy. Early TDMA trials [1, 5] used such methods to attain approximately synchronized bursts. Synchronization was then improved by closed-loop methods. These principles are used in military and maritime TDMA systems [6, 7]. Coarse sync is also applicable in TDMA data networks which have been proposed [8, 9]. For TDMA systems of the future which will be designed for data traffic only or for specialized applications, very long frames up to 200 ms can be used.

Although coarse sync is not applicable in efficient TDMA systems, this class of methods is important for its simplicity, independence and economy. Such methods are easily adapted for nonsynchronous satellites and orbiting platforms.

Clock-Controlled Methods   Conceptually, the next step from coarse sync is to have stable clocks that control burst timing. Every station works from a common timing reference based on frequency/time standards. Such **clock-controlled** concepts appear not effective by themselves, because frame efficiency and demodulation functions are only marginally improved.

Stable oscillators are available which can form the basis of nearly synchronous (also called plesiochronous) networks. Also, national standard highly stable sources can now be routinely but not yet operationally compared. Satellite **time transfers** have been carried out between the United States and Japan [10] and France and West Germany, and among Canada, the United States and France [11, 12] and Canada, France and West Germany in ongoing experiments [13]. Large synchronous networks (i.e., having symbol clocks synchronized) are now in place and will be extended to satellite systems.

Reference Burst   A very important class of open-loop sync involves the use of a reference burst. One or two can be used, but only one is in control at a given time and the other is available as a backup. A **reference burst** is a special preamble only and its purpose is to mark the start of frame with the burst codeword. The station transmitting the reference burst is called the **reference station**. The reference bursts are received by each station in the TDMA network and all transmissions are locked to the time base of the reference station. Note that the loop is open.

Reference bursts were conceived early in TDMA developments because such techniques are practical in providing accurate and precise timing information to remote stations. Since Earth station clocks are synchronized to the received clock, stable oscillators are not required. Since the reference bursts pass through the repeater, and since they usually occupy the same bandwidth as the traffic bursts, the train of bursts has information on doppler (for nonsynchronous satellites), time-delay variations due to satellite motion and channel characteristics. Although in open loop, a station can make use of this information to improve its access positioning. However, the reader should realize that only one repeater and a specific pair of uplink and downlink are included; there are difficulties in transferring such results accurately to other repeaters, other beams or stations in very different locations.

It is useful to note that reference bursts are the TDMA equivalent of pilot carriers in FDMA systems. Both are system overheads that should be kept to a minimum in duration and bandwidth, respectively; yet both are used effectively to improve performance and reduce equipment costs. We note that all forms of synchronization convey information and that we must devote some capacity to achieving synchronization.

Closed-Loop Methods

Closed-loop methods are important for their absolute accuracy and their high precision. The **closed-loop** category includes a number of techniques in which the transmitted signals are returned through the repeater to the transmitting station, the signals are compared, and the transmission is controlled by the result—hence the term "closed loop" and the description "self-locking." In most systems, the signals are bursts, which are called **sync bursts**; the time of arrival is measured relative to a time base; and early or late controls are provided to the transmitter. **Sync tracking** is this operation of periodic maintenance of closed-loop synchronization. Through superframe, frame and burst codeword designs, there is no ambiguity in the time of arrival of the return burst. The accuracy is absolute because the loop is closed, leaving no doubt as to the actual timing. Through careful signal design and processing, precisions of a few nanoseconds are possible.

Coventional Closed Loop   When every burst is received back at the transmitting station in a single hop, the most important type of synchronization scheme is indicated. This class was historically the first to be studied extensively [1, 5], was used in the TELESAT TDMA system [2, 14] and is to be used in the operational INTELSAT system in 1983 [15]. This class has also been called the INTELSAT type and classical TDMA synchronization.

Systems of this class include TDMA experiments sponsored by INTELSAT, the Japanese TTT System [16] and most domestic TDMA systems under development.

Global-Spot Class   A novel type arose out of the need to have TDMA synchronization for satellites with spot beams. In general, spot beams do not illuminate the transmitting station. Since the conventional approaches require each station to receive its own transmissions, the **global-spot** type was invented [17]. It is so identified because conventional closed-loop methods are used in the global beam and traffic is carried in the spot beams. A phase comparison method in the global beam was also proposed [18].

m-Sequence Class   A class of closed-loop TDMA sync has been called **m-sequence type**, because this technique applies the properties of maximal-length sequences. An **m-sequence** in its binary form is a train of ones and zeros which is simple to generate but has random properties, including longest length before repetitions [19]. The term "PN-sequence" (pseudo-noise sequence) is often used, although the two terms do not mean the same thing. By using wideband signals at low level and with excellent correlation properties, precise synchronization is possible with minimal interference to other signals. The concepts were applied to synchronization in the Defense Satellite Communications Systems (DSCS) [20]. Such methods are also important in acquisition, which is initial sync, treated below.

Sync Window Methods

For satellites with spot beams, connectivity among coverage zones can be achieved by having an array of programmed switches in the satellite; such methods have been called SS-TDMA (satellite switched), and also SS-TDMA/SDMA, (see Chapter 1 for a discussion of SDMA), or TDMA/SS. Earth stations must synchronize to a switching sequence being followed in the satellite. The window method is a generic form of such techniques, first proposed by Schmidt [21] and recently surveyed [22]. Table 8.2 lists some types of sync bursts for sync window methods.

A **switching satellite** consists of a number of transmitters cross-connected to receivers by a high-speed time-division switch matrix. The switch matrix connections are changed throughout the frame to produce the required interconnection of the Earth stations. A special connection at the beginning of the frame is the **sync window**, during which signals from each spot-beam zone are looped back to their originating spot beam zone; this forms the timing reference for all zones. Closed-loop synchronization can be established, but this is complicated by the necessarily short sync window.

Figure 8.3 shows a scheme for locking and tracking the sync window in the satellite switching sequence. A burst of two tones, F_1 and F_2, is transmitted by a single station. Only the portion that passes through the sync window is received back at the station. The basic concept is to measure and compare the received F_1 and F_2 sub-bursts, as illustrated. Although very narrow bandwidth and full power are used, there is need for digital averaging over many frames. The difference is used to control the position of the F_1/F_2 burst to a resolution

Table 8.2 Types of Sync Bursts for Use with Sync Windows

ASK sync burst
Coded ASK sync burst
PSK Sync burst
Coded PSK sync burst
FSK sync burst
Coded FSK sync burst
coded preamble + message + metric

where A = amplitude
P = phase
F = frequency

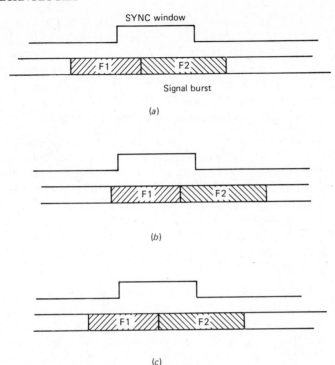

Figure 8.3 Sync window for SS-TDMA and use of bursts of two frequencies. (*a*) Signal burst too early; (*b*) signal burst too late; (*c*) synchronized to SYNC window.

of one symbol; this process is continually repeated in closed loop. The sync bursts to the network are slaved to this time base and the network is thus synchronized to the sync window and the switching sequence.

The reader should note that several repeaters are involved in SS-TDMA, that only a simple on-board function has been added and that a given link remains functionally the same as for other methods. The satellite repeaters do not initiate transmissions of their own, although the sync window is said to modulate sync signals. A TDM equivalent is a commutator.

Hybrid Types

The categories and classes of TDMA synchronization methods described readily adapt and combine to form new types. We note the evolutionary steps in random-access methods in Chapter 10. Closed-loop and open-loop methods can work together in sequential style in what are called ranging-plus-prediction methods.

Ranging and Extrapolation

With the predictability of the orbit and with due regard to propagation effects, the synchronization problem is reduced to accurate ranging to the satellite by each station. In some variations, the ranging and prediction are **sequential**. In an extrapolation scheme, ranges are predicted for future times and the station operates open loop mostly. Alternatively, an adequate model of the satellite motion can be constructed from real-time range measurement or from TTC (telemetry, tracking and control) ranging, and ranges are predicted from the model [23]. Such methods are distinguished from coarse sync in that there is a ranging function with synchronization as its purpose.

Centralized Synchronization and Ranging   The ranging and synchronization functions can be carried out **simultaneously** and in real time. The methods use forms of three-dimensional triangulation: the basic concepts are to make only necessary measurements of delays through the satellite with a minimum of three stations participating in ranging and to make this information available to all stations. Each station locks to a modulated sync burst, demodulates the data and uses its own coordinates to establish precise synchronization [24]. There are several variations possible for different geometries and accuracies can be selected during design.

Open- and Closed-Loop Selection   The foregoing classes were described as sequential or simultaneous in their application of ranging and prediction. There is another class which is **spatial** in the application of open- or closed-loop sync. In the same network, some stations can operate open loop but a few are closed loop. Medium-capacity systems can then be economical and yet flexible. TDMA applied to mobile applications is possible with reduced weight and simplicity. Also, an otherwise closed-loop system can accommodate open-loop accesses for stations with unknown locations.

Initial Acquisition Methods

Initial acquisition is defined as the process whereby a station is brought into synchronism with an operating TDMA network. Included in this concept is **reacquisition** by a station that has temporarily lost synchronism. There is a distinction from **startup**, which is a process by which a TDMA network begins to function, starting from a point in time at which no transmissions are present. Startup methods are specific to each form of sync and acquisition.

A basic method of initial acquisition can be associated with each of the synchronization classes; there are significant variations in each of the systems. Signals during initial acquisition can be at a high level and noninterfering or at a low level and possibly interfering. Levels are relative to normal transmission power according to the link equations.

Random Access

In random-access systems, by their very definition, no initial acquisition and no startup are required. When collisions of bursts occur, some systems randomize the transmission time of each access attempt, thereby reducing the probability of **permanent lockout**, which is repeated mutual interference by stations.

BCW Detection

For systems using reference bursts or sync bursts, several techniques have been developed based on precise detection of a **burst codeword** (BCW). A BCW, sometimes called a unique word (UW), is a selected set of symbols in the preamble serving to establish a known position in the burst. A station receives these frame reference bursts and establishes approximate timing. Open-loop systems continually monitor these bursts and base their transmission timing on them only. This has been called passive synchronization [3].

Orbit Prediction

Computer-prediction methods were first used for approximate acquisition [1] and extensions have been proposed [23, 25]. Orbit parameters or records of measured delays are stored for use during acquisition. Such schemes require computations to augment the stored information. Occasional station-keeping operations can cause problems and perturbations. For example, the thruster firings are constrained by factors other than the communications payload; the size and timing are somewhat random.

In closed-loop systems using high-level signals, the acquiring station transmits only the preamble in the middle of its assigned time slot. The burst is monitored by its receiver and when its own BCW is properly received in each frame, transmissions are moved to the start of the slot and a traffic portion is added. When its own BCW is missed for a specified number of frames, that station goes into the reacquisition mode.

m-Sequences

The same properties of m-sequences are used during initial acquisition as in synchronization tracking. A significant advantage is that the same hardware and similar software are used in both [20]. The INTELSAT specification calls for the use of m-sequences for initialization and recovery after significant loss of sync [26].

Wide and Narrow Carrier Pulses

Another acquistion technique [27] uses a wide carrier pulse followed by narrow carrier pulses which are slowly swept across the frame. All such pulses must be low-level signals. The wide pulse is several frames long and detections of the ending resolves the frame delay. The narrow pulse must pass through a short aperture gate (also called a window) in the receiving Earth station and fine resolution is determined. Envelope detection was proposed to keep equipment simple.

The method is directly applicable to closed-loop systems, either single hop or double hop, in which case a partner is needed. Of course, double-hop techniques are necessary for spot beams without switching. The method should also be usable in satellite-switched systems.

Sync Window Initialization

A survey paper by Carter [22] describes several methods of initial acquisition using sync windows and special signals.

8.3 ASSIGNMENT OF CAPACITY

In TDMA systems it is necessary to assign the required capacities to the slot in the frame. When a network uses only one frame, the task is simple enough, although demand assignment and its signalling remain development areas. In this section we address some of the more complex networking. Let there be n Earth stations, each within the same global or area coverage beam for the uplink, but in separate spot beams for the downlink, as illustrated in Figure 8.4. Assume that the satellite has multiple channels, m of them, each with capacity $\mathcal{C}$ units. (A unit might be one digital voice link, or 56 kbps or 10 Mbps.) Without loss of generality, let this capacity be $\mathcal{C}$ units per frame. The n stations form a network with up to $n(n-1)$ connections and the m channels are to be shared in frequency-hopping multichannel TDMA mode, called channel-hopping TDMA.

Network Constraints

Let the connection matrix for point-to-point traffic be **K**, with dimension $n \times n$.

$$\mathbf{K} = \begin{bmatrix} k_{1,1} & k_{1,2} & \cdots & k_{1,n} \\ k_{2,1} & k_{2,2} & \cdots & k_{2,n} \\ & & \vdots & \\ k_{n,1} & k_{n,2} & \cdots & k_{n,n} \end{bmatrix} \tag{8.6}$$

Element $k_{i,j}$ represents the traffic sent (originated) by station i destined for reception by the station j.

An **assignment** is an allocation of available capacity (to the uplink and to the downlink) to satisfy certain constraints of the network and its stations. Clearly, no Earth station can have more than $\mathcal{C}$ units of uplink or downlink capacity; this is because of the often tacitly accepted constraint that a station operates on only one channel at a given instant, because it only has one uplink and downlink chain. These constraints are stated as

$$S_i = \sum_{j=1}^{n} k_{i,j} \leq \mathcal{C} \qquad \text{all stations, sending} \tag{8.7}$$

$$R_j = \sum_{i=1}^{n} k_{i,j} \leq \mathcal{C} \qquad \text{all stations, receiving} \tag{8.8}$$

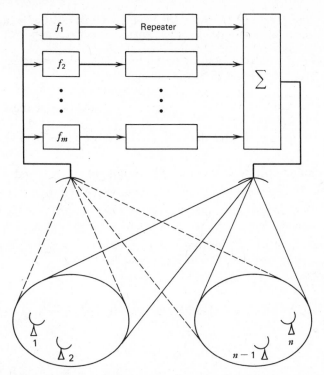

Figure 8.4 Network configuration, m channels, n stations in spot beams.

If the total of uplinks and total of downlinks are each less than $m\mathcal{C}$, an assignment can always be made. The two constraints are expressed as

$$\sum_{i=1}^{n} S_i = \sum_{i=1}^{n} \sum_{j=1}^{n} k_{i,j} = \sum_{j=1}^{n} R_j \leq m\mathcal{C} \qquad (8.9)$$

An **assignment procedure** is to allocate the required uplink and downlink capacities to the frames of the m channels. It will be evident that there is no unique assignment and that there are many equivalent.

Recursive Algorithm

It is clearly not always possible to use ad hoc assignment procedures. Formal procedures are being developed for specific applications. One particular recursive algorithm is presented here. The contents of this section are adapted from a recent paper by Acampora [28]. Define a **diagonal** of **K** as a set of nonzero elements, no two of which occupy the same row or column; let the **length** be the number of elements in such diagonals. Although often appearing as a diagonal of **K**, it may have any form.

Claim 1:   If (8.7) and (8.8) are satisfied and if (8.9) is satisfied with equality, a diagonal of length m always exists.

Claim 2:   Such a diagonal can be found to include all rows and columns that satisfy (8.7) and (8.8) with equality, if any.

A simple recursive algorithm is suggested by the proof of these claims.

1   Start with the connection matrix **K** and an empty assignment array for each channel.

2   Insert dummy traffic to satisfy (8.9) but not violate (8.7) and (8.8).

3   Sum all rows and columns, marking those which sum to $\mathcal{C}$.

4   Select the diagonal from the marked rows or columns and then select any others to make length m.

5   Make assignments in the first slot of each channel by writing down the indices of the elements. (Mark dummy traffic.)

6   Reduce by one each element of the diagonal.

The recursive nature of the algorithm is evident, since after one application of these steps,

$$\mathbf{K} \to \mathbf{K'}$$

$$\mathcal{C} \to \mathcal{C} - 1$$

and the reduced matrix satisfies the new constraints. The recursion continues with steps 3 to 6 until all assignments are made.

Example 8.2

The assignment algorithm is illustrated in Figure 8.5 for the simple case $n = 3$ stations, $m = 2$ channels and $\mathcal{C} = 12$. Uplink and downlink agilities are assumed possible and required by some stations.

Another constraint is often present: each station is permanently assigned to receive on one channel. The connection matrix is arranged into no more than m groups and this constraint is on the downlinks. In the case of spot beams, the groupings are dictated by the beam coverage.

Example 8.3

This example illustrates the application of the algorithm under all the constraints. There are $n = 7$ stations and the satellite has $m = 4$ channels, each with capacity $\mathcal{C} = 12$ units. The receive constraints are shown by grouping of the columns. A secondary condition is illustrated by reducing uplink frequency

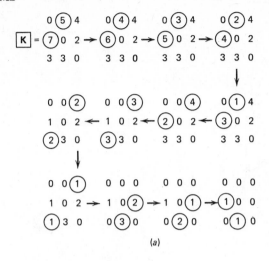

(a)

Slot	1	2	3	4	5	6	7	8	9	10	11	12
Station												
Up 1	1	1	1	1	1	1	1	1	1	2	2	2
Down	2	2	2	2	2	3	3	3	3	3	3	1
Up 2	2	2	2	2	2	3	3	3	3	3	3	3
Down	1	1	1	1	1	1	1	1	1	2	2	2

(b)

Figure 8.5 Details of the reduction of the matrix **K** for Example 8.2 and the corresponding assignments. (*a*) Matrix reduction; (*b*) traffic assignment.

agility requirements. Although a station has the capability, it is inefficient to transmit multiple bursts if not needed. The steps and recursions are detailed in Figure 8.6. Dummy traffic is shown as $+4$, $+2$, and $+1$ in the groups B, C, and D. A circled entry indicates selection on the diagonal, then reduction to zero.

8.4 TDMA SYSTEMS

This section describes some of the TDMA systems that have been developed and are now planned. The synchronization class and the initial acquisition type are indicated and special features are noted. Brief notes on implementation are given and selected performance figures are quoted from the references. Some outstanding advantages and a few problems are stated, with discussions as appropriate.

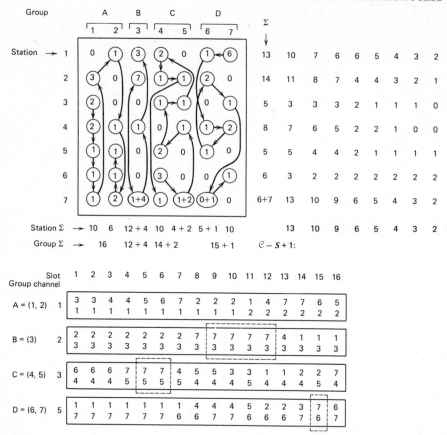

Figure 8.6 Reduction and assignment for Example 8.3; flow lines show the progress of the reduction and row sums are checked.

Experimental Systems

INTELSAT Experiments

MATE (Multiple Access Technology Experiment), the first TDMA experiment, was conducted by COMSAT in 1965 and 1966 [1, 29, 30]. Burst communication was shown to be feasible and guard times of less than 100 ns were demonstrated. Specifically, the BCW approach was proven and practical burst synchronizers were built. The frame length was 125 μs and there were three accesses.

MAT-1 or TDMA-1 had operational features for testing, including a capacity of 700 voice channels which could be demand assigned (defined in Appendix F) and reallocated during operation. It was the first 50-Mbps system and the experiment established many basic TDMA principles [5, 31].

In TDMA-2 [32], new concepts were investigated in signal processing and control equipment, such as digital speech interpolation (DSI; see Appendix E), burst formatting, minicomputer control and an FDM supergroup codec. Extensive information on INTELSAT experimental systems is to be found in Refs. 33 and 34.

SMAX

In the satellite multiple-access communication system (SMAX) [35, 36], a three-stage process is used to achieve synchronization. In step 1, a reference burst is processed to synchronize a coherent clock in the local station using conventional phase-locked-loop techniques. In the second step, the local station transmits its own burst, receives and compares it with the reference burst; there are also frequency and phase control loops. Initial acquisition is accomplished with continuous low-level m-sequences, which are reset at frame periods. Majority decision control is used in step 3 for frame sync. A guard time of one symbol is necessary. A symbol acts as a reference for differentially coherent detection and the next 12 bits complete the preamble. In this system, bursts were of fixed length and three stations were used. Various methods of channel switching and demodulation were analysed [37] and tested in field trials with ATS-1 (Application Technology Satellite) in 1968 and ATS-3 [38].

TTT

TTT is short for a TDMA system with time-preassignment and TASI (time-assigned speech interpolation) features [16, 39]. With TASI and operation at 50 Mbps, the system had a capacity of 1400 voice channels (8-bit PCM). A small computer controlled the initial acquisition, burst and channel assignment and also processed order-wire channels.

TDMA System I

The German TDMA System I and related research were described in 1971 [40]. The system has selectable rates of 100 Mbps and 50 Mbps, the latter having been used in field trials with INTELSAT III for transmission of TV signals [41].

Prototype TDMA

The INTELSAT prototype TDMA system, as it is called, is based on the preceding international studies and experiments and was specified by the TDMA working group [33, 42]. The system uses closed-loop, conventional synchronization with aperture gating; low-level initial acquisition followed by preamble placement in the assigned slot; reacquisition if required; and replacement of reference stations.

The system requirement for a reliable reference burst is critical, hence several stations assume this function, according to plan or in case of failure of the active reference station. Switchover is done automatically within prescribed limits, so as not to disturb the traffic flow.

The minimal guard time is 50 ns, so the prototype system uses three symbol times. The transmission rate is 60 Mbps and specified to be convertible to 120 Mbps. BCW detection was specifically reported [29, 43]; 20 bits for BPSK and 10 symbols for DCPSK (4-phase) are used. Other significant features are transponder hopping and use of FDM supergroup codecs.

NTT System

Nippon Telephone and Telegraph (NTT) researchers have reported a system with novel features [25]. Initial acquisition is by range prediction, based on stored information of measured ranges at 2-min intervals for the previous 2 days. Synchronization is by conventional means with provision for clock recovery from the sync burst, to have a synchronous network. The guard time is reported as 10 ns or one symbol; the modems operate at 100 Mbps and use BPSK for simplified detection.

The novel aspects include operation at 30/20 GHz, direct digital interface (DDI) at 1.544 Mbps with pulse stuffing and a switched routing system at the Earth stations.

CENSAR System

The centralized synchronization and ranging (CENSAR) system is experimental and was tested in 1976 and 1977 by Nuspl in Canada [44, 45]. Designed for spot-beam applications, synchronization is by closed-loop ranging with three cooperating stations in two-hop delay-lock loops. The central control station makes three independent range measurements, determines the four ranges to the satellite, and broadcasts the necessary portion of this information to all stations in the network. Using this information and knowing its own position, any station in the network calculates its own delay or range to the satellite. These user stations operate open loop.

The experimental system was designed for the particular constraints of the CTS satellite (also called HERMES) and available Earth stations. The control station transmitted control data on the link delays and rates of change, together with frame and bit timing, control and status signals, and an order wire; the rate was only about 500 bps and the sync burst carried the data. Each ranging station sent back a ranging burst at a time specified by the control data for that station. Sync bursts consisted of 31-bit BPSK signals at a transmission rate of 65.5 Mbps in a frame of 125 μs.

The significant results of the experiment are that centralized control of closed-loop, open-loop sync is feasible, that startup of the sync network takes less than 10 s with no prior information about the orbit, and that sync tracking is achievable to about 20 ns. This was a synchronous system: a network clock was distributed to each station.

TDMA Test Bed

INTELSAT has sponsored the development of a 120-Mbps TDMA test bed [46] to enable systematic test and evaluation of system operation in an INTELSAT V environment. Three TDMA terminals and test units were constructed. In recent tests, there was particular interest in bit-error rate (BER) and BCW error rate, also called UWER.

TDMA Field Trials

INTELSAT authorized a series of field trials with participation by five signatories [47].

Advanced TDMA System

To investigate problems of TDMA operation in the 30/20-GHz bands, Japanese researchers built a new type of TDMA system. There were comparisons of operation in this K-band and the C-band, using the CS satellite [48].

Operational Systems

TELESAT TDMA System

TELESAT Canada initiated the first commercial TDMA system in May 1976 [2, 14]. Closed-loop, conventional syncronization was used. The frames are 250 μs and the transmission rate is 61.248 Mbps; the modems are QPSK.

To achieve compatible performance in each direction of transmission in this two-access system, a rate-$\frac{3}{4}$ codec is used in the hop from Allan Park, Ontario, to Harrietsfield, Nova Scotia. The burst from Harrietsfield consists of a 12-bit guard time, a 96-bit preamble, 6435 information bits and 1 trailing bit. The burst from Allan Park consists of the same guard time and preamble (uncoded) followed by 8580 (6435 $\times \frac{4}{3}$) information bits and than a postamble of 80 symbols. The design of the burst modem, the special mastergroup PCM codec [49] and other equipment were described in Ref. 50.

This Canadian domestic service for Teleglobe Canada provided 400 two-way voice channels at a noise level of 2700 pWOp and at a voice-channel bandwidth of 3.1 kHz. The system availability is reported as greater than 99.98% with the noise objective met for more than 99.9% of the time. The system is not now in operation, but not for technical reasons.

MARISAT, TDMA Portion

The MARISAT* system [7] is a maritime mobile satellite system operating in C-band for links with shore stations and in L-band for links with ships. Voice, telex and request services are provided. Whereas the accessing plan is SCPC,

---

*MARISAT is a consortium of COMSAT General Corporation, ITT World Communications and Western Union International.

the ship-to-shore link uses TDMA for narrowband data. Synchronization is derived from the TDM continuously broadcast channel in the shore-to-ship link; a 20-bit BCW is used. The transmission rate of the TDMA bursts is 4800 bps. The guard time is 198 bits for this open-loop, reference burst class. CR, BTR and BCW take 109 bits and the remaining $12 \times 6 = 72$ bits are data, supporting 50-bps telex. The frame length is 1.74 s. This SCPC/TDMA plan can have up to 22 such TDMA channels.

DYNAC Systems

Dynamic Network Assignment Communications (DYNAC) [51] is a family of medium-capacity TDMA systems providing integrated voice and data services at rates from 268 kbps to 2 Mbps. In a STAR network, a TDM carrier and TDMA carriers form a pair; remote-to-remote circuits are by double hop through the central site. In a MESH network, full connectivity is available, with one control site. A DYNAC Earth station uses open-loop methods during initial acquisition and closed-loop in the tracking (steady-state, acquired) state. Frame lengths are selectable from 10 to 250 ms. The transmission rate is 2.048 Mbps and QPSK burst modems are used. Codecs are field selected at rates $\frac{1}{2}$ and $\frac{7}{8}$. There is provision for expansion to multiple systems by IF hopping, so that a station can transmit and receive multiple TDMA bursts per frame on different carrier frequencies.

CT-2000 TDMA

The CT-2000 [52] supports voice, data, facsimile and video services in a flexible, low-cost system of terminals for installation in user Earth stations. Frames are preselectable from 1 to 128 ms and the burst transmission rate is preselectable in options ranging from $1.5 \leq R \leq 15$ Mbps to $1.5 \leq R \leq 120$ Mbps. Burst lengths are preselectable from 1 to 348 channels (voice equivalent) and burst positions can be positioned anywhere in the frame. There are two reference bursts per superframe. CR + STR range from 16 to 64 symbols; BCW is 16 symbols. Modulation is differentially encoded coherent QPSK with spectral conditioning by means of a PN-sequence of length $2^{15} - 1$. The reported options are demand assignment, error control by FEC or ARQ, and carrier frequency hopping. CT-2000 went operational for the *Wall Street Journal* in mid-1980.

Planned Systems

INTELSAT Operational TDMA

INTELSAT has decided to have TDMA/DSI operating in 1984. The TDMA Working Group has prepared specifications [15].

ESA Computer Data Network

In a computer data network being designed [9], TDMA will be used with open-loop, coarse synchronization; the transmission rate is 4 Mbps and the frame efficiency is specified as 0.90; the following example illustrates the calculation of the frame length.

Example 8.4

$$\underset{\text{open loop}}{eff} = 0.9 = 1.0 - \frac{(3 \times 360 + 116 \times 80)/0.25 + 116 \times 240}{4000 \, F},$$

$$F_{\text{OL}} = 173.2 \text{ ms}$$

and

$$\underset{\text{closed loop}}{eff} = 0.9 = 1.0 - \frac{116 \times 5 + 116 \times 240}{4000 \, F}, \qquad F_{\text{CL}} = 71.0 \text{ ms}$$

Advanced WESTAR (AW) SS/TDMA

This system is to be operational with the K-band (14/12-GHz) payload of the Advanced WESTAR satellites [53]. The design includes a 4×4 satellite switch and four channels, each with a capacity of 250 Mbps; full interconnectivity is possible.

The beam coverages are illustrated in Figure 8.7a and the AW K-band communications payload is shown in Figure 8.7b. The four channels support seven beams, including, WEST, CENTRAL, EAST and EAST-SPOTS. New York and Miami spot beams are coupled at RF, whereas Los Angeles and San Francisco spot beams are coupled with the WEST beam at IF in the satellite. The switching sequence (mode) is discussed in Ref. 53; only reference stations use the F_1/F_2 burst through the sync window to lock to the satellite clock.

The TDMA equipment accommodates large ($112 \times T_1$), medium and small ($1 \times T_1$) capacities in a flexible, cost-effective design. The frame is 125 or 750 μs and a choice between QPSK and offset QPSK is discussed. (T_1 is a North-American hierarchial rate at 1.544 Mbps.)

In addition to the dual-band C/K operation, specialized spot beams and the 4×4 switch, other advanced features include the use of hard limiters prior to the TWTA. The all-digital links have an input dynamic range of 15 dB (-78 dBW/m² nominal, $+3$ dB to -12 dB). Frequency reuse is by vertical/horizontal and circular (cw/ccw) polarizations. Redundancy is planned for the high-capacity stations.

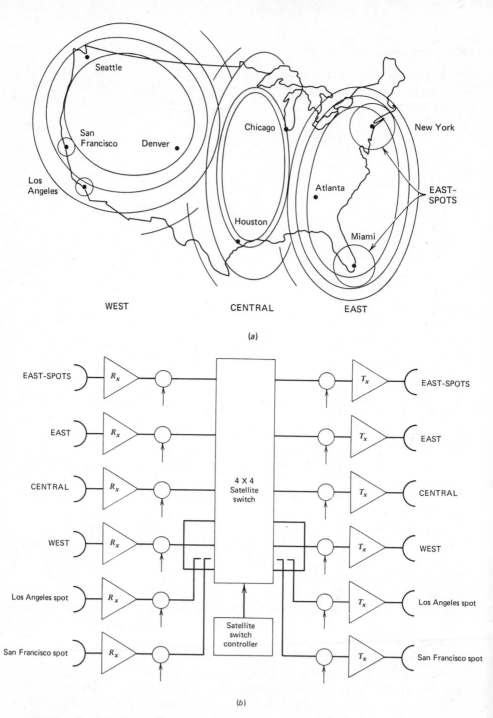

Figure 8.7 (a) Antenna coverages for Advanced WESTAR; (b) communications payload, showing seven beams and four channels through a 4×4 microwave switch matrix.

254

SBS TDMA

Satellite Business Systems (SBS)* in the United States is establishing a domestic satellite system and Earth stations operating at 14/12 GHz to provide private switched communications networks [54]. The Federal Communications Commission (FCC) of the United States approved the application in February 1977 and operations started in March 1981.

The expansive features of this system concept are noteworthy. Fully integrated voice, data and image services will be provided at a variety of data rates from 2.4 Kbps to 6.3 Mbps. Capacity will be fully demand assigned by a centralized management facility. TDMA, all-digital modulation and optional FEC will assure maximum flexibility. Earth stations will have 5- and 7-m antennas, the selection depending on the location in the coverage pattern.

Some further details illustrate the specific technologies being applied. Networking will be simplified and tailored to meet customer needs for geographical distribution, traffic mix, traffic intensity, quality of service and grade of service. The burst modems are to be QPSK type and will have a transmission rate of 43 Mbps, operate down to 1×10^{-2} BER but have 1×10^{-4} or better BER 99.5% of the time. Details on synchronization, acquisition, frame structure and the link equations are not known at this writing.

SLIM TDMA

A medium-capacity (3 Mbps) SLIM TDMA [55, 56] system is in field trials with the Canadian Satellite ANIK B; this TDMA system is compatible with other carriers in the repeater channel. Data ports operate at multiples of 2.4 to 168 kbps, at 32 kpbs for delta codec voice, and at 1.544 Mbps for simplex T_1 and slow-scan video. In open loop, a station has a 200-μs guard time and in closed loop it has 5 μs. The frame length is 20 ms and slots are preselected by an assignment. There is provision for FEC coding on selected sub-bursts and burst modem is QPSK. The baseline performance is specified as 1×10^{-4} with a margin of about 3.5 dB. Coded data will have 1×10^{-7} P(bit error) 99.9% of the time.

Scanning-Beam SS-TDMA

When a scanning beam is added to a switching satellite, a very efficient (spectral, orbital) and flexible TDMA system results [57]. The pair of scanning beams is associated with an uplink and a downlink, with both beams electronically steered (beam-formed) independently. For the example discussed, 100 locations are covered in 20-μs slots. The capacity of each beam is 600 Mbps through the use of QPSK in 500-MHz bands.

The preamble is selected at 67 symbols. Indicated frame lengths are 25 ms for small stations, 5 ms for medium, 1 ms for large and 250 μs for trunking

*SBS is a partnership sponsored by Aetna Life & Casualty, COMSAT General Corporation and IBM.

stations. Open-loop startup (cold start) and closed-loop sync tracking are planned. Demand assignment is to be used; the paper describes a signalling scheme that is centrally controlled and very efficient.

MAROTS, TDMA Portion

In the 50-kHz channels of this SCPC system, there is TDMA operation. The transmission rate is 9.6 kbps with 30 shore station bursts per frame of length 1.3 s. For the ship-to-shore links, the rate is also 9.6 kbps, but the frame length is 2.6 s to provide 120 telegraph channels. Guard times range from 40 to 23 ms, depending on the use of coarse synchronization by the ships.

8.5   TDMA SYSTEM DESIGN

This section presents the requirements and preliminary design for a hypothetical network for a computer utility; let us call it COMPUTIL for easy reference. The requirements and design procedures are selected to be instructive and to illustrate several features of TDMA systems. COMPUTIL has not been implemented, nor is it likely to be.

Requirements

The basic requirements are for a worldwide network of circuit-switched connections at rates from 2.4 kbps (for a large number of stations) in multiples of 2.4 kbps to 3 Mbps (one broadcast channel), with a total transmission rate of 4 Mbps in each zone; to provide for easy interconnection to other systems, a packet-switching network is to be included in a broadcast slot of the TDMA system. Let the BER performance be specified as 10^{-6} for each data link and suppose that the availability is to be 99% of the time.

This data network will support computer-to-computer traffic only and has no response time or setup time constraints. (This is unrealistic but reasonable for a first design.) Station capacities are to be flexible and grade of service is to be selectable for better than the nominal performance.

Some further guidelines are to provide long-term flexibility and growth potential. We seek a design that permits upgrading to higher-speed modems, which will have fixed and reconfigured assignments easily made in a first phase, and which will have demand assignment in a second phase. Consider SS-TDMA as a possibility at some future date.

The network and each station are to be commercially viable (but no cost analysis will be done here), so cost-effective designs are to be selected. Only presently available technology and components can be used, because the hypothetical requirements include operation of part of the network in one year.

Selection of Major Subsystems

Without much discussion, we arbitrarily select the major system parameters and subsystems. Direct-to-urban locations in COMPUTIL favor selection of operation at 14/12 GHz. BPSK modems at 4 Mbps are selected for economy and availability; QPSK will be used for the second phase. Error control will be with FEC for those stations wanting or needing it; a rate-$\frac{1}{2}$ system is chosen and the equipment is optional at any station. ARQ techniques are to be used for the whole network. Since DA is also in the future, the required signalling for selection of FEC and DA is to be via RequesT (RT) and AssignmenT (AT) bursts; the capacity for each is chosen as 2.4 kbps, with more capacity easily available by reassignment.

Link Calculations

The transmission rate is $R = 4$ Mbps and the required E_b/N_0 is 11.5 dB for a BER $= 10^{-6}$ with a 1.5-dB implementation margin. The bandwidth required is $4 \times 1.25 = 5$ MHz, to allow for an adequate spectral utilization efficiency of 0.8 bit/Hz.

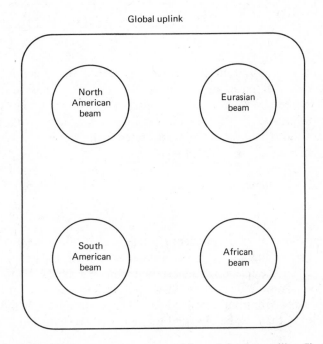

Figure 8.8 COMPUTIL antenna coverage with an Atlantic satellite. Similar coverage in Pacific, Indian regions. Uplinks are in a global beam and downlinks are in spot beams.

Table 8.3 Link Calculations for COMPUTIL

Required	E_b/N_0	11.5	dB
Transmission rate	R	+66.0	dB-Hz
Implementation	M_i	+1.5	dB
Fading	M_f	+3.0	dB
		82.0	dB-Hz
Antenna loss	L_{ANT}	+0.2	dB
Downlink	L_d	+205.8	dB
Receiver gain	$-G_r$	−57.4	dBi
System Temperature	T	+24.0	dBK
Pointing	L_{ptg}	+0.4	dB
	k	−228.6	dBW/K-Hz
Satellite	e.i.r.p.)$_s$	26.4	dBW
Small signal gain	G_{ss}	−155.0	dB
Uplink	L_u	+206.8	dB
Satellite	e.i.r.p.)$_e$	78.2	dBW
Transmit gain	$-G_t$	−58.7	dBi
Waveguide	L_{wg}	+0.7	dB
Pointing	L_{ptg}	+0.3	dB
Backoff	BO_{HPA}	+3.0	dB
Power for HPA	P_{HPA}	23.5	dBW

We assume the spot-beam zone coverages shown in Figure 8.8 for the downlinks and global beam for the uplinks. Under the constraints above, typical but hypothetical receivers are assumed to be available, since repeaters are not channelised at 5 MHz in the satellite. The stations will have the required filters and the repeaters will be operated with sufficient input backoff. Table 8.3 summarizes the link calculations.

Burst and Frame Formats

Network Size

For a respectable 0.9 frame efficiency, there can be $0.9 \times 4000/2.4 = 1500$ accesses at 2.4 kbps; this we judge as excessive and would require closed-loop operation. For an assumed average station capacity of 24 kbps and 0.9 efficiency, there can be 150 accesses. Also, there can be 7 accesses at 504 kbps or 1 access at 3.024 Mbps plus a few others. These features clearly indicate the size and flexibility that can be obtained for COMPUTIL by using TDMA.

Example 8.5 BCW Selection

For this feasibility design, we select a BCW of 31 symbols from an *m*-sequence and show its autocorrelation (truncated, not cyclic) in Figure 8.9. Its performance is analysed as follows for digital BCW detection by applying (8.1):

$$P(\text{BCW error}) = \sum_{i=3}^{31} \binom{31}{i} p^i (1-p)^{31-i}$$

$$= \binom{31}{3} p^3 (1-p)^{28} + \binom{31}{4} p^4 (1-p)^{27} + \cdots$$

$$= 4495 p^3 (1-p)^{28} + 31{,}465 p^4 (1-p)^{27} + \cdots$$

$$= 0.003392 \tag{8.10}$$

$$0.00023987$$

$$0.0000130838$$

$$+ \cdots$$

$$\overline{0.00364}$$

where $e=2$ for a design threshold of 26 and $p=0.01$ for a bad channel.

More than 26 of 31 symbols must be correct, equivalent to zero, one, or two errors. This BCW detection subsystem is to survive if the channel degrades to $p = 10^{-2}$, for which $P(\text{BCW error}) = 0.00364$, an acceptable performance. To reduce effects of malfunctioning stations, transmission is not allowed unless BCW trains have arrived correctly (details would be specified).

The BCW could also be falsely detected. This could be caused by noise or by a data pattern that was wrongly positioned and closely similar to the BCW. These events are unlikely and are further reduced by the use of an aperture gate.

Example 8.6

An aperture gate of 31 ± 4 is used and the probability of false detection is calculated.

$$P(\text{BCW falsely detected}) = P(\text{correlation} > \text{threshold} | \text{no BCW})$$

$$= P(31 \text{ bits match at } \pm 4, \pm 3, \ldots, 0)$$

$$+ P(30 \text{ bits match at } \pm 6, \ldots, 0)$$

$$+ P(29 \text{ bits match at } \pm 8, \ldots, 0)$$

$$= (9 \times 1 + 13 \times 31 + 17 \times 465) 2^{-31}$$

since $p = 1 - p = \frac{1}{2}$ for random data bits.

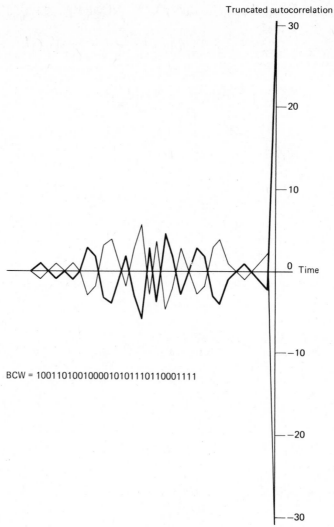

Figure 8.9 BCW of *31* symbols and its truncated autocorrelation.

Frame Format

A BroadcasT (BT) burst for packet-switched service is arbitrarily given a capacity of 24 kbps. Since only the gateway stations are frequency-agile, there are four such channels: BT1, BT2, BT3 and BT4. Two synchronization bursts are used from a gateway station and its standby, one pair for each zone; guard times of twice the nominal are to be used before and after these bursts. The BT, RT, AT and sync bursts are kept separate for more flexibility, although it would be more efficient to assemble BT and AT with S1 and have one access.

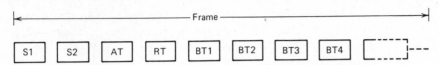

Figure 8.10 Selected frame format for COMPUTIL.

Open loops and closed loops have been compared: for simplicity, operational features and cost factors, let us choose open loop. A guard time of 200 μs is required by orbital and network geometry. The frame format is shown in Figure 8.10 for one of the zones.

Example 8.7 Frame Length

The goal of 0.9 efficiency is calculated for 150 accesses, open-loop synchronization, preambles of 60 symbols at 0.25 μs/symbol and no postambles.

$$0.90 = 1 - \frac{3 \times 400 + 5 \times 200 + 150 \times 200 + 158 \times 60 \times 0.25}{F} \tag{8.11}$$

We solve for $F = 345$ ms but choose $F = 400$ ms for convenience.

There are 1,600,000 symbols per frame; at 2.4 kbps, there are 960 bits/burst, which is the slot size for the message portions. The frame efficiency will be reduced when some stations use coding and postambles. Figure 8.11 illustrates the range of frame efficiencies for the network.

Assignments

This network uses four channels in four zones, with only gateway stations being frequency agile to transmit to all zones. Of course, the sync bursts might

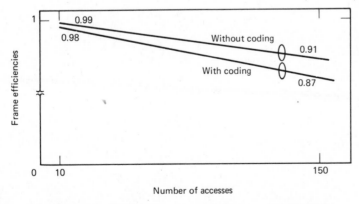

Figure 8.11 Possible frame effeciencies for COMPUTIL.

as well start the frame. It seems desirable to assign the broadcast slots next, since each station is expected to receive these. Similarly, the random access by stations needs a unique slot so that any station can make a request and not be occupied with another transmission. The assignment slot is given to one station in each of the beams and the receivers must be free.

The connection matrix **K** might appear as shown in Figure 8.12; the submatrices together with the main diagonal of **K** represent the connections within the zones and entries are nonzero; all other entries are zero, except as shown for the gateway stations. This matrix assumes that connections from local zone to remote zones have been entered as double-hop connections. **K** would be reduced for the gateway stations first and the submatrices would be reduced and assigned separately.

The assignment table is easy to reconfigure without using the RT and AT facilities, but with slow response time. For example, in phase 1 the requests could be via the packet-switched system or outside the network. The assignments would be figured out at a central location and then transmitted to each station. Zonal assignments could also be made. When DA is added, RT and AT would assure rapid response. There are many ramifications beyond the scope of this example.

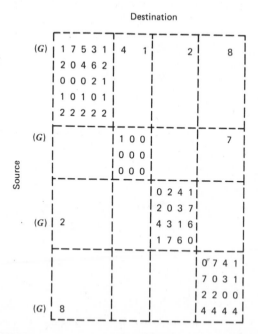

Figure 8.12 COMPUTIL connection matrix. Blanks in the matrix are 0. Many stations operate in one beam, but only gateway stations (G) access all four channels.

Block Diagram

Figure 8.13 shows the essential elements of the block diagram of a COMPUTIL station. The significant portion of this diagram, compared to Figure 8.1, is the baseband portion. Since the transmission rate is only 4 Mbps, the equivalent byte (8-bit) rate is 500 kbytes/s and the word (16-bit) rate is 250 kwords/s. This is one word every 4 μs, which is quite feasible by computers of modest capability. The buffers are therefore part of the computer. Similarly, the RequesT and AssignmenT functions are in the computer and the signals are routed between it and the burst handlers. The data flow control could be implemented by interrupt handlers in the computer.

The functions of the controller are simplified. It looks to the memory for assignments of start-burst and stop-burst events, for encode flags if the station has codecs, for frequency assignments in agile stations and other status bits. The controller sends appropriate timing signals to the burst handlers (format and unpack), to the channel select (with sufficient advance) to the codecs, to the modems and to the buffer controller in the computer. The controller derives its timing from the sync unit, which delivers start-of-frame, start-of-burst and clock signals.

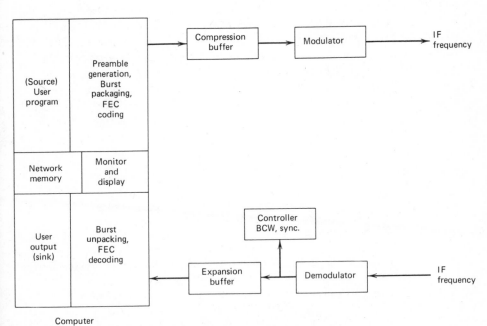

Figure 8.13 Block diagram for a COMPUTIL station.

Discussion

A few general comments can be made. In this network, the majority of the stations can be upgraded to operate closed looped, with some additional hardware to look for a station's own BCW and some software to process this measurement and control the transmission. Such open-or closed-loop operation could be selectively introduced, with proportional improvement in frame efficiency. The channel assignments would take into account the two guard-time values.

Another improvement in capacity and utilization efficiency is to introduce QPSK modems. Compatibility with existing stations could be obtained by having selectable modems, BPSK at 4 Mbps and QPSK modems at 8 Mbps, with some compromise in filter designs. Similarly, modems and codecs could be demand assigned with only a small signalling capacity.

In a third phase, when a switching satellite is available, more network connectivity could be obtained by using a 4×4 switch (MSM); there would need to be additional equipment to operate with the sync window, for the gateway stations only.

8.6   CONCLUDING REMARKS

This chapter presents a broad overview of TDMA and includes numerous references. The reader will find an interesting discussion of TDMA state of the art for 1977 in a paper by Dill [58]. Articles in the Special Issue on Satellite Communications [59] of *IEEE Proceedings* and the Special Issue on Synchronization [60] of the *IEEE Transactions on Communications* are also useful. A report by CCIR on methods of modulation and multiple access is recommended [61]. The many reprints and introductions by Van Trees in a recent book [62] are recommended; there is also an extensive bibliography.

REFERENCES

1 T. Sekimoto and J. G. Puente, "A Satellite Time-Division Multiple-Access Experiment", *IEEE Trans. Commun. Technol.*, Vol. COM-16, Aug. 1968, pp. 581–588.

2 R. K. Kwan, "The TELESAT TDMA Field Trial", in *Proc. ICDSC-3*, Kyoto, 1975, pp. 135–143.

3 P. P. Nuspl, K. E. Brown, W. Steenaart and B. Ghicopoulos, "Synchronization Methods for TDMA", *Proc. IEEE*, Vol. 65, Mar. 1977, pp. 434–444.

4 E. A. Harrington, "Issues in Terrestrial/Satellite Network Synchronization" *IEEE Trans. Commun.*, Vol. COM-27, Nov. 1979, pp. 1690–1695.

5 W. G. Schmidt, O. G. Gabbard, E. R. Cacciamani, W. G. Maillet and W. W. Wu, "MAT-1:INTELSAT's Experimental 700-Channel TDMA/DA System", in *Proc. ICDSC-1*, London, 1969, pp. 428–440.

6 J. Deal and J. Buegler, "A Demand-Assigned, Time-Division Multiple-Access System for Military Tactical Application", in *Proc. AIAA/CASI, 6th Commun. Satellite Syst. Conf.*, Montreal, Apr. 1976, paper 76-270.

7 D. W. Lipke, D. W. Swearingen, J. F. Parker, E. E. Steinbrecher, T. O. Calvit and H. Dodel, "MARISAT—A Maritime Satellite Communications System", *COMSAT Tech. Rev.*, Vol. 7, Fall 1977.

8 S. E. Dinwiddy, "A Simple TDMA System for Satellite Data Communications Networks," in *Proc. ICDSC-3*, Kyoto, 1975, pp. 378–384.

9 J. Husted and S. Dinwiddy, "Low Cost Satellite Data Transmission Networks Using Demand Assigned TDMA", in *Proc. ICDSC-4*, Montreal, 1978, pp. 8–15.

10 Y. Saburi, M. Yamamoto and K. Harada, "High-Precision Time Comparison via Satellite and Observed Discrepancy of Synchronization", *IEEE Trans. Instrum. Meas.*, Vol. IM-25, Dec. 1976, pp. 473–477.

11 C. C. Costain et al., "Two-way Time Transfer via Geostationary Satellites: NRC/NBS, NRC/USNO and NBS/USNO via HERMES and NRC/LPTF (France) via SYMPHONIE", *11th PTTI*, Washington, D.C., Nov. 1979.

12 C. C. Costain et al., "Two-way Time Transfers between NRC and Paris Observatory via the SYMPHONIE Satellite", in *Proc. 33rd Annu. Symp. Freq. Control*, Atlantic city, N.J., June 1979.

13 M. Brunet, "Two-way Time Transfers between National Research Council (Ottawa) and Paris Observatory via the SYMPHONIE Satellite", *Int. Symp. SYMPHONIE*, Berlin, Feb. 1980.

14 R. K. Kwan, "A TDMA Application in the TELESAT Satellite System", in *Proc. Natl. Telecommun. Conf.*, Atlanta, Ga., 1973, pp. 31E.1–31E.6.

15 INTELSAT, TDMA/DSI System Specifications (TDMA/DSI Traffic Terminals), Document BG-42-65E, B/6/80, 26 June 1980.

16 K. Nosaka, "TTT System—50 Mbps PCM-TDMA System with Time-Preassignment and TASI Features", in *Proc. ICDSC-1*, London, 1969, pp. 83–94.

17 "TDMA System Synchronization through a Global Beam Transponder", INTELSAT, BG/T-2-25E, 15 June 1973.

18 A. K. Jefferis and K. Hodson, "New Synchronization Scheme for Communication-Satellite Time-Division Multiple-Access Systems", *Electron. Lett.*, Vol. 9, 29 Nov. 1973, pp. 566–568.

19 Special Issue on Spread Spectrum, *IEEE Trans. Commun.*, Vol. COM-25, Aug. 1977, pp. 745–869.

20 J. Husted, A. Walker and G. Goubeaud, "A Time-Division Multiple-Access System for the Defense Satellite Communications System," in *Proc. EASCON, '70*, 1970, pp. 229–237.

21 W. G. Schmidt, "An On-Board Switched Multiple-Access System for Millimeter-Wave Satellites", in *Proc. ICDSC-1*, London, 1969, pp. 399–407.

22 C. R. Carter, "Survey of Synchronization Techniques for a Switching Satellite", *IEEE Trans. Commun.*, Vol. COM-28, Aug. 1980, pp. 1291–1301.

23 K. Hodson and S. J. Campanella, "Open Loop Acquisition and Synchronization", in *Proc. ICDSC-4*, Montreal, Oct. 1978, pp. 354–357.

24 P. P. Nuspl and R. de Buda, "TDMA Synchronization Algorithms", in *Proc. EASCON '74*, Washington, D.C., Oct. 1974.

25 Y. Watanabe and K. Izumi, "A New TDMA System for Domestic Service and Its High-Speed PSK Modem", in *Proc. ICDSC-3*, Kyoto, 1975, pp. 385–393.

26 INTELSAT, "System Specification of the INTELSAT Prototype TDMA System", BG-1-18 (Rev.2), Mar. 1974.

27 H. Kurihara, A. Ogawa and Y. Hirata, "A New Initial Acquisition Technique for PCM-TDMA Satellite Communication System", in *Proc. ICDSC-3*, Kyoto, 1975, pp. 281–287.

28 A. S. Acampora, "An Efficient TDMA Assignment Algorithm for Area Coverage Satellite Networks", to be published.

29 W. Schrempp and T. Sekimoto, "Unique Word Detection in Digital Burst Communications", *IEEE Trans. Commun. Technol.*, Vol. COM-16, Aug. 1968, pp. 597–605.

30 O. G. Gabbard, "Design of a Satellite Time-Division Multiple-Access Burst Synchronizer", *IEEE Trans. Commun. Technol.*, Vol. COM-16, Aug. 1968, pp. 589–596.

31 W. G. Maillet, "Processing of Bursts in a TDMA System", in *Proc. ICDSC-1*, London, 1969, pp. 69–82.

32 W. G. Maillet, "INTELSAT's 50 Mb/s TDMA-2 System", in *Proc. ICDSC-2*, Paris, 1972, pp. 26–32.

33 W. G. Schmidt, "The Application of TDMA to the INTELSAT IV Satellite Series", *COMSAT Tech. Rev.*, Vol. 3, Fall 1973, pp. 257–275.

34 O. G. Gabbard and P. Kaul, "Time-Division Multiple-Access", in *Proc. EASCON, '74*, Washington, D.C., 1974, pp. 179–184.

35 S. Kondo, "Clock Synchronization in TDMA Satellite Communications System-SMAX", *Rev. Elec. Commun. Lab.*, Vol. 17, Aug. 1969.

36 Y. Inoue, "Design of Acquisition System for a PCM-TDMA Satellite Communication System", *Electron. Commun. Jpn.*, Vol. 53, May 1970, pp. 99–107.

37 M. Takada, S. Nakamura, S. Kondo, Y. Inoue, M. Ono and H. Ikeda, "New PCM-TDMA Satellite Communications System and Variable Destination Channel Control Techniques", in *Proc. ICDSC-1*, London, 1969, pp. 39–50.

38 Y. Suguri, H. Doi and E. Metzger, "A TDMA/PCM Experiment on Applications Technology Satellites", *IEEE Trans. Commun. Technol.*, Vol. COM-19, Apr. 1971, pp. 196–205.

39 K. Nosaka, H. Sasaki, K. Amano and H. Michishita, "TTT System (50 Mb/s PCM-TDMA System with Time-Preassignment and TASI) and Its Satellite Test Results", *IEEE Trans. Commun. Technol.*, Vol. COM-20, Aug. 1972, pp. 820–825.

40 H. Rupp and E. Herter, "German TDMA System I-System Description", *11th Eur. Space Symp.*, Berlin, May 1971, paper 94.

41 G. Eckhardt, B. Reidel and H. Rupp, "Ein Flexibles TDMA-System für 100/50 Mbit/s" ("A Flexible TDMA System for 100/50 Mbit/s"), *Frequenz*, Vol. 25, No. 10, 1971, pp. 298–303.

42 "Statement of Work and Technical Requirements for a TDMA/SDMA Synchronization and Acquisition Study", COMSAT Tech. Rep. RFP-IS-464, 24 May 1972.

43 P. Kaul, "An All-Digital Variable Threshold Codeword Correlator", in *Proc. ICDSC-2*, Paris, 1972, pp. 33–39.

44 P. P. Nuspl, N. G. Davies and R. L. Olsen, "Ranging and Synchronization Accuracies in Regional TDMA Experiment", in *Proc. ICDSC-3*, Kyoto, 1975, pp. 292–300.

45 P. P. Nuspl and R. Mamen, "Results of the CENSAR Synchronization and Orbit Perturbation Measurement Experiments", in *Proc. ICDSC-4*, Montreal, Oct. 1978, pp. 346–353.

46 S. Tachikawa, R. Edy et al., "120-Mbit/s TDMA Test Bed", in *Proc. ICDSC-4*, Montreal, Oct. 1978, pp. 291–298.

47 S. J. Campanella and B. A. Pontano, "The INTELSAT TDMA Field Trial", in *Proc. ICDSC-4*, Montreal, Oct. 1978, pp. 299–305.

48 K. Kosaka, A. Saburi et al., "Advanced TDMA System for Experimental Study in 30/20 GHz Band", in *Proc. ICDSC-4*, Montreal, Oct. 1978, pp. 333–340.

49 H. Kaneko, Y. Katagiri and T. Okada, "The Design of a PCM Mastergroup Codec for the TELESAT TDMA System", in *Proc. Int. Conf. Commun.*, San Francisco, 1975, pp. 44.6–44.10.

50 S. Yokoyama, K. Kato, T. Noguchi, N. Kusama and S. Otami, "The Design of a PSK Modem for the TELESAT TDMA System", in *Proc. Int. Conf. Commun.*, San Francisco, 1975, pp. 44.11–44.15.

51 Technical Description, Digital Communications Corporation, Germantown, Md.

52 Technical Description, Commercial Telecommunications Corporation, Santa Maria, Calif.

53 J. Ramasastry, W. Callahan, R. Markham, P. Kaul and L. Golding, "Advanced WESTAR SS/TDMA System", in *Proc. ICDSC-4*, Montreal, Oct. 1978, pp. 36–43.

54 J. D. Barnla et al., "The SBS Digital Communications Satellite Systems", in *Proc. EASCON '77*, Washington, D.C., Sept. 1977.

55 B. C. Searle and P. P. Nuspl, "SLIM TDMA Pilot Project", in *Proc. IEEE Can. Commun. Power Conf.*, Montreal, Oct. 1980, pp. 339–341.

56 Technical Description, Miller Communications Systems, Ottawa.

57 Y. S. Yeh and D. O. Reudink, "The Organization and Synchronization of a Switched Spot-Beam System", in *Proc. ICDSC-4*, Montreal, Oct. 1980, pp. 191–196.

58 G. D. Dill, "TDMA, The State-of-the-Art", in *Proc. EASCON '77*, Washington, D.C., Sept. 1977. Also in Ref. 62.

59 Special Issue on Satellite Communications, *Proc. IEEE*, Vol. 65, Mar. 1977.

60 Special Issue on Synchronization, *IEEE Trans. Commun.*, Part I, Vol. COM-28, Aug. 1980.

61 CCIR, Report 708, *Green Book*, Vol. IV, Geneva, 1978.

62 H. L. Van Trees (Ed.), *Satellite Communications*, IEEE Press Book, New York, 1979. Also Distributed by Wiley, New York.

PROBLEMS

BCW Analysis and Design

1 (a) Find the autocorrelation function of the following BCW for QPSK transmission, including the STR sequence before BCW:
I-Channel, 1111100011011101010000100101100;
Q-channel, 0100001001011001111100011011101

(b) Calculate P (missed BCW) for a BER $= 10^{-4}$.

(c) Calculate P (false detection of BCW), without an aperture gate and with a gate of 20 ± 3 symbols.

2 For a binary BCW of length 31, processing is to be by analog correlation techniques, such as with a surface acoustic-wave (SAW) device. Using an implementation margin of 3 dB, calculate the probability of missing BCW for the COMPUTIL example (Section 8.5).

3 Verify that for $n = 31$, $e = 2$,

$$P(\text{BCW missed}) = \binom{31}{3}p^3 - 3\binom{31}{4}p^4 + 6\binom{31}{5}p^5$$

$$< \binom{31}{3}p^3$$

This is a loose bound for small p. Similar expressions are useful when investigating other n and e.

4   An ROC (Receiver Operating Characteristic) is a plot of P (detection) against P (false alarm). For a binary BCW of 15 bits and digital processing with $e = 1$ and an aperture gate of 15, plot the ROC for this BCW detector.

5   Find a shorter but adequate BCW for COMPUTIL of Section 8.5.

6   For a precise gate of 31 symbol durations, find the probability of false detection. Repeat for the case of no gate at all.

Frame Efficiencies

7   **(a)**   Show the formula for frame efficiency, using symbols as the unit of measurement.

    (b)   Calculate the frame efficiencies for each of Example 8.3.

8   Calculate the frame efficiency for the TELESAT TDMA system.

9   Calculate the frame efficiency for the SS-TDMA, scanning beam system by Bell Labs.

CHAPTER 9

Code-Division Multiple Access

In code-division multiple access (CDMA) all users operate at the same nominal frequency and simultaneously use the entire repeater bandwidth. The important feature of CDMA is that unlike FDMA and TDMA, minimal dynamic (frequency or time) coordination is needed between the various transmitters in the system. In addition, CDMA techniques can provide substantial levels of antijam (AJ) capability to combat deliberate interference and have a low probability of intercept (LPI) to reduce the probability of reception by unauthorized users. Furthermore, CDMA allows a graceful degradation of performance as the number of simultaneous users increases. Conversely, when the system is underused, the excess capacity automatically becomes increased margin.

Two most widely used CDMA techniques are:

1 Direct sequence (DS) or pseudo-noise (PN) modulation with chip time T_c. The waveform used to represent a 1 or a 0 in the PN sequence is usually referred to as a chip.
2 Noncoherent frequency hopping (FH) modulation with minimum frequency separation Δf.

These techniques use a much larger-than-required bandwidth to transmit information (typically 10^3 to 10^6 times larger than information bandwidth), and have been named spread-spectrum techniques.

In this chapter we first discuss the spread-spectrum concept [1, 2]. Since the essential ingredient of the DS and FH techniques is the PN sequence or PN code used to spread the bandwidth, we present a brief exposition of the PN sequences and their auto and cross-correlation properties [3]. It will be seen that in CDMA the utilization of sequences with low cross-correlation is important in maximizing the total number of simultaneous users. As in all

269

aspects of communications, synchronizaton represents a prominent and indis-
pensible feature in CDMA and is discussed in Section 9.3. It is absolutely
necessary that the locally generated PN sequence be in synchronism with the
received sequence to within a fraction of one chip (DS), or one bit time (FH)
[4]. We conclude the chapter by considering the performance of CDMA
satellite systems [5].

9.1 FUNDAMENTAL CONCEPTS OF THE SPREAD SPECTRUM

A basic spread-spectrum system is shown in Figure 9.1. Each user uses the
same carrier frequency ω_c and occupies the same RF bandwidth. The spectrum
-spreading part of the system is an addition to a conventional digital communication system. In many cases, it is quite difficult to distinguish between the
spreading operation and the modulation, but for analysis purposes it is
certainly accurate to treat them separately.

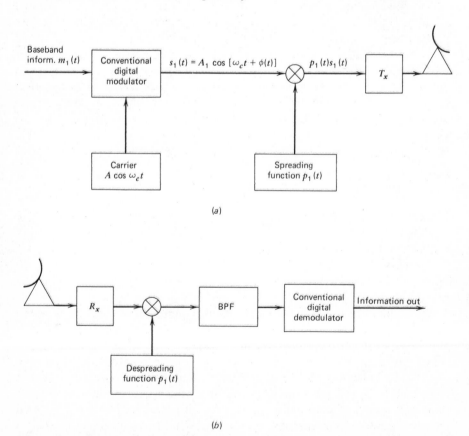

Figure 9.1 Spread-spectrum transmission/reception. (a) Spread-spectrum transmitter; (b) spread-spectrum receiver.

Generation of the transmitted signal in a multiple-access environment involves two steps. The carrier $A \cos \omega_c t$ is phase-modulated by the baseband digital information $m_1(t)$ with rate $R_i(=1/T)$, where T is the symbol duration. The modulated signal $s_1(t)$ has its bandwidth spread by a spreading function $p_1(t)$ with chip rate $R_c = 1/T_c$ just prior to transmission. The resulting signal $p_1(t)s_1(t)$ is passed through the channel, where it is combined with the other signals in the system. Thus assuming M transmitters, the received signal is given by

$$r(t) = \sum_{i=1}^{M} p_i(t)s_i(t) + I(t) + n(t)$$

Here $I(t)$ is the interference (deliberate or self-noise of CDMA system) and $n(t)$ is the additive noise. At the receiver the intended user will have $p_1(t)$ as its despreading function, which is identical to the spreading function employed to spread the baseband information $m_1(t)$. If the set of spreading waveforms is chosen cleverly, only the original, modulated waveform remains after the correlation. Other waveforms will not be correlated and are effectively spread, thus appearing as noise to the demodulator. The whole process can be viewed in a qualitative manner by examining the relationship between various spectra in a spread-spectrum system as depicted in Figure 9.2. Clearly, to ensure a given jamming or interference power density in the information bandwidth of the received signal, the jammer must increase his total power by the same amount as the bandwidth expansion of the signal.

One of the major concerns in CDMA is the amount of interference rejection possible. The most widely accepted measure of this quantity, termed the system **processing gain** G_p, is given by the ratio of RF bandwidth to the information bandwidth R_c/R_i. Typical processing gains for spread-spectrum systems run from 20 to 60 dB. Clearly, the input and output signal-to-noise ratios are related as

$$\left(\frac{S}{N}\right)_o = G_p \left(\frac{S}{N}\right)_i \tag{9.1}$$

Jamming margin M_j is the degree of interference which a spread-spectrum system can withstand while receiving a desired signal and delivering a minimum signal-to-noise ratio at its information output. Jamming margin is related to the processing gain by the relation

$$M_j = G_p - \left[\left(\frac{S}{N}\right)_o + L \right] \tag{9.2}$$

where L is the system loss. Typical system losses are in the range 1 to 3 dB and are due to imperfect generation and tracking of the spreading waveform, imperfect demodulator and so on.

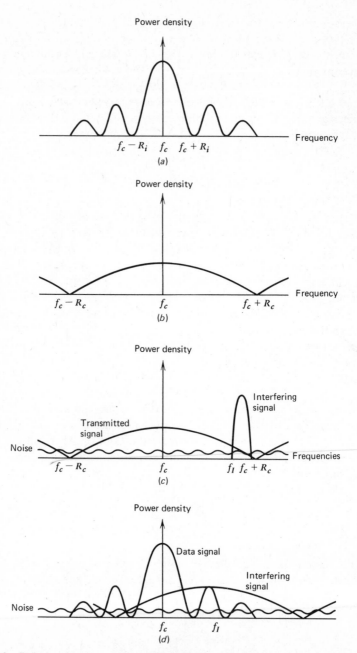

Figure 9.2 Relationship between various power spectra in a spread-spectrum system. (*a*) Spectrum of conventionally modulated signal; (*b*) transmitted spread spectrum; (*c*) received power spectrum; (*d*) despread spectrum into conventional demodulator.

Generation of the Spread-Spectrum Signal

As noted earlier, the two widely used spread-spectrum techniques are direct sequence (DS) and frequency hopping (FH). Since these two techniques are germane to the work presented here, a treatment is presented for them. Nevertheless, it should be pointed out that other spread-spectrum techniques exist. Two such techniques are time hopping (TH) and the pulsed-FM or "chirp" method. These and many other hybrid schemes are possible and are described briefly in Ref. 1.

Direct Sequence (DS) Systems

A direct sequence spread-spectrum system is shown in Figure 9.3. Although we can employ the spread-spectrum signal generation as depicted in Figure 9.1, for digital modulation it is better first to combine the information signal and the spreading sequence before phase-modulating the carrier.

The PN-code generator generates a pseudo-random binary sequence at a chip rate R_c that is some large multiple of the message bit rate R_i. This sequence is then combined with the information sequence. The combined sequence is then modulated onto a carrier. Although a variety of modulation schemes are possible, phase-shift keying (PSK) has been frequently used in

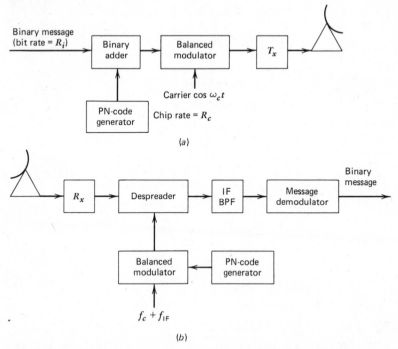

(a)

(b)

Figure 9.3   Direct sequence transmitter/receiver. (a) DS transmitter; (b) DS receiver.

spread-spectrum systems for satellite communications. Note that the bandwidth expansion (and hence the processing gain) is under the complete control of the PN-sequence, specifically the chip rate R_c.

An exact replica of the PN-sequence or code must be generated and synchronously maintained at the receiver in order that the spreading can be properly removed. Such synchronization techniques are discussed in Section 9.3. Since the recovered signal is at the same carrier frequency as the input PSK signal, a narrowband interfering signal could conceivably leak through the correlator. Hence a heterodyne correlator as depicted in figure 9.3b is used.

Frequency-Hopping (FH) Systems

Basically, an FH spread-spectrum system works similarly to the DS system, the main similarity being the correlation process required at the receiver. The main difference is the way the transmitted spectrum is generated. As depicted in Figure 9.4, the FH-PN code first goes through a code-to-frequency transformation. An FH system produces a spreading effect by pseudo-randomly hopping the final carrier frequency over a wide range of prescribed frequencies. The hopping pattern and hopping rate are determined by the PN code and code

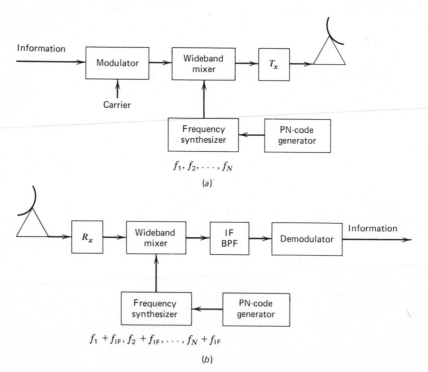

Figure 9.4 Frequency-hopping transmitter/receiver. (*a*) FH transmitter; (*b*) FH receiver.

rate, respectively. The rate of frequency hopping is tied in some way to the message rate in order that there can be coherent detection over each message bit (or *encoded chip*). At the receiver, when the local frequency synthesizer is switched with a synchronized replica of the transmitted PN code, the frequency hops on the received signal will be removed (dehopped), leaving the original modulated signal untouched. Coherent demodulation is not suitable for FH signalling because it is extremely difficult to maintain phase relationships between frequency steps and incoherent processing is used.

Let Δf be the frequency separation between discrete frequencies and N be the number of available frequency choices or channels. The processing gain is

$$G_p = \frac{\text{RF bandwidth}}{\text{message bandwidth}} = N \qquad \text{if } \Delta f = \text{message bandwidth}$$

There are some further differences between DS and FH signalling as currently practiced [6]. Code rates for DS signalling are usually much higher (near 200 Mbps) than those for FH (near 100 kbps). This is due to the relatively slow frequency synthesizers which permit code rates only up to 100 kbps. However, if cost is not a problem, one could make the frequency synthesizers as fast as desired by having several that hop ahead. Also, since the initial acquisition time is proportional to the PN chip rate R_c, the initial acquisition time for FH is much smaller than that for DS. Typical initial acquisition times are on the order of 1 ms for FH and 2 to 3 s for DS.

Because of the complementary nature of these two techniques, a combined (hybrid) FH/DS scheme has been suggested for multiple-access problems. The Joint Tactical Information Distribution System (JTIDS) [7], which is intended to be the major U.S. tactical communications system for the 1980s, is a prime example of the use of a hybrid FH/DS scheme.

The processing gain of a FH/DS system is given by the sum of the processing gains (in dB) of the FH and DS systems (Problem 1). Clearly, a DS system must have a very high chip rate R_c or the FH system must have a very large number of channels. Because of a reduction in the number of channels and in the code rate, a hybrid FH/DS system offers simpler implementation possibilities [8].

9.2 PSEUDO-RANDOM SEQUENCES AND THEIR CORRELATION PROPERTIES

Pseudo-random or pseudo-noise sequences or codes are noiselike (but deterministic) periodic sequences. Sequences that are useful for spectrum spreading have the properties described in Ref. 9: (1) denying any information about future sequence k-tuples to the unintended party, (2) practical implementation, (3) triangular autocorrelation function* having a high peak-to-side lobe ratio,

---

*Although PN-sequences have this characteristic, we do not have to have this; we just need a single autocorrelation spike.

(4) balanced k-tuple statistics and (5) for each pair of sequences, a uniformly low cross-correlation.

In much of the literature on pseudo-random sequences, the terms pseudo-noise (PN) sequence and maximal-length linear feedback shift-register sequence (m-sequences) are used synonymously. This is confusing, since pseudo-random sequence is a generic name that includes certain non-maximal-length linear feedback shift-register sequences and a host of other sequences generated by various ad hoc methods. In this section we make a clear distinction between the m-sequences and various types of related sequences, such as those described in Refs. 10 and 11. A further source of confusion is the wide variation in the literature for the notation of autocorrelation and cross-correlation of these sequences. The notation we use throughout this chapter is adapted from Ref. 12. After introducing the periodic and aperiodic correlation function and their properties, we discuss m-sequences and their properties at length. Also discussed are Gold codes [10], which are generated by modulo-2 addition of selected m-sequences.

Periodic and Aperiodic Correlation Functions

Let $\mathbf{x}^k = (x_0^k, x_1^k, \ldots, x_{N-1}^k)$, $k = 1, 2, \ldots, K$ be a set of K binary sequences of elements of $(+1, -1)$. The sequences are assumed to be periodic with period N. The **aperiodic cross-correlation** function $c_{k,i}$ for the sequence $\mathbf{x}^k$ and $\mathbf{x}^i$ is defined by

$$c_{k,i}(\tau) = \begin{cases} \displaystyle\sum_{j=0}^{N-1-\tau} x_j^k x_{j+\tau}^i, & 0 \leq \tau \leq N-1 \\ \displaystyle\sum_{j=0}^{N-1+\tau} x_{j-\tau}^k x_j^i, & 1-N \leq \tau < 0 \end{cases} \tag{9.3}$$

The terminology for τ being positive or negative is depicted in Figure 9.5. This terminology is based on the notion that N code chips $x_0, x_1, \ldots, x_{N-1}$, are arriving serially at a receiver which has a matched filter for the anticipated code sequence. τ is assumed negative before the arriving sequence fills in the matched filter; and τ is positive as the arriving sequence goes out of the matched filter.

The **periodic cross-correlation** function $\theta_{k,i}$ is defined by

$$\theta_{k,i}(\tau) = \sum_{j=0}^{N-1} x_j^k x_{j+\tau}^i, \qquad 0 \leq \tau \leq N-1$$

$$= c_{k,i}(\tau) + c_{k,i}(\tau - N) \tag{9.4}$$

The **odd cross-correlation** function $\hat{\theta}_{k,i}$ is defined by [13]

$$\hat{\theta}_{k,i}(\tau) = c_{k,i}(\tau) - c_{k,i}(\tau - N), \qquad 0 \leq \tau \leq N-1 \tag{9.5}$$

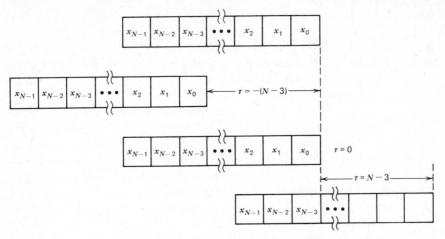

Figure 9.5 Sequence shift (τ) terminology.

The periodic, odd and aperiodic autocorrelation functions are

$$\theta_k(\tau)=\theta_{k,k}(\tau), \quad \hat{\theta}_k(\tau)=\hat{\theta}_{k,k}(\tau) \quad \text{and} \quad c_k(\tau)=c_{k,k}(\tau), \qquad \tau=0,1,\dots,N-1 \tag{9.6}$$

respectively. When only one sequence is under consideration, the subscript will be omitted.

Example 9.1

We compute the periodic autocorrelation of the sequence

$$\mathbf{x}=(-1,-1,-1,1,1,-1,1)$$

We have

$$\theta(0)=(-1)(-1)+(-1)(-1)+(-1)(-1)+1.1+1.1+(-1)(-1)+1.1$$

$$=1+1+1+1+1+1+1=7$$

$$\theta(1)=(-1)(-1)+(-1)(-1)+(-1)(1)+1.1+(1)(-1)+(-1)(1)+(1)(-1)$$

$$=1+1-1+1-1-1-1=-1$$

Similarly,

$$\theta(2)=\theta(3)=\theta(4)=\theta(5)=\theta(6)=-1$$

Thus

$$\theta(\tau)=\begin{cases} 7, & \tau=0 \\ -1, & 1\leq\tau\leq6 \end{cases}$$

is the periodic autocorrelation function of $\mathbf{x}$.

Binary Maximal-Length Sequences (m-Sequences)

As we noted in Chapter 8, m-sequences have been used extensively for initial
acquisition and synchronization tracking. The m-sequences are a special case
of binary codes which are usually produced by a linear feedback shift-register
generator (LFSRG). An enormous amount of research has gone into the
autocorrelation properties and the "noiselike" aspects of the m-sequences [14].

An LFSRG is formed, in general, by taking the binary output from some of
the m-stages of the shift register, modulo 2 adding the outputs in a connection
vector $(h_0, h_1, \ldots, h_m)$ and feeding the result back to the input of the register.
Figure 9.6 shows an m-stage LFSRG. The sequence of binary chips flowing out
of the LFSRG are determined by the feedback vector and the initial condition
or contents $(x_0, x_1, \ldots, x_{m-1})$ of the shift register. The sequence satisfies the
recurrence relation

$$x_i = \sum_{j=1}^{m} h_j x_{i-j}, \qquad i \geq m$$

Example 9.2

Consider the four-stage LFSRG with connection vector (10011) shown in
Figure 9.7. The succession of states triggered by clock pulses assuming the
initial state to be (1000) are as follows:

Clock Cycle (i)	Register Cycle	x_i
0	1000	0
1	0100	0
2	0010	0
3	1001	1
4	1100	0
5	0110	0
6	1011	1
7	0101	1
8	1010	0
9	1101	1
10	1110	0
11	1111	1
12	0111	1
13	0011	1
14	0001	1
15	1000	0
16	0100	0
17	0010	0 repeating
$\vdots$	$\vdots$	$\vdots$

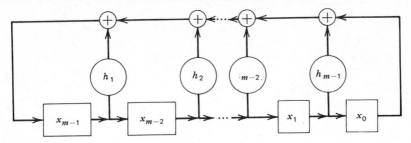

Figure 9.6 *m*-Stage linear feedback shift register. The symbol $\rightarrow h_i \rightarrow$ denotes a closed circuit if $h_i = 1$; and an open circuit for $h_i = 0$.

Note that in Example 9.2 the sequence begins to repeat itself after the first 15 terms. In general, the period N of the sequence **x** generated by an m-stage LFSRG is at most $2^m - 1$. Any sequence achieving a period $N = 2^m - 1$ is called a maximum-length sequence, or simply an m-sequence.

Whether an m-stage LFSRG will generate only one sequence with period $N = 2^m - 1$ will depend upon its connection vector. Let $h(x)$ be the mth-order polynomial given by

$$h(x) = h_0 + h_1 x \cdots h_{m-1} x^{m-1} + h_m x^m$$

We call $h(x)$ the associated polynomial of the shift register with feedback coefficients $(h_0, h_1, \ldots, h_m)$. Here $h_0 = h_m = 1$ and the other h_i's take on values 0 and 1. Thus the polynomial for the four-stage LFSRG in Figure 9.7 is given by $1 + x^3 + x^4$.

It is a well-known result of algebra that if $h(x)$ is an irreducible (not factorable) primitive polynomial (see Section 13.3) of degree m, then all sequences generated by $h(x)$ will have maximum period $N = 2^m - 1$. A detailed discussion of primitive polynomials will follow in Chapter 13, where we also establish the relationship between Hamming codes and m-sequences. In Table 9.1 we provide a short list of some binary irreducible primitive polynomials. For a more extensive listing, see Table 3.6 in Dixon [1]. Also listed in Table 9.1 is the number of different primitive polynomials of degree m (and consequently the number of m sequences) $= \phi(2^m - 1)/m$. $\phi(2^m - 1)$, the Euler phi function, is the number of integers less than and relatively prime to $2^m - 1$. As can be seen, large numbers of sequences are available for m greater than 10.

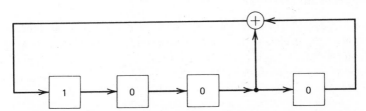

Figure 9.7 Four-stage LFSRG with connection vector (10011).

Table 9.1 Some Binary Irreducible Primitive Polynomials for m-Sequences

Number of Stages m	Sequence Length $N = 2^m - 1$	Number of m-Sequences	$h(x)$ (Example)
3	7	2	$1 + x + x^3$
4	15	2	$1 + x + x^4$
5	31	6	$1 + x^2 + x^5$
6	63	6	$1 + x + x^6$
7	127	18	$1 + x^3 + x^7$
8	255	16	$1 + x^2 + x^3 + x^4 + x^8$
9	511	48	$1 + x^4 + x^9$
10	1023	60	$1 + x^3 + x^{10}$
11	2047	176	$1 + x^2 + x^{11}$
12	4095	144	$1 + x + x^4 + x^6 + x^{12}$
13	8191	630	$1 + x + x^3 + x^4 + x^{13}$
14	16383	756	$1 + x + x^6 + x^{10} + x^{14}$
15	32767	1800	$1 + x + x^{15}$

Properties

Next, we review briefly the most important properties of m-sequences and then consider those properties that are of particular relevance in a CDMA environment.

Cyclic Property Let T denote the operator that shift n-tuples cyclically to the left by one place, that is, $T\mathbf{x} = (x_1, x_2, \ldots, x_{N-1}, x_0)$. By applying T, k times to the n-tuple, we see that

$$T^k \mathbf{x} = (x_k, x_{k+1}, \ldots, x_{N-1}, x_0, x_1, \ldots, x_{k-1}), \qquad 0 \leq k < N$$

It can now be stated that there are exactly $N = 2^m - 1$ nonzero m-sequences generated by $h(x)$: that is, $\mathbf{x}, T\mathbf{x}, \ldots, T^{N-1}\mathbf{x}$. These sequences are also called **phases** of $\mathbf{x}$.

Balance Property There are exactly $2^{m-1} - 1$ zeros and 2^{m-1} ones in one period of a maximal-length sequence.

Shift and Add Property The modulo 2 sum of an m-sequence and any of its cyclic versions is another cyclic version of the original sequence with a phase different from either version.

Periodic Autocorrelation Property The m-sequences have an interesting cyclic or periodic autocorrelation property. If we transform the binary $(0, 1)$ sequence of the shift register output to the binary $(+1, -1)$ sequence, by replacing each zero by $+1$, and each 1 by -1, then the periodic correlation function is given by

$$\theta(\tau) = N = \begin{cases} 2^m - 1, & \tau = 0 \\ -1, & \tau \neq 0 \end{cases} \tag{9.7}$$

and this is the best possible periodic correlation function in the sense that for no other binary sequence is the term $\max\limits_{\tau\neq0} \theta(\tau)$ smaller. This is one of the main reasons for the use of m-sequences in preamble design for bit timing and synchronization.

Randomness Property Since a maximal-length sequence is generated it cannot be called random, yet it has a well-defined statistical distribution for the runs of ones and zeros. In fact, in each period, one-half of the runs have length 1, one-fourth of the runs have length 2, one-eighth of the runs have length 3 and so on.

Cross-Correlation Functions

For code-division multiple-access systems, large sets of sequences with low off-peak autocorrelation (side lobes) and cross-correlation values are required. Since m-sequences possess ideal autocorrelation functions, it is of interest to examine their cross-correlation functions. Unfortunately, no general method of predicting the cross-correlation behaviour of m-sequences appears to exist at the present time. Almost all of the known results are summarized in Ref. 3, where it is shown that only very small sets of m-sequences can have good cross-correlation properties. Thus for CDMA systems in which the number of users is small (e.g., five or less) it is possible to carefully select m-sequences. Large sets of m-sequences generally have quite poor cross-correlation properties and thus are not suitable for CDMA systems with a large number of users.

Gold Sequences

Gold sequences [10] are a set of sequences generated by modulo 2 addition of selected m-sequences whose periodic cross-correlation function is bounded by

$$\max|\theta_{k,i}(\tau)| \leq 2^{(m+1)/2} + 1$$

The aperiodic cross-correlation for Gold sequences is known to be statistically as low as possible for sequences of a given length [15]. Thus Gold sequences are ideally suited for CDMA applications. Further, they may offer rapid acquisition possibilities. However, since the Gold sequences are linear, they are easily "crackable" and thus are less useful when unintended jamming parties are involved [9].

This concludes our discussion of m-sequences and Gold sequences. For many other interesting and useful pseudo-random sequences, the reader is referred to Refs. 3, 8, 13, 16 and 17.

9.3 SYNCHRONIZATION

Although a CDMA system does not require a rigid system timing, it is obviously necessary for receivers to be capable of synchronizing their reference sequences to the incoming signals. Sequence phase uncertainty due to changes in propagation path delay is the primary source of synchronization uncertainty. The PN code phase uncertainty has to be resolved within one chip duration before despreading of the received signal can result. In an ideal situation, once code phase is synchronized, it should continue to be so forever. In practice, there are two stages of synchronization: initial acquisition and tracking. Once initial synchronization has been achieved, one must hold it in alignment and the system enters the tracking mode. Feedback tracking loops of the types delay lock [18, 19] or dither [20] are employed to track the PN code. They are actually the same but are implemented somewhat differently in that the delay-lock loop requires two correlators while the dithering loop requires a single correlator.

Initial Synchronization

As stated by Dixon [2], little information is available in the literature that discusses the specific problems of gaining initial synchronization in spread-spectrum systems. The simplest method of obtaining initial synchronization is the sliding correlator [1]. Here we adjust the rate of generation of the locally generated PN-sequence and correlate it with the incoming code as shown in Figure 9.8. The two PN-sequences slip in phase with respect to each other, stopping only when the despreader produces a satisfactory result. In practice, the sliding correlator is often preceded by some other method for reducing the area of search, such as the transmission of a synchronization preamble, which is a short PN-sequence itself. In this situation the time uncertainty is just the period of the short sequence.

Another technique for rapid lockup that has been used is to employ **acquirable codes** [21]. These codes consist of a number of shorter codes which are combined so as to maintain some correlation with the component codes. The best known codes of this type are the JPL codes [1]. For example, consider

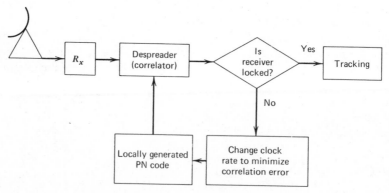

Figure 9.8 Sliding-correlator synchronizer.

the sequence $x = x_1 \oplus x_2$, where x_1 and x_2 are m-sequences with periods N_1 and N_2 chosen to be relatively prime. The acquirable code x (which is not an m-sequence) has period $N_1 N_2$. In this situation, the search time is proportional to $N_1 + N_2$ rather than $N_1 N_2$.

Another technique for acquisition is the sequential estimation [22]. Here the incoming signal is demodulated and stored in a shift register. This code is then compared to all the stored codewords to activate the correct PN-sequence generator. Obviously, this technique does not enjoy the benefit of the processing gain and is thus more susceptible to interference and noise.

Passive Correlation Technique

Theoretically, for passive correlation, which is an ideal technique, the receiver employs tapped delay-line (TDL) matched filters to recognize a code sequence in only one specific configuration. The processing gain or the time–bandwidth product determines the complexity of the matched filters, which can be implemented in many ways. It is claimed [23] that for bandwidths in excess of 20 MHz, surface acoustic wave (SAW) matched filters are simpler than CCD (charge-coupled devices) or digital electronics implementation of the TDL matched filter, and for bandwidths in excess of 50 MHz, SAW matched filters are probably the only realistic choice. In practice, time–bandwidth products of 1000 to 10,000 are achievable, which limits the maximum processing gain (the level at which sync can be achieved) to 30 to 40 dB.

Tracking

Delay-Lock Tracking Loop

The basic delay-lock loop is depicted in Figure 9.9, in which coherent demodulation to baseband will be assumed. The results directly apply to an IF loop (see Problem 11, Chapter 5). An alternative form of the delay-lock loop has been described by Gill [24]. The incoming signal is applied to two multipliers

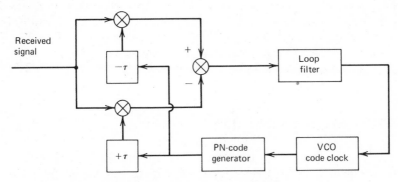

Figure 9.9 Delay-lock tracking loop.

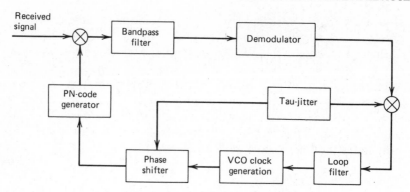

Figure 9.10 Dithering tracking loop.

(correlators), where it is multiplied by an early and a late version of the locally generated PN code. The difference signal is used to control the VCO code clock. Note that it is necessary that the transfer functions of the two paths be identical. An extensive analysis of the delay-lock tracking loop performance as well as optimization of delay-lock correlator spacing has been outlined in Ref. 25.

Dithering (Tau-Jitter) Tracking Loop

An alternative form of tracking loop called the dithering loop is shown in Figure 9.10. The received signal, just as in the case of a delay-lock loop, is correlated with an early and late version of the locally generated PN code, the difference being that the correlation is done by a single correlator on an alternating basis. As shown in Figure 9.10, the PN-code generator is driven by a clock signal whose phase is "dithered" back and forth in accordance with the binary signal; this is called tau-jitter. This eliminates the necessity of ensuring identical transfer function of the two paths. However, the signal-to-noise performance is about 3 dB worse than the delay-lock loop [20].

Because of its implementation simplicity, the tau-jitter loop has been recommended for PN code tracking in the TDRSS (Tracking and Data Relay Satellite System)-to-Shuttle Ku-band communication link [26]. The PN code used is an m-sequence of length 1023 ($=2^{10}-1$) with a chip clock rate of 3.03 MHz. The acquisition time is estimated to be between 5 and 8 s at $C/N_0 = 60.2$ dB-Hz.

9.4 PERFORMANCE OF CDMA SYSTEMS

The application of CDMA is usually not attractive from a capacity viewpoint alone. One or more of the following desirable objectives can be accomplished with CDMA:

1 Reduce the spectral density.
2 Provide protection against jamming.
3 Provide message privacy (i.e., low probability of intercept). (Note that since signals may be recorded and processed off-line, interception is easier than jamming.)
4 Combat multipath in mobile communication systems.
5 Conduct precise range measurements (e.g., CENSAR, described in Section 8.4).

Because of these features, CDMA techniques have been used for a variety of satellite systems, such as NASA TDRSS [27], satellite systems to provide communications for mobile users [28], air traffic control systems [29] and military satellite systems [30].

In TDRSS, the use of CDMA techniques accomplished many objectives [31]:

1 It enables discrimination between 20 simultaneous users operating at the same frequency.
2 The processing gain provides some protection against narrowband interference.
3 Range measurements are made by comparing the transmitted and received code phase.
4 Reduced flux density at the surface of the Earth is achieved.

In the late 1960s, a number of researchers [32–34] analyzed the performance of CDMA with a hard-limiting satellite repeater. Most of this work assumed time synchronization between transmitter and receiver and thus dealt with periodic cross-correlation properties of the sequence. In Ref. 33 it has been shown that the maximum number of identical users M in a band B which can be supported by the satellite downlink is given by

$$M = \frac{\pi/4}{R_c(E_b/N_0)} \cdot \frac{P_r}{N_r + N_j} \tag{9.8}$$

where P_r = total satellite power arriving at the receiver
$\quad R_c$ = chip rate assumed to be the same for all transmitters
$\quad N_r$ = noise density at the receiver
$\quad N_j = \dfrac{P_r}{B}\left(\dfrac{M-1}{M}\right) \cong \dfrac{P_r}{B}$ = equivalent noise density generated by the
$\quad\quad (M-1)$ users to which the given receiver is not tuned

The factor $\pi/4$ accounts for the intermodulation loss in a hard limiter for a signal in Gaussian noise [21]. The signal-to-noise ratio at the output of the correlation receiver has also been derived [33].

Equation (9.8) may be written as

$$M = \frac{(\pi/4)B}{R_c(E_b/N_0)} \cdot \frac{P_r/N_rB}{1+(P_r/N_rB)} \tag{9.9}$$

If the bandwidth is large enough for N_r to dominate N_j such that (P_r/N_rB) $\ll 1$, we are in the "power-limited" region. In this situation the maximum number of users is the same as in an orthogonal system operating at the same E_b/N_0. Conversely, for B small such that $(P_r/N_rB) > 1$, we are in the "bandwidth-limited" region. The multiple-access noise density N_j dominates and CDMA is rather inefficient.

Anderson and Wintz [34] in their investigation assumed no synchronization between the transmitter and the receiver and clearly demonstrated the effects of aperiodic cross-correlation. They obtained the following bound on the signal-to-noise ratio at the output of the correlation receiver for CDMA with a hard-limiting satellite repeater:

$$(SNR)_o \geq \frac{0.88}{M} \cdot \frac{N(SNR)_f}{1+2\pi_c^2 N(SNR)_f} \tag{9.10}$$

where N = length of the PN code
$\quad M$ = number of users
$\quad \pi_c$ = peak of the periodic and aperiodic cross-correlations of the M PN-sequences
$(SNR)_f$ = [statistical average of the (energy per chip) SNR in fundamental at the output of the hard limiter]

Anderson and Wintz also showed that for sufficiently small $(SNR)_f$ the additive noise dominates. On the other hand, if the number of users is sufficiently large, the cross-correlation terms dominate. Note that we can solve (9.10) in terms of M to obtain a bound on the number of users.

Recently, Pursley [12] analysed the performance of CDMA systems and the effects of aperiodic cross-correlation functions. His results were derived for a linear channel only but are included here for the sake of completeness. His first result is in the form of an upper bound on the worst-case bit error probability P_{max}, which may be useful for certain CDMA systems where the length N of the PN codes used is much larger than the number of users M. Let C_{max} denote the maximum magnitude of the aperiodic cross-correlation function as defined in Section 9.2. Then

$$P_{max} \leq 1 - Q\left\{\left[1-(M-1)\left(\frac{2C_{max}}{N}\right)\right]\sqrt{\frac{2E_b}{N_0}}\right\} \tag{9.11}$$

Often the large cross-correlation values arise for only a few values of the delay parameter τ. In this situation it is more meaningful to consider the average performance rather than the worst case. Two measures of average performance are the average error probability, which is described in Ref. 35, and the average signal-to-noise ratio, which is discussed in Ref. 12.

9.5 CONCLUDING REMARKS

Synchronization and antijamming are the principal problems in CDMA. Rapid acquisition implies short sequences, while longer sequences with ever-increasing time–bandwidth product are required to combat jamming. To satisfy these conflicting requirements, the use of subsequences of a long sequence to provide synchronization has been proposed [8]. The problem of constructing sequences with good aperiodic cross-correlation functions is also undergoing further investigation.

Although CDMA has been used primarily in military systems, there is a renewed interest in it for commercial applications due to new technologies [36]. The hardware associated with CDMA-code generators, correlators and matched filters are within the state of the art and are now not significant extra burdens. As users of satellite systems become concerned with interference and un-authorized reception, the substantial levels of antijam (AJ) and low probability of intercept (LPI) provided by CDMA techniques will become more important.

In our treatment of CDMA, we have not considered the antijam and anti-intercept performance criteria. These are described in some detail by Ristenbatt and Daws [9]. We have not considered error-control coding aspects. Viterbi [37] has shown that coding is always beneficial and sometimes crucial in CDMA systems.

REFERENCES

1   R. C. Dixon, *Spread Spectrum Systems*, Wiley, New York, 1976.

2   R. C. Dixon, *Spread Spectrum Techniques*, IEEE Press, New York, 1976.

3   D. V. Sarwate and M. B. Purseley, "Crosscorrelation Properties of Pseudorandom and Related Sequences", *Proc. IEEE*, Vol. 68, May 1980, pp. 593–619.

4   J. K. Holmes and C. C. Chen, "Acquisition Time Performance of PN Spread-Spectrum Systems", *IEEE Trans. Commun.*, Vol. COM-25, Aug. 1977, pp. 778–783.

5   Special Issue on Spread Spectrum Communications, IEEE *Trans. Commun.*, Vol. COM-25, Aug. 1977, pp. 745–869.

6   R. A. Yost, "Susceptibility of DS/FH, *M*-ary DPSK to Partial- and Full-Band Noise, CW-Tone, and Periodic FM Jamming", Interim Tech. Rep., School of Elec. Eng., Georgia Inst. of Technol., Atlanta, Ga., for HqUSAF Contract No. F30602-75-0118, Feb. 1978.

7   "JTIDS/TIES Consolidated Tactical Communications", *Electron. Warfare*, Sept. 1977, pp. 45–53.

8   I. F. Blake and J. W. Mark, "CDMA Sequences and Techniques", Tech. Rep., Dept. of Elect. Eng., Univ. of Waterloo, Waterloo, Canada, for DSS Project No. 808-01, Dec. 1978.

9   M. P. Ristenbatt and J. L. Daws, Jr., "Performance Criteria for Spread Spectrum Communications", *IEEE Trans. Commun.*, Vol. COM-25, Aug. 1977, pp. 756–763.

10 R. Gold, "Optimal Binary Sequences for Spread Spectrum Multiplexing", *IEEE Trans. Inf. Theory*, Vol. IT-13, 1967, pp. 619–621.

11 R. J. McEliece, "Correlation Properties of Sets of Sequences Derived from Irreducible Cyclic Codes", *Inf. Control*, Vol. 45, Apr. 1980, pp. 18–25.

12 M. B. Pursley, "Performance Evaluation for Phase-Coded Spread-Spectrum Multiple-Access Communications—Part I: System Analysis", *IEEE Trans Commun.*, Vol. COM-25, Aug. 1977, pp. 795–799.

13 J. L. Massey and J. J. Uhran, "Sub-band Coding", in Proc. 13th Annu. Allerton Conf. Circuits Syst. Theory, Oct. 1975, pp. 539–547 (see also "Final Report for Multipath Study", Dept. of Elec. Eng., Univ. of Notre Dame, Notre Dame, Ind., 1969).

14 S. W. Golomb, *Shift Register Sequences*, Holden-Day, San Francisco, 1967.

15 R. Gold, "Code Synthesis Study", Final Report, Vol. II, Robert Gold Assoc., Los Angeles, June 1973.

16 F. J. MacWilliams and N. J. A. Sloane, "Pseudorandom Sequences and Arrays", *Proc. IEEE*, Vol. 64, 1976, pp. 1715–1719.

17 L. R. Welch, "Lower Bounds on the Maximum Cross-Correlation of Signals", *IEEE Trans. Inf. Theory*, Vol. IT-17, 1974, pp. 397–399.

18 J. J. Spilker, "Delay-Lock Tracking of Binary Signals", *IRE Trans. Space Electron. Telem.*, Vol. SET-9, Mar. 1963, pp. 1–8.

19 J. J. Spilker and D. T. McGill, "The Delay-Lock Discriminator—An Optimum Tracking Device", *Proc. IRE*, Vol. 49, Sept, 1961, pp. 1403–1416.

20 H. P. Hartmann, "Analysis of a Dithering Loop for PN Code Tracking", *IEEE Trans. Aerosp. Electron. Syst.*, Vol. AES-10, Jan. 1974, pp. 2–9.

21 N. G. Davies, "Performance and Synchronization Considerations", in L. A. Gerhardt (Ed.), *Spread Spectrum Communications*, AGARD NATO Lecture Series (58), 1973, pp. 4.1–4.24.

22 R. B. Ward, "Acquisition of Pseudonoise Signals by Sequential Estimation", *IEEE Trans. Commun. Technol.*, Vol. COM-13, Dec. 1965, pp. 475–483.

23 L. B. Milstein and P. K. Das, "Surface Acoustic Wave Devices", *IEEE Commun. Mag.*, Vol. 17, Sept. 1979, pp. 25–33.

24 W. J. Gill, "A Comparison of Delay-Lock Loop Implementations", *IEEE Trans. Aerosp. Electron. Syst.*, Vol. AES-2, July 1966, pp. 415–426.

25 C. R. Cahn, "Spread Spectrum Applications and State-of-the-Art Equipment", in L. A. Gerhardt (Ed.), *Spread Spectrum Communications*, AGARD NATO LEcture Series (58), 1973, pp. 6.1–6.111.

26 S. Udalov, "Shuttle Ku-Band Forward Link Signal Design and Performance Considerations", in *Proc. Natl. Telecommun. Conf.*, Dec. 1977, pp. 27.6.1–27.6.9.

27 R. A. Stampfl and A. E. Jones, "Tracking and Data Relay Satellite Systems", *IEEE Trans. Aerosp. Electron Syst.*, Vol.. AES-6, May 1970, pp. 276–289.

28 I. L. Lebow, K. L. Jordan and P. R. Drouilhet, Jr., "Satellite Communications to Mobile Platforms, *Proc. IEEE*, Vol. 59, Feb. 1971, pp. 139–159.

29 I. G. Stiglitz, "Multiple-Access Considerations—A Satellite Example", *IEEE Trans. Commun.*, Vol. COM-21, May 1973, pp. 577–582.

30 A. N. Ince, "Code Division Multiplexing for Satellite Systems" in J. K. Skwirzynski (Ed.), *Communication Systems and Random Theory*, Darlington, U.K., 1978, pp. 821–857.

31 C. C. Chen and J. W. Burnett, "TDRS Multiple Access Channel Design", in *Proc. Natl. Telecommun. Conf.*, Dec., 1977, pp. 19.2.1–19.2.7.

32 J. M. Aein, "Multiple Access to a Hard-Limiting Communication Satellite Repeater", *IEEE Trans. Space Electron. Telem.*, Vol. SET-10, Dec. 1964, pp. 159–167.

33 J. W. Schwartz, J. M. Aein and J. Kaiser, "Modulation Techniques for Multiple Access to a Hard-Limiting Satellite", *Proc. IEEE*, Vol. 54, May 1966, pp. 763–777.

34 D. R. Anderson and P. A. Wintz, "Analysis of a Spread-Spectrum Multiple-Access System with a Hard Limiter", *IEEE Trans. Commun. Technol.*, Vol. COM-17, Apr. 1969, pp. 285–290.

35 K. Yao, "Error Probability of Asynchronous Spread Spectrum Multiple Access Communication Systems", *IEEE Trans. Commun.*, Vol. COM-25, Aug. 1977, pp. 803–809.

36 G. R. Cooper and R. W. Nettleton, "A Spread-Spectrum Technique for High-Capacity Mobile Communications", *IEEE Trans. Veh. Technol.*, Vol. VT-27, Nov. 1978, pp. 264–275.

37 A. J. Viterbi, "Spread Spectrum Communications—Myths and Realities", *IEEE Commun. Mag.*, Vol. 17, May 1979, pp. 11–18.

PROBLEMS

1 **(a)** Draw the block diagram of a hybrid FH/DS transmitter/receiver.

(b) Show that

$$G_p(\text{FH/DS}) = G_p(\text{FH}) + G_p(\text{DS}) \qquad \text{dB}$$

$$= 10\log_{10}(\text{number of channels})$$

$$+ 10\log_{10}\left(\frac{R_c}{R_i}\right) \qquad \text{dB}$$

(c) What would be the processing gain of a FH/DS system with a chip rate of 30 Mcps, a data rate of 4 kbps and 150 frequency channels?

(d) Why does FH/DS offer simpler implementation possibilities?

2 **(a)** Show that the normalized power density of a PN code of length N and chip duration T_c is given by

$$\frac{N+1}{N^2}\left[\frac{\sin(\omega T_c/2)}{\omega T_c/2}\right]^2 \sum_{\substack{n=-\infty \\ n\neq 0}}^{\infty}\left(\omega - \frac{2\pi n}{T_c}\right) + \frac{1}{N^2}\delta(\omega)$$

(b) Show that by doubling N, the lines in the spectrum are twice as dense but the power in each line is reduced by a factor of 2.

(c) For a fixed T_i, why is the processing gain of a DS system determined solely by T_c?

(d) A truly random waveform has no DC term. Why is there a DC term in the power density of the PN code?

3 **(a)** What are the sequences generated by the polynomials $1+x^2+x^3+x^4+x^5$ and $1+x+x^2+x^4+x^5$? Assume an initial condition of (11111) in both cases.

(b) Compute and plot the periodic and aperiodic cross-correlation function for the sequences obtained in part (a).

4 Let $\{\mathbf{u}, \mathbf{v}\}$ denote a "preferred pair" of m-sequences of period $N = 2^m - 1$ generated by the primitive binary polynomials $h(x)$ and $\hat{h}(x)$, respectively. Then $G(u, v)$, a set of Gold sequences is defined as

$$G(\mathbf{u}, \mathbf{v}) = \{\mathbf{u}, \mathbf{v}, \mathbf{u} \oplus \mathbf{v}, \mathbf{u} \oplus T\mathbf{v}, \mathbf{u} \oplus T^2 \mathbf{v}, \dots, \mathbf{u} \oplus T^{N-1} \mathbf{v}\}$$

(a) Show that every sequence in $G(\mathbf{u}, \mathbf{v})$ can be generated by the polynomial $f(x) = h(x)\hat{h}(x)$.

(b) Show that the non-maximal-length sequences belonging to $G(\mathbf{u}, \mathbf{v})$ can be generated by term-by-term modulo 2 addition of the outputs of the shift registers corresponding to $h(x)$ and $\hat{h}(x)$.

(c) Show that the values taken on by the "preferred" three-valued cross-correlation function are -1, $-t(m)$ and $t(m) - 2$, where

$$t(m) = 1 + 2^{(m+2)/2}$$

5 The JPL ranging codes are constructed by modulo-2 addition of two or more m-sequences with relatively prime lengths.

(a) Show that the length of the JPL code is equal to the product of the length of the constituent m-sequences.

(b) Explain the use of these codes for initial synchronization.

6 Show that the rms tracking error $\sigma_{\Delta\tau}$ for the delay-lock tracking loop is given by

$$\sigma_{\Delta\tau} = \frac{1}{\sqrt{S/2N_0 B_L}} \frac{\sqrt{R(0)_{BL} - R(2\tau_d)_{BL}}}{2R'(\tau_d)_{BL}}$$

where S = average signal power

B_L = one-sided loop noise bandwidth

N_0 = one-sided noise power spectral density

τ_d = delay or advance of the reference PN code

$R(x)_{BL}$ = bandlimited autocorrelation function for the code evaluated at x

7 The sum signal $s(t) + i(t)$ is applied to a hard limiter whose output has a constant amplitude 1, to produce a signal with phase modulation only. Assuming that the interference is of constant amplitude and is large with respect to the signal, show that the signal power-to-interference ratios at the input and the output of the hard limiter are related by

$$\left(\frac{S}{I}\right)_o = \left(\frac{S}{I}\right)_i - 6 \text{ dB}$$

8 What form does (9.11) take for the simple case of two users?

9 In FH with M-ary FSK, the jammer spaces a series of jamming tones across the bandwidth. Show that

$$P(\text{hop-decision error}) = (M-1)\frac{N_j}{N_f}$$

where N_j = number of jamming tones
N_f = number of orthogonal frequencies

10 In Problem 9 we have seen that the signalling efficiency of FH with M-ary FSK degrades as M increases, which is the reverse of what happens in white Gaussian noise. Show that the use of a Reed–Solomon code (see Chapter 13) will yield dramatic performance improvements for frequency-hopped waveforms in a multitone interference environment.

CHAPTER 10

||||||||||||||||||||||||||||||||||||||||||||||||||||

Packet Satellite Networks

This chapter treats some important areas of packet switching as applied to satellite services. As a TDM technique for accessing a satellite channel, packet switching is a true form of multiple access and closely related to TDMA; the similarities and the distinguishing differences are discussed. The most common name is packet satellite networks (PSN). This chapter is a concise survey of an evolving and rapidly changing field; hence it is more descriptive than analytic. Some concepts and definitions of packet switching are given and a review of protocols for satellite networks is presented. Experimental and operational systems are briefly described.

Packet satellite networks are based on packet communications networks which have existed since 1965 [1–16]. The best-known packet network is ARPANET [17] and there were announcements of packet-switching services in the United Kingdom, The United States, France and Canada from 1972 to 1974. The first public network, TELENET, began service in the United States in August 1975. A comprehensive overview of the evolution of packet switching was given by Roberts [1].

10.1 PACKET-SWITCHING CONCEPTS

We begin with some concepts that are applicable to all packet-switching applications, whether they use wires, cables, radio or satellites. Then some special ideas and methods are introduced to take advantage of the wide-area, broadcast features of satellites, but which exhibit much longer delays than for terrestrial networks.

Why Packet Switching?

Computer communication networks are the primary reason for the existence of packet concepts. Such networks have a distributed network structure and use store-and-forward switching in the provision of services which are interactive but not simultaneously two-way. A typical service is a transaction that might consist of a terminal sending a few characters to a computer, which then sends back (echoes) these characters. This natural packetizing in the sources and the fundamental bursty (high peak-to-average usage) nature of the transmissions are the basic reasons for packet networks, which are combined software and hardware implementations of communication networks. In contrast to circuit switching, in which circuits are actually or virtually set up for the duration of a call, then disconnected, packet switching uses existing circuits to convey packets that contain necessary information to direct the packet to its destination. In packet communications series of packets may be regarded as moving independently in the network. Hence there are queues formed by packets waiting for processing and transmission; queueing theory thus plays a major role in the analysis and design of packet networks [18]. Analogies to postal concepts (packets, addresses, routing, delays, etc.) are quite useful but should not be taken to extremes.

Whereas circuit-switched services are simply characterized by rate and error performance, packets and packet-switched services have many descriptors. We present a brief discussion of the more common descriptors [19–24]; a packet network does not necessarily possess all these features.

Priority

Users can put a priority on each of their messages to satisfy a variety of considerations. Priority is a desirable feature for military and private networks, and is associated with different costs in commercial contexts. The use of priority significantly affects the capacity allocations.

Delay Class

There are significant cost and performance differences associated with meeting different user delay requirements. In a packet network, voice traffic usually cannot tolerate more than a single-hop delay (0.25 s), while some data traffic may be satisfied with delays of more than 1 s. Data traffic can also require fast, moderate or slow service response times.

Holding Time

A user can define a holding time for a message, with the implication that a message can be discarded by the network rather than be delivered late. Maximum flexibility is achieved by allowing the user to choose such holding times and by permitting protocols to choose retransmission times.

Reliability

Some traffic has inherent redundancy, such as in 8-bit PCM voice; some is tolerant of high error rates, such as delta-coded voice; but data traffic cannot tolerate errors and hence error control coding may have to be used. In packet switching, optimization of the overall path can be achieved by allowing users to select the level of error control appropriate to their traffic.

Length

The length of a packet is an important parameter, since it directly affects internal network buffer sizes. Long lengths of 8000 to 16,000 bits reduce the need for fragmenting messages into smaller segments and for keeping header and processing overheads down. Smaller lengths are easier to encode and buffer, and can be transmitted faster. Some systems have different packet lengths, depending on purpose.

Conferencing

This is the network capability to form a mininetwork for a specific purpose or time period. A group addressing service is needed to allow users to define dynamically the set of conference participants. Membership authentication, directory services and demand assignment signalling in PSNs are related features.

Duplicates

Since packets are usually repeated, there is the problem of how to handle duplicate packets. Duplicate detection requires that each station retain information to allow unique identification of each message. The numbering scheme must allow for the maximum holding time, maximum retransmission rate and minimum packet size.

Sequencing

There is also the problem of whether the network should deliver all messages in the order received from the source. Sequencing imposes significant increases in buffering and processing at destination, particularly in PSNs with long delays.

Definitions

A formal definition of a packet is in order [7]. A **packet** of information is a finite sequence of bits, divided into a control portion and a data portion. A header in the control portion contains enough information (addressing) for the packet to be routed (switched) to its destination; the header might also contain very specific descriptors of the packet. There are usually some checks on each packet, so that any node through which the packet passes may exercise error control or other network control.

A **packet switch** at a node or station is the collection of hardware and software resources which implement all procedures within the network, such as routing, resource allocation and error control, and which provides network services through a host.

A **host** is the generic name in computer communications for the collection of hardware and software that uses the packet-switching services to support user services. In data communications, the source and the sink reside in or are the host, which may be a computer.

A **protocol** is a set of communications conventions, including formats and procedures, which allow two or more end points to communicate. In general, the **end points** may be packet switches, hosts, terminals, data bases, people and so on.

In the analysis of packet-switching networks, two significant measures or indicators of performance are throughput and delay. The ratio of the useful traffic to the total number of bits is the **throughput**; in this normalized form, throughput $\eta < 1$. It is measured as the ratio of successful packets to the total that were sent. Chapter 11 contains a detailed discussion of error control, throughput and delay as applied to satellites. The **delay** of a packet is simply the time between start and successful completion of transmission and delivery to the user. But there are many detailed definitions of delay, so comparisons can be difficult for different systems. The average message delay in a system is given by [8, 9]

$$T = K + \sum_{i=1}^{L} \left(\frac{\lambda_i}{\gamma} \cdot \frac{\lambda_i \alpha'/C_i}{C_i/\alpha' - \lambda_i} + \frac{1}{C_i/\alpha} + P_i + K \right) \qquad (10.1)$$

where K = nodal processing time
    L = total number of links in the path
    λ_i = packet arrival rate, independent packets
    γ = total traffic entering network
    α = average length of data packet
    α' = average length of all packets (data, control)
    C_i = transmission rate of link i
    P_i = propagation delay

This equation makes the independence assumption on each queue, and the evaluated delay is generally longer than the real delays in a network with packet lengths that are almost constant.

10.2   SERVICES AND ACCESS METHODS

We now focus on services and access methods in packet satellite networks which take advantage of the wide-area coverage of satellites, the broadcast capability (or point-to-multipoint services) and the mesh topology, whereby pairs of nodes are connected directly by a single hop. The term "packet satellite" is used to distinguish from packet radio, surveyed by Kahn et al. [11].

Service Objectives

The following types of services are possible in packet networks:

1 Interactive terminal/computer
2 Distributed computers and resources
3 Batch operations
4 File transfers
5 Datagrams
6 Streams (voice, facsimile)
7 Data from remote platforms

A **datagram** is a self-contained packet with all appropriate descriptors, including destination addresses; it is particularly useful in broadcast applications. A **stream** is a service characterized by volatile periodic traffic for which the maximum acceptable delay is only slightly larger than the minimum propagation time, or for which the allowable variance in the packet delays is small [10]. This is typified by voice and some forms of facsimile. Datagrams and streams are to be distinguished from virtual circuits, which are typical in distributed networks.

PSN Types of Access

In all forms of packet switching, access protocols are the key to the performance of the network. There are several classifications: distributed control, central control, combination of both; random or deterministic. The distinction between circuit switching and packet switching has already been made. The following categories of access protocols now exist:

1 Fixed assignments
2 Random access
3 Implicit reservation
4 Explicit reservation
5 Hybrid circuit/packet switching

We present the categories in the order of their development in parallel with TDMA demand assignment techniques. Table 10.1 summarizes the categories and classes described in the literature.

Fixed Assignment

F-TDMA is the simplest protocol of those implemented in SATNET (discussed below). In fixed assignment the frame time is divided into slots of fixed

Table 10.1 Access Protocols for Packet Satellite Network

	Network	Characteristics
Fixed assignments	F-TDMA	Fixed, no DA
Random access	Pure Aloha	Unslotted
	S-Aloha	Slotted
	CSMA and others	Carrier sense
Implicit reservation	R-Aloha	Reservation via S-Aloha
Explicit Reservation	R-TDMA	Reservation
	C-PODA	Contention during requests
Hybrid circuit/packet switching	PODA	Priority-oriented DA
	TBACR	Transmit before assignment, collision requests
	Flexible hybrid + many others	

duration with the slots equally divided among the stations. The assignment is permanent and is an example of a fixed-assignment access method. F-TDMA as described is similar in frame structure to conventional TDMA treated in Chapter 8, with the important distinctions (a) that F-TDMA does not have network synchronization and packets are sent asynchronously, and (b) there are no frame sync signals.

Random Access

Pure Aloha Aloha is a packet radio system that was extended to packet satellite applications. What is described here is known as pure-Aloha or **unslotted Aloha**. Terminals of the Aloha system transmit packets randomly, and hence the packets sent from the different terminals may overlap and result in transmission errors. When they overlap, a situation called packet **collision**, the terminals retransmit the packets until they are free from overlap. To avoid repeated overlap, the interval of packet retransmission is randomized in each terminal. The analysis of the Aloha system was made by Abramson and others [8, 9, 25], and is summarized as follows.

Let us assume that a common radio channel is randomly shared by M terminals, and that the message traffic of the terminal i, as normalized by the link capacity, is $S_i (i = 1, \ldots, M)$. Packet retransmission takes place when packet collision occurs; thus the number of packets from terminal i is larger than the message traffic measured in packets. Let us denote the actual traffic of terminal i as normalized by the link capacity by G_i. The probability that a packet from terminal i does not overlap with a packet from other terminals is given by [8]

$$P(i \text{ not overlap}) = \prod_{\substack{j=1 \\ j \neq i}}^{M} (1 - 2G_j) \tag{10.2}$$

The normalized message traffic of the terminal i is therefore expressed as

$$S_i = G_i \prod_{\substack{j=1 \\ j \neq i}}^{M} (1 - 2G_j) \tag{10.3}$$

If all the terminals have identical amounts of message traffic, the normalized total message traffic S can be expressed as

$$S = \sum_{i=1}^{M} S_i = S_i M = G_i M \left(1 - \frac{2G_i M}{M} \right)^{M-1}$$

$$= G \left(1 - \frac{2G}{M} \right)^{M-1} \tag{10.4}$$

where $S = S_i M$ is called the message traffic rate and $G = G_i M$ is called the link traffic rate. When the number of terminals M is infinitely large, (10.4) becomes

$$\lim_{M \to \infty} S = G \lim_{M \to \infty} \left(1 - \frac{2G}{M} \right)^{M-1} = G \exp(-2G) \tag{10.5}$$

S-Aloha To reduce the probability of collisions, time slots were introduced and transmissions could begin only at the start of slots. In the S-Aloha protocol each station maintains two output queues: the new queue (for new packets) and the retransmit queue (for packets that need to be retransmitted because of a previous conflict). All stations follow the same transmission rules. At the beginning of a slot a station transmits a packet from the retransmit queue with probability P_r (retransmit gate). Only if the retransmit queue is empty will the station then transmit a packet from the new queue with probability P_n (new gate). Furthermore, a packet arriving at an empty station (i.e., both queues empty) is transmitted with probability $P_n = 1$. If two or more stations transmit in the same slot, their packets collide and each is received unreliably. The senders detect the conflict after a round-trip time by monitoring the downlink, and promptly return a copy of the collided packet to the retransmit queue.

The probability that a packet from terminal i is free from collision is now given by

$$P(i \text{ not overlap}) = \prod_{\substack{j=1 \\ j \neq i}}^{M} (1 - G_j) \tag{10.6}$$

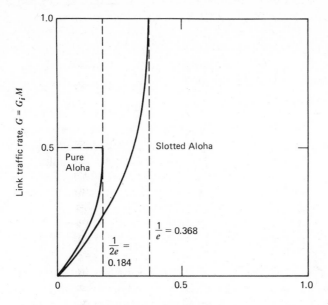

Figure 10.1 Relationship between link and message traffic rates. Note asymptotic maxima of $1/2e$ and $1/e$.

and (10.5), therefore, is replaced by

$$\lim_{M \to \infty} S = G \exp(-G) \tag{10.7}$$

The curves in Figure 10.1 illustrate the relation between S and G for pure Aloha and for S-Aloha. For pure Aloha the maximum utilization of the link is $1/2e$; for S-Aloha it is $1/e$, or about 36% of the available capacity is used. It is noteworthy that these bounds apply to large uniform networks and can be exceeded in real networks with unequal packet sizes and arrival rates.

Others There are other techniques of random access. The most notable for radio packets is carrier-sense multiple access (CSMA) which has maximum usage exceeding 80% if propagation delay is negligible compared to the packet transmission time [26]. For geostationary satellite links this is not practical, since the frames would be required to be several seconds long. Any other form of monitoring of the link status has the same problems.

Implicit Reservation

These methods are called **implicit** because they involve reservation by use.

R-Aloha Another approach to packet demand assignment is the use of a framing concept to permit implicit reservations with the S-Aloha method; this is called R-Aloha. The frame provides a basis for allowing stations with high traffic rates to have one or more slots in each frame for their exclusive use (reservation). These stations are removed from the random-access contention occurring in the remaining slots. Control is distributed in that each station executes an identical assignment algorithm based on global information from the network. Stations maintain a record of slot utilization for one frame; when a station uses a contention slot successfully, the slot is assigned to that station in each successive frame until it stops using it. The frame duration must be at least the time for a single hop; otherwise, there are severe instabilities. Lam showed that R-Aloha can achieve throughput ranging from S-Aloha to F-TDMA, depending on the traffic conditions [37].

Explicit Reservation

An **explicit reservation** is a distinct and unique assignment of slots to a user by the network scheduler.

R-TDMA The R-TDMA protocol establishes a permanent association (ownership) between slots and stations similar to F-TDMA. Unlike F-TDMA, however, the slots not claimed by the original owner may be reassigned on a round-robin* basis to the stations that have traffic to send. Thus it is a dynamic reservation scheme.

In R-TDMA a frame consists of M reservation slots (where M is the number of stations) and a number of data slots. Reservation slots are smaller than regular slots. When the frame size is chosen to balance the loading for reservations and message packets, the maximum channel usage is given by

$$\max S = \frac{AMB}{AMB + 2e} \tag{10.8}$$

where B is the average number of packets per reservation and A is a constant near unity, depending on the retransmission strategy.

Each station declares its backlog (i.e., the number of packets awaiting transmission) using its reservation slot. Reservations are monitored by all stations and synchronized reservations tables are maintained in all stations showing the outstanding transmission requirements. The reservation table is used by the channel scheduler (a distributed algorithm that runs synchronously and identically in all stations) to assign future slots to users in a demand-access round-robin fashion.

C-PODA PODA (Priority Oriented Demand Assignment) techniques are presented below. C-PODA (contention based) is a satellite channel access scheme [10] which allocates slots to users based on prior reservations. Users contend for reservations by making repeated requests using a S-Aloha channel,

---

*Round robin is used in the sense of each station getting access in an ordered, sequential manner but not necessarily predetermined.

or they may piggyback reservation requests in the header of ongoing traffic. When a reservation request is honored, the message is scheduled for transmission according to priority, type and desired delivery time. The C-PODA protocol handles both data and stream traffic and allows for the existence of large and small stations.

The C-PODA frame consists of an information subframe I, used for sending messages, and a control subframe C, used for sending control packets that carry acknowledgements and reservation requests. The information subframe is partitioned into variable-size slots equal to the size of the packets to be transmitted, in contrast with the previous access schemes, which used fixed slot sizes. The control subframe is divided into smaller uniform time slots called virtual slots, and accessed by means of a S-Aloha protocol. Each station chooses randomly with uniform probability at most one virtual slot in each subframe in which to send its control packets. The boundary between the data subframe and the control subframe is dynamically adjusted based on load fluctuations. However, to guarantee a certain performance, the control subframe is not permitted to be reduced below a minimum size.

Hybrid Circuit/Packet Switching

Hybrid networks listed in Table 10.1 take advantage of features of both circuit switching and packet switching [27, 28]. As shown in the table, hybrid schemes include generalized packet satellite networks (GPSN), TBACR and flexible schemes.

PODA The concept of a general-purpose packet satellite network (GPSN) was introduced by Jacobs et al. [10]. This is also called generalized PSN. The requirements of a GPSN include:

1 Efficient use of satellite bandwidth
2 Satisfaction of multiple delay constraints
3 Multiple priority levels
4 Variable message lengths
5 Stream traffic
6 Different transmission rates
7 Fairness
8 Efficient message acknowledgements
9 Robustness

No one system will satisfy all these requirements and significant tradeoffs are needed. Shorter delays must be achieved at the expense of lower throughput (server efficiency) and conversely. This is complicated by the necessity to use the satellite links with their 250-ms delays. Minimum delays of less than 1s are expensive to achieve. Jacobs et al. discuss these requirements in some detail.

PODA is a protocol designed to satisfy efficiently the requirements of a GPSN, including support of datagram, stream and burst traffic, multiple delay classes, levels of priorities, and variable message lengths. Its design is an integration of circuit and packet-switching assignment and control techniques [10]. Its features are summarized here.

1 For datagram messages, explicit reservations are used, generally resulting in double-hop delays.

2 A single explicit reservation is used to set up each stream, with subsequent packets automatically scheduled as TDMA bursts, thus experiencing one-hop delays.

3 Very high reliability, when required, is provided by the use of scheduled acknowledgements and packet retransmissions.

4 High availability and improved performance are achieved by the use of distributed control techniques.

5 Robustness and mixed transmission rates are achieved by the integrated use of distributed and centralized control techniques.

TBACR Schemes   Bose and Rappaport [29] have described and analysed two variations of Transmit Before Assignment using Collision Request (TBACR) schemes. Scheme A uses three types of forward channels. The single request channel is used for random transmissions of requests for assignments. There are C_p pooled channels which are used in demand-assigned mode with a central station making the assignments. There are also C_t trailer channels used for random transmissions of packets while waiting for assignments. The reverse direction has two channels: one for assignments from the central station; the other for acknowledgements from the destination stations; these reverse channels are free of contention.

Scheme B has a single request channel and C_m message channels. Random transmissions are allowed on unassigned channels. Of course, all users must monitor the assignments from central control and there are provisions for inhibition and preemption of random transmissions.

These schemes were analysed for throughput and average delay, with bounding techniques used where exact analysis was not possible. The published curves in Ref. 29 show low delay at low traffic demand, similar or better than pure Aloha, and up to 70% throughput with reasonable delays, which is better than S-Aloha. With suitable selection of parameters these schemes are applicable to PSNs.

Flexible Hybrid   Yet another scheme emphasizes the integration of voice and data services [30]. Voice and interactive data are packet switched but bulk data are circuit-switched. The goals of this scheme include stable and consistent system performance over a wide range of user demands and yet sufficient flexibility to accommodate new technology and new approaches.

Other Protocol Methods and Analysis

Some other approaches to protocols, assignments and analysis techniques are briefly mentioned and referenced. Tree search algorithms, first studied extensively by Capetenakis [31], are deterministic assignment schemes. In packet satellite applications, there is another level of control required to coordinate the concurrent conduct of several tree searches. For a discussion and survey of techniques, the reader is referred to Ref. 32.

Comparisons of Protocols

Figure 10.2 compares the delay versus throughput curves for the protocols discussed above. C-PODA exhibits a negative slope since reservations can piggyback on regular packets as traffic increases: the reservation delay is reduced and the average packet delay is lower.

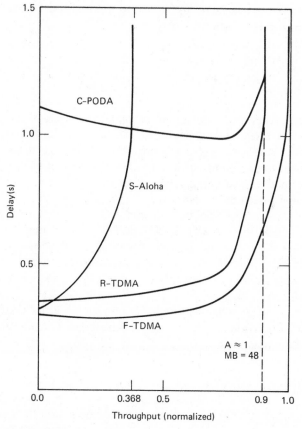

Figure 10.2 Delay versus throughput in a three-station configuration for various channel access protocols.

10.3   EXAMPLES OF PACKET SATELLITE NETWORKS

Aloha

The Aloha PACSAT experiments were the first to apply packet switching with satellites.

SATNET

SATNET is an experimental PSN for voice and data. It was the most extensive field trial as of 1979 of packet satellite concepts, designs, development and international cooperation. The objectives of SATNET are to

1   Provide hardware and software checkout for various protocol designs.
2   Study the behaviour of various protocols from channel, user and application viewpoints.
3   Select system parameters to optimize desired performance measures (throughput, delay, fairness, robustness, etc.).
4   Support the development of reliable host-SATNET protocols, interfaces and internetworking requirements.
5   Assist in the development of efficient voice conferencing protocols.

The international aspects of this experiment are noteworthy. Commercial, educational and research organizations from the United States, the United Kingdom, and Norway are participants. INTELSAT provides the satellite capacity at a new tariff approved in late 1974 for a new kind of service known as Multi-Destination Half-Duplex (MDHD). The significance of this and some implications have been discussed by Kahn [19].

The capacity for SATNET is permanently assigned to a SPADE channel (see Chapter 7) of the Atlantic INTELSAT IV-A satellite. This single RF channel is shared by all stations in SATNET and provides a nominal bandwidth of 38 kHz; it supports 64 kbps using QPSK modems. This is an interesting application of hybrid TDMA and PSN techniques in a SCPC/FDMA channel.

Figure 10.2 portrays the measured delay versus total throughput for a test network consisting of three stations with balanced traffic on each. An artificial traffic generator at each station was generating R packets per slot, where R was varied from 0 to 1.

WDPSN

ARPA (Advanced Research Project Agency) plans to use the SATNET technology in a U.S. domestic PSN called Wideband Domestic Packet Satellite Network (WDPSN). Initially, there will be five locations using a 3-Mbps channel and integrated voice/data is the major interest.

Data Retransmission

Random-access or round-robin protocols are used in what are commonly called **data retransmission platforms** (DRPs) or **remote monitoring systems** (RMS). These are systems for the collection of data via satellite from land-based monitoring sites. Existing systems include LANDSAT, NOAA/GOES and several experimental systems; these are surveyed and an international experiment in RMS is described in Ref. 33. Such systems are characterized by low data rates, low power and low cost, often with transmit-only stations. Low-altitude polar-orbiting satellites are used for total Earth coverage and for lower path losses.

Electronic Mail

INTELSAT and some signatories are operating Intelpost, an international system for electronic mail services.

SARSAT

Search and Rescue Satellite Aided Tracking (SARSAT) is an international project with participation by the United States, Canada and France, and possibly Norway and others. The USSR is planning COSPAS, which will be compatible. There are now about 250,000 Emergency Locator Transmitters (ELTs) deployed primarily in commercial aircrafts and some marine vessels. A polar-orbiting satellite picks up the ELT signals and relays them to a central processing site. Based on doppler processing of the received signal, in a frequency–time plot the slope of the curve has information on the longitude, and the zero crossing contains latitude data. Experiments have verified the feasibility of timely location determination to within a 15-km radius, so that rescue units can be dispatched quickly and have a small area to search. Signals from ELTs are now analog; developments by the SARSAT project will permit digital modulation with identification and status messages. Also, on-board processing is being developed in France [34–36].

REFERENCES

1 L. G. Roberts, "The Evolution of Packet Switching", Special Issue on Packet Communication Networks, *Proc. IEEE*, Vol. 66, Nov. 1978, pp. 1307–1313.

2 J. Lederberg, "Digital Communications and the Conduct of Science: The New Literacy", Special Issue on Packet Communication Networks, *Proc. IEEE*, Vol. 66, Nov. 1978, pp. 1314–1319.

3 L. Kleinrock, "Principles and Lessons in Packet Communications", Special Issue on Packet Communication Networks, *Proc. IEEE*, Vol. 66, Nov. 1978, pp. 1320–1329.

4 J. C. R. Licklider and A. Vezza, "Applications of Information Networks", Special Issue on Packet Communication Networks, *Proc. IEEE*, Vol. 66, Nov. 1978, pp. 1330–1346.

5 L. Pouzin and H. Zimmerman, "A Tutorial on Protocols", Special Issue on Packet Communication Networks, *Proc. IEEE*, Vol. 66, Nov. 1978, pp. 1346–1370.

6 R. F. Sproull and D. Cohern, "High-Level Protocols", Special Issue on Packet Communication Networks, *Proc. IEEE*, Vol. 66, Nov. 1978, pp. 1371–1386.

7 V. G. Cerf and P. T. Kirstein, "Issues in Packet-Network Interconnection", Special Issue on Packet Communication Networks, *Proc. IEEE*, Vol. 66, Nov. 1978, pp. 1386–1409.

8 H. Inose and T. Satio, "Theoretical Aspects in the Analysis and Synthesis of Packet Communications Networks", Special Issue on Packet Communication Networks, *Proc. IEEE*, Vol. 66, Nov. 1978, pp. 1409–1422.

9 F. A. Tobagi, M. Gerla, R. W. Peebles and E. G. Manning, "Modeling and Measurement Techniques in Packet Communication Networks", Special Issue on Packet Communication Networks, *Proc. IEEE*, Vol. 66, Nov. 1978, pp. 1423–1447.

10 I. M. Jacobs, R. Binder and E. V. Hoversten, "General Purpose Packet Satellite Networks", Special Issue on Packet Communication Networks, *Proc. IEEE*, Vol. 66, Nov. 1978, pp. 1448–1467.

11 R. E. Kahn, S. A. Gronemeyer, J. Burchfiel and R. C. Kunzelman, "Advances in Packet Radio Technology", Special Issue on Packet Communication Networks, *Proc. IEEE*, Vol. 66, Nov. 1978, pp. 1468–1496.

12 D. D. Clark, K. T. Progran and D. P. Reed, "An Introduction to Local Area Networks", Special Issue on Packet Communication Networks, *Proc. IEEE*, Vol. 66, Nov. 1978, pp. 1497–1517.

13 J. M. McQuillan, "Enhanced Message Addressing Capabilities for Computer Networks", Special Issue on Packet Communication Networks, *Proc. IEEE*, Vol. 66, Nov. 1978, pp. 1517–1527.

14 S. L. Mathison, "Commercial, Legal, and International Aspects of Packet Communication", Special Issue on Packet Communication Networks, *Proc. IEEE*, Vol. 66, Nov. 1978, pp. 1527–1539.

15 P. T. F. Kelly, "Public Packet Switched Data Networks, International Plans and Standards", Special Issue on Packet Communication Networks, *Proc. IEEE*, Vol. 66, Nov. 1978, pp. 1539–1549.

16 I. Gitman and H. Frank, "Economic Analysis of Integrated Voice and Data Networks: A Case Study" Special Issue on Packet Communication Networks, *Proc. IEEE*, Vol. 66, Nov. 1978, pp. 1549–1570.

17 L. G. Roberts, "Multiple Computer Networks and Intercomputer Communication", *ACM Symp. Oper. Syst. Princ.*, Gatlinburg, Tenn., Oct. 1967.

18 L. Kleinrock, *Queueing Systems*, Vols. I and II, Wiley-Interscience, New York, 1976.

19 R. E. Kahn, "The Introduction of Packet Satellite Communications", *Packet Satellite Communication Techniques and Experience*, Natl. Telecommun. Conf., Washington, D.C., Nov. 1979.

20 I. M. Jacobs, E. V. Hoversten, N. L. Abel, R. Binder, R. D. Bressler, W. B. Edmond and E. A. Killian, "Packet Satellite Network Design Issues", *Packet Satellite Communication Techniques and Experience*, Natl. Telecommun. Conf., Washington, D.C., Nov. 1979. pp. 45.2.1–45.2.12.

21 D. A. McNeill, J. G. Cole, J. H. Malman, W. C. Milliken, S. Rothschild, W. A. Redman, E. V. Hoversten and S. Blake, "SATNET Monitoring and Control", *Packet Satellite Communication Techniques and Experience*, Natl. Telecommun. Conf., Washington, D.C., Nov. 1979, pp. 45.3.1–45.3.4.

22 W. W. Chu, M. Gerla, W. E. Naylor, S. Treadwell, D. Mills, P. Spilling and F. A. Aagesen, "Experimental Results on the Packet Satellite Network", *Packet Satellite Communication Techniques and Experience*, Natl. Telecommun. Conf., Washington, D.C., Nov. 1979, pp. 45.4.1–45.4.12.

23 P. T. Kirstein, C. Bradbury, R. C. Cringle, H. R. Gamble, R. Binder, D. McNeill, J. Cole, R. Bressler and D. L. Mills, "SATNET Applications Activities", *Packet Satellite Communication Techniques and Experience*, Natl. Telecommun. Conf., Washington, D.C., Nov. 1979, pp. 45.5.1–45.5.7.

24 D. L. Mills, "SATNET Demonstrations", *Packet Satellite Communication Techniques and Experience*, Natl. Telecommun. Conf., Washington, D.C., Nov. 1979, p. 45.6.1.

25 N. Abramson, "The Aloha System", *Computer Communication Networks*, Prentice-Hall, Englewood Cliffs, N.J., 1973.

26 L. Kleinrock and F. Tobagi, "Random Access Techniques for Data Transmission over Packet-Switched Radio Channels", *Proc. NCC'75*, May 1975, pp. 187–201.

27 D. Weese, "A Satellite Data Network Model Employing Packet and Circuit Switching", in *Proc. 3rd Int. Conf. Comput. Commun.*, Toronto, 1976, pp. 111–116.

28 A. Ephremides and O. Mowafi, "Analysis of a Hybrid Scheme for Buffered Users— Probabilistic Time Division", in *Proc. Natl. Telecommun. Conf.*, Houston, Tex., Dec. 1980.

29 S. Bose and S. S. Rappaport, "A Multiple Access Scheme for Packet Radio Channels", in *Proc. Natl. Telecommun. Conf.*, Washington, D.C., Nov. 1979, pp. 11.5.1–11.5.5.

30 M. J. Bose and C. M. Sidio, "Approaches to the Integration of Voice and Data Telecommunications", in *Proc. Natl. Telecommun. Conf.* Washington, D.C., Nov. 1979, pp. 46.6.1–46.6.8.

31 J. Capetenakis, "The Multiple Access Broadcast Channel: Protocol and Capacity Considerations", Report ESL-R-806, Elect. Syst. Lab., MIT, Cambridge, Mass., Mar. 1978.

32 C. Meubus and M. Kaplan, "Protocols for Multi-access Packet Satellite Communications", in *Proc. Natl. Telecommun. Conf.*, Washington, D.C., Nov. 1979, pp. 11.4.1–11.4.7.

33 K. Singh and G. P. Forcina, "Satellite Remote Monitoring System: General Requirements and a Proposed New Approach", in *Proc. ICDSC-4*, Montreal, 1978, pp. 16–22.

34 W. N. Radisch and B. J. Trudell, "The Search and Rescue Satellite Mission—a Basis for International Co-operation", Position Location and Navigation Symposium (PLAN-78), San Diego, Nov. 1978.

35 Ozurabor et al., "COSPAS Project—A Satellite Aided Experimental System for SAR Applications", *Congr. Int. Astr. Fed.*, Munich, Sept. 1979, paper IAF-79-A-33,

36 H. L. Werstiuk and A. E. Winter, "The Search and Rescue Satellite (SARSAT) System Project", *Operational Modelling of the Aerospace Propagation Enviornment*, AGARD-CPP-238, Apr. 1978, pp. 20.1–20.5.

37 S. S. Lam, "Packet Switching in a Multi-access Broadcast Channel with Application to Satellite Communication in Computer Network", Ph.D. dissertation, Dept. of Comp. Sci. Univ. of Calif., Los Angeles, Mar. 1974; also in Univ. Calif., Los Angeles Tech. Rep. UCLA-ENG-7429, Apr. 1974.

PROBLEMS

1 For a fully connected mesh of N nodes, verify that the number of one-way connections or links is

$$N(N-1)$$

2 For a network of N nodes and L links, verify that the number of two-way connections is

$$\frac{\left[\dfrac{N(N-1)}{2}\right]!}{\left[\dfrac{N(N-1)}{2}-L\right]!\,L!}$$

PART 3

Coding

Fundamentals of Error-Control Coding: Forward Error Correction and Automatic Repeat Request

In previous chapters we treated the modulation, signal processing and multiple-access problems related to digital communications by satellite. We now turn our attention to the problem of error-control coding on these channels.

As mentioned in Chapter 1, error-control coding can be achieved by **forward error correction (FEC), automatic repeat request (ARQ)** or by a combination of FEC and ARQ. The applications of these techniques to satellite communications are considered first. Following the presentation of the fundamental properties, potential gains and limitations of coding in general, ARQ as it applies to satellite channels is discussed in detail. Specific and powerful coding/decoding techniques especially attractive for satellite communications are examined thoroughly in Chapters 12, 13 and 14.

11.1 INTRODUCTION: THE CODING PROBLEM

In the general communication system using FEC depicted in Figure 11.1, the source generates data bits or messages that must be transmitted to a distant user over a noisy channel. Generally speaking, a specific signal is assigned to each of the possible M messages that can be transmitted by the system, and the selection rule that assigns a transmitted signal to each possible message is the code. Because of the channel noise, the transmitted signals do not arrive at the receiver exactly as transmitted, and hence errors are made in conveying the source information to the user. A natural objective of any system is to achieve error-free transmission in the most economic manner.

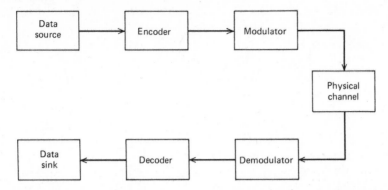

Figure 11.1   Block diagram of a digital communication system using FEC.

The coding theorem of Shannon [1] demonstrates the existence of codes that achieve reliable communication if and only if the information transmission rate is smaller than some maximum rate $\mathcal{C}$ called the **channel capacity**. This remarkable result indicates that the ultimate limit of performance set by the noise on the channel is not the accuracy, as generally believed before Shannon's work, but by the rate at which the data can be reliably transmitted. The concept of channel capacity is fundamental to communication theory and is surprisingly powerful and general since it can be applied to large classes of channel models, whether memoryless or with memory, discrete or nondiscrete. In this chapter we investigate the implications of the channel capacity for the space channel modelled as a discrete memoryless Gaussian channel. Detailed treatment of the channel capacity concept may be found in communication and information theory texts [2–5].

Shannon's celebrated coding theorems are only existence theorems; they give no clue as to the actual construction of these promising coding schemes. Consequently, since publication of Shannon's result, a considerable amount of work has been concerned with the search and analysis of practical coding and decoding techniques that permit reliable communication at the high data rate promised by the theory. Some of these techniques particularly suited to satellite communications are presented in subsequent chapters.

The first codes investigated were called **block codes**. In these codes a sequence of k information symbols are encoded in a block of n symbols, $n>k$, to be transmitted over the channel. For a data source delivering the information bits at the rate R bps, every T seconds the encoder receives a sequence of $k=RT$ bits which defines the message. After these k information bits have entered the encoder, the encoder generates a sequence of **coded symbols** of length n symbols to be transmitted over the channel. In this transmitted sequence or **codeword**, n must be greater or equal to k in order to guaranty a unique relationship between each codeword and each of the possible 2^k messages. Such a code which maps a block of k information symbols into a block of n coded symbols (also called channel symbols) is called an (n, k)

block code. The code rate is $r=k/n$ bits/symbol, and n is called the **block length**.* Block codes have been studied extensively and there are many good (n,k) block codes [4–7]. However, although Shannon showed that block codes exist for which the error probability decreases exponentially with the block length n, the long blocks which are easy to decode, yield a significantly larger error performance than predicted by the theory.

Another important class of codes are the **convolutional codes** [8]. These codes are particularly suitable when the information symbols to be transmitted arrive serially in long sequences rather than in blocks. That is, in convolutional codes long sequences of information symbols are encoded continuously in a serial form. This is achieved by entering these symbols one at a time into the encoder, which has some finite memory capacity. The information symbols are sequentially shifted through a K-stage shift register, and following each shift some number v of coded symbols are generated and transmitted. These v coded symbols are obtained by parity checking, that is, by modulo-2 addition of the contents of various stages of the shift register according to the specific code. The length K of the shift register is called the **constraint length** of the code, and the code rate is $r=1/v$ bit per transmitted symbol. Randomly chosen convolutional codes are known which achieve error probabilities decreasing exponentially with the constraint length of the code. Moreover, it has been shown [8–10] that when used in conjunction with a good decoder, convolutional codes outperform block codes with the same order of complexity, or the same encoder memory. Convolutional encoding is treated in detail in Chapter 12.

11.2 CODING FOR THE SATELLITE CHANNEL: CAPACITY, CODING GAIN AND BANDWIDTH EFFICIENCY

Having introduced the coding problem for arbitrary channels, we now apply coding to a satellite channel and determine the potential benefit and communication efficiency improvement to be gained by using coding.

The two principal characteristics of the somewhat idealized satellite channel model are the following:

1   The principal disturbance is additive wideband white Gaussian noise.
2   The transmission delay is large, about 250 ms for a synchronous geostationary satellite.

Furthermore, up to the present and the foreseeable future, satellite repeaters are more power limited than bandwidth limited and hence sufficient

---

*In what follows we will make the distinction between the information symbols delivered by the source and the coded or channel symbols which are transmitted in the channel. A binary information symbol carries 1 bit of information, whereas a binary channel symbol carries r, $r<1$, bits of information.

bandwidth is available to allow some bandwidth expansion. In this environment the improvement of the communication efficiency by coding becomes particularly attractive.

A block diagram of a model of a data communication system using a satellite is given in Figure 11.2. Let the system transmit the information bits at the rate R bps over the channel where the only disturbance is additive white Gaussian noise of power spectral density N_0 W/Hz, and let the received modulated-signal power be P watts. The received energy per bit-to-noise ratio (also called the signal-to-noise ratio per information bit) E_b/N_0 required to achieve a given bit-error probability serves as a figure of merit for different combinations of coding and modulation schemes. Writing E_b/N_0 in terms of the received signal power and information data rate, we have

$$\frac{E_b}{N_0} = \frac{P}{N_0 R} \tag{11.1}$$

Clearly, a coding or modulation scheme that reduces the E_b/N_0 required for a given bit-error probability leads to either an increase in the allowable information rate R in bps, or a decrease in the necessary P/N_0, or a combination of both. A fundamental problem is then determining the practical system that will operate at the lowest E_b/N_0 with a given error performance.

Following the block diagram of Figure 11.2, information bits produced by the source enter the encoder at a rate R bps. Assuming that block coding is performed with a code of code rate $r = k/n$, the encoder inserts $(n-k)$ redundant symbols for every block of k information bits, resulting in blocks n coded symbols long. Now since n coded symbols must be transmitted in the time it takes the source to send k information symbols, the transmission speed

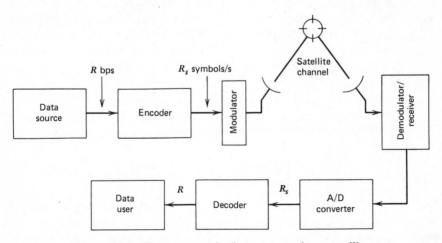

Figure 11.2 Data communication system using a satellite.

R_s in the channel is increased to the value

$$R_s = R\frac{n}{k} = \frac{R}{r} \qquad \text{symbols/s} \tag{11.2}$$

Consequently, if E_b is the received energy per bit, then the corresponding received energy per coded or channel symbol is

$$E_s = E_b\frac{k}{n} = E_b r \qquad \text{J/channel symbol} \tag{11.3}$$

Hence the use of coding entails an increase in the signalling speed on the channel and an increase in the required bandwidth together with a decrease in the received energy per symbol.

At the receiving end, as decoding is usually performed by a digital machine, the demodulator-receiver output for each received symbol enters an analog-to-digital converter (A/D), resulting in either a hard decision (or hard quantization, i.e., binary), or a soft decision (or soft quantization, i.e., A-ary). The decoder observes these decisions, and knowing the particular coding and quantization schemes used, it attempts through a decoding algorithm to correct the transmission errors and then extracts the information bits that may have been sent by the source. These estimates of the information bits are then delivered to the user.

It is worth mentioning that since $E_s/N_0 < E_b/N_0$, more symbols will be received in error with a coded system than with a direct noncoded system. However, in a well-designed coded system, the larger number of channel-induced errors will be compensated by the error-correcting capability of the coding/decoding scheme resulting in a superior overall error performance than with an uncoded system.

The satellite channel is modelled by the well-known and much studied Gaussian memoryless channel. We now specialize the channel capacity discussion to such a channel and examine the implications and the potential benefit to be gained by using coding [11, 12]. Recall that Shannon's capacity theorem establishes the maximum rate at which information can be reliably transmitted through a channel but does not indicate how to construct practical coding/decoding schemes that will achieve capacity.

For a Gaussian memoryless channel of nominal bandwidth B Hz, received signal power P and noise spectral density N_0, it can be shown [2] that the capacity is given by

$$\mathcal{C} = B\log_2\left(1 + \frac{P}{N_0 B}\right) \qquad \text{bps} \tag{11.4}$$

Whenever the information rate R is less than $\mathcal{C}$, then some hypothetical coding–modulation–demodulation–decoding scheme exists which yields an arbitrarily small error probability. Conversely, if R is larger than $\mathcal{C}$, then regardless of the modulation or coding scheme, the error probability will be bounded away from zero. Equation (11.4) indicates that the capacity increases with both the available bandwidth and the signal-to-noise ratio. For an infinite bandwidth the asymptotic value of the capacity is

$$\mathcal{C}_\infty = \lim_{B \to \infty} \mathcal{C} = \lim_{B \to \infty} B \log_2 \left(1 + \frac{P}{N_0 B} \right) \tag{11.5}$$

Using the relation $\lim_{n \to \infty} \ln(1 + x/n)^n = x$, we have

$$\mathcal{C}_\infty = \lim_{B \to \infty} \frac{1}{\ln 2} \ln \left(1 + \frac{P}{N_0 B} \right)^B = \frac{P/N_0}{\ln 2} \tag{11.6}$$

Equation (11.6) shows that as the bandwidth becomes infinitely large, the channel capacity is not infinite and is ultimately bounded by the signal-to-noise ratio. The corresponding limiting value of the received energy per bit-to-noise ratio E_b/N_0 can readily be obtained by replacing in (11.1) the information rate R by its ultimate value $\mathcal{C}_\infty$.

$$\frac{E_b}{N_0} = \frac{P}{N_0 \mathcal{C}_\infty} = \ln 2 \tag{11.7}$$

Under no bandwidth limitations, the minimum value of E_b/N_0 referred to as $E_{b\,\min}/N_0$ is called the **Shannon limit**, and approaches a limit of $\ln 2 = 0.693$, or -1.6 dB. Naturally, because available power is at a premium aboard a satellite, it is desirable to design communication systems that operate with an E_b/N_0 as close as possible to this limiting value of -1.6 dB. However, the selection of the particular modulation–coding–demodulation–decoding techniques to be used in a given application is an engineering problem that depends on both theoretical and practical factors such as available bandwidth, implementation cost and complexity, and so on.

The capacity expression (11.4) shows that the bandwidth B and the power signal-to-noise ratio P/N_0 can be traded off in achieving a desired capacity. That is, at a given data rate, a given level of error performance can be achieved by the right combination of bit energy and bandwidth, and by expanding the bandwidth the required E_b/N_0 can be reduced to its minimum possible value. This exchange of bandwidth with signal-to-noise ratio is especially interesting in communication satellites, where the available power is rather limited but the bandwidth is readily available. By comparison, channel bandwidth is restricted in terrestrial systems, and thus, over such bandlimited channels, higher data rates with a given error performance are usually achieved by increasing the transmitted power. Consequently, another useful parameter in establishing the efficiency of a communication system is the **bandwidth efficiency** defined as the number of information bits that are transmitted per cycle of bandwidth used (or, equivalently, in bps per hertz of bandwidth). The maximum value of the bandwidth efficiency is reached at channel capacity since it corresponds to the maximum information rate within a given bandwidth. Assuming transmission at capacity, the maximum bandwidth efficiency for a Gaussian memoryless channel of bandwidth B is readily obtained from the capacity expression (11.4):

$$\frac{\mathcal{C}}{B} = \log_2\left(1 + \frac{P}{N_0 B}\right) \qquad \text{bps/Hz} \tag{11.8}$$

Using (11.1) for $R = \mathcal{C}$, (11.8) becomes

$$\frac{\mathcal{C}}{B} = \log_2\left(1 + \frac{P}{N_0 \mathcal{C}} \cdot \frac{\mathcal{C}}{B}\right) = \log_2\left(1 + \frac{E_b}{N_0} \cdot \frac{\mathcal{C}}{B}\right) \tag{11.9}$$

or, equivalently,

$$\frac{E_b}{N_0} = \frac{1}{\mathcal{C}/B}\left(2^{\mathcal{C}/B} - 1\right) \tag{11.10}$$

Using the bandwidth efficiency as a criterion, different modulation and coding schemes may be compared to each other and to the ultimate limit given by (11.9).

The maximum bandwidth efficiency curve as a function of E_b/N_0 is shown in Figure 11.3, and for convenience the relation is plotted in two parts: Figure 11.3a for $\mathcal{C}/B > 2$, and Figure 11.3b for $\mathcal{C}/B < 2$, corresponding approximately to the **bandwidth-limited** and the **power-limited** regions of operations, respectively. As expected, Figure 11.3 indicates that capacity can be achieved in either the bandwidth-limited or the power-limited region.

In the bandwidth-limited region, capacity can be reached together with a high bandwidth efficiency, but at the expense of ever larger values of E_b/N_0, which increase without bounds. As coding requires a bandwidth expansion, then clearly in this region coding has not much to offer with respect to reducing the required E_b/N_0. This is to be contrasted with the power-limited region of operation, where the bandwidth efficiency is rather poor, but where capacity can be achieved with decreasing E_b/N_0 (up to the Shannon limit of -1.6 dB) as the bandwidth is increased. In fact, Figure 11.3b shows that in attempting to reach the Shannon limit, the law of diminishing returns applies: at a bandwidth efficiency $\mathcal{C}/B = 0.5$ bit per cycle (corresponding to a coding rate of $\frac{1}{4}$ with ideal QPSK), E_b/N_0 is within 0.8 dB of the ultimate limit of -1.6 dB which is theoretically obtained with an infinitely large bandwidth.

When power is readily available but bandwidth strictly limited, multilevel signalling or polyphase signalling may be utilized to have more information bits represented by each transmitted symbol. In multilevel signalling each transmitted symbol may assume A possible levels instead of 2, leading to a coding rate of

$$r_A = \log_2 A \qquad \text{bits/transmitted symbol} \tag{11.11}$$

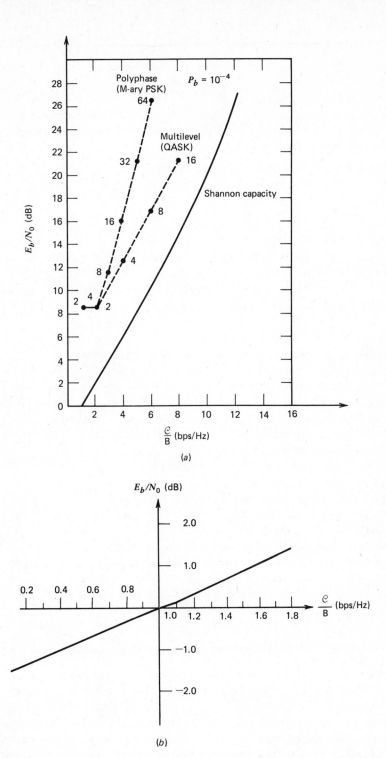

Figure 11.3 Maximum bandwidth efficiencies for the white Gaussian noise channel.

For multilevel signalling doubling the number of amplitudes while maintaining the same probability of error requires maintaining the same level
separation, that is, quadrupling the average power. Therefore, as the available
power increases, capacity can be reached by going deeper into the bandwidth-
limited region, which means a rather inefficient use of the available power.

In polyphase signalling or M-ary phase-shift keying (M-ary PSK), M
phasors are uniformly spaced on a circle of radius $\sqrt{E}$, where E is the signal
energy. These systems have a good bandwidth efficiency given by

$$\frac{R}{B} = \log_2 M, \qquad M > 2 \tag{11.12}$$

but as M increases, a given error probability can be maintained only by
increasing the signal separation, that is, the average power [11]. The bandwidth
efficiency of these two multilevel signalling schemes is shown in Figure 11.3a
for a bit-error probability of 10^{-4}, and it can be seen that these practical
systems perform well below Shannon's capacity limit.

We now investigate the benefit to be gained by using coding in the
power-limited region of Figure 11.3b. In order to compare the performance of
coding with traditional noncoded schemes, the curve of the ideal coherent PSK
modulation error performance over a white Gaussian noise channel is repeated
for convenience in Figure 11.4, where the Shannon limit of -1.6 dB is shown.
The other capacity value ($+0.4$ dB) indicated in the figure takes into consideration the 2-dB loss incurred by the receiving system when hard decisions are
made on the received data [2].

From Figure 11.4 an E_b/N_0 equal to 9.6 dB is required to obtain a bit-error
probability of 10^{-5}. Since $E_{b\,\min}/N_0$ is -1.6 dB, then a potential gain of 11.2
dB is theoretically possible for this error performance. In fact, all systems that
operate with an error performance curve located between that of the ideal PSK
modulation and the Shannon limit will enjoy a so-called coding gain. The
coding gain of a coding scheme is the difference expressed in dB in the required
E_b/N_0 for a given error performance, between the ideal PSK and the particular
coding scheme. This gain is especially attractive in a satellite environment,
since it may be translated either as a reduced transmitted power or reduced
antenna size, or as an increased data rate for the same error performance.

To illustrate the use of coding on digital satellite links, some coding
schemes particularly suitable on the satellite channel are now briefly mentioned. (A detailed presentation of these schemes is given in Chapters 12, 13
and 14.) Since the trend for satellite repeaters is toward larger transmission
power without a corresponding increase in available bandwidth, only codes
that require a modest bandwidth expansion will be considered.

For discrete memoryless channels such as satellite channels, systems employing convolutional encoding at the input and **probabilistic decoding** at the
output are among the most attractive means of approaching the reliability of
communication predicted by Shannon's coding theorems. Probabilistic decoding refers to techniques where the decoded message is obtained by probabilistic

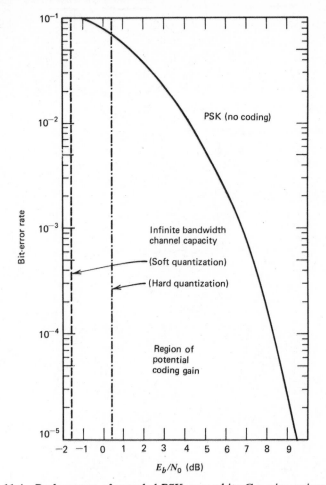

Figure 11.4 Performance of uncoded PSK over white Gaussian noise channel.

considerations rather than by a fixed set of algebraic operations. Moreover, no particular algebraic structure is imposed on the code, which may be chosen at random. The two principal probabilistic decoding techniques are sequential decoding [13] and Viterbi decoding [8]. The applicability of these decoding schemes to the satellite channel has been widely demonstrated [9, 10]. These techniques are examined in Chapter 12.

Many other good nonprobabilistic decoding techniques exist for memoryless channels, and several of them are being used or contemplated for satellite communications. For convolutional codes there is Massey's threshold decoding scheme [14] used on the SPADE system of INTELSAT [15], and for block codes efficient algorithms are known for the Golay and Bose–Chaudhuri–Hocquenghem (BCH) codes [16]. In particular the double-error-correcting, triple-error-detecting (128,112) extended BCH code has been selected for high-speed data transmission with the INTELSAT V satellite [17].

To give some insight into the performance of error-correcting codes on random error channels, the performance of several codes is shown in Figure 11.5. From this figure we observe that the constraint length $K=7$ Viterbi decoder provides superior performance to that of both block codes and the threshold decoder. Furthermore, by using a 3-bit quantization on the channel output rather than a hard decision of the binary symmetric channel, the performance of either Viterbi or sequential decoder may be further improved by approximately 2 dB [10].

Another point of interest is the steep slope of the sequential decoding curve: a relatively small variation in E_b/N_0 results in a very large opposite variation of the bit-error probability (roughly an order of magnitude for each 0.2-dB variation in E_b/N_0). Such behaviour is an indication of the high

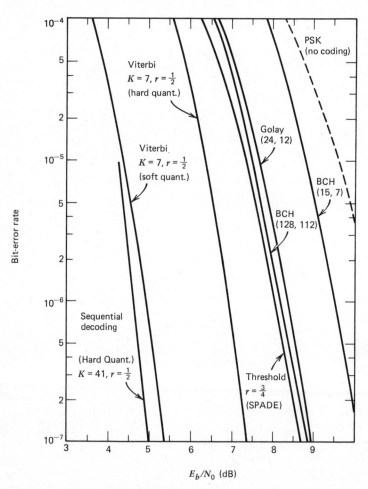

Figure 11.5 Performance of several coding schemes.

sensitivity of this powerful decoding scheme, which would then require accurate control of the transmission and reception of the coded signals.

The effectiveness of communication systems using the foregoing coding techniques may be measured by the coding gain over ideal noncoded PSK signalling at a given error probability. Table 11.1 lists the coding gain at the 10^{-5} bit-error probability for the coded systems of Figure 11.5.

For example, the hard-quantized sequential decoder requires an E_b/N_0 equal to 4.4 dB, whereas the PSK system requires 9.6 dB. This 5.2-dB coding gain afforded by the sequential decoder can be translated as either a 5.2-dB reduction in the transmitter power for the same data rate as the PSK system, or as an increased data rate equal to $10^{0.52} = 3.3$ times the uncoded data rate. For the rate-$\frac{3}{4}$ threshold decoder, the coding gain is a more modest 1.2 dB. This gain is quite sufficient for the application of this relatively simple coding scheme to the satellite transmission of data over regular voice channels [18]. For lower error probability, the gain of the coded systems relative to the uncoded system is even larger.

It must be stressed that this coding gain is obtained at the expense of an increase in the necessary transmission bandwidth. The **bandwidth expansion** is the reciprocal of the coding rate, that is, 2 for the sequential decoding example and $\frac{4}{3}$ for the threshold decoder. Consequently, the transmitted symbol rate is increased by a factor of 2 and $\frac{4}{3}$, respectively, over the information data rate. This is a moderate price to pay and may be within the means of the available bandwidth of the satellite.

Without error correction coding, efficient communication through minimization of E_b/N_0 may be achieved, and even capacity can be approached by selecting efficient modulation or signalling techniques. In particular, the class of M-ary orthogonal, M-ary biorthogonal and M-ary simplex signals [2, 19] can achieve error-free transmission (i.e., capacity) as M becomes infinitely large. It

Table 11.1 Coding Gain for Several Coded Systems

System	Required Value of E_b/N_0 at $P_b = 10^{-5}$ (dB)	Coding Gain (dB)
Ideal PSK (no coding)	9.6	—
BCH (15, 7)	8.7	0.9
Golay (24, 12)	7.6	2.0
BCH (128, 112)	7.5	2.1
Threshold ($r = \frac{3}{4}$)	7.4	2.2
Viterbi $K = 7$, $r = \frac{1}{2}$		
Hard quantization	6.5	3.1
Soft quantization	4.5	5.1
Sequential $K = 41$, $r = \frac{1}{2}$		
Hard quantization	4.4	5.2

must be noted that the exchange between the bandwidth and signal-to-noise-ratio still applies for these modulation schemes, and thus the lower values of E_b/N_0 required for a given error probability will be achieved at the expense of the bandwidth occupancy or bandwidth efficiency. Thus as capacity is reached with these systems, the necessary bandwidth approaches infinity [19, 20]. The bandwidth efficiency of the modulation schemes [21] are given by

Orthogonal

$$\frac{R}{B} = \frac{2\log_2 M}{M} \tag{11.13}$$

Biorthogonal

$$\frac{R}{B} = \frac{4\log_2 M}{M} \tag{11.14}$$

Simplex

$$\frac{R}{B} = \frac{2\log_2 M}{M-1} \tag{11.15}$$

Table 11.2 lists the required value of E_b/N_0 for a bit error probability of 10^{-5}

Table 11.2 Required Values of E_b/N_0 and Bandwidth Efficiency to Achieve $P_b = 10^{-5}$

Modulation System	M	E_b/N_0 (dB)	Bandwidth Efficiency (bps/Hz)
M-orthogonal	2	12.60	1.
	8	8.35	0.75
	16	7.36	0.50
	32	6.58	0.31
	64	6.0	0.19
	128	5.61	0.11
	512	4.95	0.035
	1024	4.59	1.95×10^{-2}
	32768	3.61	9.1×10^{-4}
M-biorthogonal	2	9.56	2
	8	8.12	1.5
	16	7.36	1.0
	32	6.53	0.62
	64	5.93	0.38
	128	5.61	0.22
	512	4.95	0.07
	1024	4.59	3.91×10^{-2}
	32768	3.60	1.83×10^{-3}

(assuming perfect coherent demodulation) together with the corresponding bandwidth efficiency of the orthogonal and biorthogonal systems. (For $M > 8$ the simplex and orthogonal systems are essentially the same.) Table 11.2 shows that biorthogonal signalling is twice as bandwidth efficient as orthogonal signalling for essentially the same performance. For these two systems as well as for the simplex system the required E_b/N_0 decreases relatively slowly as the dimensionality M of the signalling increases, eventually reaching capacity as $M \to \infty$. However, the cost, complexity and bandwidth requirements all increase very rapidly with M, so that these systems are practical only with an M value of the order of 32. In fact, an $M = 32$ biorthogonal signalling scheme was used in the Mariner 6 and 9 spacecraft telemetry systems for the transmission of pictures of Mars to the Earth [19].

As a final point it may be interesting to compare the cost of each decibel gained by modulation or coding with that obtained by other means in satellite communication. For example, to improve the data returns from the Mariner Jupiter/Saturn fly-by mission, the National Aeronautics and Space Administration (NASA) considered the following modifications to increase the gain of the antenna system of its Deep Space Network [22]:

1   Antenna upgrading from 64 to 70 m: gain = 0.8 dB.
2   Telescope resurfacing for the larger antennas: gain = 0.7 dB.
3   Secondary cones' optimization: gain = 0.7 dB.
4   Using the 34-m antennas in phased arrays: gain = 0.8 dB.

In such an environment the equivalent gain that could be provided by coding may indeed be a rather attractive and elegant alternative. Further discussion on the applications, gains and tradeoffs afforded by coding may be found in Ref. 35.

11.3   ERROR CONTROL BY DETECTION–RETRANSMISSION METHODS

In the preceding section we showed that no practical coding or modulating scheme achieves the elusive capacity and error-free transmission. Even the most powerful FEC techniques have inherent limitations and hence convey erroneous data to the user. In certain applications, for example in text or message containing internal redundancy, these transmission errors may be ignored or tolerated. However in many applications, such as computer-to-computer communication, transmission errors cannot be tolerated; for these applications the most widely used error-control technique for improving the reliability of the systems involves an error detection and retransmission procedure known as **automatic repeat request**, ARQ. The essential characteristics of ARQ systems are now presented in detail together with their applicability to satellite communication.

Fundamental Concepts of ARQ Systems

In ARQ systems shown schematically in Figure 11.6, data originating from a controllable source is delivered to a sending terminal. The sending terminal partitions the information bits in packets or blocks of length k, buffers the blocks, inserts the proper control and synchronization bits and delivers these blocks to the encoder (which may also be part of the sending terminal). The encoder performs block encoding on the data blocks by inserting $(n-k)$ redundant parity symbols for error-control purposes, resulting in blocks of length n coded symbols. The code is thus an (n, k) block code with code rate $r=k/n$ bits/symbol. Following encoding, the blocks are transferred to the modem and transmitted over the channel.

At the receiving end the block is demodulated and passed on to the decoder, which attempts to *detect* the presence of errors in the block. This error-detecting operation is easily performed by reconstructing the parity symbols from the received data and comparing them with the actually received parity symbols. The difference between the reconstructed and received parity symbols is called the **syndrome** of the block and indicates the presence of erroneous symbols in the received blocks. (The concept of syndrome is treated in detail in Chapter 13.) If there are no discrepancies, the syndrome is zero and the block is assumed to be error-free. It is then delivered to the data sink through the receiving terminal, which notifies the sending terminal that the block has been correctly received. Upon reception of this **acknowledgement (ACK)** the sending terminal proceeds to transmit a new block.

Should discrepancies exist, the syndrome would be nonzero, indicating the presence of errors in the block. The block is not accepted and a request for repeat (**NACK**) is generated by the receiving terminal and transmitted back to the sending terminal over the reverse channel. The same block is then retransmitted. It is therefore always assumed that a transmitted block is stored at the sending terminal and is not discarded until the corresponding acknowledgement is received. Clearly, with such a system erroneous information bits are delivered to the user only if the error-detecting capability of the code is exceeded. It will be shown in Chapter 13 that by making n sufficiently larger than k (i.e., by reducing the code rate), any desired degree of protection against undetected errors can be obtained. However, a larger block length entails large

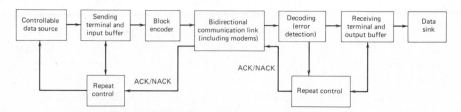

Figure 11.6 Block diagram of an ARQ system.

time delays, larger buffers, increased hardware complexity and reduced link efficiency.

The two principal performance criteria to evaluate ARQ systems are:

1   The **undetected block error probability** P_u, that is, the probability of delivering erroneous data to the user.

2   The **throughput** η, defined as the ratio of the information bits delivered to the user to the total number of symbols transmitted in the channel. This is also referred to in the literature as **throughput rate** and **throughput efficiency**.

In evaluating the performance of ARQ systems, several factors must be taken into consideration, in particular:

1   Channel error characteristics and error statistics (random, burst or compound error patterns).

2   Code selection to detect the vast majority of all error patterns at the highest possible code rate.

3   System parameters such as modem signalling rate, propagation delay, turnaround time for half-duplex mode of transmission and so on.

4   Specific techniques and protocols used for acknowledgements and block transmissions.

Powerful error-detecting codes exist [6, 7, 13]. Some are even standardized [23], and hence a code can easily be chosen to detect nearly all types of error patterns, regardless of how these errors occur on the channel. Therefore, the problem of undetected errors is for all practical purposes nonexistent in ARQ systems. The inherent high reliability of ARQ systems, together with the insensitivity of this reliability and robustness to varying conditions on the channel, is a striking characteristic of ARQ systems, which, unfortunately, is not shared by FEC systems. Consequently, the key figure of merit in the evaluation of ARQ systems is the throughput.

Types of ARQ Systems

There are three principal ARQ schemes, designated as follows:

1   Stop-and Wait ARQ
2   Continuous ARQ
3   Selective-Repeat ARQ

These techniques are described below and their throughput efficiency evaluated. The dependency of the throughput upon system parameters such as channel error statistics, code rate, signalling speed, block length and retransmission protocols are examined [24–27]. The applicability of these different ARQ schemes to satellite communication will be considered [28–31].

Stop-and-Wait ARQ Systems

The Stop-and-Wait ARQ, also called **idle-RQ** [24], is the simplest and the most widely used ARQ system over terrestrial systems. In these systems the blocks are transmitted one at a time, and following the transmission of a block, no further block is transmitted by the sending terminal until an acknowledgement is received from the receiving terminal. If the acknowledgement is positive (ACK), the next data block is transmitted and a copy of it preserved in the input buffer. If the acknowledgement is negative (NACK), the same block is repeated by transmitting the contents of the input buffer. The copy of a block residing in the input buffer may be discarded only after receiving the corresponding ACK, and clearly, the size of that buffer must be equal to the length of the largest block to be sent.

With such a system, transmission on the forward channel is halted during the time interval separating two consecutive block transmissions, that is, the round-trip delay, plus transmission time of the control signals, plus modem turnaround time in half-duplex mode of operation. It is therefore expected that as the transmission loop delay increases and becomes larger than the block duration, the transmission efficiency of these systems will degrade rapidly.

Assuming that the acknowledgement messages are received error-free, the throughput of Stop-and-Wait ARQ systems is given by

$$\eta = \frac{k}{n + R_s \Delta T} \overline{N} \qquad \text{bits/symbol} \qquad (11.16)$$

where $\overline{N}$ = average number of blocks delivered per transmission

k = number of information bits per block

n = block length, symbols

R_s = signalling speed on the channel, symbols/sec

ΔT = overall round-trip delay: propagation and operational delays

Ignoring the undetected block error probability, it is easily shown that

$$\overline{N} = 1 - P_B \qquad (11.17)$$

where P_B is the **block error probability**. Using (11.17), the throughput is then

$$\eta = \frac{k(1 - P_B)}{n + R_s \Delta T} \qquad \text{bits/symbol} \qquad (11.18)$$

Defining $D = R_s \Delta T$ as the overall delay in transmitted binary symbols, the throughput can be expressed as

$$\eta = r\frac{n(1-P_B)}{n+D} \tag{11.19}$$

$$= r\varepsilon \tag{11.20}$$

where ε is the retransmission channel efficiency and r is the code rate, $r = k/n$, in bits/symbol. To examine the dependency of the system throughput on the channel error rates and other system parameters, we first note that the throughput is given in terms of the block error probability P_B. For general channels where errors occur in bursts, burst statistics are required in order to relate P_B to the average bit or raw error rate on the channel. However, since on a satellite channel the errors tend to be produced randomly rather than in bursts, the channel may be reasonably well modelled by a binary symmetric channel, that is, a binary input and binary output channel having a crossover probability or raw error rate p. Therefore, assuming that all errors are detected, the block error probability for such a channel is

$$P_B = 1 - (1-p)^n$$

$$\simeq 1 - e^{-np} \qquad \text{for } np \ll 1 \tag{11.21}$$

Using (11.19), the throughput is given by

$$\eta = r\frac{ne^{-np}}{n+D} \tag{11.22}$$

The block length n_{opt} that maximizes the throughput could then be obtained by differentiating (11.22) with respect to n and setting to zero. Assuming a constant code rate r, the optimum block length is then given by

$$n_{\text{opt}} = \frac{D}{2}\left(-1 + \sqrt{1 + \frac{4}{Dp}}\right) \tag{11.23}$$

and the maximum value of the throughput is given by

$$\eta_{\text{opt}} = r\frac{n_{\text{opt}}e^{-n_{\text{opt}}p}}{n_{\text{opt}} + D} \tag{11.24}$$

or equivalently

$$\eta_{\text{opt}} = r\varepsilon_{\text{opt}} \tag{11.25}$$

where ε_{opt} is the maximum retransmission efficiency.

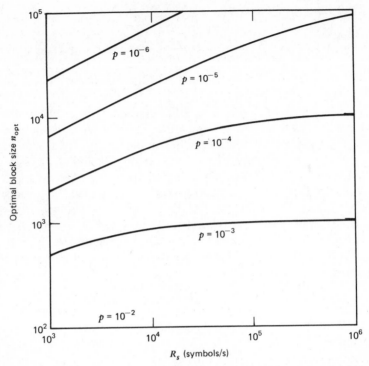

Figure 11.7   Optimal block size as a function of the signalling speed for Stop-and-Wait ARQ systems over a satellite link.

In Figures 11.7 and 11.8 we plot the optimal block lengths n_{opt} and maximum retransmission efficiency ε_{opt} as a function of the signalling rate R_s for a satellite channel, with raw channel error rates p varying from 10^{-2} to 10^{-6}. Operational delays were assumed very small in comparison with the round-trip propagation delay of 512 ms, and thus have been neglected.

Figure 11.7 shows that the optimal block length n_{opt} is monotonically increasing with the signalling rate on the channel, and tends to level off to a value that gets larger as the channel-error rate p becomes smaller. For example, at a signalling rate of 50 ksymbols/s, the optimum block length varies from 964 to 39,390 symbols as the channel error rate varies from 10^{-3} to 10^{-5}.

Even in applications where very large blocks may be acceptable, finding good error-detecting codes for such large blocks may present some difficulties. The problem may be circumvented by partitioning the long blocks into a sequence of several subblocks, each of which is protected by its own error-detection scheme.

Stop-and-Wait ARQ systems are quite attractive because of the simplicity of their operation and the small amount of buffering they require. Although inherently inefficient due to the idle time spent waiting for acknowledgements, they are widely used on terrestrial circuits, where the round-trip delay is small.

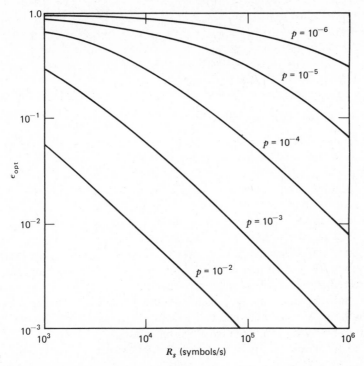

Figure 11.8 Maximum retransmission efficiency as a function of the signalling speed for Stop-and-Wait ARQ systems over a satellite link.

However, this inefficiency becomes unacceptable on systems that employ high-speed modems and where the round-trip delay is high like on satellite links. As shown in Figure 11.8, the maximum retransmission efficiency ε_{opt} falls off rapidly as the signalling rate increases, so that for satellite channels the loss in throughput incurred by the 512-ms delay really offsets the simplicity advantage of the system. Stop-and-Wait systems are therefore not well suited to satellite links.

Continuous ARQ Systems

A method of reducing the idle time between the transmitted blocks is to send the blocks continuously over the forward channel, without waiting for the acknowledgement signals which are transmitted on the reverse channel. Whenever a NACK is received at the sending terminal, the transmitter backs up to the erroneous block and repeats that block and all subsequent blocks in order to preserve the natural ordering of the blocks. Upon receiving a NACK, the transmitter completes transmission of the current block, then goes back and repeats the block detected in error and all the following blocks, including the one being sent when the NACK signal was received. Such a scheme is often

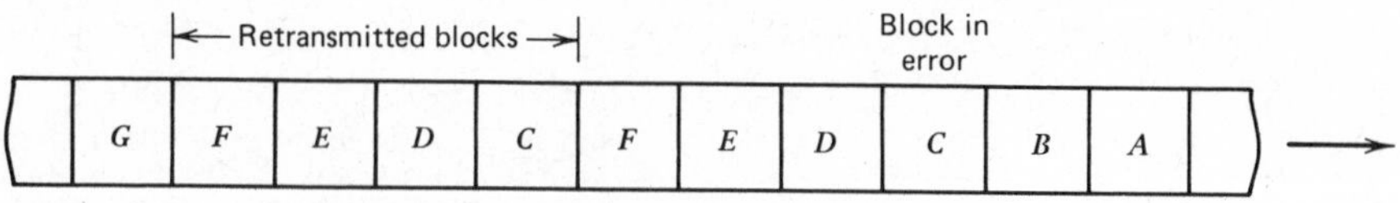

Figure 11.9 Block sequencing following a block error in a go-back-4 Continuous ARQ system.

called **go-back-N ARQ**, where N refers to the number of blocks repeated. Figure 11.9 shows the block sequencing for a go-back-4 ARQ.

With these ARQ systems, buffering must be provided at the sending terminal and the buffer size depends on the transmission rate and round-trip delay, with a minimum of two block lengths. At the receiving terminal buffering is simplified since no block needs to be preserved unless it is free of errors and has the proper sequence number. Further operational details of Continuous ARQ systems may be found in Refs. 23 to 27. We may expect Continuous ARQ to be very efficient when the channel is good and few blocks are detected in errors. Its principal drawback is the buffering requirement, but it offers substantial advantages over Stop-and-Wait ARQ when the round-trip delay is large. For satellite channels, Continuous ARQ is far more efficient than Stop-and-Wait, and in fact Recommendation V.41 of the CCITT was amended to permit the use of go-back-2 ARQ systems over satellite circuits [32].

Neglecting second-order effects due to acknowledgement errors, the throughput of Continuous ARQ systems is given by

$$\eta = r \frac{n(1 - P_B)}{n + DP_B} \qquad \text{bits/symbol} \qquad (11.26)$$

$$\eta = r\varepsilon \qquad (11.27)$$

where the same definitions as the Stop-and-Wait ARQ apply, and where $D = R_s \Delta T$ is the overall delay in symbols between the end of transmission of a block and the beginning of the retransmission of that same block if a NACK is received. Using $r = k/n$ the throughput is also given by

$$\eta = k \frac{(1 - P_B)}{n + DP_B} \qquad \text{bits/symbol} \qquad (11.28)$$

For a satellite link modelled as a binary symmetric channel with raw error rate p, and block error probability $P_B \simeq 1 - e^{-np}$, the throughput is

$$\eta \simeq k \frac{e^{-np}}{n + D(1 - e^{-np})} = \frac{rne^{-np}}{n + D(1 - e^{-np})} \qquad (11.29)$$

and the channel retransmission efficiency is expressed as

$$\varepsilon \simeq \frac{ne^{-np}}{n+D(1-e^{-np})}$$ (11.30)

Assuming a constant code rate for the error-detecting code, the optimal block length is again obtained by setting $d\eta/dn=0$. The optimal block length is found here to be the minimum, $n_{opt}=1$. The maximum throughput is therefore upper bounded by

$$\eta_{opt} = \frac{r}{1+Dp} = \frac{r}{1+pR_s\Delta T}$$ (11.31)

and the retransmission efficiency is bounded by

$$\varepsilon_{opt} = \frac{1}{1+pR_s\Delta T}$$ (11.32)

Both the round-trip delay (expressed in number of transmitted symbols) and the channel raw error rate limit the performance of Continuous ARQ systems. As the transmission speed R_s on the channel increases, the retransmission efficiency can only be maintained by upgrading the channel raw error rate. This is indicated in Figure 11.10, where ε_{opt} is plotted for a satellite link as a function of the signalling speed, with channel error rates p varying from 10^{-3} to 10^{-7}. Figure 11.10 shows that for small values of the pR_s product, the retransmission efficiency may be increased to 1, but as the product increases, this efficiency becomes severely limited by the satellite round-trip delay. Therefore, Continuous ARQ systems offer reasonable retransmission efficiencies over satellite links.

Reference 30 reports tests conducted by COMSAT Laboratories using the CCITT recommendation V.41 go-back-2 ARQ with blocks of data of length 3860 symbols over an intercontinental data circuit consisting of a single-hop satellite link and terrestrial circuits. An excellent throughput efficiency of 98.7% was obtained over a raw channel error rate of 3.4×10^{-5} at a modem rate of 4800 symbols/s.

From a general practical point of view, although (11.31) and (11.32) give a good measure of the expected efficiency of the system, these expressions provide no help in determining the actual block length to use. In fact, contrary to the assumption that the code rate is maintained constant regardless of the block length, it is well known that for small block lengths the code rate is actually low and may vary widely.

As an example, consider the error-detecting code used by IBM in their computer-communication systems [33]. The code uses $(n-k)=16$ parity symbols and is capable of detecting all blocks with three of fewer errors, or with

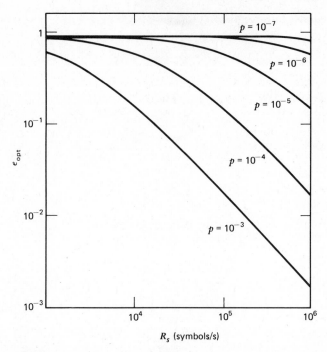

Figure 11.10 Maximum retransmission efficiency as a function of the signalling speed for Continuous ARQ systems over a satellite link (two-way transmission delay, 0.512 s, no operational delay).

bursts spanning at most 16 symbols. This code may be used in conjunction with blocks of length up to 32,767 symbols.

When the number of parity symbols of the code $(n-k)$ is maintained constant regardless of the code rate, an optimum value of the block length may be easily obtained.

Solving $d\eta/dn = 0$ for $np \ll 1$, the optimal block length is found to be

$$\hat{n}_{\text{opt}} = \frac{n-k}{2} + \frac{1}{2}\sqrt{(n-k)^2 + \frac{4(n-k)}{p}(1+Dp)} \qquad (11.33)$$

Whenever $Dp = R_s p \Delta T \ll 1$, and $(n-k)/p \gg (n-k)^2$, this expression simplifies to

$$\hat{n}_{\text{opt}} \simeq \frac{n-k}{2} + \sqrt{\frac{n-k}{p}} \qquad (11.34)$$

Substituting either (11.33) or (11.34) in (11.29) will yield the corresponding throughput value $\hat{\eta}_{\text{opt}}$.

For example, consider a Continuous ARQ scheme used over a satellite link with the following parameters:

Modem signalling rate: $R_s = 10^4$ symbols/s
Overall propagation delay: $\Delta T = 0.512$ s
Number of parity symbols per block: $n - k = 16$

Table 11.3 lists the values of the optimal block length $\hat{n}_{\mathrm{opt}}$, the corresponding system throughput $\hat{\eta}_{\mathrm{opt}}$ and the upper bound η_{opt} for channel error rates p varying from 10^{-7} to 10^{-3}.

Table 11.3 demonstrates that when the channel-error rate p is small and the Dp product negligible compared to 1, the throughput upper bound is actually reached with long blocks. However, as the channel degrades and the Dp product becomes larger than 1, the discrepancies between the calculated throughput and its upper bound become significant, indicating the greater importance played by the code rate and the block length on the system throughput. Consequently, in the process of determining the optimum block length through maximization of the system throughput, the code rate must be taken into account. Furthermore, from a user point of view, in addition to the throughput, the undetected block error probability P_u is also an important factor in evaluating the system performance since it relates to the reliability of the delivered data. Therefore, meeting the P_u requirement should also be taken in consideration when choosing the code.

As an illustrative example of Continuous ARQ performance over a satellite link with round-trip delay of 0.512 s, let the error-detecting code be a (1023, 913) BCH block code, which can detect up to 22 random errors in the block. The system throughput as a function of the signalling rate for channel error rates p varying from 10^{-3} to 10^{-7} is shown in Figure 11.11.

In Figure 11.12 the throughputs for Stop-and-Wait and Continuous ARQ systems using the same error-detecting code (1023, 913) over a channel with $p = 10^{-4}$ are compared to each other and to their corresponding upper bounds, given by (11.24) and (11.31), respectively. These results demonstrate the superiority of Continuous over Stop-and-Wait ARQ, and show that high throughput values may be achieved whenever the pR_s product is relatively low. However, whenever the channel quality degrades below a raw error rate of 10^{-3}, the system throughput decreases drastically, making the system totally inefficient.

Table 11.3   Values of $\hat{n}_{\mathrm{opt}}$, $\hat{\eta}_{\mathrm{opt}}$ and η_{opt} for Different Values of p

Channel error rate, p	10^{-7}	10^{-6}	10^{-5}	10^{-4}	10^{-3}
Block length, $\hat{n}_{\mathrm{opt}}$	12650	4008	1273	500	321
Throughput, $\hat{\eta}_{\mathrm{opt}}$	0.997	0.986	0.927	0.614	0.128
Upper bound, η_{opt}	0.998	0.995	0.951	0.661	0.163

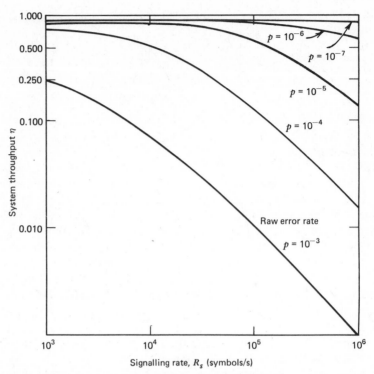

Figure 11.11 Throughput versus signalling rate for a Continuous ARQ system over a satellite link. Propagation delay, 0.512 s; BCH error-detecting code (1023,913).

Selective-Repeat ARQ Systems

The Selective-Repeat ARQ scheme is a variation of Continuous ARQ systems whereby only the particular block received in error is retransmitted. The number of retransmitted blocks is therefore lower than with the go-back-N ARQ, and the system performs better at higher channel error probabilities. However, as the blocks are not always received in serial order at the receiving terminal, more complex logic and larger buffers are required. Furthermore, the acknowledgement signal must permit the identification of the block it is associated with, resulting in a significantly more complex overall system than go-back-N ARQ.

Since the channel retransmission efficiency is

$$\varepsilon = 1 - P_B \tag{11.35}$$

the throughput is thus given by

$$\eta = r(1 - P_B) = \frac{k}{n}(1 - P_B) \qquad \text{bits/symbol} \tag{11.36}$$

where the same notation as earlier is used.

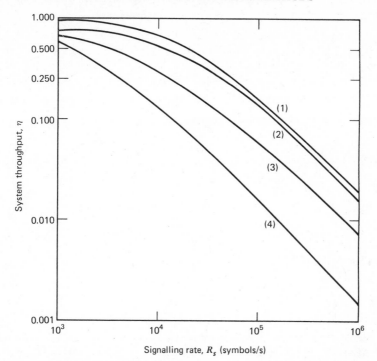

Figure 11.12 Throughputs and upper bounds as a function of the signalling rate for Continuous and Stop-and-Wait ARQ systems over a satellite link with raw error rate $p = 10^{-4}$. (1), Continuous ARQ, upper bound on throughput (optimum η); (2), Continuous ARQ, BCH code (1023,913); (3), Stop-and-Wait ARQ, upper bound on throughput (optimum η); (4), Stop-and-Wait ARQ, BCH code (1023,913).

The most striking feature of the Selective-Repeat ARQ scheme is the total independence of the throughput on both the overall transmission delay and the signalling rate on the channel, which makes it very attractive for satellite applications.

In Figure 11.13 the retransmission channel efficiency ε given by (11.35) is plotted as a function of the raw error rate of the channel, for a Selective-Repeat ARQ System using a block length $n = 1023$. Figure 11.13 shows that as long as the channel is good and few blocks are detected in error, the channel efficiency is close to 100% and the throughput of the system is practically equal to the rate of the error-detecting code. However, as the channel degrades and more retransmissions occur, then, as expected, the retransmission efficiency falls sharply; that is, like in any ARQ scheme, the blocks keep on being repeated with hardly any error-free blocks delivered to the user.

A measure of comparison between Selective-Repeat ARQ and both Continuous and Stop-and-Wait ARQ systems transmitting at 5 ksymbols/s over a satellite link is also provided in Figure 11.13. The superiority of the selective scheme over all other ARQ schemes is quite apparent, making this error-control technique the best one for satellite channels.

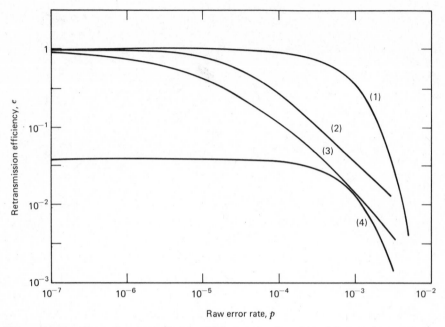

Figure 11.13   Channel retransmission efficiency as a function of the raw error rate for several ARQ schemes over a satellite channel. (1), Selective ARQ ($n=1023$); (2), Continuous ARQ upper bound ($R_s = 50$ ksymbols/s); (3), Stop-and-Wait upper bound ($R_s = 50$ ksymbols/s); (4), Stop-and-Wait ($n = 1023$, $R_s = 50$ ksymbols/s).

Variations of ARQ Systems

We have shown that Continuous ARQ schemes whether go-back-N or Selective-Repeat are in general more efficient than Stop-and-Wait schemes, and for satellite links Selective-Repeat ARQ is the most attractive system although the most complex to implement. As the error probability on the channel increases, all ARQ systems suffer from a rapidly falling throughput, a behaviour that is further aggravated (except for Selective schemes) by the long transmission delays of satellite links. However, even during periods of low throughput values, the reliability of the delivered blocks is usually maintained.

Several variations of basic ARQ techniques have been proposed with the objective of satisfying specific systems constraints of particular applications or alleviating some inherent drawbacks of a given ARQ system.

Modifications of both Stop-and-Wait and Continuous ARQ schemes have been proposed by Sastry [29] in order to improve the throughput over satellite links under high-error-rate conditions. In the modified Stop-and-Wait ARQ, whenever a request for repeat is received, a finite number ℓ of retransmissions of the erroneous block are sent repeatedly before stopping and waiting for the acknowledgement. Only if none of the ℓ retransmissions are received correctly will the system request a repeat for that block. In this variant, the throughput is improved because the delay between retransmissions is distributed over the ℓ retransmissions. The improvement will depend on the expected number of retransmissions and hence on the pR_s product.

The throughput of this variant of Stop-and-Wait ARQ [29] is given by

$$\eta_\ell = \frac{k}{(n+D)+(\ell n+D)\left[P_B/\left(1-P_B^\ell\right)\right]} \qquad (11.37)$$

which reduces to the usual Stop-and-Wait system of (11.18) when $\ell=1$.

Using essentially the same approach, the Continuous ARQ scheme is modified as follows: whenever a repeat is requested, the same required block is transmitted repeatedly until it is received correctly. This is to be contrasted with the usual Continuous ARQ method, which sends, after each request for repeat $(N-1)$ subsequent blocks which are ignored at the receiving terminal. The throughput improvement over the conventional Continuous ARQ scheme is brought about by a reduction in the time occupied by the total number of retransmissions for a block. It can be shown [29] that the throughput is significantly improved for high channel error rates and whenever the overall delay D in number of transmitted symbols is much larger than the block length (i.e., for satellite links and high transmission rates).

Several other variants of the basic ARQ techniques consist of adapting either the length of the transmitted blocks or the transmission rate with the actual error rate on the link. **Adaptive ARQ** schemes attempt to keep the link efficiency as high as possible by using two or more block lengths. The receiving terminal monitors the rate of the transmission repeats, and notifies the sending terminal to send short blocks during a period of high noise, and long blocks during a period of good channel conditions. Adapting the block length to channel conditions reduces the block error probability and thus improves the system throughput. However, the incremental advantage over fixed-block ARQ is limited and this advantage is obtained at a cost of more complex transmit and receive terminals.

In **Variable Rate** systems the return channel is used to adaptively control the forward transmission rate and prevent it from unacceptable degradation. This variant of ARQ may be attractive to satellite links where the feedback delays are large and where the changes in channel characteristics such as deep fades are slow. The principal drawback of Variable Rate schemes is the significant increase in cost and complexity, since each allowed transmission rate may require a distinct modulator and demodulator.

Implementation of Selective-Repeat ARQ

The Selective-Repeat ARQ performs better than any other form of ARQ and may be seen as an upper bound on any ARQ-alone error-control technique. However, its implementation is significantly more involved than either Stop-and-Wait or Continuous ARQ systems. A relatively simple method has been proposed to implement a Selective-Repeat ARQ scheme for a high-speed bidirectional satellite data link randomly accessed by a large number of low-speed users [34]. To fully utilize the satellite link, the blocks are transmitted without interruption and only those blocks received in error are retransmitted. The technique consists of multiplexing the satellite channel into a finite number L of logical subchannels on which a simple Stop-and-Wait ARQ scheme is implemented. Clearly, in order to keep the satellite link fully occupied, the minimum number of these subchannels, L_{min}, must be equal to the number of blocks that can be transmitted during a round-trip delay ΔT, that is,

$$L_{min} = \frac{R_s \Delta T}{n} = \frac{D}{n} \qquad (11.38)$$

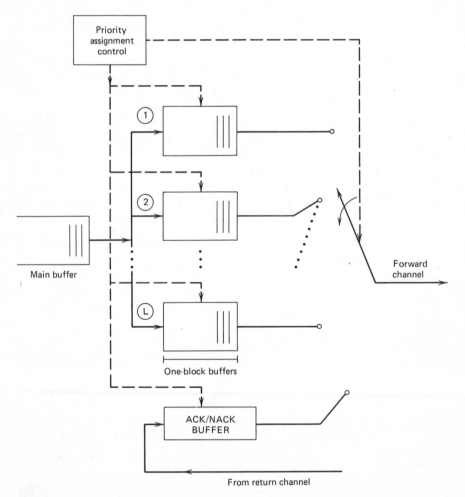

Figure 11.14   Schematic diagram of a variant of the Selective Repeat ARQ scheme.

As shown in Figure 11.14, upon arriving at the sending terminal, a block enters a queue in the main buffer, which feeds L one-block buffers according to a first-in first-out mode of operation. The L buffers are connected to the channel through a logical switch controlled by a priority assignment protocol. Regardless of the particular assignment protocol, following transmission of a block, a copy of the block is kept in its buffer until the corresponding ACK is received. Therefore, each of the L one-block buffers may be accessible to the main buffer only after its own ACK has been received. To free the logical subchannel buffers promptly, the transmission priorities listed in decreasing order are as follows: control blocks, retransmitted blocks and blocks transmitted for the first time.

From a computer simulation of this variant it is shown [34] that the observed throughput closely matches that of a theoretical Selective-Repeat ARQ Scheme, and furthermore this particular access method utilizes efficiently the satellite link while serving rapidly and evenly each of the particular users.

11.4   CODE RATE AND UNDETECTED ERROR PROBABILITY

In the analysis of ARQ systems the delivery of erroneous blocks to the users was always considered a second-order effect and hence negligible. In fact, the undetected block error probability P_u of an ARQ system may be used as a performance criterion; in computer-to-computer communication, P_u is often required to be less than 10^{-10}. Although a detailed treatment of block codes is given in Chapter 13, we now briefly address the general problem of selecting an error-detecting code that must meet both requirements of high efficiency and low undetected error probability.

The probability of undetected block errors, P_u, depends on the error-detection capability of the error-detecting code. It is shown in Chapter 13 that the error-detecting capability of a block code is approximately twice that of its error-correction capability, and within a given class of error-correcting codes, this capability is a function of the coding rate r. Consequently, there is a functional relationship between P_u and r, and this relationship necessarily involves the block length n of the code. Therefore, it is conceptually possible to determine the code rate $r(n, P_u)$ as a function of the block length and undetected error probability. However, the required amount of computations is so large as to make the method worthless.

A more rewarding approach consists of determining the required error-detecting capability t_d of a code of block length n (or equivalently, the ratio t_d/n) given the value of the undetected error probability P_u. Having obtained t_d and n, a corresponding code with the appropriate rate may then be selected from tables of codes readily available in the literature (see, e.g., Table 9.1 and Appendix D in Ref. 6).

For example, consider an ARQ system used over a satellite link modelled as a binary symmetric channel with crossover probability p. Let t_d denote the error-detecting capability of the code and n the block length. P_u is then given by

$$P_u = 1 - \sum_{i=0}^{t_d} \binom{n}{i}(1-p)^{n-i}p^i \qquad (11.39)$$

In Figure 11.15 the ratio t_d/n is plotted as a function of the block length n for $P_u = 10^{-10}$ and $p = 10^{-3}$ and 10^{-4} [28]. Figure 11.15 shows that for fixed p and P_u the required error-detecting capability of the code decreases as the block length increases, indicating the need of less powerful codes, that is, codes with higher code rates r.

Figure 11.16 shows the variations of the required code rate r as a function of the block length n for different combinations of P_u and p. As expected from Figure 11.16, for fixed values of P_u and p the code rate increases toward 1 with the block length. For moderate values of n (a few thousand digits) the code rate is substantial (approximately 0.9) and not very sensitive to variations of the block length. However, for block length smaller than 100, the required code rate is strongly dependent on the block length and falls rapidly below 0.6, supporting our earlier observation that in ARQ systems, maximization of the

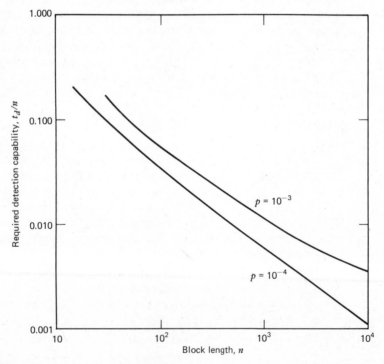

Figure 11.15 Required code capability t_d/n as a function of block length n for an undetected block error probability of 10^{-10}.

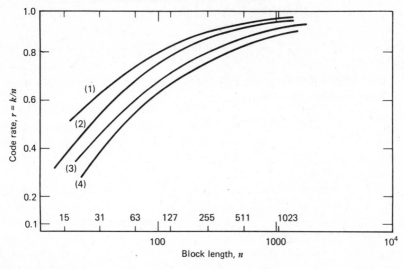

Figure 11.16 Required code rate versus block length. (1), $P_u = 10^{-8}$, $p = 10^{-4}$; (2), $P_u = 10^{-10}$, $p = 10^{-4}$; (3), $P_u = 10^{-8}$, $p = 10^{-3}$; (4), $P_u = 10^{-10}$, $p = 10^{-3}$.

throughput via block length optimization must take into consideration the rate of the code. Finally, it may be mentioned that from a theoretical point of view, the law of diminishing returns apply to the known classes of suitable block codes: keeping the rate fixed, the ratio t_d/n approaches zero as n increases, indicating that the codes are weak for very large n. In attempting to keep r fixed while maintaining a given P_u, one should not let the block length grow indefinitely.

11.5   COMPARISON BETWEEN FEC AND ARQ TECHNIQUES

In this chapter the two principal error-control techniques, FEC and ARQ, have been presented and their merits analyzed. Each technique has its own advantages and disadvantages which must be properly weighed for each application when deciding which one to choose.

Given a good knowledge of the channel error mechanisms and statistics, an FEC technique with an adequately chosen code can improve the error performance of a link in which transmission power is limited but excess bandwidth is available. The ensuing coding gain may be substantial at the low error rates required for data transmission; this gain is especially interesting for satellite links. The steep response of the bit-error rate to E_b/N_0 makes FEC systems very sensitive to channel degradations such as fading, and in general, unlike ARQ systems, FEC systems are not very robust to a wide range of channel performance variations. However, the rise of the error rate during channel degradations is compensated by several advantages of FEC systems. Unlike

ARQ schemes, in general FEC systems do not require the use of a feedback link or of a controllable source, nor do they introduce interruptions or delay variations in the transfer of the information to the users. Moreover, FEC systems enjoy a constant throughput and do not require extensive buffering (except for coding and decoding purposes) of the incoming data.

In ARQ systems knowledge of the nature of the errors in the channel is, by far, not as critical as for FEC systems in selecting the code to use. In general, error-detecting codes used in ARQ systems require less redundant symbols than do error-correcting codes, but most important, the hardware implementation for the error detection is considerably less complex than for the error correction needed in FEC.

On digital terrestrial communication links, ARQ is the most common form of error control. For satellite data link even though the long round trip transmission delay reduces the throughput and may necessitate large buffers, well-designed ARQ systems may be very attractive.

In general, it has been shown [36] that for independent channel errors, FEC systems enjoy a higher throughput than ARQ when the required error probability is moderate, and a lower throughput when the error probability is very low. However, for dependent and bursty channel errors, ARQ systems outperform FEC systems in both throughput and probability of undetected errors.

11.6 HYBRID FEC/ARQ

The main drawback of ARQ systems is their rapidly falling throughput when the channel degrades and/or where the signalling speed increases. Since this inefficiency is caused by the channel error-induced retransmissions, for a given signalling speed a throughput improvement will come from a reduction in the number of retransmissions (i.e., by upgrading the channel). On a satellite link this could be achieved by a brute-force increase of the repeater radiated power and/or the receiving antenna G/T. However, improvement of the throughput may also be achieved by means of coding that is by combining the advantages of both FEC and ARQ techniques in a hybrid FEC/ARQ scheme.

A FEC/ARQ hybrid system consists of a FEC system contained within an ARQ system: the outer ARQ system provides the very low undetected error probability performance required, while the function of the inner FEC system is to reduce the number of retransmissions by correcting as many blocks in errors as possible.

As shown in Figure 11.17, for the outer ARQ system the coder–raw channel–decoder may be regarded as a super channel whose error rate is the error probability of the inner FEC system, which is hopefully smaller than for the raw channel alone. Since only the erroneous blocks that cannot be corrected by the FEC system are subjected to retransmission requests, clearly under conditions of high channel noise and frequent retransmissions, the throughput is expected to improve with such a scheme. However, since the

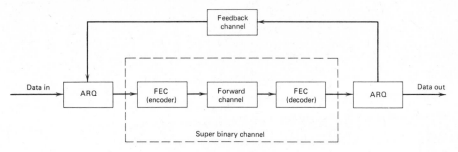

Figure 11.17   FEC/ARQ hybrid scheme.

overall coding rate is the product of the coding rates of both the error-correcting and error-detecting codes, then for good channels the throughput of a hybrid scheme may be smaller than that of the corresponding ARQ scheme alone.

In designing hybrid systems, consideration must be given to the nature of the errors on the super channel. Because the error statistics at the output of the FEC system are substantially different from those of the raw channel, the characteristics of the super and raw channels differ considerably. Consequently, the error-detecting code of the ARQ system must be chosen accordingly if a given undetected error performance is to be maintained. When the FEC system fails to correct an erroneous block, it may introduce additional errors to the channel-induced errors. Furthermore, these errors often appear in bursts. Therefore, the ARQ system sees a highly bursty error channel even if the raw channel introduces only random errors; provisions (such as interleaving) must thus be taken to circumvent this difficulty.

As an example, consider the following hybrid FEC/ARQ scheme proposed for satellite links using a time-division multiple-access system [28,31]. In this system, depicted schematically in Figure 11.18, at each Earth station data arrive asynchronously from different low-speed users and enter the ARQ sending terminals, where they are coded and processed according to the particular ARQ scheme employed. To each of the L low-speed user input ports there corresponds an ARQ terminal. Following ARQ processing the blocks enter the main TDMA terminal equipment, where they are interleaved, multiplexed and properly formatted. The output sequence of the TDMA system is then encoded for error-correction purposes by the FEC system prior to transmission in the channel as a high-speed burst.

At the receiving end the process is repeated in reverse. Demultiplexing and deinterleaving at the TDMA receiving terminal will recreate the blocks for the different users. Should any of the blocks be in error, a retransmission for these blocks will be requested through the return channel.

Let the TDMA high-speed burst transmission rate on the forward channel be R_S symbols/s. Then, as shown in Chapter 1 for a given satellite e.i.r.p. and a given receiving station G/T, the symbol energy-to-noise ratio E_s/N_0 in dB

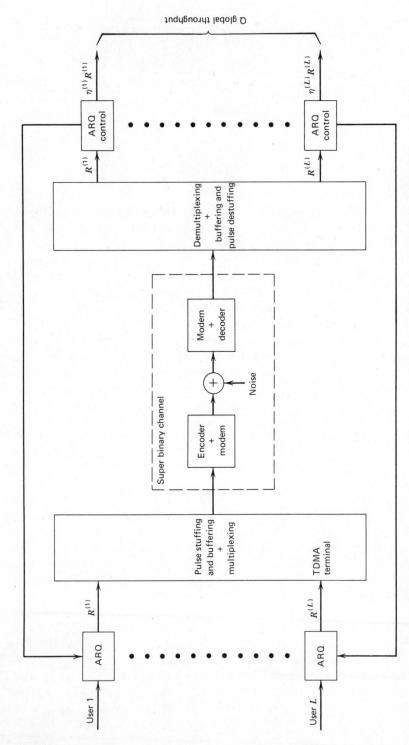

Figure 11.18 Block diagram of the hybrid ARQ/FEC scheme used in conjunction with TDMA.

344

on the forward channel, as given by the link equation, may be written as

$$\frac{E_s}{N_0} = \text{e.i.r.p.} + \frac{G}{T} + \text{gains} - \text{losses} - 10\log_{10} R_s \qquad (11.40)$$

or

$$\frac{E_s}{N_0} = A + \frac{G}{T} - 10\log_{10} R_s \qquad (11.41)$$

where the constant A encompasses the e.i.r.p. value and all gains and losses of the link.

Assuming that the TDMA scheme provides perfect interleaving, the combination of the encoder, raw channel, and decoder may be seen as a super memoryless binary channel whose transition probability p is the output bit-error rate of the inner FEC system. Throughput computation of the system is then straightforward.

Referring to Figure 11.18, let $R^{(j)}$ be the signalling speed of the jth user, and let $\eta^{(j)}$ be its throughput. The information throughput corresponding to this jth user is then given by

$$q^{(j)} = \eta^{(j)} R^{(j)} \qquad \text{bps} \qquad (11.42)$$

Provided that $\sum_{j=1}^{L} R^{(j)} \leq R_s$, the overall information throughput Q for the entire system is

$$Q = \sum_{j=1}^{L} q^{(j)} = \sum_{j=1}^{L} \eta^{(j)} R^{(j)} \qquad \text{bps} \qquad (11.43)$$

Assuming for simplicity that all L users are identical, the overall information throughput may be written as

$$Q = \eta^{(j)} R^{(j)} \left[\frac{R_s}{R^{(j)}} \right] \simeq \eta^{(j)} R_s \qquad \text{bps} \qquad (11.44)$$

where $[x]$ represents the integer part of x.

Comparison of the computed information throughput between a Continuous ARQ-alone scheme and a hybrid FEC/ARQ scheme over a TDMA system is provided in Figure 11.19. In this scheme the ARQ system utilizes a (1023,913) BCH code and the FEC uses rate-$\frac{1}{2}$ convolutional coding with optimum Viterbi decoding [28, 31]. The receiving antenna G/T is 28 dB/K, the overall constant A of (11.41) is 47.6 dBW and the users' rate is 2400 symbols/s.

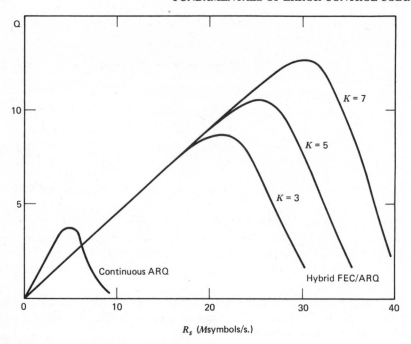

Figure 11.19 Overall information throughput as a function of burst rate for Continuous ARQ and hybrid FEC/ARQ schemes. G/T, 28 dB/K; ARQ, BCH (1023,913); FEC, $r=\frac{1}{2}$ Viterbi; user rate R, 2400 symbols/s.

From Figure 11.19 it can be seen that for relatively low burst rates and thus good channel conditions, the ARQ system alone is superior to the hybrid scheme, indicating that very few retransmissions occur. Hence the FEC system needlessly introduces a throughput reduction. However, as the transmission burst rate increases, the channel degrades, more and more retransmissions are requested and consequently the throughput decreases. With the hybrid scheme, as long as the FEC system can correct the channel errors, no retransmissions are requested and thus the information throughput increases in proportion with the burst rate. However, as the error-correcting capability of the code is exceeded, the retransmission frequency increases, yielding the usual reduction of the throughput. Throughput curves are plotted for three convolutional codes, having constraint lengths equal to 3,5 and 7. As expected, the most powerful of the three codes ($K=7$) reaches the largest throughput at the highest burst rate. Clearly, then, with this system, for burst transmission rates R_s lower than 6.5 Msymbols/s, an ARQ scheme alone will give a superior performance at a smaller cost than will a hybrid FEC/ARQ scheme. For higher burst rates, hybrid systems outperform ARQ systems, increasing the information throughput by a factor of 3, at the expense of a larger bandwidth requirement and a more complex implementation. Whether this advantage is warranted in actual systems is an engineering problem and will depend on the particular application.

REFERENCES

1   C. E. Shannon, "A Mathematical Theory of Communication". *Bell Syst. Tech. J.*, Vol. 27, 1948, pp. 379–423 (Part I), pp. 623–656 (Part II).

2   J. M. Wozencraft and I. M. Jacobs, *Principles of Communication Engineering*, Wiley, New York, 1965.

3   R. G. Gallager, *Information Theory and Reliable Communication*, Wiley, New York, 1968.

4   A. J. Viterbi and J. K. Omura, *Principles of Digital Communication and Coding*, McGraw-Hill, New York, 1979.

5   R. J. McEliece, *The Theory of Information and Coding*, Addison-Wesley, Reading, Mass., 1977.

6   W. W. Peterson and E. J. Weldon, Jr., *Error Correcting Codes*, 2nd ed., MIT Press, Cambridge, Mass., 1972.

7   F. J. MacWilliams and N. J. A. Sloane, *The Theory of Error-Correcting Codes, I and II*, North-Holland, New York, 1977.

8   A. J. Viterbi, "Convolutional Codes and Their Performance in Communication Systems", *IEEE Trans. Commun. Technol.*, Vol. COM-19, Oct. 1971, pp. 751–772.

9   I. M. Jacobs, "Sequential Decoding for Efficient Communication from Deep Space", *IEEE Trans. Commun. Technol.*, Vol. COM-15, Aug. 1967, pp. 492–501.

10 J. A. Heller and I. M. Jacobs, "Viterbi Decoding for Satellite and Space Communication", *IEEE Trans. Commun. Technol.*, Vol. COM-19, Oct. 1971, pp. 835–848.

11 G. D. Forney, Jr., "Coding and Its Application in Space Communications", *IEEE Spectrum*, Vol. 7, June 1970, pp. 47–58.

12 R. M. Gagliardi, *Introduction to Communications Engineering*, Wiley, New York, 1978.

13 J. M. Wozencraft and B. Reiffen, *Sequential Decoding*, MIT Press, Cambridge, Mass., 1961.

14 J. L. Massey, *Threshold Decoding*, MIT Press, Cambridge, Mass., 1963.

15 J. G. Puente, W. G. Schmidt and A. M. Werth, "Multiple-Access Techniques for Commercial Satellites", *Proc. IEEE*, Vol. 59, Feb. 1971, pp. 218–229.

16 S. Lin, *An Introduction to Error-Correcting Codes*, Prentice-Hall, Englewood Cliffs, N.J., 1970.

17 T. Muratani, H. Saitoh, K. Koga, T. Mizuna, Y. Yasuda and J. S. Snyder, "Application of FEC Coding to the INTELSAT TDMA System", in *Proc. ICDSC-4* Montreal, Oct. 1978, pp. 108–115.

18 W. W. Wu, "New Convolutional Codes—Part 1", *IEEE Trans. Commun.*, Vol. COM-23, Sept. 1975, pp. 942–955.

19 W. C. Lindsey and M. K. Simon, *Telecommunications Systems Engineering*, Prentice-Hall, Englewood Cliffs, N.J., 1973.

20 A. J. Viterbi, *Principles of Coherent Communications*, McGraw-Hill, New York, 1966.

21 I. Jacobs, "Comparison of M-ary Modulation Systems", *Bell Syst. Tech. J.*, Vol. 46, May–June 1967, pp. 843–864.

22 *Electron. Des.*, Feb. 1977.

23 J. E. McNamara, "Technical Aspects of Data Communication", Digital Equipment Corporation, Bedford, Mass., 1977.

24 R. J. Benice and A. H. Frey, "An Analysis of Retransmission Systems", *IEEE Trans. Commun. Technol.*, Vol. COM-12, Dec. 1964, pp. 135–154.

25 H. O. Burton and D. D. Sullivan, "Errors and Error Control", *Proc. IEEE*, Vol. 60, Nov. 1972, pp. 1293–1301.

26 M. D. Balkovic and P. E. Muench, "Effect of Propagation Delay, Caused by Satellite Circuits, on Data Communication System That Use Block Retransmission for Error Correction", in *Proc. Int. Conf. Commun.*, Boulder, Colo. 1969, pp. 29–36.

27 D. R. Doll, *Data Communications; Facilities, Networks, and Systems Design*, Wiley, New York, 1978.

28 J. Conan, D. Haccoun and H. H. Hoang, "Error Control Techniques for Data Transmission over Satellite Channels", Technical Report EP-75-R-32, Dept. of Elec. Eng., École Polytechnique de Montréal, Montréal, 1975.

29 A. R. K. Sastry, "Improving ARQ Performance on Satellite Channels under High Error Rate Conditions", *IEEE Trans. Commun.*, Vol. COM-23, Apr. 1975, pp. 436–439.

30 A. Gatfield, "Error Control on Satellite Channels Using ARQ Techniques", *COMSAT Tech. Rev.*, Vol. 6, Spring 1976, pp. 179–188.

31 J. Conan and D. Haccoun, "High-Speed Transmission of Reliable Data on Satellite Channels", in *Proc. ICDSC-4*, Montreal, Oct. 1978, pp. 269–274.

32 Comité Consultatif International sur le Téléphone et le Télégraphe (CCITT), "Code-Independent Error Control System", Vth Plenary Assembly, *Green Book*, Vol. VIII, Rec. V.41, Int. Telecommun. Union, Geneva, 1972.

33 *General Information—Binary Synchronous Communications*, 2nd ed., IBM Syst. Ref. Library, File TP-09, Order GA27-3004-1, Dec. 1969.

34 D. Haccoun, J. Conan and G. Golly, "Node to Node Protocols on a High Speed Full-Duplex Satellite Link", in *Proc. Natl. Telecommun. Conf.*, Dallas, Tex., Dec. 1978, pp. 28.1.1–28.1.5.

35 I. M. Jacobs, "Practical Applications of Coding", *IEEE Trans. Inf. Theory*, Vol. IT-20, May 1974, pp. 305–310.

36 R. J. Benice and A. H. Frey, Jr., "Comparisons of Error Control Techniques", *IEEE Trans. Commun. Technol.*, Vol. COM-12, Dec. 1964, pp. 146–154.

PROBLEMS

1 A digital signal is received from a satellite with a signal-to-noise ratio $P/N_0 = 20$ dB.

   (a) What is the maximum data transmission rate possible with an FSK modulation if the bit-error rate is required to be 10^{-5}?

   (b) Compare your result with the capacity of this system if a bandwidth of 100 Hz is available. What is the required E_b/N_0 at capacity? What is the bandwidth efficiency of this system?

   (c) If an arbitrarily large bandwidth and an arbitrarily complex system are available, what is the maximum transmission rate possible? Compare with part (a).

2 Consider transmitting sequences of $k=4$ bits of information through an additive white Gaussian noise channel with a noise spectral density N_0 W/Hz.

   (a) Assuming an ideal PSK modulation scheme, what is the required value of E_b/N_0 if the bit-error rate is $p=10^{-5}$? What is the uncoded word-error probability for this system?

(b) A (7,4) single-error-correction block coding scheme is now utilized. Assuming that the E_b/N_0 value of part (a) is unchanged, find the coded word-error probability. Compare with part (a).

(c) Repeat parts (a) and (b) if $p=10^{-4}$ and $p=10^{-3}$. What can you conclude about coding systems in general as the noise becomes increasingly severe?

3 (a) Plot the capacity versus the bandwidth of a satellite channel with additive white Gaussian noise of spectral density N_0 and average signal power P. What is the asymptotic value of the capacity?

(b) Let $P/N_0 = 20$ dB. Find the maximum information transmissions rate R in bps if ideal PSK modulation is used and an error probability of 10^{-5} is required. Repeat for an error probability of 10^{-4}.

(c) Repeat part (b) if DPSK modulation is used.

(d) Let $P/N_0 = 20$ dB and assume a channel bandwidth of 1000 Hz. What is the maximum transmission rate R in bps if an arbitrarily complicated system may be used and an arbitrarily small error probability is required? Compare with parts (b) and (c).

(e) Repeat part (d) if the bandwidth may be as large as desired. Compare your results with those of part (b).

4 Digital communication from a satellite in geostationary orbit (36,000 km from the Earth) to an Earth station is to be investigated. The transmission power is 10 W, the frequency is 250 MHz and the antenna gains of the satellite and Earth station are 10 and 30 dB, respectively. The additive white Gaussian noise has a spectral density $N_0 = 10^{-20}$ W/Hz, and an error probability of 10^{-5} is desired.

(a) Assuming ideal PSK modulation, what is the maximum data rate R this system allows?

(b) The binary data uses M-ary orthogonal modulation. Find the maximum binary rate for $M=32$.

(c) The error probability is desired to be vanishingly small. If an arbitrarily complex encoding scheme and an arbitrarily large bandwidth are allowed, what is the maximum transmission rate possible? Compare with parts (a) and (b).

5 Digital pictures are to be transmitted for a deep-space probe using DPSK modulation at a transmission rate of 4.8×10^6 bps. The received signal-to-noise ratio is $P/N_0 = 73$ dB and the bit-error probability is to be less than 10^{-4}.

(a) Determine whether the link can support this transmission rate.

(b) If your answer to part (a) is negative, determine the necessary improvement in P/N_0, and suggest a coding scheme that may allow this improvement.

6 Aboard a deep-space probe a source transmits at a rate of 1000 times per second an information digit chosen in a ternary alphabet (e.g., -1, 0, or $+1$). The additive white Gaussian noise has a spectral density $N_0 = kT_r$, where k is the Boltzmann constant, equal to 1.38×10^{-23} J/K and T_r is the noise temperature of the system. The received power P is equal to 4×10^{-18} W.

   (a) What is the data rate in bps?

   (b) Removing any restrictions on the allowed bandwidth, what is the temperature T_r beyond which no coding technique will give an error probability arbitrarily small for this system?

7 In a Stop-and-Wait ARQ system erroneous blocks are retransmitted with probability P_B and delivered in error to the user with probability P_u. Assuming that the blocks may be cycling indefinitely between receiver and transmitter, find an expression for:

   (a) The block error probability of the entire system.

   (b) The average number of blocks transmitted in the system per block delivered. Using this result, verify your answer to part (a).

8 **(a)** Consider a binary transmission system whereby each bit received in error is automatically retransmitted. If p is the bit-error probability on the forward channel, what is the maximum average rate (in bits per channel use) that can be used over this channel with no errors delivered to the user?

   (b) Consider a Stop-and-Wait ARQ system with block error probability P_B. What is the average number of blocks delivered per transmitted block?

9 **(a)** Show that in Stop-and-Wait ARQ systems the average time to transfer a block of size n from the source to the user is given by

$$\overline{\Delta T} = \frac{n + R_s \Delta T}{R_s (1 - P_B)}$$

where ΔT is the overall round-trip delay in seconds, R_s is the transmission speed on the channel in symbols/s, and P_B is the block error probability.

   (b) Using this result, derive the throughput of Stop-and-Wait ARQ systems.

10 **(a)** For Continuous ARQ systems, express the average number of symbols $\overline{D}$ separating two consecutive error-free blocks in terms of the block length n, the overall delay D (in symbols) and the block error probability P_B.

   (b) Using your result of part (a), determine the throughput of Continuous ARQ systems.

11 **(a)** Establish expression (11.33), giving the optimum block length for continuous ARQ system using a fixed number of parity symbols in the error-detecting code.

(b) Determine the optimal block length $\hat{n}_{opt}$ and the corresponding throughput value $\hat{\eta}_{opt}$ for a satellite link using a code with 12, 18 and 24 parity symbols over a channel with raw error rate p varying from 10^{-7} to 10^{-3}. Compare your results with the throughput upper bound η_{opt} given by (11.31).

12 **(a)** Show that for the variation of Stop-and-Wait systems involving sets of ℓ consecutive retransmissions of the erroneous blocks, the expected number of sets of retransmissions is given by

$$E = \frac{P_B}{1 - P_B^{\ell}}$$

where P_B is the block error probability.

(b) Plot E as a function of ℓ for blocks of length $n=256$ and satellite raw error rates $p=10^{-3}$, 10^{-4} and 10^{-5}.

(c) Show that the throughput of this scheme is given by (11.37).

(d) Comparing the throughputs of this variant and of conventional Stop-and-Wait systems, determine an upper bound on the number of retransmissions ℓ, so that an improvement is obtained from the conventional system.

(e) Find an expression for the optimal value of ℓ for which the improvement is maximum. Plot the improvement for $\ell=2$ as a function of p, $10^{-2} \leq p \leq 10^{-6}$, for a system with $n=1000$, $k=960$, $R_s=2400$ symbols/s and $\Delta T=600$ ms.

13 Blocks of data of length $n=1000$ bits are transmitted over a satellite link using noncoded PSK modulation with ideal coherent demodulation. The uplink is assumed error-free and the downlink budget calculation yields the following expressions for the receiving signal-to-noise ratio (in dB):

$$\frac{P}{N_0} = 47.6 + \frac{G}{T}$$

(a) Assuming independent errors in the blocks, plot the allowable data rates R in Mbps as a function of the block error probability P_B, $10^{-1} \leq P_B \leq 10^{-8}$, for different receiving antenna systems having $G/T=40.7$, 39.0, 34 and 27 dB. Comment on the slope of the curves for the different antenna systems.

(b) A rate-$\frac{1}{2}$ coding scheme is used to improve the link. Assuming that the coding/decoding scheme provides a coding gain of 5 dB, and assuming independent errors in the blocks, repeat the plots of R versus P_B of part (a). Comment on the additional costs in bandwidth, complexity and so on entailed by this system over noncoded PSK.

14 An ARQ system over a satellite link uses the BCH code (1023, 913) for the error detection of the blocks. The downlink budget has given $P/N_0 = 69.6$ dB for the received signal-to-noise ratio. The system uses PSK modulation with ideal coherent demodulation. Assume that the errors occur independently in the blocks and that all ACK/NACK traffic is error-free.

(a) Plot the output data rate R in Mbps versus the signalling rate in the channel R_s, $10^5 \leq R_s \leq 10^7$ symbols/s, for both Continuous and Selective ARQ schemes.

(b) The system is to be improved by using a FEC system within the ARQ scheme. Two FEC systems are considered: a rate-$\frac{3}{4}$ coding scheme yielding a constant coding gain of 2 dB, and a rate-$\frac{1}{2}$ coding scheme yielding a constant coding gain of 4.5 dB. Again assuming the errors in the blocks to be error-free, plot the output data rate R versus the signalling rate R_s for each of these four possible FEC/ARQ schemes. Compare with your curves of part (a) and determine the rates over which each system is the most suitable to be used.

CHAPTER 12

||||||||||||||||||||||||||||||||||||||||||||||||||||

Convolutional Encoding; Viterbi and Sequential Decoding

In Chapter 11 we saw that in block coding a sequence of k information bits is encoded into a block of n symbols to be transmitted over the channel. The block encoder accepts k-bit information blocks and generates codewords of length n symbols. Therefore, the codewords are produced in a block-by-block fashion and the encoder must buffer the entire information block before generating a codeword. In some applications this may be undesirable.

Convolutional codes are codes that are particularly suitable when the information symbols to be transmitted arrive serially rather than in long blocks. A convolutional encoder processes the information bits continuously in a serial fashion, a few bits at a time, and very often only one bit at a time.

After presenting some basic properties of convolutional codes we will present and analyse two interesting and powerful decoding techniques especially suitable for satellite communications: Viterbi decoding and sequential decoding. Threshold decoding, which can be used with either convolutional or block coding, is examined in Chapter 14.

12.1 STRUCTURE OF CONVOLUTIONAL CODES

In this section the fundamental structures and basic properties of convolutional codes are presented. The different graphic representations of convolutional codes are examined with enough details to facilitate the analysis of their performance and the understanding of the techniques used for their decoding.

Convolutional Encoder

A binary convolutional code of rate $1/v$ bits/symbol may be generated by a linear finite-state machine consisting of a K-stage shift register, v modulo 2 adders connected to some of the shift register stages, and a commutator that

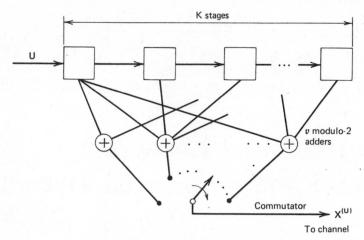

Figure 12.1 Convolutional encoder.

scans the output of the modulo 2 adders. The machine is called a convolutional encoder and is sketched in Figure 12.1.

Let us assume that the information to be encoded is the information sequence, or message

$$\mathbf{U} = (u_1, u_2, u_3, \ldots, u_L) \tag{12.1}$$

of binary digits 0 or 1. Assuming the shift register to be initially filled with an all-zero sequence (or any other known sequence), the first binary digit u_1 is fed into the first stage of the shift register. The modulo 2 adders are then sampled in sequence by the commutator, producing v output coded symbols

$$x_1^{(1)}, x_1^{(2)}, \ldots, x_1^{(v)}$$

which are transmitted in the channel. In modulo 2 addition, the output of an adder is "1" if and only if the total number of 1's in the shift register stages connected to the adder is odd. Otherwise, the adder output is "0". After the vth output symbol $x_1^{(v)}$ is obtained, the second input u_2 is shifted into the first stage of the shift register, causing the contents of all other stages to move one step to the right. The rightmost digit leaving the encoder is lost. The v modulo 2 adders are sampled again, yielding v new output symbols

$$x_2^{(1)}, x_2^{(2)}, \ldots, x_2^{(v)}$$

which are again transmitted. This procedure continues until the last input digit u_L enters the shift register. Now in order to return the shift register to its initial zero state, with each shift of u_L a "0" symbol is fed in until u_L leaves the shift register. This terminating sequence of $(K-1)$ zeros is called the **tail of the message**.

The L-bit message sequence $\mathbf{U}$ of (12.1) produces the output coded sequence or codeword

$$\mathbf{X}=\left(x_1^{(1)}, x_1^{(2)},\ldots, x_1^{(v)}, x_2^{(1)}, x_2^{(2)},\ldots, x_2^{(v)},\ldots, x_{K+L-1}^{(1)},\ldots, x_{K+L-1}^{(v)}\right) \quad (12.2)$$

of length $(K+L-1)v$ binary digits. The **rate of the code** is therefore

$$r=\frac{L}{(L+K-1)v} \qquad \text{bits/symbol} \qquad (12.3)$$

and for $L\gg K$ the rate is written as

$$r\simeq\frac{1}{v} \qquad \text{bits/symbol} \qquad (12.4)$$

In order to describe the code, let the row vector called the **connection vector**

$$\mathbf{G}_j=(g_{1j}, g_{2j},\ldots, g_{vj}), \qquad j=1,2,\ldots, K \qquad (12.5)$$

specify the connections between stage j of the shift register and the v modulo 2 adders. The component g_{ij} is equal to 1 if the ith modulo 2 adder is connected to stage j; otherwise, it is equal to 0. For example, the rate-$\frac{1}{3}$ convolutional encoder shown in Figure 12.2 has the connection vectors

$$\mathbf{G}_1=(1,1,1)$$

stage

$$\mathbf{G}_2=(1,1,0)$$

$$\mathbf{G}_3=(1,1,1)$$

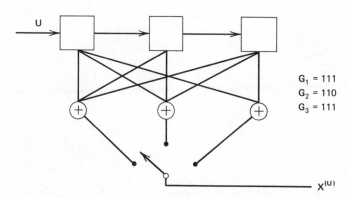

Figure 12.2 Convolutional encoder with $K=3$, $v=3$.

Assuming that the shift register is initially filled with zeros, as an example, let the encoder of Figure 12.2 be used to encode the information sequence

$$\mathbf{U}=(u_1, u_2, u_3, u_4, u_5, \ldots)=(1,0,1,1,0,\ldots) \qquad (12.6)$$

Denoting the modulo 2 addition by the sign $\oplus$, the coded symbols corresponding to the first four information bits are evaluated as follows:

$$\mathbf{x}_1 = u_1\mathbf{G}_1 = 111 \qquad (12.7a)$$

$$\mathbf{x}_2 = u_2\mathbf{G}_1 \oplus u_1\mathbf{G}_2 = 110 \qquad (12.7b)$$

$$\mathbf{x}_3 = u_3\mathbf{G}_1 \oplus u_2\mathbf{G}_2 \oplus u_1\mathbf{G}_3 = 000 \qquad (12.7c)$$

$$\mathbf{x}_4 = u_4\mathbf{G}_1 \oplus u_3\mathbf{G}_2 \oplus u_2\mathbf{G}_3 = 001 \qquad (12.7d)$$

The encoding steps are illustrated in Figure 12.3, and another representation of a convolutional encoder is given in Figure 12.4.

From the operation of the encoder, in general, the v-symbol output sequence $\mathbf{X}_i = (x_i^{(1)}, x_i^{(2)}, \ldots, x_i^{(v)})$ produced when the input binary digit u_i is first shifted in the shift register is given by the expression

$$\mathbf{X}_i = u_i\mathbf{G}_1 \oplus u_{i-1}\mathbf{G}_2 \oplus \cdots \oplus u_{i-K+1}\mathbf{G}_K \qquad (12.8)$$

Equation (12.8) is recognized as the convolution of the vectors $\mathbf{G}$ and the

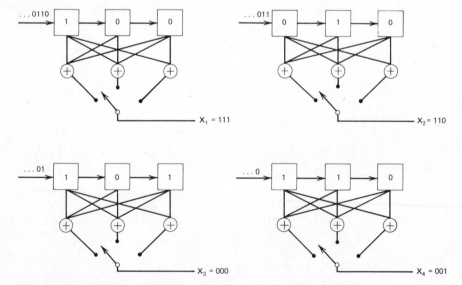

Figure 12.3 Steps in encoding the sequence $\mathbf{U}=(10110\ldots)$ with the coder of Figure 12.2.

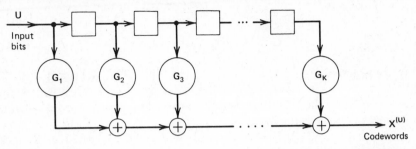

Figure 12.4 Binary convolutional encoder of constraint length K and generator $\mathbf{G}^{*}=(\mathbf{G}_{1},\mathbf{G}_{2},\ldots\mathbf{G}_{K})$.

K-bit-long input sequence $u_{i}, u_{i-1}, \ldots, u_{i-K+1}$. The term "convolutional" is taken from the form of (12.8) and clearly, the vectors $\mathbf{G}_{i}$, $i=1,2,\ldots,K$, specify the code. We observe that the v output symbols corresponding to a particular input digit depend upon that digit and the $(K-1)$ preceding digits that entered the encoder. The constraint length of the code expressed in information bits is defined as the number of shifts over which a single information bit can influence the encoder output. For the simple binary convolutional code, the constraint length is equal to K, the length of the shift register.* Since each output symbol of the encoder depends on the present input information bit and the $(K-1)$ preceding ones, the **memory** of the encoder is equal to $(K-1)$ information bits.

Considering the input binary sequence as an L-dimensional row vector $\mathbf{U}$, the $(L+K-1)\times v$-component output codeword $\mathbf{X}$ can be written, using modulo 2 arithmetic, as

$$\mathbf{X}=\mathbf{U}[\mathbf{G}] \tag{12.9}$$

where

$$\overleftarrow{\hspace{2cm}(L+K-1)v \text{ columns}\hspace{2cm}}$$

$$[\mathbf{G}]=\begin{bmatrix} \mathbf{G}_1 & \mathbf{G}_2 & \mathbf{G}_3 & \cdots & \mathbf{G}_K & \mathbf{0} & \mathbf{0} & \cdots & \mathbf{0} \\ \mathbf{0} & \mathbf{G}_1 & \mathbf{G}_2 & \mathbf{G}_3 & \cdots & \mathbf{G}_K & \mathbf{0} & \cdots & \mathbf{0} \\ \mathbf{0} & \mathbf{0} & \mathbf{G}_1 & \mathbf{G}_2 & \mathbf{G}_3 & \cdots & \mathbf{G}_K & \mathbf{0}\cdots & \mathbf{0} \\ \vdots & \vdots & & \ddots & \ddots & \ddots & & \ddots & \vdots \\ \mathbf{0} & \mathbf{0} & \cdots & \mathbf{0} & \mathbf{G}_1 & \mathbf{G}_2 & \mathbf{G}_3 & \cdots & \mathbf{G}_K \end{bmatrix} \begin{array}{l} \\ \\ L \text{ rows} \\ \\ \end{array} \tag{12.10}$$

The matrix $[\mathbf{G}]$ is called a **generator matrix** of the code, and the Kv-component

---

*The constraint length of a code is sometimes expressed in coded symbols, as the number of output coded symbols that are affected by a single information bit. It is then equal to Kv.

vector

$$G^* = (G_1, G_2, G_3, \ldots, G_K) \qquad (12.11)$$

is called a **generator** of the code. The first row of matrix $[G]$ is the generator G^* followed by $(L-1)v$ zeros, and each succeeding row is the previous row shifted v places to the right, with all elements to the left of G_1 equal to zero. The number of rows is the length L of the input sequence, and the number of columns is $(K+L-1)v$. Every codeword X can be expressed as a linear combination of the rows of $[G]$. For the example of Figure 12.2, the generator of the code is

$$G^* = (111 \quad 110 \quad 111)$$

and for the 4-bit-long input sequence $U = (1011)$ followed by a tail of two zeros, the output sequence is

$$X = (111 \quad 110 \quad 000 \quad 001 \quad 001 \quad 111)$$

Systematic Convolutional Codes

A convolutional code is said to be **systematic** if the first of each v coded symbols is the information symbol that generated these v symbols. A convolutional encoder is thus systematic if the first modulo 2 adder is connected only to the first stage of the shift register. Therefore, the output of the first modulo 2 adder is the information bit itself, and hence this adder may be eliminated, simplifying the encoder somewhat. Unlike nonsystematic codes, the recovery of the information bits from the encoded sequences is trivial with systematic codes. However, in general, systematic codes do not perform as well as nonsystematic codes with the same rate and same constraint length.

Impulse Response

Since the convolutional encoder is a linear finite-state machine, it may be described by its impulse response or its transfer function.

The **impulse response** of a convolutional encoder is the output X_I corresponding to the impulse input sequence $I = (1 \quad 0 \quad 0 \quad 0 \quad \cdots)$. From the discussion above, the impulse response is recognized as the generator of the code followed by a string of zeros:

$$X_I = (G_1, G_2, \ldots, G_K, 0, 0, \ldots) \qquad (12.12)$$

For example, the impulse response of the encoder of Figure 12.2 is $(111 \quad 110 \quad 111 \quad 000 \quad \cdots)$. Now since any input sequence can be expressed as a linear combination of the impulse input sequence and its delayed versions, by linearity it follows that the corresponding output sequence can be expressed by the same linear combination of delayed versions of the impulse response of the encoder. Naturally, this amounts to performing the matrix multiplication given by (12.9) with all operations performed in modulo 2 arithmetic.

Output-Symbol Representation

The previous description of the encoder yields the output sequence $\mathbf{X}$ of the encoder given an input sequence $\mathbf{U}$. Since the input sequence may be arbitrarily long, the description of the entire output sequence of the encoder by (12.9) using the generator matrix (12.10) becomes rather cumbersome. A more attractive approach is to describe and evaluate the encoder output symbol by symbol.

Keeping the description of the code by its connection vectors of (12.5), for an input information bit u_i the output symbol $x_i^{(j)}$ of the jth modulo 2 adder is given by

$$x_i^{(j)} = u_i g_{j1} \oplus u_{i-1} g_{j2} \oplus \cdots \oplus u_{i-K+1} g_{jK}, \qquad j=1,2,\ldots,v \quad (12.13a)$$

or equivalently,

$$x_i^{(j)} = \sum_{\ell=1}^{K} u_{i-\ell+1} g_{j\ell}, \qquad j=1,2,\ldots,v \quad (12.13b)$$

where $\sum$ represents the modulo 2 sum. The v-tuple output $\mathbf{X}_i = (x_i^{(1)} x_i^{(2)} \cdots x_i^{(v)})$, corresponding to the input information bit u_i is then given by the encoding equation

$$\left(x_i^{(1)}, x_i^{(2)}, \ldots, x_i^{(v)} \right) = \left(u_i, u_{i-1}, \ldots, u_{i-K+1} \right) \begin{bmatrix} g_{11} & g_{21} & \cdots & g_{v1} \\ g_{12} & g_{22} & \cdots & g_{v2} \\ \vdots & \vdots & & \vdots \\ g_{1K} & g_{2K} & \cdots & g_{vK} \end{bmatrix}$$

$$(12.14)$$

Using this description, a rate-$\frac{1}{2}$ convolutional encoder is schematically represented in Figure 12.5.

It is worthwhile to note that in general the number of modulo 2 adders is substantially smaller than the constraint length. Consequently, instead of describing the code by K different v-dimensional connection vectors of the form of (12.5), it may be more convenient to describe it by v connection vectors, each of dimension K, that is, using the same notation as in (12.5),

$$\mathbf{G}_j = (g_{j1}, g_{j2}, \ldots, g_{jK}), \qquad j=1,2,\ldots,v \quad (12.15)$$

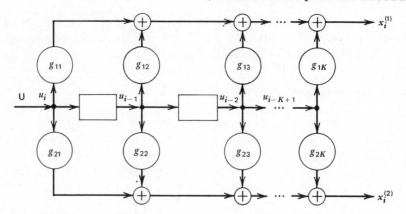

Figure 12.5   Rate-$\frac{1}{2}$ convolutional encoder of constraint length K.

For example, the connection vectors for the encoder of Figure 12.2 become

$$\mathbf{G}_1 = 111$$

$$\mathbf{G}_2 = 111$$

$$\mathbf{G}_3 = 101$$

It is readily verified that this code description does not alter the encoding equation given by (12.14).

Polynomial Form Representation

Denoting a **time unit delay** by D and n unit delays by D^n, the information sequence **U**, output sequence **X** and connection vectors $\mathbf{G}_j$ can all be represented by polynomials of the form

$$U(D) = u_0 D^0 \oplus u_1 D \oplus u_2 D^2 \oplus \cdots \tag{12.16}$$

$$G_j(D) = g_{j1} D^0 \oplus g_{j2} D \oplus \cdots g_{jK} D^{K-1}, \qquad j = 1, 2, \ldots, v \tag{12.17}$$

The polynomials $G_j(D)$ are called **generator polynomials**. From (12.13) it is easily shown that the outputs of the modulo 2 adders are given by

$$X_j(D) = U(D)G_j(D), \qquad j = 1, 2, \ldots, v \tag{12.18}$$

which suggests that the generator polynomials can be interpreted as the v transfer functions of the encoder.

    Finally, the output codeword or output polynomial $X(D)$ is the single sequence obtained by multiplexing the v output sequences $X_j(D)$ prior to transmission. Observing that even-numbered output $X_j(D)$ are sampled at odd time intervals, and that odd-numbered outputs $X_j(D)$ are sampled at even time intervals, the encoder output polynomial is expressed as

$$X(D) = X_1(D^v) \oplus DX_2(D^v) \oplus \cdots \oplus D^{v-1}X_v(D^v) \qquad (12.19)$$

or, equivalently, as

$$X(D) = U(D^v)[G_1(D^v) \oplus DG_2(D^v) \oplus \cdots \oplus D^{v-1}G_v(D^v)] \qquad (12.20)$$

For example, the generator polynomials of the encoder of Figure 12.2 are

$$G_1(D) = 1 \oplus D \oplus D^2 \qquad (12.21a)$$

$$G_2(D) = 1 \oplus D \oplus D^2 \qquad (12.21b)$$

$$G_3(D) = 1 \oplus D^2 \qquad (12.21c)$$

Using the same input sequence $\mathbf{U} = (1 \quad 0 \quad 1 \quad 1)$ as earlier, the corresponding information polynomial is $U(D) = 1 \oplus D^2 \oplus D^3$ and the output polynomials are calculated as follows:

$$X_1(D) = U(D)G_1(D) = (1 \oplus D^2 \oplus D^3)(1 \oplus D \oplus D^2)$$
$$= 1 \oplus D \oplus D^5$$
$$X_2(D) = U(D)G_2(D) = (1 \oplus D^2 \oplus D^3)(1 \oplus D \oplus D^2)$$
$$= 1 \oplus D \oplus D^5$$
$$X_3(D) = U(D)G_3(D) = (1 \oplus D^2 \oplus D^3)(1 \oplus D^2)$$
$$= 1 \oplus D^3 \oplus D^4 \oplus D^5$$

We then have

$$X_1(D^3) = 1 \oplus D^3 \oplus D^{15}$$

$$DX_2(D^3) = D \oplus D^4 \oplus D^{16}$$

$$D^2X_3(D^3) = D^2 \oplus D^{11} \oplus D^{14} \oplus D^{17}$$

Finally, using (12.19), the encoder output polynomial is given by

$$X(D) = 1 \oplus D \oplus D^2 \oplus D^3 \oplus D^4 \oplus D^{11} \oplus D^{14} \oplus D^{15} \oplus D^{16} \oplus D^{17}$$

which corresponds, as expected, to the same codeword obtained earlier,

$$\mathbf{X} = (111 \quad 110 \quad 000 \quad 001 \quad 001 \quad 111)$$

Generalization

One can generalize the binary convolutional encoder by allowing b, $b \geq 1$, information symbols at a time to enter the shift register, and also by allowing these information symbols to be q-ary rather than binary. Hence in a generalized convolutional encoder the K-stage shift register receives groups of b q-ary information symbols of an input sequence whose components are elements of a finite Galois field $GF(q)$. Then the input to the encoder is the information sequence

$$U = (\mathbf{u}_1, \mathbf{u}_2, \mathbf{u}_3, \dots, \mathbf{u}_L) \tag{12.22}$$

where

$$\mathbf{u}_i = \left(u_i^{(1)}, u_i^{(2)}, \dots, u_i^{(b)} \right)$$

$$u_i^{(j)} \in GF(q); \qquad j = 1, 2, \dots, b; \, i = 1, 2, \dots, L$$

The v modulo 2 adders, $v > b$ and their connections, are replaced by v inner product computers, each of which computes the inner product in $GF(q)$ of the shift register contents and some specified vector. In other words, the elements of the generator matrix $[G]$ belong to $GF(q)$, and the operations of (12.9) are performed by arithmetic operations in $GF(q)$. The length K is chosen to be a multiple of the integer b, $K = kb$, and the constraint length of the code is K, the number of bits which can influence the encoder output. The rate of the code is then

$$r = b \frac{\log_2 q}{v} \qquad \text{bits/symbol} \tag{12.23}$$

Following the encoding of the L groups of b information symbols, a tail of $(k - 1)$ groups of b zeros is shifted in to clear the shift register.

In practice it may be simpler to consider the b-tuple input sequences entering the encoder in parallel. The K-stage shift register is then replaced by a stack of b k-stage shift registers, each with its own kv-component generator vector. The generator of the whole code is then a $b \times kv$ matrix

$$[G^*] = \begin{bmatrix} G_1^* \\ G_2^* \\ \vdots \\ G_b^* \end{bmatrix} \tag{12.24}$$

where G_i^*, $i = 1, 2, \dots, b$, is the generator corresponding to the ith shift register. The generator matrix $[G]$ of the whole code has the same form as that of the

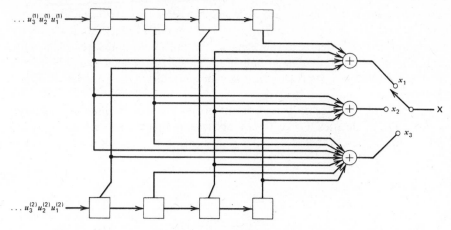

Figure 12.6 Rate-$\frac{2}{3}$ convolutional encoder with $K=8$, $v=3$, $b=2$.

simple binary code of (12.10), but with each row replaced by the matrix $[\mathbf{G}^*]$ of (12.24). For example, the rate-$\frac{2}{3}$ code with generator vectors

$$\mathbf{G}_1^* = (111 \quad 011 \quad 001 \quad 100)$$

$$\mathbf{G}_2^* = (101 \quad 001 \quad 111 \quad 011)$$

is shown in Figure 12.6. The input sequences starting with $11001001\ldots$, will be encoded as $010 \quad 010 \quad 001 \quad 001 \quad \cdots$.

In summary, from the foregoing description of the operation of a convolutional encoder, we observe that

1 The coded symbols depend on present and past values of the inputs.
2 The past dependency does not extend to the infinite past but is limited to the length of the shift register.
3 The code is **linear** since the output symbols are linear combinations of past inputs.

We now show that the first two observations lead to two different representations of convolutional codes. In the following sections, unless specified, we consider only binary symbols, although the generalization to q-ary symbols is straightforward.

Tree Structure

The fact that at any time an input may take in general 2^b different values yielding 2^b output sequences suggests representing the output of a convolutional encoder by a **tree** with $2^{b\ell}$ possible sequences corresponding to ℓ b-inputs.

The **root node** of the tree has no predecessor, whereas every other node has one predecessor. There are 2^b branches stemming out of each node, with the exception of the nodes corresponding to the tail of the message, which have only a single branch extension. All branches carry v coded symbols. Figure 12.7 depicts the tree representation for the outputs of the encoder of Figure 12.2.

A **path** in the tree is traced from left to right according to the input sequence that specifies it. For the binary tree of Figure 12.7, a 0 input means taking the upper branch leaving a node, and a 1 means taking the lower

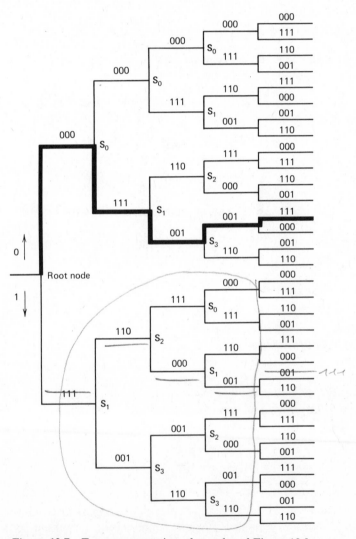

Figure 12.7 Tree representation of encoder of Figure 12.2.

branch. Hence a message sequence $u_1, u_2, u_3, \ldots$ traces a path through the tree, and the corresponding coded symbols on the branches of the path are the channel inputs that are transmitted. In Figure 12.7, the input message starting with the sequence 01100 determines the path indicated by the thick line having the encoded sequence 000 111 001 001 111. The tree of Figure 12.7 is said to have five levels, one for each branching along a possible path. The number of levels in a tree can be extended indefinitely, and naturally there is a one-to-one correspondence between the set of all 2^L information sequences of length L bits and the set of paths through the L levels of the tree. A code that can be represented by a tree is called a **tree code**. In general, tree codes are codes where coded symbols depend on past input sequences and where paths may be represented as branch choices on a tree. Not all tree codes are convolutional codes.

State Diagram

Since a convolutional encoder is a linear finite-state machine, it can be described by a state diagram. In the causal transformation of the input information sequence $\mathbf{U}$ into the coded sequence $\mathbf{X}$, at any given time the v output digits are completely determined by the most recent K information bits that are residing in the shift register. (In what follows we will assume that $b=1$ without loss of generality.) Now, since a finite-state machine changes from one state to another according to the input it receives, the **state** of a convolutional encoder is then simply the $(K-1)$ input symbols preceding the current input. That is, the state is the contents of the first $(K-1)$ register stages, and hence the encoder state together with the new input symbols uniquely specify the v output symbols. We can label the states by the sequence

$$\mathbf{S} = (s_1, s_2, \ldots, s_{K-1}) \tag{12.25}$$

K=stages in the register

where s_i are the input to the encoder and where input s_{i+1} entered the encoder before s_i.

The total number of states is then 2^{K-1} for single-input encoders (2^{K-b} for shifts of b inputs) and from one state the machine may move to two other states (2^b other states for b inputs). Naturally, some input symbols may cause the machine to stay in the same state, as indicated in Table 12.1.

This description suggests representing the encoder by a **state diagram** with 2^{K-1} states or nodes with two branches leaving and entering each state. These branches represent the transitions of the encoder from one state to another, and therefore they carry the v coded symbols delivered by the encoder in the transition. For example, the state-diagram representation of the encoder of Figure 12.2 is shown in Figure 12.8. It has four states and each branch is labelled by the corresponding input bit (above) and output v-tuples (below).

The state diagram is a very compact representation of the encoder. Starting from the initial state $\mathbf{S}_0 = (0, 0, \ldots, 0)$, the coded output corresponding to the

Table 12.1   State Transitions of the Encoder of Figure 12.2

Old State	Input Bit	New State	Output Symbols
00	0	00	000
00	1	10	111
10	0	01	110
10	1	11	001
01	0	00	111
01	1	10	000
11	0	01	001
11	1	11	110

input sequence $\mathbf{U} = (u_1, u_2, u_3, \ldots)$ is readily obtained by writing the branch symbols corresponding to the transitions due to $u_1, u_2, u_3, \ldots$. For example, from Figure 12.8, the output to the message $110100\cdots$ is 111, 001, 001, 000, 110, 111. A potentially infinite tree has been reduced to a mere four-state diagram.

The state diagram can be regarded as a signal flow graph, and particular codes can be analysed via the transfer function of their state diagram [1]. However, the great difficulty in finding the transfer function for long-

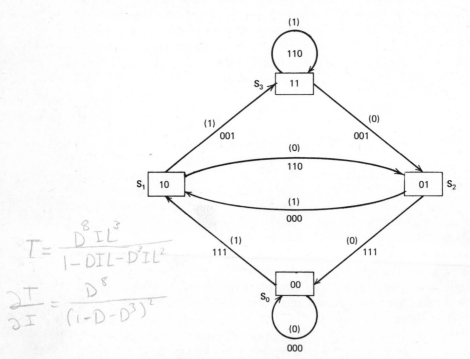

$$T = \frac{D^8 I L^3}{1 - DIL - D^3 I L^2}$$

$$\frac{\partial T}{\partial I} = \frac{D^8}{(1 - D - D^3)^2}$$

Figure 12.8   State diagram for the encoder of Figure 12.2.

constraint-length codes limits the use of this technique to very short constraint-length codes.

Trellis Structure

To each node of the tree there corresponds an encoder state. For an input sequence of L bits, there are 2^L terminal nodes in the L-level tree, but only 2^{K-1} different states. Therefore, for $L \geq K$, the number of nodes in the tree exceeds the number of states of the encoder and hence several nodes must correspond to the same encoder state. Clearly, after the input symbols have been identical for $(K-1)$ consecutive symbols, the encoded symbols are the same: the paths are said to **remerge**. However, this fact is obscured in the tree representation of the code since the tree keeps on branching off with two branches from each node, as if the states were all different. This behaviour is equivalent to considering an infinite number of states or a dependence over the infinite past. Therefore, we see that the tree contains much redundant information which can be eliminated by merging together, at any same level, all nodes corresponding to the same encoder state. The redrawing of the tree with merging paths is called a **trellis**.

Taking the example of Figure 12.7, let the four states be $S_0 = (0,0)$; $S_1 = (1,0)$; $S_2 = (0,1)$; $S_3 = (1,1)$. After the third level, the information sequences $100\, u_4 u_5 u_6 \ldots$ and $000\, u_4 u_5 u_6 \ldots$ generate the same code symbols, and hence both nodes in Figure 12.7 labelled S_0 can be joined together. The same reasoning applies for the other nodes, and the tree collapses into the trellis of Figure 12.9. More insight about the remerging of the tree paths in general is given by the following properties.

1   If two sequences of information symbols are identical except for n consecutive information symbols, the corresponding codewords will be distinct over $(n+K-1)$ consecutive branches.

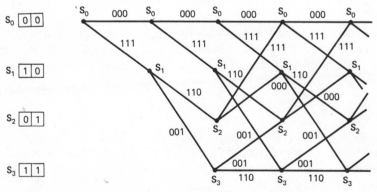

Figure 12.9   Trellis representation of the encoder of Figure 12.2.

$$\frac{\partial T}{\partial I} = \frac{D^8}{(1-D-D^2)^2}$$

2 If the information symbols of two paths agree at some point in $(K-1)$ branches, the subtrees extending from these two paths thereafter must be identical.

The trellis representation of a convolutional code is more instructive than the tree representation, for it explicitly uses the fact that the encoder is a finite-state machine. Each state of a binary trellis has two possible successors, two possible predecessors, and from level to level the structure is repetitive.

This reduction of the tree structure of the code into a trellis is credited to Forney [2] and the trellis structure is at the heart of an optimum decoding technique for convolutional codes due to Viterbi [1, 3]. This decoding technique, which has been introduced in Chapter 6, is presented in detail in Section 12.4.

12.2 DISTANCE PROPERTIES OF CONVOLUTIONAL CODES

The minimum distance properties of convolutional codes play an important role in determining the error-correcting capability of the code. However, unlike block codes, several distance measures have been proposed for convolutional codes, and each one of them is important and useful for particular decoding techniques.

Let $U = (u_1, u_2, u_3, \dots)$ and $V = (v_1, v_2, v_3, \dots)$ be two input information sequences, and let $X^{(U)}$ and $X^{(V)}$ be the corresponding codewords. In the sequel, the first n branches of a codeword $X^{(\cdot)}$ will be denoted $X_n^{(\cdot)}$.

The **Hamming distance** $d^{(U,V)}(n)$ between two codewords of length n branches, $X_n^{(U)}$ and $X_n^{(V)}$, is equal to the number of positions where $X_n^{(U)}$ and $X_n^{(V)}$ differ. That is,

$$d^{(U,V)}(n) = d_H\left(X_n^{(U)}, X_n^{(V)}\right) = W_H\left(X_n^{(U)} \oplus X_n^{(V)}\right) \tag{12.26}$$

where $d_H(A, B)$ represents the Hamming distance between A and B, and where $W_H(A \oplus B)$ is the **Hamming weight**, that is, the number of 1's in the sum $(A \oplus B)$.

The nth-order **column distance function** $d_c(n)$ of a convolutional code is the minimum Hamming distance between all pairs of codewords of length n branches which differ in their first branch of the coded tree. It is given by

$$d_c(n) = \min_{U_1 \neq V_1} d_H\left(X_n^{(U)}, X_n^{(V)}\right), \qquad X_1^{(U)} \neq X_1^{(V)} \tag{12.27}$$

The column distance function is a nondecreasing function of n, and assumes two particular values of special interest: $d_{\min}$, the **minimum distance** of the code when $n = K$ the constraint length of the code; and d_{free}, the **free distance** of the code when $n \to \infty$ [4]. That is,

$$d_{\min} = d_c(K) \tag{12.28}$$

and

$$d_{\text{free}} = \lim_{n \to \infty} d_c(n) \tag{12.29}$$

Since the column distance function $d_c(n)$ is nondecreasing with n, then obviously we must have

$$d_{\min} \le d_{\text{free}} \tag{12.30}$$

However, for many codes $d_{\text{free}} = d_{\min}$.

The minimum distance of a convolutional code is the important parameter for determining the error probability of the code when used with threshold decoding, or some other decoding algorithms that operate over one constraint-length segment of the codewords. However, as the free distance is the minimum distance between arbitrarily long encoded sequences, it is the principal distance parameter in determining the code performance with a decoding algorithm than can observe the encoded sequences over more than one constraint length. Two such decoding algorithms are Viterbi decoding and sequential decoding.

Finally, the **distance profile** [5] of a convolutional code is the K-tuple of column distance functions

$$\mathbf{d} = \left(d_c(1), d_c(2), d_c(3), \dots, d_c(K) \right) \tag{12.31}$$

This distance profile is a measure of the rate of growth of the column distance functions and plays an important role in the computational effort of sequential decoders. In this type of decoder the column distance function determines the decoding speed. A code with a good distance profile will have a rapidly increasing column distance function, and a fast and early growth of the column distance function will usually guarantee quick decoding [6].

In determining the different distances of a convolutional code, it must be observed that because of the linearity of the code, the minimum Hamming distance between all pairs of codewords is equal to the minimum Hamming weight of the nonzero codewords. Hence setting $\mathbf{X}_n^{(V)} = \mathbf{0}$ in (12.27), and using (12.26), we obtain

$$d_c(n) = \min d_H\left(\mathbf{X}_n^{(U)}, \mathbf{0} \right) = \min_{\mathbf{X}_n \ne \mathbf{0}} W_H(\mathbf{X}_n) \tag{12.32}$$

Consequently, to determine $d_{\min}$ or d_{free}, one has to consider only the codewords lying at the lower half of the coded tree since they all begin with a nonzero branch. This lower half of the coded tree is frequently called the **incorrect subset** of codewords and is denoted by S_I. Therefore, we can write

$$d_{\min} = \min_{\mathbf{X} \in S_I} W_H(\mathbf{X}_K) \tag{12.33}$$

and

$$d_{\text{free}} = \lim_{n \to \infty} \min_{\mathbf{X} \in \mathbb{S}_I} W_H(\mathbf{X}_n) \tag{12.34}$$

For example, let us find $d_{\min}$ for the $K=3$ code of Figure 12.2. Using the tree representation of Figure 12.7, it is necessary only to determine the Hamming weights of all the three-branch paths lying at the lower half of the tree. These Hamming weights are easily found to be 8, 5 and 6. Thus $d_{\min} = 5$ and corresponds to the codewords (111, 110, 000) and (111, 001, 001).

Determining d_{free} is more involved, because in principle, arbitrarily long paths in the incorrect subset must be examined. However, since d_{free} is finite and since the Hamming weight of a path stops growing after remerging with the all-zero path, then it follows that the path corresponding to d_{free} must eventually remerge with the all-zero path. Consequently, d_{free} may be determined from the state diagram or the trellis representation of the code by considering all the paths leaving the zero state and returning thereafter to that state. Naturally, the lengths of these paths may be arbitrarily large, but for most codes the minimum path corresponding to d_{free} is only a few constraint lengths long, and very often just one constraint length long (K branches).

For example, for our code of Figure 12.2, from either the state diagram of Figure 12.8 or the trellis representation of Figure 12.9, it is easy to see that $d_{\text{free}} = 8$. Here two paths of lengths 3 and 4 branches yield d_{free}: (111, 110, 111) and (111, 001, 001, 111). Pursuing this example, the entire column distance function of the code of Figure 12.2 may be easily determined. It is plotted in Figure 12.10, showing that d_{free} is reached at the eighth order of the column distance function.

Depending on the application and type of decoder utilized, one should select an optimum minimum distance code, or an optimum free distance code,

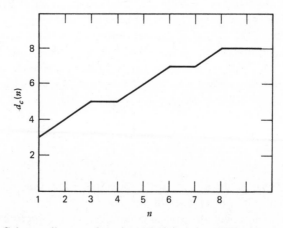

Figure 12.10 Column distance function of the code of Figure 12.2: $d_{\min} = 5$ and $d_{\text{free}} = 8$.

or an optimum distance profile code. The search for these codes is usually performed by computer, but since the number of sequences to be examined grows exponentially with K, the procedure is very time consuming, and even for small constraint lengths an exhaustive search is impossible. Therefore, computer search techniques have been developed to reject as early as possible those codes that cannot be optimum. Using these techniques, optimum codes with respect to several distance measures have been found for different coding rates and constraint lengths [5–16]. Some of these codes are tabulated in Appendix 12.1.

12.3 MAXIMUM LIKELIHOOD DECODING OF CONVOLUTIONAL CODES

In this section basic principles of maximum likelihood decoding over discrete memoryless channels are presented. These concepts are fundamental to the understanding of the two powerful decoding techniques for convolutional codes, the Viterbi and sequential decoding, which are presented in subsequent sections of this chapter.

In a terminated convolutional code, the message input sequence U is encoded as a sequence X and is represented by a particular path through the tree or trellis of the code. This codeword X (also called the transmitted sequence) is sent through a noisy memoryless channel to a decoder and a user.

A **discrete memoryless channel** (DMC) is an idealized model of a noisy channel with digital input and quantized or digital outputs. The input to a DMC is a sequence of binary symbols and the output is a sequence of letters from a J-symbol alphabet. During each channel use, a symbol from the channel input sequence is transmitted, and the corresponding channel output symbol is received. A received symbol j, $j=1, 2,\ldots, J$ is assumed to be statistically dependent only on the corresponding input and is determined by a fixed conditional probability assignment $P(j|i)$. If $J = 2$, the output of the DMC is a binary sequence, and the channel is called a **binary symmetric channel** (BSC) or **hard-quantized** channel. If $J>2$, the channel output, or equivalently, the demodulator output, is said to be **soft-quantized**. It is well known [17, Chap. 6] that for additive white Gaussian noise 3-bit (soft) quantization (i.e., $J=8$), yields a 2-dB performance improvement over hard quantization.

The decoder observes the output sequence of the DMC corresponding to the transmitted sequence, and determines which of the possible 2^L data sequences of length L entered the encoder. Figure 12.11 shows the model of a communication system employing a convolutional encoder and a DMC.

When a message m enters the encoder, the codeword $X^{(m)}$ is transmitted, and on the basis of the corresponding received sequence Y, the decoder produces a decoded message m'. Decoding errors occur if $m \neq m'$.

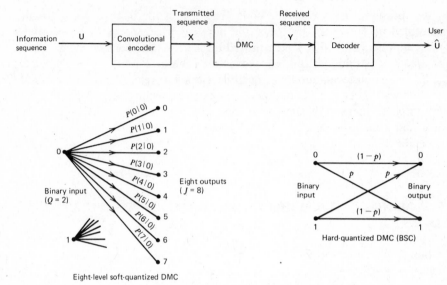

Figure 12.11   Models of communication system and discrete memoryless channels.

It is well known [18, Chap. 5] that for completely general channels, the decoder, which, given **Y** chooses m' for which

$$P\big(\mathbf{Y}|\mathbf{X}^{(m')}\big) \geq P\big(\mathbf{Y}|\mathbf{X}^{(m)}\big), \quad \text{all } m \neq m' \tag{12.35}$$

minimizes the error probability of the sequence, if all input data sequences are equally likely. This decoder, which is optimum, is called a **maximum likelihood sequence decoder**, and the conditional probabilities $P(\mathbf{Y}|\mathbf{X}^{(\cdot)})$ are called the **likelihood functions**. Hence a maximum likelihood decoder will choose the message that maximizes the likelihood function.

For binary convolutional codes of rate $1/v$, there are two branches leaving each node, with v transmitted symbols per branch. A node on tree level (or tree depth) ℓ is uniquely specified by the data sequence $\mathbf{U}_\ell = (u_1, u_2, \ldots, u_\ell)$, which locates the encoder at that node. The v symbols on the last branch reaching node $\mathbf{U}_\ell$ are denoted by

$$\mathbf{x}_\ell^{(\mathbf{U})} = \big(x_{\ell 1}^{(\mathbf{U})}, x_{\ell 2}^{(\mathbf{U})}, \ldots, x_{\ell v}^{(\mathbf{U})} \big) \tag{12.36}$$

Similarly, let

$$\mathbf{y}_i = (y_{i1}, y_{i2}, \ldots, y_{iv}) \tag{12.37}$$

represent the v received symbols when $\mathbf{x}_i^{(\mathbf{U})}$ is transmitted. As the channel is memoryless, the likelihood function for the ith branch of that path is given by

$$P\big(\mathbf{y}_i|\mathbf{x}_i^{(\mathbf{U})}\big) = \prod_{j=1}^{v} P\big(y_{ij}|x_{ij}^{(\mathbf{U})} \big) \tag{12.38}$$

Generally, it is more convenient to use the logarithm of the likelihood function because the logarithm is a monotonically increasing function, and therefore does not alter the final result. Defining the **log-likelihood function** $\gamma_i^{(U)}$ for the ith branch on the path specified by the input sequence **U** as

$$\gamma_i^{(U)} \triangleq \log P(\mathbf{y}_i | \mathbf{x}_i^{(U)}) = \sum_{j=1}^{v} \log P(y_{ij} | x_{ij}^{(U)}) \tag{12.39}$$

the log-likelihood function $\Gamma_N^{(U)}$ for the first N branches of that path is thus

$$\Gamma_N^{(U)} = \sum_{j=1}^{N} \gamma_j^{(U)}, \quad N = 1, 2, 3, \dots \tag{12.40}$$

(In what follows, whenever there is no ambiguity as to which path is being considered, the superscript **U** will be removed.)

The log-likelihood function (or simply the likelihood) is used as a **metric,** and for memoryless channels the metric is additive over the received symbols. Because a metric can be computed for each path in the tree or trellis, maximum likelihood decoding of convolutional codes may be regarded simply as the finding of the path with the largest accumulated metric, given the received sequence. However, depending on whether the tree or trellis structure of the code is utilized, the determination of this most likely path will lead to either an impractical or practical decoding technique.

Likelihood Function and Hamming Distance over the BSC

For a binary symmetric channel (BSC) it is easy to show that maximizing the log-likelihood function log $P(\mathbf{Y}|\mathbf{X})$ is equivalent to minimizing the Hamming distance between the two sequences **X** and **Y**.

Following Viterbi [1], let **X** be a transmitted codeword in a BSC with transition probability p, and let **Y** be the corresponding received sequence. Suppose that **X** and **Y** are N coded symbols long and that z transitions occurred in the transmission of **X**. Hence **X** and **Y** differ in z positions; that is, their Hamming distance is z. As the channel is memoryless, the log-likelihood function or metric is given by

$$\log P(\mathbf{Y}|\mathbf{X}) = \log\left[p^z (1-p)^{N-z} \right] \tag{12.41}$$

Upon expanding, we obtain

$$\log P(\mathbf{Y}|\mathbf{X}) = N \log(1-p) - z \log\left(\frac{1-p}{p} \right)$$

$$= -A - Bz \tag{12.42}$$

where A and B are positive constants, since $p < 0.5$. Therefore, minimizing the

Hamming distance z corresponds to maximizing the metric. Consequently, over a BSC the metric is conveniently replaced by the Hamming distance, and a maximum likelihood decoder will choose in the tree or trellis the path $\mathbf{U}$ whose corresponding encoded sequence $\mathbf{X}^{(\mathbf{U})}$ is closest to the received sequence $\mathbf{Y}$ in Hamming distance.

Using the Tree or Trellis Structure for Maximum Likelihood Decoding of Convolutional Codes

By definition, maximum likelihood decoding implies comparing the received sequence with all possible transmitted sequences before making a decision. Hence in general for memoryless channels, the optimal decoding of an L-bit-long binary sequence requires comparing the 2^L accumulated metrics of the 2^L different codewords that could have been transmitted, and selecting the best one. Because of this exponential increase of the decoding effort with the length L of the sequence, maximum likelihood decoding is usually difficult to implement and therefore rarely used in practice. However, its importance lies in its use as a standard with which other practical suboptimum decoding techniques may be compared. Moreover, it is used to determine the performance of codes, since a measure of performance for any code is the probability of error with the optimum decoder.

For the decoding of convolutional codes, either the tree or trellis structure of the code may be used. In the tree representation of the code, the fact that the paths remerge is entirely ignored, and hence the optimal decoding of an L-bit sequence requires the exhaustive comparison of 2^L accumulated metrics. Consequently, over the tree structure of the code, optimal decoding is not practical, and hence the estimation of the most likely data sequence would call for a suboptimum but more practical decoding technique. One of these suboptimum techniques is sequential decoding.

Consider now the trellis structure of the code where the redundancy of the tree was eliminated by merging. In principle this redundancy could be exploited by a clever decoder which would consider only those paths that could ever maximize the likelihood function over the whole set of paths. In the decoding process, if at some point it is realized that a path cannot possibly yield the largest metric, then that path would be discarded by the decoder. The decoded path would then be chosen from among the set of remaining or "surviving" paths that reached the last level L. Such a decoder would still be optimum in the sense that the decoded path would be the same as the decoded path obtained from a "brute-force" maximum likelihood decoder, but the early rejection of unlikely paths would reduce the decoding effort. The objective is, naturally, to find a procedure that breaks the exponential message sequence-length dependency of the decoding effort and yet yields maximum likelihood decisions. We now show that the trellis structure of the code allows such a decoding procedure.

Consider the set of all tree paths lying at level (or depth) ℓ, $K \leq \ell \leq L$, whose end node correspond to the same encoder state $\mathbf{S}_i^{(\ell)}$, $i = 1, 2, \ldots, 2^{K-1}$, where the superscript refers to the level of the tree. These paths are distinct in the tree but converge into a single node in the trellis. In this set, the path having the largest accumulated metric is called the **survivor** at state $\mathbf{S}_i^{(\ell)}$, and all other paths constitute the nonsurvivor set as $\mathbf{S}_i^{(\ell)}$. It is obvious that an optimum decoder can discard the entire nonsurvivor set of paths without altering the optimum decision, since a maximum likelihood path cannot belong to any nonsurvivor set. This set can consequently be discarded.

It is clear that a decoder that would compare the metrics of all paths converging into a node, keeping only the survivor at that node, will yield a maximum likelihood decision if the operation is repeated for all distinct states at each level. The natural structure to use is of course the trellis. There are 2^{K-1} states per level, hence 2^{K-1} survivors must be determined for each level, yielding a decoding effort that would vary as 2^{K-1} for a code of constraint length K. Therefore, the exponential growth of the decoding effort is in the constraint length of the code, *not* in the length of the input sequence as in maximum likelihood decoding. Whether or not this effort is tolerable will depend upon the particular application, but it may already be concluded that maximum likelihood decoding of convolutional codes using their trellis structure will be limited in practice to codes of short constraint lengths.

A very practical optimum decoding technique is the Viterbi decoding algorithm invented by Viterbi in 1967 [3]. This decoding procedure, which has been introduced in Chapter 6, uses the trellis structure of the code and is very efficient in the way the periodic remerging of the paths is systematically exploited to determine the survivors. The optimality of the Viterbi decoder was demonstrated by Forney [2] and Omura [19]. Forney showed that for any convolutional code, the output of the decoder is a sequence of estimated information digits which is maximum likelihood conditioned upon the received sequence, and Omura showed that the Viterbi algorithm was in fact a forward dynamic programming technique.

12.4 VITERBI DECODING ALGORITHM

The Viterbi decoding algorithm is a powerful and practical technique for decoding convolutional codes, which has been shown to be particularly suitable for satellite channels [20].

We recall that of all the paths remerging at a given node, only the survivor at that node needs to be retained by the decoder. In the trellis structure, a node is entirely specified by its state and its level. That is, $\mathbf{S}_i^{(n)}$ specifies a node at level n, having state $\mathbf{S}_i$, $i = 1, 2, \cdots, 2^{K-1}$. For a binary trellis, the essence of the Viterbi decoding algorithm consists of examining each of the two paths leading to a given node $\mathbf{S}_i^{(j+1)}$ and selecting that path which is most likely to have been taken by the encoder. This most likely path (the survivor) is retained and the other is discarded. The procedure is repeated for all states at each trellis level.

The Viterbi decoding algorithm operates as follows, where we may assume a binary trellis without loss of generality. Given the first $(K-1)$ branches of the received sequence, $\mathbf{Y}_{K-1}=(\mathbf{y}_1,\mathbf{y}_2,\ldots,\mathbf{y}_{K-1})$, start by examining, from the origin out, all 2^{K-1} paths of length $(K-1)$ branches and compute their total metrics $\Gamma_{K-1}^{(i)}$, $i=1, 2,\ldots, 2^{K-1}$. These metrics may be either the log-likelihood functions of (12.40) or the Hamming distances if a BSC is used. As no merging takes place for the first $(K-1)$ consecutive branches of the tree, those 2^{K-1} paths are distinct and identical for both tree and trellis. Upon reception of the Kth branch $\mathbf{y}_K$, each path extends its two branches to level K, where two branches converge into each node. The total metrics of the two paths converging into each node are compared, and only the path with the largest metric is retained; the other path is discarded. All 2^{K-1} survivors are determined in the same way. Now finding the survivors at level $(K+1)$ involves only the extension of each survivor into its two successors, computing for each the branch metrics $\gamma_{K+1}^{(i)}$, $i=1, 2,\ldots, 2^K$ (given the received branch $\mathbf{y}_{K+1}$), and comparing pairwise the total metrics Γ_{K+1}'s of the converging paths.

· The mechanics of the decoding are now apparent: at each step the 2^{K-1} surviving paths are extended and the comparison is made among the paths which were generated by input sequences not previously discarded. Out of each comparison a single path is chosen, and this survivor together with its metric is stored. Hence at each step the extensions increase the number of paths by a factor of 2 while the comparisons reduce that number by a factor of 2, resulting in a constant number of survivors. From our earlier discussion we know that the paths eliminated in the reduction from 2^K new paths to 2^{K-1} new survivors do not affect the optimality of the final decision.

A difficulty may arise in the case of ties: the survivor at a node is not unique; that is, two or more paths yield the same highest accumulated metric. Keeping all the contenders would not in any way help solve the ambiguity later, since thereafter further received symbols would yield identical metrics. Therefore, in the case of ties the sole survivor is chosen at random.

To help make a final decision as to which survivor should be chosen as the decoded path, the trellis is terminated by a tail of $(K-1)$ known information symbols. In the tail, branching ceases, since only the branch corresponding to the known transmitted symbols is extended from each state. Therefore, the number of survivors is reduced by a factor of 2 by the comparison at each step. Consequently, after the $(K-1)$ tail branches are received and decoded, there is only a single path left in the entire trellis: this path is accepted as the decoded path, and given the received sequence, it corresponds to the most likely transmitted sequence. No other path has a larger accumulated metric.

The operations of the Viterbi decoder are not complex, and the motion of the decoder is always forward with no backing up. A decoding step involves only the determination of the branch metrics, the total accumulated metric and the pairwise comparison and proper path selection. These operations are identical from level to level, and as they must be performed at every state, the complexity of the decoder is proportional to the number of states, and hence grows exponentially with the constraint length. To decode a block of L information bits the total number of Viterbi operations is $L2^{K-1}$, and for $L>K$ it is far smaller than the total number $(2^L - 1)$ of operations that would be required to perform maximum likelihood decoding on a tree. However, as a practical decoding technique, clearly the exponential growth of the number of states with K will limit Viterbi decoding to convolutional codes of short constraint lengths ($K \le 8$).

As an illustrative example, consider Viterbi decoding over a BSC for the $K=3$, rate-$\frac{1}{3}$ convolutional code of Figure 12.2, whose trellis is repeated for convenience in Figure 12.12. Let the 5-bit information sequence be $\mathbf{U}=(1, 0, 1, 1, 0, 0, 0)$, where the last 2 bits correspond to the tail of the message. As shown in Figure 12.12, the corresponding transmitted codeword is $\mathbf{X}=(111, 110, 000, 001, 001, 111, 000)$. Let us assume that six encoded symbols at positions 3, 9, 10, 14, 18 and 21 are received in error, so that the received sequence is $\mathbf{Y}=(110, 110, 001, 101, 011, 110, 001)$. Given the received sequence $\mathbf{Y}$ and using the Hamming distance between $\mathbf{Y}$ and all distinct paths in the trellis as the metric, we now use the Viterbi decoding algorithm to determine the most likely transmitted path. The decoder is assumed to have a replica of the encoder and hence it can generate the encoding trellis. $\quad$ *Vitali*

Following the algorithm, the metrics of the eight first three-branch-long paths are calculated and compared pairwise at each of the four states. At each remerging node the path having the smallest Hamming distance of the pair (i.e., the survivor) is kept. The other path (the nonsurvivor), shown in dashed lines in Figure 12.13, is discarded. The survivors at level 4 are determined by extending the two branches emerging from each state, calculating the Hamming distance increments and adding them to the survivor's Hamming distance at level 3, from which the branches emerge. The survivors are again the best

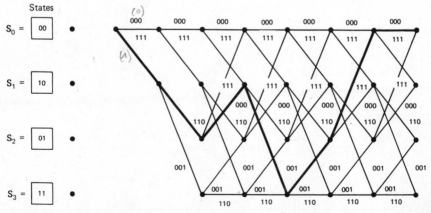

Figure 12.12 Trellis representation of the code of Figure 12.2 showing the path corresponding to $\mathbf{U}=(1011000)$.

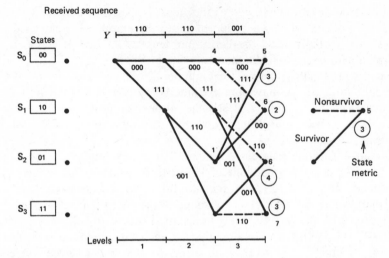

Figure 12.13 Trellis pattern for the first three trellis levels, showing the survivors (continuous line) and nonsurvivors (dashed lines). (Hamming distances are indicated next to the nodes.)

paths of each converging pair. The procedure is repeated for each succeeding level until the last level L is reached. In our example, $L=5$ and the survivors for levels 4 and 5 are shown in Figures 12.14 and 12.15, respectively. At level 4 we observe that two ties occur at states $S_0^{(4)}$ and $S_2^{(4)}$, and at level 5 two ties occur at states $S_1^{(5)}$ and $S_3^{(5)}$. Therefore, the survivor at each of these states is determined at random, with a 50% chance of selecting the wrong survivor.

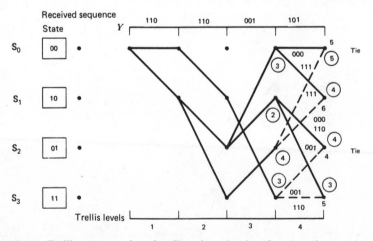

Figure 12.14 Trellis pattern for the first four levels, showing the surviving and nonsurviving paths. (Two ties occur at nodes $S_0^{(4)}$ and $S_2^{(4)}$; their survivors are chosen at random.)

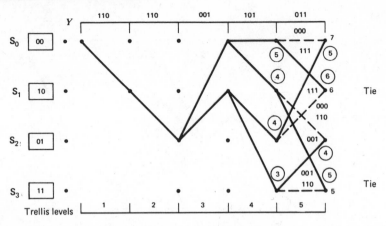

Figure 12.15 Trellis pattern for the first five levels showing the surviving and non-surviving paths. (Two ties occur at nodes $S_1^{(5)}$ and $S_3^{(5)}$.)

The decoding of the last two received branches is simplified as they correspond to the tail of the message. Since here the tail of the message consists of two "0's", only the branch corresponding to a "0" will be extended by the decoder. Consequently, as shown in Figures 12.16 and 12.17, the number of survivors is reduced to two at level 6 and to only one at level 7. This last and unique survivor is the decoded sequence. As shown in Figure 12.17, it is at distance 6 from the received path, and as no other path is closer to the received sequence, it is the most likely path given that received sequence. Tracing this path back to the origin and deleting the tail, the decoded information sequence that will be delivered to the user is thus $\hat{U} = (10110)$, indicating no error. All six channel errors that occurred during the transmission have been corrected by the algorithm.

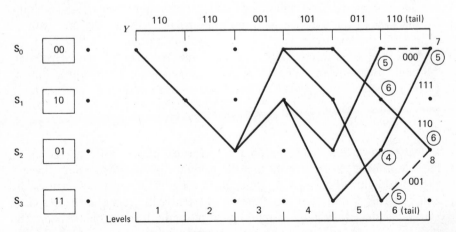

Figure 12.16 Trellis pattern at the first tail branch. (Only one branch is extended from each state.)

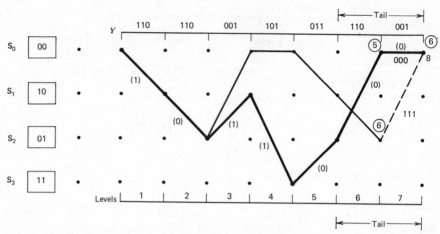

Figure 12.17 Trellis pattern for the last level of the received sequence, showing the single survivor, that is, the finally decoded path $\mathbf{U}=(10110)$.

Interesting observations may be drawn from this simple example. Examination of Figures 12.14 and 12.15 shows that none of the extensions of the survivors at nodes $\mathbf{S}_2^{(3)}$ and $\mathbf{S}_3^{(3)}$ yields a survivor at level 4. Therefore, the path sections terminating at $\mathbf{S}_2^{(3)}$ and $\mathbf{S}_3^{(3)}$ are eliminated since they cannot be part of the most likely path. Consequently, as shown in Figure 12.15, no surviving path goes through nodes $\mathbf{S}_0^{(1)}$, $\mathbf{S}_0^{(2)}$, $\mathbf{S}_1^{(2)}$, $\mathbf{S}_3^{(2)}$, $\mathbf{S}_2^{(3)}$, and $\mathbf{S}_3^{(3)}$. This illustrates the general observation that a surviving path at any level is always a surviving path at all of its preceding levels, but is not necessarily a part of a surviving path at any subsequent levels of the trellis. Therefore, as the decoding progresses through the trellis, earlier states may be vacant, and most interestingly, from some trellis level to the origin, only one state may be occupied at each level. Consequently, from the root node to that level, a unique survivor path exists and this path cannot be altered by subsequent decoding. Hence it must be the starting part of the decoded path and may be delivered to the user without waiting for the last trellis level to be reached. For example, in Figure 12.15, from the root node to level 2 of the trellis, only the path $\mathbf{S}_0^{(0)}$, $\mathbf{S}_1^{(1)}$, $\mathbf{S}_2^{(2)}$ exists and could therefore be delivered without further delay. Such a procedure is usually embedded in practical Viterbi decoders to simplify their operation.

Metric Quantization

We have shown that if the decoding is performed over a BSC, the Hamming distance is used as the metric. For the general discrete memoryless channel with soft quantization ($J>2$), although the log-likelihood function must be used, in terms of implementation the use of positive integers as metrics is more convenient since they must be represented in digital form in the decoder. The rounding off of the metric can be performed by replacing $\log P(y_i|x_i)$ by

Table 12.2 Set of Symbol Metrics for Eight-level Quantized Channel

Channel output y_i		0	1	2	3	4	5	6	7
Channel input x_i	0	7	6	5	4	3	2	1	0
(Coded symbols)	1	0	1	2	3	4	5	6	7

$\alpha[\log P(y_i|x_i)+\beta]$, where α is a real number and β a positive real number. In practice it has been found that α and β may be chosen in such a way that the smallest symbol metric is zero and all other symbol metrics are positive integers.

Naturally, depending on the choice of α and β, several integer metric tables are possible, and clearly, owing to the metric quantization, the Viterbi decoder becomes slightly suboptimum. However, it has been shown that for satellite channels the decoder performance is quite insensitive to the particular choice of the set of integer metrics [20]. In fact, the use of integers as symbol metrics, such as those shown in Table 12.2 for a 3-bit quantized DMC, results in a negligible performance degradation. Naturally, as the channel is more finely quantized, more bits will be needed to represent the metric symbols in the decoder. As mentioned in Chapter 1, it has been verified [20] that the performance of the system is about 2 dB better with eight levels of channel quantization than it is with only two levels (i.e., BSC). In fact, there is little to gain to increase the number of channel quantization levels beyond 8; the performance is nearly as good with eight quantization levels as with an arbitrarily larger number of levels. As for the best spacing between the quantized levels, it is approximately $\sqrt{N_0/2}$, where N_0 is the received noise spectral density.

Practical Simplifications of the Algorithm

The great advantage of the Viterbi decoder is that the number of decoding operations performed per received branch is constant. These 2^{K-1} operations are always of the same nature and do not require a sophisticated logic. The main disadvantage is the huge memory (2^{K-1} L-bit survivors) necessary to store all the paths. Moreover, forcing the encoder into a known final state by using a tail of a known sequence of information bits is equivalent to waiting until all L information symbols have been decoded *before* starting the delivery of the *first* symbol to the user. In a real system this delay may be operationally undesirable.

Through extensive simulations, it has been observed that all the surviving paths do not stay distinct over their entire length but have a tendency to stem from a single node several constraint lengths earlier [20]. Hence, as illustrated in our example, it is not necessary to wait until the unique decoded path has been obtained before starting to deliver the decoded bits. A final decision may be made on these branches prior to the point of common convergence, and the decoder can safely deliver the information with a small delay as it progresses in the trellis. With this procedure, clearly the decoder can eliminate the tail altogether and operate continuously.

Getting rid of the tail in this fashion leads to a very attractive situation with respect to the memory requirement of the decoder. Clearly, instead of storing the entire path history over the total length L, for each survivor the decoder needs only to store the history of the paths up to and including the point of common convergence. If the point of common convergence lies M levels back, then the amount of information bits path history storage required for each state is M, and the total amount of storage is equal to $M2^{K-1}$ bits. The path history length is called the **memory of the decoder**. As for the information bit decisions, they consist simply in deciding on the oldest information bit on an arbitrary one of the 2^{K-1} surviving paths. Each time the decoder penetrates the trellis one level deeper, it delivers this oldest bit to the user.

An actual decoder has a given amount of storage available and the past history storage is a substantial part of it. To minimize the required decoder memory compatible with a given error performance, the method described above is usually refined to deliver the oldest information bit on the surviving path with the highest metric rather than the oldest bit on any surviving path [20]. It has been shown theoretically by Forney [2] that with this procedure, asymptotically the expected value of the error probability on that oldest bit is not increased if the memory M is large enough. It has been found through simulation [7, 20] that a memory of four or five constraint lengths is sufficient to ensure a negligible degradation from the error probability of the optimum decoder. Should an error occur by this method it does not lead to catastrophe, for although the Viterbi decoder was assumed to start operating from a known state, it has been found through simulation that the Viterbi decoder can resynchronize itself; that is, it may start decoding from any state. The first three or four constraint lengths of information bits may be unreliable but normal operations will prevail thereafter. Therefore, for a practical decoder the tail can be eliminated and the memory reduced without altering the performance of a (theoretical) optimum decoder. Such a practical decoder is sometimes called a **truncated Viterbi decoder**.

12.5 MULTIPROCESSOR-BASED IMPLEMENTATION OF VITERBI DECODERS

In this section we consider practical methods for implementing Viterbi decoders either in hardware or software. We shall consider only rates $1/v$ binary convolutional codes, which imply restricting the discussion to binary input channels. However, the channel output need not be binary like the BSC, but may take any number of quantization levels, because Viterbi decoders can easily operate with almost any number of channel output quantization levels.

From our earlier discussion we saw that for each surviving path the decoder must store the path information bit history over a length corresponding to the decoder memory. In addition, the cumulative path metric of the survivor at each state must be stored by the decoder as a state metric. At any trellis depth n, $n \geq K$, for each of the 2^{K-1} states the following processing must be performed:

1 Compute the total metric for the two paths entering the state by adding the previous state metrics to the branch metrics converging into that state.
2 Compare the two path metrics and store the largest as the new state metric.
3 Update the path information bit history corresponding to the state by adding the information bit of the surviving branch at that state to the path history of the originating state at the preceding level.

These relatively simple tasks may be accomplished in parallel using a separate processor for each of the 2^{K-1} distinct states. Indeed, since at any trellis level the branch symbols are uniquely determined by the state and the information bit corresponding to that branch, then clearly the coded symbols for the branches entering or leaving a given state are always the same. Consequently, given y_n the received branch at level n, each of the 2^{K-1} processors can immediately determine the appropriate branch metrics by looking up a metric table. These metric tables can conveniently be stored in a read-only memory (ROM), and hence changing metric tables will consist only of changing the ROM chip. The rest of the elementary operations, such as adding the previous survivor's metric, comparing the metrics and selecting the best path together with the path history update, are rather easy to perform. Therefore, each elementary processor is a relatively simple digital machine.

These basic operations and the required transfer of information are schematically illustrated in Figure 12.18, where each box represents such an elementary processor.

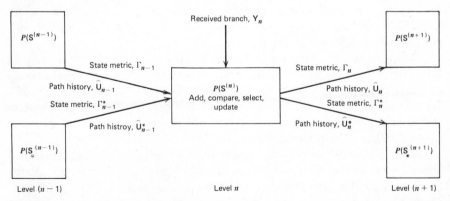

Figure 12.18 Information transfer between elementary processors of a Viterbi decoder.

Let $P(\mathbf{S}^{(n)})$ denote the processor residing at a given state $\mathbf{S}^{(n)}$ and operating on the received branch $\mathbf{y}_n$ at level n. This processor is logically connected (according to the trellis diagram) to two processors denoted $P(\mathbf{S}^{(n-1)})$ and $P(\mathbf{S}_*^{(n-1)})$ operating at the preceding level $(n-1)$, and to two processors denoted $P(\mathbf{S}^{(n+1)})$ and $P(\mathbf{S}_*^{(n+1)})$ operating at the following level $(n+1)$. In order to perform its decoding operations at level n, the processor $P(\mathbf{S}^{(n)})$ receives from processors $P(\mathbf{S}^{(n-1)})$ and $P(\mathbf{S}_*^{(n-1)})$ the corresponding state metrics denoted Γ_{n-1} and Γ_{n-1}^*, together with the paths history $\hat{\mathbf{U}}_{n-1}$ and $\hat{\mathbf{U}}_{n-1}^*$ for the two surviving paths, which are residing in $P(\mathbf{S}^{(n-1)})$ and $P(\mathbf{S}_*^{(n-1)})$, respectively. Given the received branch $\mathbf{y}_n$, $P(\mathbf{S}^{(n)})$ can therefore immediately determine the branch metric increments and hence the total metrics of the two remerging paths at state $\mathbf{S}^{(n)}$. Comparison of these paths yields the new survivor at $\mathbf{S}^{(n)}$ together with its state metric Γ_n. The processor $P(\mathbf{S}^n)$ then updates the new path history $\hat{\mathbf{U}}_n$ by adding the appropriate information bit, and transfers it along with the new state metric Γ_n to the processors $P(\mathbf{S}^{(n+1)})$ and $P(\mathbf{S}_*^{(n+1)})$. This same cycle is identical for each state and is repeated at every trellis level. Obviously, in addition to these operations the decoded information bits must be retrieved. However, determining these decoded bits does not increase the decoder complexity very much. Further details concerning multiprocessor-based Viterbi decoders are given in Refs. 21 and 22.

For the first $(K-1)$ branches of the trellis, and also in the tail, not all the states are occupied. Moreover, the decoding of the tail involves only one branch extension instead of two. Therefore, some of the elementary processors should in principle be deactivated. However, to maintain the uniformity of the decoding operations throughout and to simplify the controlling logic, the following initialization procedure may be performed [21]. At step 0, all processors except the one corresponding to state zero, $\mathbf{S}_0^{(0)}$, are initialized with a very large negative metric value. The initial metric value for state zero may be set to zero. It is easy to verify that this initialization procedure effectively deactivates the idle states at the beginning of the trellis without altering the regular operation of the decoder. Now if the decoder utilizes a tail, the decoded path will correspond to the survivor at state $\mathbf{S}_0^{(L+K-1)}=(0, 0, 0,\ldots, 0)$. Therefore, one simply ignores the output of all the processors except the one corresponding to the all-zero state.

Finally, although there are 2^{K-1} elementary processors, they are not all different. Regardless of the value of K, the number of different elementary processors is equal to the number of possible different encoded branches of v binary digits: that is, 2^v. For example, a practical realization of a Viterbi decoder for a rate-$\frac{1}{2}$ code with $K=8$ would necessitate $2^7=128$ elementary processors but of only four distinct types.

When a Viterbi decoder is implemented in software or simulated on a general-purpose computer, the operations at each of the 2^{K-1} states are usually performed serially. Two sets of registers containing the survivors' path history for two consecutive trellis levels are required, and at each new penetration in the trellis, the program updates and swaps the path history information bits between the two registers. The overflow bits of these registers may conveniently be used as the decoder output.

Regardless of the implementation, constraint lengths on the order of 7 or 8 constitute the limit of practicality for Viterbi decoders. In fact, for moderate bit error probability (10^{-4} to 10^{-5}), Viterbi decoders are among the best practical coding schemes to date and are widely used in space communications [20]. Extensive parallel operations allow decoding speeds of several Mbps. Furthermore, as mentioned in Chapter 15, it appears that Viterbi decoders will be increasingly used in satellite and terrestrial communication systems not only to correct noise-induced errors, but also as a very effective means to combat intersymbol interferences in digital communication [23, 24].

12.6 ERROR PERFORMANCE ANALYSIS OF VITERBI DECODING OVER DMC

In this section we analyse the error performance of convolutional codes with an optimum Viterbi decoder over discrete memoryless channels. Some general error probability bounds based on the ensemble of convolutional codes are presented first. Because these bounds apply to the average error probability for the ensemble of codes, they may not be very useful for specific, short-constraint-length codes that are used with a Viterbi decoder. Therefore, the analysis for specific codes is carried out using the generating function approach introduced by Viterbi [1].

General Error Probability Bounds

For discrete memoryless channels Viterbi [3] has shown that over the ensemble of randomly chosen convolutional codes, the error probability is bounded by

$$P(E) < L2^{-KR_0/r}; \qquad r < R_0 \tag{12.43}$$

where L is the information sequence length, r the code rate, K the constraint length, and R_0 is a parameter that depends only on the channel. For a general DMC having Q inputs and J outputs, it can be shown [17, 18] that R_0 is given by

$$R_0 = -\log_2 \min_{\{q\}} \sum_{j=1}^{J} \left[\sum_{i=1}^{Q} q_i \sqrt{P(j|i)} \right]^2 \tag{12.44}$$

where the $\{q_i\}$ represent the channel input probabilities, $P(j|i)$ the channel transition probabilities, and the minimization is performed over the channel input distribution.

For the DMC with equally likely inputs, the calculation of R_0 is straightforward since equally likely inputs correspond to the minimizing distribution. In particular, for the BSC with transition probabilities p and $q_0 = q_1 = \frac{1}{2}$,

(12.44) yields

$$R_0 = 1 - \log_2 \left[1 + 2\sqrt{p(1-p)} \right] \qquad (12.45)$$

The R_0 parameter is a fundamental quantity in information theory, where it is used to express error probability bounds over the DMC. In particular, it can be shown [17, 18] that over the ensemble of block codes of code rate r having 2^L codewords of length L/r, the average word error probability with maximum likelihood decoding is bounded by

$$\overline{P_w(E)} \le 2^{-L\left(\frac{R_0}{r} - 1\right)}; \qquad r < R_0 \qquad (12.46)$$

When the ensemble of codes consists of only two codewords the bound is

$$\overline{P_w(E)} \le 2^{-LR_0/r} \qquad (12.47)$$

which led to calling R_0 the **two-codeword exponent for the DMC**.

We shall see that R_0 also plays a fundamental role in sequential decoding, where it is called the **computational cutoff rate** and is usually denoted R_{comp}.

Restricting the discussion to convolutional codes, the bound (12.43) which was derived for the ensemble of codes may not necessarily yield accurate results when applied to a specific code, especially when the constraint length is short. However, the bound is very instructive and useful for long-constraint-length codes, say $K > 20$, for it shows that the error probability is exponentially decreasing with the constraint length of the code, and can in principle be made arbitrarily small by increasing K and/or decreasing the code rate r. Therefore, for a given application the designer should trade off the performance and complexity of the decoder in choosing the code, as both parameters are exponentially dependent on the constraint length.

To derive useful error probability bounds applicable to short-constraint-length codes, the performance analysis of Viterbi decoding over DMC using a specific code is presented in the next subsections. The presentation follows closely Viterbi's analysis using the generating function approach of the code [1].

Generating Function of a Convolutional Code

For analysing the performance of Viterbi decoding using a specific convolutional code over a DMC, Viterbi has shown that the state diagram is the effective tool [1]. Because a convolutional code is a group or linear code, we may assume that the all-zero coded sequence was transmitted. Then an error may occur whenever the metric of any other path that reaches state $S_0 = (0, 0, \ldots, 0)$ at time j is greater than that of the all-zero path up to time j. In other words, an error may occur whenever the all-zero path is not the survivor at time j. To analyse this event we must determine the Hamming distances between *all* such incorrect paths and the all-zero or correct path. Once the

Hamming distances are obtained for the specific code, a union bound on the error probability can easily be derived.

To illustrate the technique, following Viterbi [1] we will consider the nonsystematic $r = \frac{1}{2}$, $K = 3$ convolutional code depicted in Figure 12.19 together with its state diagram. Selecting a code with a lower rate or a longer constraint length will unnecessarily complicate the derivations without adding much to the understanding of the technique.

To obtain a complete description of the Hamming weights of all nonzero codewords, the state diagram of Figure 12.19 is modified as shown in Figure 12.20. The all-zero state is split open into an initial state and a final state, and the code symbols for each branch are replaced by D^i, where i is the Hamming

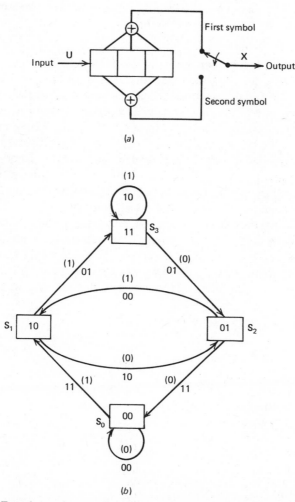

Figure 12.19 Encoder and state diagram for the rate-$\frac{1}{2}$, $K = 3$ code with $\mathbf{G}_1 = 111$, $\mathbf{G}_2 = 101$. (a) Encoder; (b) state diagram.

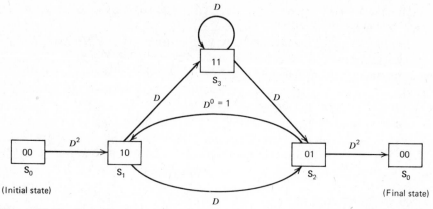

Figure 12.20 State diagram of the encoder of Figure 12.19 labelled with branch gains (Hamming distances from the all-zero path).

weight of the coded branch symbols. In Figure 12.20, $i=2$ for the coded symbols 11; $i=1$ for the coded symbols 01 and 10; and $i=0$ for the coded symbols 00. The D^i are sometimes called the **branch gains.** Each path connecting the initial state to the final state is a nonzero codeword with a generating function of the form D^j, where j is the total Hamming weight of the path, or equivalently is the Hamming distance from the correct (all-zero) path. For such paths the total weight is called the **path gain**, and clearly, it is the product of all the branch gains along the path. For example, the path represented by the state sequence $\mathbf{S}_0$, $\mathbf{S}_1$, $\mathbf{S}_3$, $\mathbf{S}_2$, $\mathbf{S}_0$ has path gain or **generating function** D^6, and hence its weight or Hamming distance from the all-zero path is 6.

Now, finding the set of weights of *all* nonzero codewords of the state diagram consists simply in determining the overall transfer function or generating function $T(D)$ of the directed graph. This generating function can be determined by elementary signal flow graph techniques, as illustrated in Figure 12.21. For our example, referring to Figures 12.20 and 12.21, we can write

$$\mathbf{S}_3 = D\mathbf{S}_3 + D\mathbf{S}_1$$

or

$$\mathbf{S}_3 = \frac{D}{1-D}\mathbf{S}_1 \qquad\qquad (12.48)$$

Using (12.48), the forward path function between $\mathbf{S}_1$ and $\mathbf{S}_2$ is given by

$$D \cdot \frac{D}{1-D} + D = \frac{D}{1-D}$$

and hence

$$\mathbf{S}_2 = \frac{D/(1-D)}{1-[D/(1-D)]}\mathbf{S}_1$$

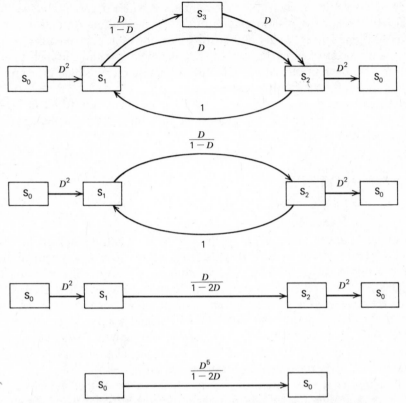

Figure 12.21 Signal flow graph of the state diagram of Figure 12.20.

or

$$S_2 = \frac{D}{1-2D} S_1 \tag{12.49}$$

Therefore, the overall generating function is

$$T(D) = D^2 \cdot \frac{D}{1-D} \cdot D^2$$

or

$$T(D) = \frac{D^5}{1-2D} = D^5(1 + 2D + 4D^2 + \cdots + 2^\ell D^\ell + \cdots) \tag{12.50}$$

This expression indicates that there is one incorrect path of weight 5, two of weight 6, four of weight 7, and in general there are 2^ℓ paths of weight $(5+\ell)$. From our discussion on the free distance of the code, clearly the free distance is the Hamming weight of the lowest-order term in the expansion of $T(D)$. In this example $d_{\text{free}} = 5$, indicating that any pair of channel errors can be corrected.

Now, as we are interested in determining the bit error probability, we must also find the number of information bits in error resulting from an incorrect path decision. This is readily obtained by labelling each branch corresponding to a nonzero information bit with B. For example, the three-branch path of weight 5 is now represented by D^5B, indicating that only a single "1" enters in its information sequence (in the transition from state 00 to state 10). The augmented state diagram of Figure 12.20 is shown in Figure 12.22. Using the same technique as above, the generating function of the augmented state diagram becomes

$$T(D, B) = \frac{D^5B}{1-2DB} = D^5B(1+2DB+4D^2B^2 + \cdots + 2^\ell D^\ell B^\ell + \cdots)$$

$$(12.51)$$

Hence we read that there is one path of weight 5 resulting in 1 bit in error, there are two paths of weight 6 resulting in 2 bits in error, and in general there are 2^ℓ paths each of weight $(\ell+5)$ resulting in $(\ell+1)$ bits in error. As the weights of all nonzero codewords and their corresponding number of bits in error are given by the generating function, an upper bound on the sequence and bit error probability can easily be determined for the specific convolutional code of interest. However, it must be stressed that this technique is severely limited by the number of states in the encoder, and becomes rather cumbersome when the number of states exceeds 16 ($K \geq 5$).

Finally, the state diagram of Figure 12.22 can be further augmented by explicitly indicating the length (in branches) of the different paths. As shown in Figure 12.23 each branch contains a multiplier coefficient $\mathcal{L}$, so that the exponent of the $\mathcal{L}$ factor will be increased by one every time a branch is passed through on a path. The length of a path is thus given by the exponent of the $\mathcal{L}$

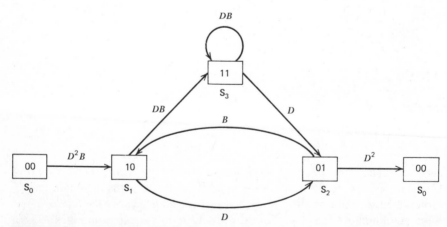

Figure 12.22 State diagram of the encoder of Figure 12.19 labelled with branch gains and number of inputs "1".

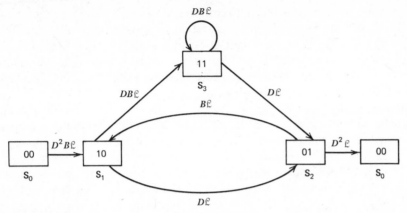

Figure 12.23   Augmented state diagram of the encoder of Figure 12.19 labelled with Hamming weights, lengths, and number of inputs "1".

term. For our example the generating function becomes

$$T(D, B, \mathcal{L}) = \frac{D^5 B \mathcal{L}^3}{1 - D\mathcal{L}(1 + \mathcal{L})B}$$

$$= D^5 B \mathcal{L}^3 + D^6 B^2 \mathcal{L}^4 (1 + \mathcal{L}) + D^7 B^3 \mathcal{L}^5 (1 + \mathcal{L})^2$$

$$+ \cdots + D^{5+\ell} B^{1+\ell} \mathcal{L}^{(3+\ell)} (1 + \mathcal{L})^\ell + \cdots \qquad (12.52)$$

Thus there is one path of length 3, having distance 5 from the all-zero path, and which carries a single 1 in its information sequence. There are two paths of distance 6, one four branches long and the other five branches long, and they both correspond to information sequences with two 1's, and so on.

Performance Analysis of Specific Codes

Having obtained both the weight distribution of all nonzero codewords and their corresponding information sequences, the error performance analysis of specific codes can be carried out. For convenience and without loss of generality we again assume that the all-zero codeword is transmitted over a BSC, and we will take as an example the rate-$\frac{1}{2}$, $K=3$ convolutional code of Figure 12.19. We first obtain a bound on the probability of a first-error event. Then using a union bound argument the sequence and bit-error probabilities are derived.

First-Error-Event Probability

We say that a first-error event occurred at an arbitrary trellis level j if the correct path is eliminated for the first time at the jth step. That is, the correct path is not a survivor at the all-zero state at depth j. Now for this event to occur, clearly an incorrect path must be closer (in Hamming distance) to the received path than the correct path. It is well known [17, 18] that over the BSC with transition probability p, an error will occur between two codewords with Hamming distance d if there are more than $d/2$ symbol errors, or channel transitions, among the d symbols in which the two paths differ. For our example of the code of Figure 12.19, the incorrect path of length three branches is at distance 5 from the all-zero correct path (see Figure 12.24a). A first-error event will be made if in these five positions where the correct and incorrect paths differ, the received sequence agrees with the incorrect path in three or more positions. Clearly, the errors occurring on the positions where the correct and incorrect paths agree have no consequence in determining the survivor, since they increase the metric of both paths by the same amount. For this distance-5 comparison, the first-error-event probability denoted P_5 is thus given by

$$P_5 = \sum_{i=3}^{5} \binom{5}{i} p^i (1-p)^{5-i} \qquad (12.53)$$

Now some incorrect paths may have an even weight, such as the distance-6

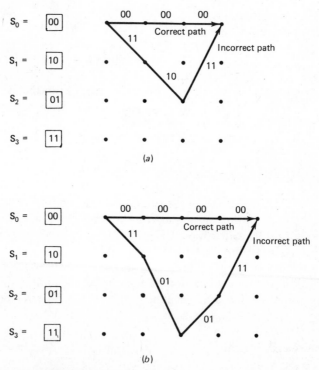

Figure 12.24 First-error event for the code of Figure 12.19. (a) Odd weight in correct paths; (b) even weight in correct paths.

path of length four branches shown in Figure 12.24*b*. Here if exactly three errors are made among the positions where the correct and incorrect paths disagree, then the metrics are tied at the remerging node. Since the survivor is selected at random, an error is then made with probability $\frac{1}{2}$. Obviously, no ambiguity exists if four or more errors occur. Consequently, for this distance-6 path the first-error-event probability is given by

$$P_6 = \frac{1}{2}\binom{6}{3}p^3(1-p)^3 + \sum_{i=4}^{6}\binom{6}{i}p^i(1-p)^{6-i} \tag{12.54}$$

Therefore, in general, if the incorrect path is at distance d from the correct path, the corresponding first-error-event probability P_d is

$$P_d = \begin{cases} \displaystyle\sum_{i=\frac{d+1}{2}}^{d}\binom{d}{i}p^i(1-p)^{d-i}, & d \text{ odd} \\[4mm] \displaystyle\sum_{i=\frac{d}{2}+1}^{d}\binom{d}{i}p^i(1-p)^{d-i} + \frac{1}{2}\binom{d}{\frac{d}{2}}[p(1-p)]^{d/2}, & d \text{ even} \end{cases} \tag{12.55}$$

In evaluating the overall first-error-event probability at any step, denoted $P(E)$, we observe that all incorrect paths of length ℓ branches or less can cause a first-error event at level ℓ. Hence a bound on $P(E)$ can be obtained by summing the first-error-event probabilities for each of all the possible incorrect paths that may have caused a first-error event at step ℓ. Since for a code with free distance d_{free} the generating function

$$T(D) = \sum_{\ell=d_{\text{free}}}^{\infty} a_\ell D^\ell \tag{12.56}$$

gives the total number of incorrect paths a_ℓ at distance ℓ, $\ell \geq d_{\text{free}}$, from the correct path, the probability of a first-error event at any step is bounded by

$$P(E) < \sum_{\ell=d_{\text{free}}}^{\infty} a_\ell P_\ell \tag{12.57}$$

where P_ℓ is given by (12.55).

For the code of Figure 12.19, using the expansion of (12.50), we obtain

$$P(E) < P_5 + 2P_6 + 4P_7 + \cdots + 2^\ell P_{\ell+5} + \cdots \tag{12.58}$$

Now it is not difficult to show [18] that for d even or odd, (12.55) can be bounded by the expression

$$P_d < 2^d[p(1-p)]^{d/2} \tag{12.59}$$

Therefore, using (12.56), (12.57) and (12.59), for any convolutional code with generating function $T(D)$, the probability of first-error event is bounded by

$$P(E) < \sum_{\ell=d_{\text{free}}}^{\infty} a_\ell 2^\ell [\, p(1-p)]^{\ell/2} \tag{12.60}$$

or

$$P(E) < T(D)|_{D=2\sqrt{p(1-p)}} \tag{12.61}$$

Applying this result to our example, from (12.58) and (12.60) we obtain

$$P(E) < \sum_{\ell=5}^{\infty} 2^{\ell-5} 2^\ell [\, p(1-p)]^{\ell/2}$$

$$= \frac{2^5 [\, p(1-p)]^{5/2}}{1-4\sqrt{p(1-p)}} \tag{12.62}$$

For reasonably good channels (i.e., for small p), the bound of (12.60) is dominated by the term corresponding to the path with free distance d_{free} from the correct path (i.e., the first term), so that

$$P(E) \simeq a_{d_{\text{free}}} \Big[2\sqrt{p(1-p)} \Big]^{d_{\text{free}}} \tag{12.63}$$

and since $p \ll 1-p$,

$$P(E) \simeq a_{d_{\text{free}}} 2^{d_{\text{free}}} p^{(1/2)d_{\text{free}}} \tag{12.64}$$

As an example, for the code of Figure 12.19, $d_{\text{free}} = 5$ and $a_{d_{\text{free}}} = 1$. Hence, for a BSC with a transition probability $p = 10^{-3}$, using (12.64) the probability of first-error event is

$$P(E) \simeq 2^5 (0.001)^{5/2} = 1.01 \times 10^{-6}$$

whereas using the bound of (12.62), it is found to be

$$P(E) < 1.155 \times 10^{-6}$$

As mentioned earlier, we see that the free distance is the most important single measure of the convolutional code's error performance. Not surprisingly, finding good codes for Viterbi decoding consists in finding codes with large free distances.

Bit-Error Probability Bounds

The foregoing analysis gave upper bounds on the sequence error probability for Viterbi decoding of convolutional codes. However, from the user point of view a more important and practical performance measure is the number of information bits decoded in error, that is, the bit-error probability, denoted P_b. As for the probability of first-error event, the generating function approach will be used to derive useful bounds on P_b.

We recall that the transfer function $T(D, B)$ explicitly gives the number of information bits "1" corresponding to each of the incorrect paths, and hence it can be used to determine the number of information bits in error for each incorrect path. Therefore, an upper bound on the bit-error probability can easily be obtained by weighting each term of the bound on the probability of the first error event by the corresponding number of erroneous information bits, that is, the number of output bit errors.

For our example, rewriting $T(D, B)$ in expression (12.51) as

$$T(D, B) = \frac{D^5 B}{1 - 2DB} = D^5 B + 2D^6 B^2 + 4D^7 B^3 + \cdots + 2^\ell D^{\ell+5} B^{\ell+1} + \cdots$$

we see that the exponent of B in each term gives the number of bits in error for the paths corresponding to that term. Therefore, the number of bits in error for each path can easily be computed by differentiating $T(D, B)$ with respect to B and then setting $B = 1$. That is,

$$\left. \frac{dT(D, B)}{dB} \right|_{B=1} = D^5 + 2.2D^6 + 3.4D^7 + \cdots + (\ell+1)2^\ell D^{\ell+5} + \cdots$$

$$= \frac{D^5}{(1 - 2D)^2} \tag{12.65}$$

Again using the union bound, the bit-error probability will thus be bounded by

$$P_b < P_5 + 2.2P_6 + 3.4P_7 + \cdots + (\ell+1)2^\ell P_{\ell+5} + \cdots \tag{12.66}$$

where P_ℓ is given either by (12.55) or the bound (12.59). For the latter the summation yields

$$P_b < \sum_{\ell=5}^{\infty} (\ell-4)2^{\ell-5} 2^\ell \left[p(1-p) \right]^{\ell/2}$$

$$= \frac{\left[2\sqrt{p(1-p)} \right]^5}{\left[1 - 4\sqrt{p(1-p)} \right]^2} \tag{12.67}$$

More generally, for any convolutional code with a generating function $T(D, B)$, we can write

$$\left.\frac{dT(D, B)}{dB}\right|_{B=1} = \sum_{\ell} c_{\ell} D^{\ell} \tag{12.68}$$

so that

$$P_b < \sum_{\ell} c_{\ell} P_{\ell} \tag{12.69}$$

Therefore, using bound (12.59) on P_{ℓ} the general bound on the bit-error probability is

$$P_b < \left.\frac{dT(D, B)}{dB}\right|_{B=1, D=2\sqrt{p(1-p)}} \tag{12.70}$$

Again, for reasonably good BSC, p is small and the bound is dominated by the path at free distance d_{free} from the correct path, that is, the first term of (12.69). We then obtain

$$P_b \simeq c_{d_{\text{free}}} \left[2\sqrt{p(1-p)}\right]^{d_{\text{free}}} \simeq c_{d_{\text{free}}} 2^{d_{\text{free}}} p^{d_{\text{free}}/2} \tag{12.71}$$

For the code of Figure 12.19, $c_{d_{\text{free}}} = 1$, and for $p = 0.001$, we obtain $P_b \simeq 1.01 \times 10^{-6}$, which is the same as $P(E)$. This indicates that for this code the error events are caused typically by the weight-5 path resulting in one output bit error.

The foregoing expression for the error probability can be directly applicable to other memoryless channels [1]. For example, for coherent PSK modulation over an additive white Gaussian noise (AWGN) channel with hard quantization, we recall from Chapter 2 that the crossover probability is given by

$$p = Q\left(\sqrt{\frac{2rE_b}{N_0}}\right) \tag{12.72}$$

where E_b/N_0 is the bit energy-to-noise ratio, r the coding rate, and $Q(x)$ is as defined in Section 2.3. For large x a useful approximation is

$$Q(x) \simeq e^{-x^2/2} \tag{12.73}$$

Therefore, for large values of E_b/N_0, substituting $p \simeq e^{-rE_b/N_0}$ in (12.64) and (12.71), respectively, yields

$$P(E) \simeq a_{d_{\text{free}}} 2^{d_{\text{free}}} e^{-(rE_b/2N_0)d_{\text{free}}} \tag{12.74}$$

$$P_b \simeq c_{d_{\text{free}}} 2^{d_{\text{free}}} e^{-(rE_b/2N_0)d_{\text{free}}} \tag{12.75}$$

We see again the important part played by the free distance in determining the error probability.

The general expressions of the probability of first-error event and bit-error probability over an AWGN can be obtained in terms of the free distance. It is easy to show [1] that the pairwise error probability for two codewords that differ in ℓ positions is given by

$$P_\ell = Q\left(\sqrt{\frac{2\ell r E_b}{N_0}}\right)$$

$$= Q\left(\sqrt{\frac{2(d_{\text{free}}+m)r E_b}{N_0}}\right)$$

where $\ell = (d_{\text{free}}+m)$, $m=0,1,2,\ldots$. Using the inequality

$$Q\left(\sqrt{x+y}\right) \leq e^{-y/2}Q\left(\sqrt{x}\right) \tag{12.76}$$

the pairwise error probability bound becomes

$$P_\ell \leq e^{-(\ell-d_{\text{free}})r E_b/N_0}Q\left(\sqrt{\frac{2r E_b d_{\text{free}}}{N_0}}\right) \tag{12.77}$$

Substituting in (12.57), the bound on the probability of first-error event becomes

$$P(E) < \sum_{\ell=d_{\text{free}}}^{\infty} a_\ell P_\ell \leq Q\left(\sqrt{\frac{2r E_b d_{\text{free}}}{N_0}}\right) \sum_{\ell=d_{\text{free}}}^{\infty} a_\ell e^{-(\ell-d_{\text{free}})r E_b/N_0}$$

$$\tag{12.78}$$

or, equivalently,

$$P(E) < Q\left(\sqrt{\frac{2r E_b d_{\text{free}}}{N_0}}\right)e^{r E_b d_{\text{free}}/N_0}T(D)\Big|_{D=e^{-r E_b/N_0}} \tag{12.79}$$

Similarly, the probability of bit error for Viterbi decoding of binary convolutional codes over AWGN channels can be obtained in the same way. From (12.68), (12.69) and (12.77) we obtain

$$P_b < Q\left(\sqrt{\frac{2r E_b d_{\text{free}}}{N_0}}\right)e^{r E_b d_{\text{free}}/N_0}\frac{dT(D,B)}{dB}\Big|_{B=1,\, D=e^{-r E_b/N_0}} \tag{12.80}$$

As an illustrative example, for the code of Figure 12.19, where $r=\frac{1}{2}$ and

$d_{\text{free}} = 5$, the bound on P_b is

$$P_b < Q\left(\sqrt{\frac{5E_b}{N_0}}\right) e^{(5/2)(E_b/N_0)} \frac{e^{-5E_b/2N_0}}{[1-2e^{-E_b/2N_0}]^2}$$

that is,

$$P_b < \frac{Q\left(\sqrt{5E_b/N_0}\right)}{[1-2e^{-E_b/2N_0}]^2} \tag{12.81}$$

At this point it may be instructive to compare the bit-error probability performance of convolutional coding and Viterbi decoding with a noncoded modulation scheme such as coherent PSK over an AWGN channel. From Chapter 2 we recall that for the noncoded scheme the bit-error probability is

$$P_b = Q\left(\sqrt{\frac{2E_b}{N_0}}\right) \tag{12.82}$$

which for large values of E_b/N_0 simplifies to

$$P_b < e^{-E_b/N_0} \tag{12.83}$$

The performance advantage of convolutional coding with Viterbi decoding over uncoded transmission for the same E_b/N_0 can be obtained by comparing (12.80) with (12.82) or (12.75) with (12.83) when E_b/N_0 is large.

Simple upper and lower bounds on the coding gain achieved on AWGN channels by convolutional codes can readily be obtained. Comparing (12.75) with (12.83) and considering only the exponents of the expressions, we see that for a given E_b/N_0, the exponent with convolutional coding is larger than the exponent with no coding by at least a factor of $(\frac{1}{2})rd_{\text{free}}$. This factor may thus be taken as a lower bound on the coding gain of the convolutional code. Now an upper bound on the coding gain is provided by (12.80), where we can write

$$P_b > Q\left(\sqrt{\frac{2rE_b d_{\text{free}}}{N_0}}\right) \tag{12.84}$$

Comparing with (12.82), an upper bound on the coding gain is thus rd_{free}. Therefore,

$$\tfrac{1}{2}rd_{\text{free}} \leq \text{coding gain} \leq rd_{\text{free}} \tag{12.85}$$

indicating that a 3-dB margin separates these two bounds.

Table 12.3 Coding Gain Upper Bounds for Some Convolutional Codes

	Rate-$\frac{1}{2}$ Codes			Rate-$\frac{1}{3}$ Codes	
		Upper Bound			Upper Bound
K	d_{free}	(dB)	K	d_{free}	(dB)
3	5	3.97	3	8	4.26
4	6	4.76	4	10	5.23
5	7	5.43	5	12	6.02
6	8	6.00	6	13	6.37
7	10	6.99	7	15	6.99
8	10	6.99	8	16	7.27
9	12	7.78	9	18	7.78

Table 12.3 lists the upper bound on the coding gains for several maximum free distance convolutional codes with constraint lengths varying from 3 to 9, indicating that fairly simple convolutional codes may provide a significant coding gain.

The actual coding gain experienced with a Viterbi decoder varies with the required bit-error probability P_b, with a tendency to increase as P_b is decreased. Coding gains for a number of codes with rates varying from $\frac{1}{3}$ to $\frac{3}{4}$ are presented in Table 12.4 together with the theoretical upper bound. These coding gains have been achieved with actual hardware implementation or by computer simulation [25]. For example, for the $r=\frac{1}{2}$, $K=7$ code and soft-decision Viterbi decoding, a coding gain of 3.8, 5.1 and 5.8 dB has been achieved at a bit-error probability of 10^{-3}, 10^{-5} and 10^{-7}, respectively, whereas the asymptotic upper bound value is about 7 dB. Also, examination of Table 12.4 reveals that at $P_b = 10^{-7}$, the shorter-constraint-length codes are

Table 12.4 Basic Coding Gain (dB) for Soft-Decision Viterbi Decoding

E_b/N_0 Uncoded	r	$\frac{1}{3}$		$\frac{1}{2}$			$\frac{2}{3}$		$\frac{3}{4}$	
(dB)	P_b \ K	7^a	8	5	6^a	7^a	6	8^a	6	9^a
6.8	10^{-3}	4.2	4.4	3.3	3.5	3.8	2.9	3.1	2.6	2.6
9.6	10^{-5}	5.7	5.9	4.3	4.6	5.1	4.2	4.6	3.6	4.2
11.3	10^{-7}	6.2	6.5	4.9	5.3	5.8	4.7	5.2	3.9	4.8
	Upper bound	7.0	7.3	5.4	6.0	7.0	5.2	6.7	4.8	5.7

Source: Ref. 25: © 1974 IEEE; reprinted, with permission, from "Practical Applications of Coding", by Irwin M. Jacobs, from *IEEE Trans. Inf. Theory*, Vol. IT-20, May 1974, p. 306.
[a]Presently implemented.

closer to the upper bound than are the more powerful codes with longer constraint lengths. However, the derivation of the bound is such that it may not be tight for any practical values of E_b/N_0. Further details concerning coding gains of implemented coding systems may be found in Ref. 25.

12.7   SIMULATION AND EXPERIMENTAL RESULTS

Viterbi decoding has been extensively simulated on a computer and tested on actual hardware implementations [25]. Several alternatives for applying high-speed and high-performance Viterbi decoding to TDMA satellite communications have been analysed and extensively simulated on a computer [26]. In such systems the decoder may be operating either on the TDMA burst or at the level of each one of the low-speed users.

In general, Viterbi decoding is used over binary input channels with either hard or 3-bit soft quantized outputs. The constraint lengths vary between 3

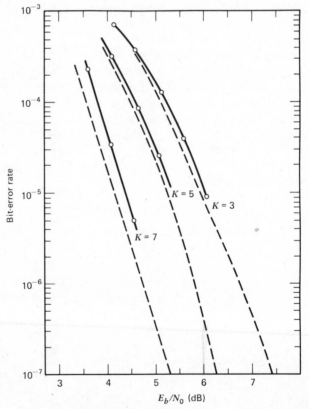

Figure 12.25   Bit-error performance for rate-$\frac{1}{2}$ codes and Viterbi decoding with soft-quantized channel and 32-bit path memory. —O—O—, simulation; --------, upper bound.

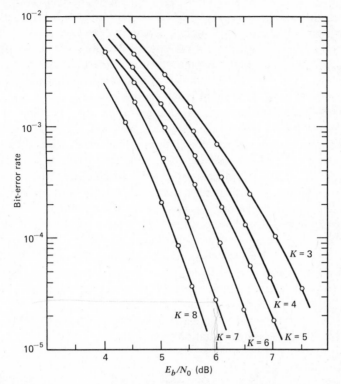

Figure 12.26 Bit-error performance for rate-$\frac{1}{2}$ codes and Viterbi decoding over BSC and 32-bit path memory.

and 8, the coding rate is rarely smaller than $\frac{1}{3}$, and the path memory of the decoder is usually a few constraint lengths [20]. Typical error performance curves are given in Figures 12.25 and 12.26. As expected, it is always found that the performance is improved with the constraint length, where each increment in K improves the required E_b/N_0 by a factor of 0.5 dB at $P_b = 10^{-5}$. It is also found that a moderate path memory length (in the vicinity of 30 to 40 branches) is quite adequate and a further increase of the decoder memory does not lead to any practical performance improvement. Also as expected, it is observed that a 3-bit quantization of the channel output provides a 2-dB gain over the hard-quantized BSC, although the use of soft quantization entails only a slightly more complex decoder: principally an analog-to-digital converter at the channel output (matched filter), a larger metric table and a corresponding increase in the memory required to store the path metrics.

Using $P_b = 10^{-5}$ as a reference, the coding gains of the decoders could easily be evaluated from Figures 12.25 and 12.26. In particular, for $K=7$ a coding gain of over 5 dB for a soft-quantized channel and over 3 dB for a hard-quantized channel is afforded by Viterbi decoding. In certain applications such as satellite communication where transmission power is at a premium, this coding gain provided by Viterbi decoding is thus rather attractive. For example, rate-$\frac{1}{2}$, $K=7$, soft-quantized Viterbi decoding was successfully used on two-way digital video link using the Canadian experimental satellite Hermes, where the 5-dB coding gain was translated into a substantial reduction of the antenna size and a smaller overall Earth station cost [27].

In all our discussion and analysis of Viterbi decoding, the received signals were supposed to be coherently detected. However, in actual systems the carrier phase is not known exactly at the receiver, and hence the inaccurate carrier phase reference will degrade the system performance. In particular, a constant phase error ϕ will decrease the received energy-to-noise ratio by the factor $\cos^2\phi$. Since the performance curve of coded systems is very steep, a small reduction of E_b/N_0 due to phase uncertainty will be translated as a comparatively larger degradation of the error probability. Analysis and simulation of Viterbi decoding in the presence of phase uncertainty has shown that although more severe than for uncoded systems, the error probability degradation does not vary too abruptly and can easily be brought under control [20].

One last point worth mentioning is that Viterbi decoding does not perform very well in a bursty channel. In those channels, interleaving of the data may thus have to be considered to obtain low correlation between noise samples. The ensuing performance degradation would then be minimal.

12.8   SEQUENTIAL DECODING TECHNIQUES

In the preceding section we saw that since the number of states increases exponentially with the constraint length, Viterbi decoding becomes rapidly impractical as K exceeds about 10. Therefore, if the desired performance requires the use of codes with larger free distances and larger coding gains, a somewhat suboptimal decoding technique must be used instead of Viterbi decoding.

Given the received sequence, the search for the most likely transmitted path can, in principle, be attempted sequentially, one branch at a time, making tentative decisions (or hypotheses) on successive branches of the explored paths, in such a way that the actual path being followed is the most likely among the *subset* of paths currently examined. Whenever a subsequent choice indicates an earlier incorrect decision, this decision can be properly modified, and hence at each decoding step, the current most likely path may be chosen among different paths lying at different levels in the tree (or trellis). Clearly, the explored path is then only **locally most likely**; that is, the procedure is *suboptimum*.

Such a step-by-step procedure is facilitated by the tree (or trellis) structure of the code, and by following only the current most likely path, unlikely sequences can be eliminated until a single sequence of length L remains; this sequence is then accepted as the decoded path.

The elimination of an unlikely path based on the observation of its first ℓ branches, $\ell < L$, is of course equivalent to the elimination of every path beginning with these ℓ branches [i.e., a whole subtree of length $(L-\ell)$]. The earlier the rejection, the larger (exponentially) the number of paths discarded. Hence the decoding effort will be reduced but clearly the procedure will be suboptimum. This idea of extending only the current most likely path together with the concomitant path rejection is credited to Wozencraft, who in 1957 exploited it in a particular decoding procedure for tree codes called **sequential decoding** [28].

One could improve on this suboptimal search procedure by extending not just the most likely path, but the set of the M most likely paths [29, 30]. By increasing the set of paths considered in the determination of the most likely sequence, such a procedure "reduces" the suboptimality of sequential decoding at a cost of a larger decoding effort. Moreover, instead of the tree, the trellis structure of the code could be used by allowing the decoder to recognize and exploit the remerging of paths just like an optimum Viterbi decoder.

Sequential decoding is the most powerful and practical decoding technique available, in terms of error performance on the DMC. Like Viterbi decoding, it is a **probabilistic** decoding technique because the decoded path is obtained by probabilistic considerations rather than by a set of fixed sequence of algebraic operations. Moreover, the decoding operations are independent of specific properties of the code which may be chosen at random. Thus the decoding technique is applicable in principle to any convolutional code.

There exists also a number of nonprobabilistic (or algebraic) suboptimum decoding techniques for convolutional codes, for example Massey's **threshold decoding** [31] suitable for codes having certain algebraic (orthogonal) properties. However, as shown in Chapter 14, these techniques yield inferior performance compared to Viterbi or sequential decoding, but are easy to implement.

Tree Search: Sequential Decoding Principles

Just as for Viterbi decoding, the sequential decoder is assumed to have a replica of the encoder, and hence is able to duplicate all possible transmitted sequences that constitute the code. Successive sequences of v channel output symbols (the **received branches**) are received from the channel, and from these the corresponding branch metrics can be computed. Assuming a DMC, the metric along the paths is cumulative and therefore, for both sequential and Viterbi decoding, the objective of the decoder is to find the path that yields the largest accumulated metric, given the received sequence. However, Viterbi (or optimum) decoding represents an exhaustive search that takes advantage of the code topology, whereas the suboptimal sequential decoding uses an intuitive trial-and-error search method to reduce the average decoding effort.

Intuitively, one could suspect that most of the time, the correct path has a larger metric than the incorrect paths diverging from it, and hence it could be estimated by considering only a few branches. Starting from the origin of the tree, the path selected to be searched one step further (i.e., extended one level deeper in its two branches) is the path that has the largest accumulated metric among those paths already examined. Therefore, by extending only the path that appears to be the most promising, most of the computations necessary for an optimum decoding can be avoided. This idea is common to various algorithms known as **sequential decoding algorithms**, the specific method of searching and selecting the path to be extended depending on the particular algorithm. Among the subset of explored paths, the path that reaches the last level of the tree (including the tail) with the highest metric is accepted as the decoded path.

Jacobs and Berlekamp [32] have given two conditions that a decoding algorithm must satisfy in order to be a sequential decoding algorithm.

1 The branches of the explored part of the tree are examined sequentially, and the decision on which path to extend is based only on those paths already examined. Each new path is thus an extension of a previously examined one.

2 The decoder performs at least one computation for each of every examined path.

Algorithms that do not have these two properties are not considered to be sequential decoding algorithms.

Fano Metric

As the decoding proceeds, the number of paths in the subset of explored paths grows, and occasionally the decoder goes back in the tree and extends earlier and possibly incorrect paths. A fundamental idea of sequential decoding is to *bias* the metrics in such a way that the backing up and extension of unlikely paths is reduced to a minimum. For a memoryless channel, the branch metric $\gamma_j^{(U)}$ of the branch $\mathbf{x}_j^{(U)} = (x_{j1}^{(U)}, \ldots, x_{jv}^{(U)})$ of a path $\mathbf{U}$ is generally defined as

$$\gamma_j^{(U)} = \sum_{i=1}^{v} \left[\log_2 \frac{P\left(y_{ji} \mid x_{ji}^{(U)}\right)}{f(y_{ji})} - B \right] \qquad (12.86)$$

where $\mathbf{y}_j = (y_{j1}, y_{j2}, \ldots, y_{jv})$ is the sequence of v received symbols corresponding to the transmitted branch $\mathbf{x}_j^{(U)}$, and where $f(j)$ is the nominal probability of having output j for a DMC with an input probability assignment $q(i)$, $i = 1, 2, \ldots, Q$, and transition probabilities $P(j \mid i)$, $j = 1, 2, \ldots, J$, $J > Q$. That is,

$$f(j) = \sum_{i=1}^{Q} q(i) P(j \mid i) \qquad (12.87)$$

For example, for the BSC with transition probability p, $J = Q = 2$. Assuming

$q(1)=q(2)=0.5$, $f(j)$ is given by

$$f(j)=\frac{1}{2}p+\frac{1}{2}(1-p)=\frac{1}{2}$$ (12.88)

The bias term B is chosen in such a way that the branch metrics are positive along the correct path and negative along all incorrect paths. Just as in Viterbi decoding, the metric assigned to a node $\mathbf{U}_N$ on a path of length N branches is

$$\Gamma_N^{(U)}=\sum_{j=1}^{N}\gamma_j^{(U)}, \qquad N=1,2,3,\ldots$$ (12.89)

The metric of (12.86) is a "tilting" of a modified form of the log-likelihood function of (12.39), and as for Viterbi decoding, it is an integer-valued metric. This biasing or "tilting" of the likelihood function is common to all sequential decoding algorithms, and the value of the bias that minimizes the average computational effort per decoded bit is equal to r, the rate of the code [33,34]. This bias value was first proposed by Fano [33] and the metric (12.86) is usually called the **Fano metric.** The consequence of biasing the likelihood function is that along the correct path the accumulated metric tends to increase, whereas along any incorrect path it tends to decrease. Therefore, although the objective of any sequential decoder is to find the path that yields the largest accumulated metric, the strategy is to search and follow the path of increasing metric value.

As an example, let us evaluate the branch metrics for a rate-$\frac{1}{2}$ convolutional code over a BSC with transition probability p, $p<0.5$. Observing that either zero, one, or two agreements may exist between the symbols on the received branch $\mathbf{y}_j$ and the code symbols on the tree branch $\mathbf{x}_j$, then three branch-metric values are possible. Using (12.86), those branch metrics are

$$\gamma_j=1+2\log_2 p,$$

$$\gamma_j=1+\log_2 p(1-p),$$

and $$\gamma_j=1+2\log_2(1-p)$$ (12.90)

which yield the values $+0.88$, -3.70 and -8.29, respectively, for $p=0.04$. Rounding off to the next integer, the branch metric values are then $+1$, -4 and -9.

For more general DMC, determining the different values of the branch metrics may be somewhat cumbersome. For example, for a rate-$\frac{1}{2}$ code, a binary input and J-output DMC, there may be as many as $[J+\frac{1}{2}J(J-1)]$ different values of γ_j if the channel is symmetric from the input. Consequently,

Table 12.5 Channel Transition Probabilities

Transition from Input 0 [a]	Transition Probability	
	$rE_b/N_0 = 0.0\,\mathrm{dB}$	$rE_b/N_0 = 0.5\,\mathrm{dB}$
0	0.465818	0.499207
1	0.194822	0.191554
2	0.159057	0.150101
3	0.101653	0.092072
4	0.050853	0.044208
5	0.019912	0.016613
6	0.006102	0.004886
7	0.001783	0.001359

[a] The transitions from input 1 are the complements of the transitions from input 0.

in practice it may be simpler to construct a table of **symbol metrics**

$$\left[\log_2 \frac{P(y_{ji}|x_{ji})}{f(ji)} - r \right]$$

from which the branch metrics are easily obtained. For example, for a binary input, eight-output DMC whose transition probabilities are shown in Table 12.5, such a metric table is shown in Table 12.6.

Sequential Decoding Algorithms

There are two principal sequential decoding algorithms: the Fano algorithm [17,33] and a search algorithm introduced independently by Jelinek [35] in 1969 and by Zigangirov in 1966 [36]. Although seemingly different, both algorithms search the tree from its root node out, trying to "match" sequentially the beginning segment $Y_i, i = 1, 2, \ldots, L,$ of the received sequence Y_L to the corresponding initial segments of the various paths of the tree. The Fano

Table 12.6 Symbol Metric Table for the Channel of Table 12.5

rE_b/N_0	x_{ji} \ y_{ji}	0	1	2	3	4	5	6	7
0.0 dB	0	4	4	3	−1	−9	−21	−36	−60
	1	−60	−36	−21	−9	−1	3	4	4
0.5 dB	0	4	4	3	−1	−9	−23	−39	−64
	1	−64	−39	−23	−9	−1	3	4	4

algorithm [33] introduced by Fano in 1963 is a modification of the original algorithm invented by Wozencraft in 1957 [28]. This algorithm is quite complex, the simplicity of the decoding principles being somewhat lost in the details of the rules that govern the motion of the decoder. It is nevertheless quite popular, and has been used to derive most of the theoretical results on sequential decoding. As for the Zigangirov–Jelinek (Z-J) algorithm, it is far simpler to understand and will be presented first.

Zigangirov–Jelinek (Z-J) Stack Algorithm

The Z-J algorithm is a very simple algorithm in which the essential concepts of sequential decoding are readily apparent. The decoder consists of a **list** or **stack** of the already searched paths ordered in *decreasing* order of their metric values. The **top of the stack** is the path with the largest accumulated metric among the paths in the stack, and will be searched further (extended one level further in its two branches) since it is the one that is the most likely to be the correct path. The stack is reordered after each extension, so that a path whose metric is ever increasing will continue to be searched further. Should its metric decrease and drop from the top position, that path will be properly stored in the stack and the new top node will be extended.

Denoting each explored path by the node of its extremity, the stack can equivalently be considered as a list of nodes ordered according to their metric values. The objectives of the decoder are thus the finding of the top node, the extension of its successors, and the proper reordering of the stack. After initially loading the stack with the origin node whose metric is taken as zero, the decoding algorithm consists of the following steps:

1 Compute the metrics of all successors of the top node and enter the new paths in their proper place in the stack.

2 Remove from the stack the node whose successors were just inserted.

3 If the new top node is the final node, stop. Otherwise, go to step 1.

When the algorithm gets out of the loop the top node is the end node of the decoded path. The whole path is then easily recovered from the information stored in the stack.

A decoding example for a binary tree is illustrated in Figure 12.27. The paths are associated with their terminal nodes, which are numbered in the figure. The numbers written on top of the branches represent the corresponding branch metrics for some received sequence Y_4 of length 4 branches. Ordering the nodes from left to right, the contents of the stack during decoding is then:

Step Number	Stack Contents: Node (Metric)
Initialization 1	0(0)
2	2(-1), 1(-7)
3	5(-5), 6(-5), 1(-7)

4　　$6(-5), 11(-6), 1(-7), 12(-12)$

5　　$11(-6), 1(-7), 13(-9), 14(-9), 12(-12)$

6　　$1(-7), 13(-9), 14(-9), 23(-10), 24(-10), 12(-12),$

7　　$4(-8), 13(-9), 14(-9), 23(-10), 24(-10), 12(-12),$
　　　$3(-14)$

8　　$10(-8), 13(-9), 14(-9), 23(-10), 24(-10), 9(-10),$
　　　$12(-12), 3(-14)$

9　　$21(-6), 13(-9), 14(-9), 23(-10), 24(-10), 9(-10),$
　　　$12(-12), 3(-14), 22(-18)$

The decoded path is thus specified by the data sequence terminating at node ㉑, that is, $(0, 1, 1, 0)$.

With the help of this example, the Z-J algorithm may be regarded as the most natural way to extend the explored path having the largest biased metric. By always extending the most likely path among those already explored (i.e., stored in the stack), the algorithm maximizes the probability that the next step will be taken along the correct path in the encoded tree.

The mechanics of the search are quite simple in principle. After the top node is eliminated, the two new successors are inserted in the stack at the place assigned by their total metrics. The stack grows by one entry at each decoding step, so that after j steps, it contains $(j+1)$ paths of various lengths, terminating in different nodes in the tree. Defining a computation as the execution of step 1 of the algorithm, the total number of computations to decode a binary tree is equal to $(W-1)$, where W is the size of the stack when decoding stops.

A graph of the explored paths metric values, called the **received value tree**, may be constructed as in Figure 12.28 for the example of Figure 12.27. At the fourth and last level, obviously the decoded path has the highest metric value. However, this is not necessarily the case for all intermediate nodes, and by the rules of the algorithm all incorrect paths whose metric is higher than that of the current correct path must be extended. A typical example of the behaviour of the correct and incorrect path metrics is shown in Figure 12.29. A span of severe noise causes the metric to drop, and hence some incorrect paths may temporarily occupy the top of the stack and be extended. Therefore, some nodes on the correct path will require a single computation to be decoded, and some other nodes will require several computations. It is the occasional drop in the correct path metric value and the concomitant extensions of incorrect paths that are responsible for the random nature of the decoding effort of sequential decoders.

Let $\mathbf{U}_j$ be a node lying at level j on the correct path $\mathbf{U}$. If

$$\Gamma_j^{(\mathbf{U})} \leq \Gamma_i^{(\mathbf{U})}, \qquad j \leq i \leq L \tag{12.91}$$

then $\mathbf{U}_j$ is called a **breakout node** [18]. If inequality (12.91) is strict, then $\mathbf{U}_j$ is called a **strict** breakout node. Breakout nodes play a central role in the

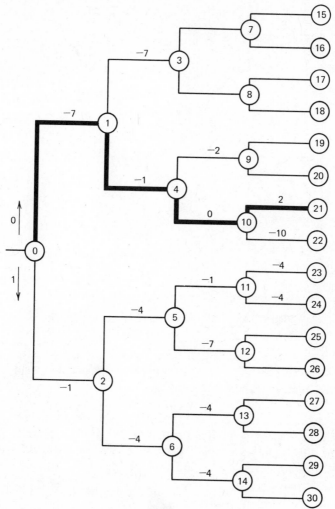

Figure 12.27 Branch metrics on a tree. The heavy line corresponds to the decoded path.

computational behaviour of any sequential decoding algorithm. With the help of Figure 12.29 it can be seen that the only incorrect paths the decoder will ever extend emerge from nonbreakout nodes on the correct path. Furthermore, it is easy to show that a sequential decoder never backs up beyond the level of the last decoded breakout node [29].

The discussion above shows that decoding proceeds smoothly when the decoder moves along breakout nodes on the correct path, and each decoded breakout node sets a limit on the depth of any future back-search. The decoding effort increases and becomes variable only when the decoder enters a series of nonbreakout nodes. Any increase of the number of consecutive

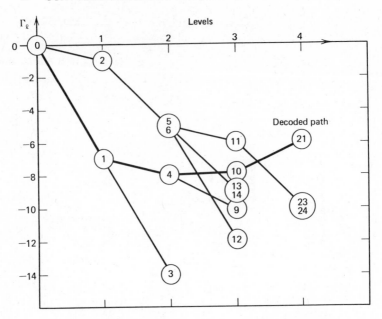

Figure 12.28   Metric values of the explored paths.

nonbreakout nodes and increase of the correct path metric dip (i.e., the largest metric difference between two consecutive nonadjacent breakout nodes) will increase the number of incorrect paths that must be explored (i.e., the decoding effort).

Using a Markov chain model introduced by Massey et al. [37], it has been shown and verified by computer simulation [38] that the cumulative distribution of the dips is an exponentially decreasing function of the dip value. As we shall see, this property is fundamental to the computational behavior of sequential decoding.

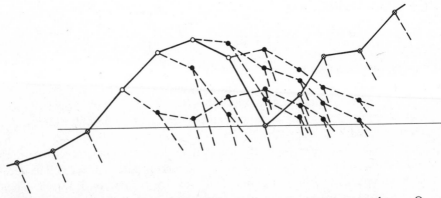

Figure 12.29   Correct and incorrect path metrics. --●--, incorrect paths; —⊗—, breakout nodes; —O—, nonbreakout nodes.

Quantized Z-J Stack Algorithm

The major problem with the Z-J stack algorithm is keeping the stack exactly ordered after each step. The exact ordering of the nodes is so time consuming that the algorithm becomes practically worthless. To overcome this difficulty, Jelinek [35] proposed a quantized version of the algorithm in which the nodes are placed at random into substacks (also called "**bins**") according to their metric values. In each substack are stored all those nodes whose metric value lies within a certain range. That is, a node m of metric Γ_m is inserted at random into substack Q if

$$QH \leq \Gamma_m < (Q+1)H \tag{12.92}$$

where H is the substack spacing in metric value.

In this quantized version, the search for the top node reduces to the search for the highest nonempty substack, clearly a much simpler task. The node that is to be extended is then taken at random from the top substack, usually in a last-in first-out mode. This quantized version of Z-J algorithm is sometimes called the **Jelinek algorithm**, where the only modifications to the original decoding rules are at step 1, which now reads: "Compute the metrics of the successors of any node from the highest nonempty substack and enter them in their proper substack". The rest of the algorithm remains unchanged.

With this quantized form, the Z-J algorithm becomes practical and competitive with the older Fano algorithm, described next.

Fano Algorithm

We have seen that the stack algorithm will require a considerable amount of storage to "remember" all terminal nodes of the paths examined during the decoding process. The fact that the next path to be searched further is always readily available in the stack is the main reason for the simplicity of the algorithm. However, this advantage is paid for by the large storage that may be required in some applications.

By contrast, the Fano algorithm [33] uses hardly any storage and hence does not "remember" the metrics of the paths that have been previously explored and rejected. Consequently, in order to know whether the current path being explored has the highest metric and thus must be explored further, the decoder uses a set of threshold values to test the acceptability of the paths. As long as the current threshold value is satisfied, that is, the metric of the currently explored path has a higher metric than the current threshold value, the decoder assumes to be moving along an acceptable path and proceeds forward. However, as soon as the current threshold is violated, the algorithm stops from going farther ahead along that path, and goes into a search for a better path. Because of the lack of storage, the search for a better path cannot be performed by jumping directly to the next-best node as in the stack algorithm, but must be performed on a branch-per-branch basis. From its current node, the decoder retraces its way back and attempts to find another path which does not violate the current threshold value. This new path is then taken as the new best path to be explored further. The details of the rules that govern the motion of the decoder constitute the Fano algorithm and are best explained by an example. The flowchart of the Fano algorithm is given in Figure 12.30, where the forward loop and the search loop are identified.

We illustrate next the different operations of the algorithm by finding the best path through the received distance tree shown in Figure 12.31. But first, some explanatory notes are necessary:

1 A particular node is said to satisfy any threshold smaller or equal to its metric value, and to violate any threshold larger than its metric value.

2 Starting at value zero, the threshold T changes its value throughout the algorithm by multiples of increments Δ, a preselected constant. For example, in Figure 12.31 nodes C, I, J and N satisfy threshold $T=\Delta$, whereas nodes A, G, E and K violate threshold $T=0$.

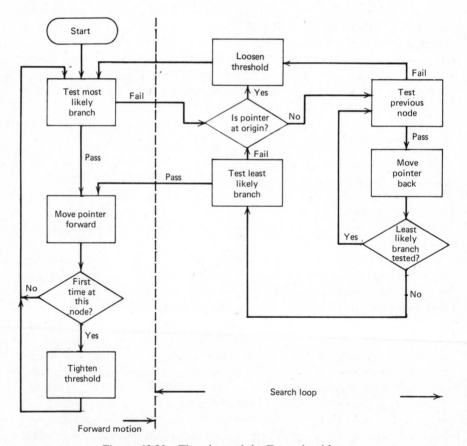

Figure 12.30 Flowchart of the Fano algorithm.

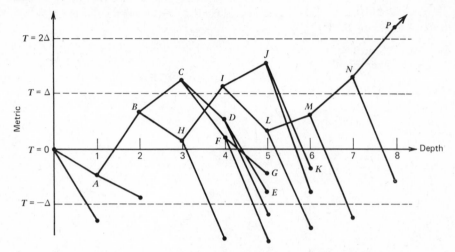

Figure 12.31 Received distance tree searched by the Fano algorithm.

3 The threshold is said to have been *tightened* when its value is increased by as many Δ increments as possible without being violated by the current node's metric.

4 The node being currently examined by the decoder is indicated by a search pointer.

5 In the flow diagram, when a node is *tested*, the branch metric is computed and the total accumulated metric for that node is evaluated and compared to the threshold. A node may be tested in both a forward and a backward move.

6 The decoder never moves its search pointer to a node that violates the current threshold.

7 The threshold is tightened only when the search pointer moves to a node never before visited.

Table 12.7 gives all the operations performed by the algorithm in decoding the tree of Figure 12.31. In the 21 steps necessary to find the best path at node depth 8, every possible action of the algorithm has been used.

Following the different steps we note that in searching for a better path the decoder always examines all accessible nodes lying above a given threshold before decreasing that threshold. This repeated examination of the same nodes (which is not encountered in the stack algorithm) overtaxes the decoding effort during periods of high noise. However, when there is little noise, back-searching is reduced to a minimum and decoding speeds of several Mbps are possible.

In a comparative study between the Fano and Z-J algorithms it has been shown that although quite different, both algorithms follow essentially the same rules of path extension and arrive at the same decoded path [39, 40].

Table 12.7   Details of the Operations of the Fano Algorithm for Decoding the Tree of Figure 12.31

Step	Search Pointer Position	Threshold Value	Action					
1	0	0	Test A	Fail	Set $T=-\Delta$			
2	0	$-\Delta$	Test A	Pass	Move to A			
3	A	$-\Delta$	Test B	Pass	Move to B		Set $T=0$	
4	B	0	Test C	Pass	Move to C		Set $T=\Delta$	
5	C	Δ	Test D	Fail	Test B	Fail	Set $T=0$	
6	C	0	Test D	Pass	Move to D			
7	D	0	Test E	Fail	Test C	Pass	Move to C	
8	C	0	Test F	Pass	Move to F			
9	F	0	Test G	Fail	Test C	Pass	Move to C	
10	C	0	Test B	Pass	Move to B			
11	B	0	Test H	Pass	Move to H			
12	H	0	Test I	Pass	Move to I		Set $T=\Delta$	
13	I	Δ	Test J	Pass	Move to J			
14	J	Δ	Test K	Fail	Test I	Pass	Move to I	
15	I	Δ	Test L	Fail	Test H	Fail	Set $T=0$	
16	I	0	Test J	Pass	Move to J			
17	J	0	Test K	Fail	Test I	Pass	Move to I	
18	I	0	Test L	Pass	Move to L			
19	L	0	Test M	Pass	Move to M			
20	M	0	Test N	Pass	Move to N		Set $T=\Delta$	
21	N	Δ	Test P	Pass	Move to P		Set $T=2\Delta$	

Considering the decoding time as a measure of the decoding effort, it was observed by simulation that for cases of low noise, the Fano decoder performs better than the stack decoder, but the advantage goes rapidly to the stack decoder as the noise increases. Hence the stack decoder would be superior to the Fano decoder for periods of high channel noise, and vice versa for periods of low channel noise. The choice of the particular decoder will depend on a tradeoff among storage, speed, software sophistication and cost.

Computational Problem of Sequential Decoding

Regardless of the algorithm used, sequential decoding involves a random motion into the tree, where the exact moves of the decoder are determined by the received sequence and the particular algorithm. As any decoder move implies a computation, the number of computations to decode a given block of information bits is a random variable. Consequently, the analysis of sequential decoding is concerned not only with the error probability but also with the distribution of the computational effort. The variability of the computation is the principal drawback of sequential decoding. However, typically the average number of computations per decoded information bit is in general very small, much smaller than the fixed number 2^{K-1} required for Viterbi decoding. Furthermore, unlike Viterbi decoding, the average number of computations for sequential decoding is insensitive to the constraint length of the code.

A combination of asymptotic results by many authors, including Jacobs and Berlekamp [32], Savage [41], Jelinek [42], Forney [43] and Falconer [44], shows that the number of computations required to decode one information bit for any sequential decoding algorithm has asymptotically a **Pareto distribution**, that is

$$P(C \geq N) < \beta N^{-\alpha}, \qquad N \gg 1 \tag{12.93}$$

where α and β depend only on the channel transition probabilities and the code rate r. The exponent α is called the **Pareto exponent** and is a very important parameter in determining the performance of sequential decoders.

Examination of (12.93) shows that the distribution of the computational effort decreases only algebraically rather than exponentially with N, which is clearly an undesirable property. Therefore, the questions of how to ensure that the average number of computations is finite and how to accommodate the variable decoding delay become important.

The answer to the second question lies with the use of input and output buffers, whereas the answer to the first question has been determined by analysis and by both computer simulation and hardware measurement. When there is little noise, the correct path is readily followed with very few back-searches, and about one computation per bit is required. However, as the channel error rate increases, more searching is performed by the decoder and the average number of computations increases drastically; eventually, it may even become unbounded.

The analysis of the moments of the Pareto distribution shows that if the Pareto exponent α of (12.93) is less than 2, the variance of the decoding effort is unbounded, and for $\alpha \leq 1$, the mean does not even exist: the average number of computations to decode one bit becomes theoretically infinite. In practice, this is observed as very erratic decoding with a dramatic increase in computations, long searches, number of errors and overflows. The coding rate for which $\alpha = 1$ is denoted R_{comp} and is called the **computational cutoff rate** of sequential decoding. Clearly, R_{comp} is a very important parameter for sequential decoding systems since it is the limiting code rate for these decoders; practical systems usually tend to operate with code rates within a few percent of R_{comp} from below.

The theoretical value of the Pareto exponent is given by the parametric equation

$$r = \frac{E_0(\alpha)}{\alpha} \tag{12.94}$$

where r is the code rate and $E_0(\alpha)$ is the so-called **Gallager function** [18]. For a

DMC with Q inputs having a distribution $\{q(i), i = 1, 2, \ldots, Q\}$, J outputs, and transition probabilities $P(j|i)$, $E_0(\alpha)$ is given by

$$E_0(\alpha) = -\log_2 \sum_{j=1}^{J} \left[\sum_{i=1}^{Q} q(i) P(j|i)^{1/(1+\alpha)} \right]^{1+\alpha} \qquad (12.95)$$

$E_0(\alpha)$ is a concave monotonic nondecreasing function of α and has the following properties:

$$E_0(0) = 0 \qquad (12.96)$$

$$E_0(1) = R_{comp} = R_0 \qquad (12.97)$$

$$\lim_{\alpha \to 0} \frac{d(E_0(\alpha))}{d\alpha} = \mathcal{C}_s \qquad (12.98)$$

where $\mathcal{C}_s$ is the channel capacity in bits/symbol.

For $\alpha = 1$ it is easy to verify that (12.95) yields the R_0 expression encountered earlier in (12.44). Therefore, R_0 of a general DMC is R_{comp} of sequential decoding, and for the BSC R_{comp} is given by (12.45), while for the binary input and continuous output channel it is given by [17]

$$R_{comp} = 1 - \log_2(1 + e^{-rE_b/N_0}) \qquad (12.99)$$

where rE_b/N_0 is the energy-to-noise ratio of each channel digit.

Such continuous output channels are equivalent to unquantized receivers, and for the same R_{comp} they require an (rE_b/N_0) value some 2 dB smaller than for the hard-quantized BSC. This soft-quantization gain may be very attractive for satellite channels even though it involves a more complex decoder.

Table 12.8 lists a few values of R_{comp} for the BSC. For (E_b/N_0) values smaller than 4.5 dB, Table 12.8 indicates that considerable computational difficulties may be encountered with sequential decoding if the code rate is $\frac{1}{2}$, since R_{comp} would be less than $\frac{1}{2}$. However, if the code rate is $\frac{1}{3}$, an E_b/N_0 as small as 4.2 dB will present no serious computational problem.

Table 12.8 R_{comp} for the BSC

rE_b/N_0 (dB)	p	R_{comp}	rE_b/N_0 (dB)	p	R_{comp}
−0.5	0.0909	0.344	2.0	0.0375	0.535
0.0	0.0786	0.379	2.5	0.0296	0.578
0.5	0.0670	0.415	3.0	0.0228	0.622
1.0	0.0563	0.453	3.5	0.0172	0.666
1.5	0.0464	0.493	4.0	0.0125	0.711

Since $E_0(\alpha)$ is monotone nondecreasing, from (12.94) we can write

$$\alpha \le \frac{R_{comp}}{r} \qquad \text{for } r \ge R_{comp} \qquad (12.100)$$

and

$$\alpha \ge \frac{R_{comp}}{r} \qquad \text{for } r \le R_{comp} \qquad (12.101)$$

Given the function $E_0(\alpha)$ for a DMC, the values of α and R_{comp}/r may be easily obtained by graphical construction. In Figures 12.32 and 12.33 the Pareto exponent is plotted as a function of E_b/N_0 for the hard- and soft-quantized channels, respectively. One could also use the approximation

$$\alpha \simeq \frac{R_{comp}}{r} \qquad (12.102)$$

which is exact only if $r = R_{comp}$ or $\alpha = 1$. However, as α increases or decreases from this value, the approximation becomes progressively worse. From (12.93) we observe that as the Pareto exponent becomes larger, the tail of the distribution of the computation decreases more rapidly. This is a very desirable situation which can be obtained either by decreasing the code rate or by increasing the energy-to-noise ratio.

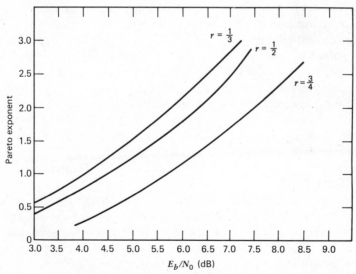

Figure 12.32 Pareto exponent as a function of E_b/N_0 for the BSC.

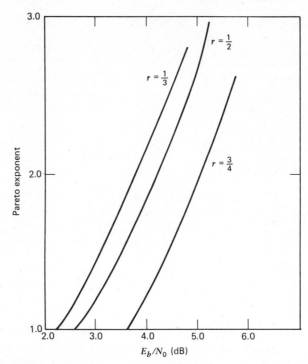

Figure 12.33 Pareto exponent as a function E_b/N_0 for the 3-bit soft-quantized AWGN ($0.5\sqrt{N_0/2}$ quantization levels).

Many experimental investigations [45] confirm the Pareto nature of the computation for the coherent and incoherent discrete memoryless channels [46, 47]. A simple intuitive argument may explain this behaviour. When the noise causes the correct path metric to dip, the decoder goes into a back-search and enters a subset of incorrect paths. As the number of paths in the incorrect subset grows exponentially with the penetration into the subset, the number of required computations will also grow exponentially with the length of the correct path metric dip. However, on the discrete memoryless channel, the interval of high noise that caused the correct path metric to dip occurs with a probability that is exponentially decreasing with its length. The combined effect of these two exponential behaviours results in a distribution that decreases at most algebraically. A typical distribution of computation curves is given in Figure 12.34.

It is worth mentioning that (12.93) has been derived for the ensemble of randomly chosen codes, and hence for a specific code only the slope of the tail that is the Pareto exponent of the distribution will be given by (12.93). Modelling the incorrect paths explored by a sequential decoder as a branching process, accurate expressions for the average number of computations and the distribution of computation for specific codes and branch metrics have been

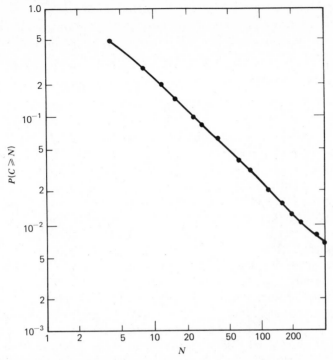

Figure 12.34 Typical distribution of computations for sequential decoding over the BSC. E_b/N_0, 4.45 dB; K, 36; r, $\frac{1}{7}$; r/R_{comp}, 0.98; 100 blocks, 735 bits/block.

obtained by Haccoun [48] and Johannesson [49], respectively. Furthermore, Chevillat and Costello [50] have expressed the distribution of computations in terms of the distance properties of the specific codes used, showing that a rapidly increasing column distance function of the code will yield a rapidly decreasing computational effort. Consequently, good convolutional codes for sequential decoding must have both a large free distance and a rapidly increasing column distance function. Some of these codes are listed in Appendix 12.1.

Buffer Overflow

Another unpleasant consequence of the variability of the computation is that incoming data may not always be immediately processed as they arrive. The question of how to accommodate the variable decoding delay is simply resolved by the use of a buffer to store the incoming data waiting to be processed. Regardless of the size of the buffer, there is a nonzero probability that it may fill up, leading to an overflow and possible communication breakdown. However, recovering procedures are usually applied during buffer overflows, which occur with a relatively large (typically 10^{-3} to 10^{-4}) probability. This overflow problem is the most serious problem with sequential decoding and the probability of overflow, also called **erasure probability**, has been bounded by Savage [41]. For an L-branch tree, the probability of overflow is given by

$$P(\text{overflow}) \simeq L(SB)^{-\alpha} \qquad (12.103)$$

where S is the **decoder speed factor** in computations per information bit interarrival time, B the size of the buffer, and α the Pareto exponent. The probability of overflow is rather difficult to combat because of its relative insensitivity to buffer size and decoder speed.

To overcome this major difficulty, data are sent in blocks, usually on the order of 500 to 1000 branches, and known sequences (the tail of the message) clear the shift register of the convolutional encoder after each block and resynchronize the system.

In case of an overflow with no recovering procedure, the entire block is lost, but the buffer is cleared and decoding can be resumed in the following block. A simple recovery procedure consists in storing the data of the blocks about to overflow and either request a repeat of these blocks or process them off-line at a later time. In the case of systematic codes or quick-look-in codes [14], one can even directly deliver an estimate of the data to the user, without decoding.

In some applications the disadvantage of overflows or erasures of sequential decoding may turn out to be useful. When a very noisy sequence is encountered, most probably it will be incorrectly decoded even by an optimum decoder. Since the sequential decoder will overflow, erasures rather than errors will be delivered to the user; hence the decoder may trade errors that are not detected for erasures that are easily detected.

Some new methods have been proposed to alleviate the computational difficulty of sequential decoding. These methods are based on the Z-J algorithm rather than the older Fano algorithm, which, because of its rules of motion, leaves little room for modification. A simple variation of the Z-J stack algorithm **called multiple-path stack algorithm** consists in extending several paths simultaneously from the top of the stack [29, 30]. Furthermore, the remerging of paths may even be exploited to eliminate unnecessary computations and stack storage. With these modifications it is shown that the variability of the computational effort is reduced, reducing with it both the error and erasure probability at a cost of a modest increase of the average number of computations.

Another method, called the **multiple stack algorithm**, has been recently developed by Chevillat and Costello [51]. It is also a modification of the Z-J algorithm and uses several stacks, which are filled one after another, but in such a way that erasures are practically eliminated at the expense of a substantial increase in stack and buffer storage.

Error Performance

The sequential decoder is said to have made an error if it makes an incorrect decision on an information bit and never returns to correct it. The error probability in sequential decoding decreases exponentially with the constraint length of the code and hence can be made arbitrarily small. Sequential decoding being a suboptimal procedure, a lower bound on its error probability is obviously the lower bound on the error probability for the optimal Viterbi decoding. For code rates $r < R_{comp}$, the error probability can be upper bounded by

$$P(E) < Ae^{-KR_{comp}/r} \qquad (12.104)$$

where A is a constant and K the constraint length of the code.

Long constraint lengths present no problem in actual sequential decoders because the coder complexity varies only linearly with the constraint length and the probability of overflow is insensitive to it. (Typical values of code constraint lengths range from 20 to 60.) The real problems with sequential decoding are the variability of the computational effort and the buffer overflow.

Regardless of the particular sequential decoding algorithm, the choice of the code and the choice of the set of metrics affect both the error and the computational performance of sequential decoders. In general, the achievement of low error probabilities, fast decoding speeds and low overflow probabilities requires the selection of codes of long constraint lengths having a large free distance d_{free} and good distance profile, that is, a rapidly increasing column distance function. Some of these codes are given in Appendix 12.1.

12.9   COMPARISONS AND LIMITATIONS OF SEQUENTIAL AND VITERBI DECODING

To summarize, the two decoding techniques presented in this chapter attempt to find the shortest path through a graph. The Viterbi algorithm was shown to be an exhaustive search growing naturally out of the code topology, whereas sequential decoding appears heuristically to be a natural method of reducing the average number of computations (but not the maximum) by trial-and-error search in the tree. Among the several sequential decoding algorithms, the stack algorithm of Zigangirov and Jelinek is the simplest and most natural way to utilize the key concepts of path rejection of sequential decoding.

As discussed previously, the computational cutoff rate R_{comp} limits the code rates at which sequential decoders can be used, but a Pareto distribution prevails regardless of the code rate. Storage requirements and decoding speed must be traded in such a way that a decrease in buffer storage must be compensated for by an increase of the decoding speed over the bit rate (i.e., the speed factor of the decoder), thus limiting the maximum bit rate capability. Below R_{comp}, if the number of computations that the decoder can perform per unit of time is greater than the required mean value of the corresponding computational effort, then on the average the decoder can keep up with the data, although buffering is necessary. Regardless of the buffer storage provided and even for rates well below R_{comp}, occasional long searches do occur, resulting in a possible buffer overflow and the consequent erasing of long sequences. For sequential decoders using long constraint lengths the probability of undetected error is indeed very small, and usually the main contribution of errors comes from the buffer overflows.

For the Viterbi decoder such a situation obviously does not occur, owing to the fixed nature of the computational effort. Storage path history and decoding speed requirements may definitely be set for a particular application.

Since both the number of operations per decoded branch and the total storage are proportional to the number of states, the complexity of the Viterbi decoder is also proportional to the number of states, and hence it increases exponentially with the constraint length K of the code. Although the error probability decreases exponentially with the constraint length, clearly any improvement on the error performance by increasing K is expensive in terms of computational effort and storage. Therefore, Viterbi decoders are limited to small ($K \leq 8$)-constraint-length codes, and consequently are used where a moderate error probability is sufficient (10^{-4} to 10^{-5}). However, the very simple nature of a Viterbi operation and its identical repetition from level to level permits a parallel implementation of a decoder and hence decoding at very high data rates, in the tens of Mbps range.

Sequential decoding is finding application in severe environments such as deep space and satellite links requiring maximum coding gains. This coding gain is theoretically in excess of 7 dB at a 10^{-5}-bit error probability. In principle, sequential decoding could be performed with soft quantization. However, for reasons of complexity, buffer size requirements, computation speed and metric sensitivity, most of the implementations have used hard quantization (with a 2-dB loss in E_b/N_0) [25, 45, 52, 53]. Indeed, the use of a hard-quantized receiver instead of an eight-level receiver reduces the effective buffer storage by a factor of 3 and simplifies the logic requirements. Since typical buffer sizes vary between 10^4 and 10^5 branches, the savings may be substantial.

The error performance of a sequential decoder is exponentially decreasing with the constant length K of the code, whereas the computational effort is almost independent of K and the complexity of the machine is only linearly dependent on K. Therefore, long-constraint-length codes are indicated with sequential decoding, and practical implementations have used rate-1/2 codes with constraint lengths larger than 45 at data rates up to 5 Mbps, achieving coding gains of 4.2 dB at 10^{-5} and 6.5 dB at 10^{-7} [25].

Decoding delays are inescapable for both sequential and Viterbi decoders. For sequential decoders the delay is a direct consequence of the variability of the computational effort, and for Viterbi decoders it corresponds to the path memory storage of each state. Sequential decoders tend to have buffers considerably longer (tens of kilobits) than the path memory storage of Viterbi decoders. However, in a real-time situation where a fixed time delay is required for delivering the data, the total number of computations required in one time delay grows only linearly with the constraint length K for sequential decoders, whereas it grows exponentially with K for Viterbi decoders.

In general, performance and complexity combine in such a way that sequential decoding is chosen over Viterbi decoding when high performance is required ($P_b < 10^{-5}$), provided that the code rate is sufficiently low that operation below R_{comp} is assured. Moreover, since sequential decoders will be extremely sensitive to bursty noise patterns, they will usually be restricted to well-behaved memoryless channels such as the space and satellite channels. Otherwise, they must be used in conjunction with interleaving, or be modified to use different metric tables according to the state of the channel [54].

Viterbi decoders may be preferred when the performance demanded is more modest but the data rate very high. The crossover point occurs for bit-error probability in the 10^{-3} to 10^{-4} range and constraint lengths between 6 and 8.

Most of the drawbacks of sequential decoding are the consequences of the variability of the computational effort and improvements should be directed at reducing this variability without undue increase in decoder complexity. In that respect the Z-J stack algorithm offers far more versatility than the Fano algorithm. This versatility comes from the possibility of exploring several paths simultaneously, even dynamically adapting the decoding effort to the requirements imposed by the actual noise in the channel [30]. Furthermore, stack storage and input buffer storage may be traded for a given probability of overflow. By controlling the stack size in such a way that overflows occur in the stack first, difficult blocks can be erased relatively early, hence reducing the storage requirements for the input buffer. Therefore, although up to the present only the Fano algorithm has been implemented, as the cost for memory drops and refinements develop, sequential decoding with the stack algorithm will become attractive for satellite and other well-behaved channels. A crypto-system based on convolutional encoding and sequential decoding has even been proposed for secure communications over noisy channels [55].

APPENDIX 12.1 BEST KNOWN CONVOLUTIONAL CODES

In this appendix tables of the best known convolutional codes with respect to several distance measures are provided. Only noncatastrophic codes are listed.*

The generators are the coefficients of the connections vectors (12.15)

$$\mathbf{G}_j = \left(1, g_{j2}, g_{j3}, \ldots, g_{jK}\right), \qquad j = 1, 2, \ldots, v$$

With the exception of Tables A12.4 and A12.5, the generators are expressed

---

*A *catastrophic code* is a code that propagates the errors, that is, for which a finite number of channel errors may cause an infinite number of bit errors. Nonsystematic codes may be catastrophic, whereas systematic codes are never catastrophic.

in octal notation. However, care should be exercised in interpreting the notation, as the original authors of the tables (and discoverers of the codes) did not use the same notation. In Tables A12.1 to A12.3 the last octal digit corresponds to $[g_{j(K-2)}, g_{j(K-1)}, g_{jK}]$; that is, 35 corresponds to [11101] and 75 corresponds to [111101]. In Tables A12.6 to A12.12 the first octal digit corresponds to $[1, g_{j1}, g_{j2}]$, so that 72 becomes [11101] and 54 corresponds to [1011].

Table A12.1   Rate-$\frac{1}{2}$ Codes with Maximum Free Distance

K	Generators		d_{free}	Upper Bound
3	5	7	5	5
4	15	17	6	6
5	23	35	7	8
6	53	75	8	8
7	133	171	10	10
8	247	371	10	11
9	561	753	12	12
10	1167	1545	12	13
11	2335	3661	14	14
12	4335	5723	15	16
13	10533	17661	16	16
14	21675	27123	16	17

Source: Ref 8: © 1973 IEEE; reprinted, with permission, from "Short Convolutional Codes with Maximal Free Distances for Rate $\frac{1}{2}$, $\frac{1}{3}$, and $\frac{1}{4}$", by K. J. Larsen, from *IEEE Trans. Inf. Theory*, Vol. IT-19, May 1973, p. 371.

Table A12.2   Rate-$\frac{1}{3}$ Codes with Maximum Free Distance

K	Generators			d_{free}	Upper Bound
3	5	7	7	8	8
4	13	15	17	10	10
5	25	33	37	12	12
6	47	53	75	13	13
7	133	145	175	15	15
8	225	331	367	16	16
9	557	663	711	18	18
10	1117	1365	1633	20	20
11	2353	2671	3175	22	22
12	4767	5723	6265	24	24
13	10533	10675	17661	24	24
14	21645	35661	37133	26	26

Source: Ref. 8: © 1973 IEEE; reprinted, with permission, from "Short Convolutional Codes with Maximal Free Distances for Rate $\frac{1}{2}$, $\frac{1}{3}$, and $\frac{1}{4}$", by K. J. Larsen, from *IEEE Trans. Inf. Theory*, Vol. IT-19, May 1973, p. 372.

Table A12.3 Rate $\frac{1}{2}$ Complementary Codes with Maximum Free Distance[a]

K	Generator (Octal)	d_{free}	$d_{\min}$
3	5	5	3
4	13	6	3
5	31	7	4
6	61	8	4
7	121	9	5
8	211	10	5
9	503	11	6
10	1065	12	6
11	2415	13	7
12	5121	14	7
13	12043	15	7
14	24421	16	8
15	51303	17	7
16	120643	18	8
17	346411	18	9
18	425551	20	8
19	1411041	20	9
20	2734605	20	10
21	5011303	22	9
22	11047441	22	10
23	22517023	24	10
24	51202215	24	10

Source: Ref. 9: © 1971 IEEE; reprinted, with permission, from "Rate $\frac{1}{2}$ Convolutional Codes with Complementary Generators", by L. R. Bahl and F. Jelinek, from *IEEE Trans. Inf. Theory*, Vol. IT-17, Nov. 1971, p. 724.

[a] In a complementary code of rate $\frac{1}{2}$ the first and last state of the shift register are always connected to the modulo-2 adders. The remaining coefficients of one generator are the complement of the coefficients of the other generator. That is,

$$g_{11} = g_{21} = g_{1K} = g_{2K} = 1$$

and

$$g_{1j} = g_{2j} \oplus 1; \qquad j = 2, 3, \ldots, k-1$$

Hence one generator description is sufficient.

Table A12.4 Generators for Rate-$\frac{2}{3}$ Convolutional Codes

K	G_0	G_1	G_2	G_3	G_4	G_5	d_{free}	Upper Bound
2	101	111					3	4
	011	100						
3	101	011	000				4	4
	011	001	101					
4	101	100	110				5	6
	011	101	011					
5	101	111	011	000			6	6
	011	001	101	101				
6	101	111	010	101			7	8
	011	111	101	011				
7	101	110	011	011	000		8	8
	011	001	101	111	110			
9	101	001	101	011	110	000	9	10
	011	010	011	100	001	101		
10	101	100	010	011	101	110	10	9
	011	111	100	010	100	011		

Source: Ref. 10: © 1974 IEEE; reprinted, with permission, from "Short Binary Convolutional Codes with Maximal Free Distances for Rates $\frac{2}{3}$ and $\frac{3}{4}$", by E. Paaske, from *IEEE Trans. Inf. Theory*, Vol. IT-20, Sept. 1974, p. 688.

Table A12.5 Generators for Rate-$\frac{3}{4}$ Convolutional Codes

K	G_0	G_1	G_2	G_3	d_{free}	Upper Bound
	1111	0000	0000			
3	0101	0110	0000		4	4
	0011	0100	0011			
	1001	1111	0000			
5	0101	0101	1001		5	6
	0011	0100	0011			
	1001	1001	0101			
6	0101	1001	1010		6	6
	0011	1110	0110			
	1001	1110	1100	0000		
8	0101	0000	1101	1001	7	8
	0011	0010	0110	1010		
	1001	0011	0110	0110		
9	0101	0111	0001	1100	8	8
	0011	1011	1000	1001		

Source: Ref. 10: © 1974 IEEE; reprinted, with permission, from "Short Binary Convolutional Codes with Maximal Free Distances for Rates $\frac{2}{3}$ and $\frac{3}{4}$", by E. Paaske, from *IEEE Trans. Inf. Theory*, Vol. IT-20, Sept. 1974, p. 688.

Table A12.6 Rate-$\frac{1}{2}$ Optimum Distance Profile Systematic Convolutional Codes Which Are Also Optimum Minimum Distance

K	Generator	d_{min}	# paths[a]	d_{free}	# paths[a]
2	6	3	2	3	1
3	7	3	1	4	2
4	64	4	3	4	1
5	72	4	1	5	2
6	73	5	5	6	3
7	730	5	2	6	3
	734	5	3	6	1
8	714	6	11	6	2
9	715	6	5	7	2
	671	6	6	7	1
10	6710	6	1	7	1
	7154	6	3	8	4
11	6710	7	12	7	1
	7152	7	13	8	3
12	6711	7	5	8	2
	7153	7	6	9	3
13	67114	8	29	9	1
14	67114	8	12	9	1
15	67115	8	6	10	4

Source: Ref. 5: © 1975 IEEE; reprinted, with permission, from "Robustly Optimal Rate One-Half Binary Convolutional Codes", by R. Johannesson, from *IEEE Trans. Inf. Theory*, Vol. IT-21, July 1975, p. 465.
[a] # paths is the number of paths yielding the distance shown

Table A12.7 Optimum Distance Profile Systematic Codes with Rate $\frac{1}{2}$[a]

M	$g^{(2)}$	d_M	#paths	d_∞	#paths
15	714474	8	1	10	1
16	714476	9	18	10	1
	671166	9	22	12	13
17	671145	9	7	11	1
	671166	9	13	12	13
18	6711454	9	3	12	4
19	7144616	10	31	12	3
20	7144616	10	13	12	3
	7144761	10	18	12	1
21	67114544	10	4	12	1
22	71446162	10	1	13	2
	71446166	10	6	14	6
23	67114543	11	27	14	6
	67115143	11	32	14	2
24	714461654	11	11	15	5
	671151434	11	16	15	4
25	714461654	11	5	15	5
	671145536	11	9	15	3
26	671145431	11	1	15	1
	671151433	11	4	16	8
27	7144616264	12	21	14^L	1
	7144760524	12	26	16	7
28	6711454306	12	8	16	4
	6711514332	12	13	16	3
29	7144616573	12	2	17^L	3
	7144760535	12	6	18	22
30	71446162654	13	43	16^L	2
	67114543064	13	44	16^L	1
31	71446162654	13	15	16^L	2
	67114543066	13	24	18	11
32	71446162655	13	4	17^L	2
	71447605247	13	13	18^L	2
33	714461626554	13	1	18^L	5
	671145430654	13	4	18^L	1
34	714461626554	14	34	18^L	5
	714461625306	14	42	18^L	1
35	714461625313	14	14	18^L	3
	714461626555	14	19	19^L	2

Note: L denotes that this number is actually d_{71} which is a lower bound on d_∞.

Source: Ref. 5: © 1975 IEEE; reprinted, with permission, from "Robustly Optimal Rate One-Half Binary Convolutional Codes", by R. Johannesson, from *IEEE Trans. Inf. Theory*, Vol. IT-21, July 1975, p. 465.

[a]Constraint length, $K = M+1$; d_∞, free distance d_{free}; d_M, minimum distance d_{min}; # paths, number of paths yielding the distance shown.

Table A12.8 Optimum Distance Profile systematic Codes with Rate $\frac{1}{2}$ [a]

M	$G^{(2)}$	d_M	# paths
36	6711454544704	14	5
37	6711454544676	14	2
38	6711454575564	15	31
39	71446165734534	15	12
40	67114545755712	15	3
41	71446165734537	15	1
42	67114545755464	16	31
43	714461626554012	16	14
44	714461626554427	16	5
45	7144616265544274	16	1
46	6711454575564666	17	39
47	6711454575564667	17	13
48	67114545755646674	17	4
49	67114545755646676	17	1
50	67114545755646676	18	38
51	671145457556466760	18	16
52	671145457556466760	18	7
53	714461626553260462	18	2
54	714461626556137204	19	43
55	714461626556137206	19	20
56	714461626556137206	19	7
57	71446162655561372064	19	2
58	71446162655561372064	20	60
59	67114545755646670367	20	25
60	671145457556466703670	20	10

Source: Ref. 12: © 1976 IEEE; reprinted, with permission, from "Some Long Rate One-Half Binary Convolutional Codes with an Optimum Distance Profile", By R. Johannesson, from *IEEE Trans. Inf. Theory*, Vol. IT-22, Sept. 1976, p. 630.

[a] Constraint length, $K = M + 1$; d_M, minimum distance d_{min}: # paths, number of paths yielding the distance shown.

Table A12.9 Optimum Distance Profile Quick-Look-In Codes with Rate $\frac{1}{2}$ [a]

M	$G^{(1)}$	$G^{(2)}$	d_M	#paths	d_∞	#paths
1	6	4	3	2	3	1
2	7	5	3	1	5	1
3	74	54	4	3	6	1
4	72	52	4	1	6	1
5	71	51	5	5	7	1
	75	55	5	6	8	2
6	704	504	5	2	7	1
	714	514	5	3	8	1
7	742	542	6	11	9	1
8	742	542	6	5	9	1
9	7404	5404	6	1	9	1
	7434	5434	6	2	10	2
10	7406	5406	7	12	10	1
	7422	5422	7	13	11	2
11	7421	5421	7	5	11	1
	7435	5435	7	6	12	5
12	74044	54044	8	29	11	1
13	74042	54042	8	12	11	1
	74046	54046	8	17	13	2
14	74042	54042	8	6	11	1
	74047	54047	8	7	14	2
15	740414	540414	8	1	13	1
	740470	540470	8	3	14	2
16	740416	540416	9	18	14	1
	740462	540462	9	22	15	3
17	740415	540415	9	7	15	3
	740463	540463	9	9	16	2
18	7404244	5404244	9	3	15	1
	7404634	5404634	9	4	16^L	1
19	7404242	5404242	10	31	15	1
20	7404241	5404241	10	13	14^L	1
	7404155	5404155	10	18	18^L	2
21	74042404	54042404	10	4	15	1
	74041550	54041550	10	8	18^L	2
22	74041566	54041566	10	1	18	1
	74042436	54042436	10	8	19^L	2
23	74042417	54042417	11	27	18^L	1
	74041567	54041567	11	32	19^L	1

Note: L denotes that this number is actually d_{71} which is a lower bound on d_∞.

Source: Ref. 5: © 1975 IEEE; reprinted, with permission, from "Robustly Optimal Rate One-Half Binary Convolutional Codes", by R. Johannesson, from *IEEE Trans. Inf. Theory*, Vol. IT-21, July 1975, p. 465.

[a] A quick-look-in (QLI) rate-$\frac{1}{2}$ code is a nonsystematic convolutional code where the information bits can easily be retrieved by adding modulo-2 the output sequences of the two modulo-2 adders of the encoder.

Table A12.10 Optimal Distance Profile Quick-Look-In Codes with Rate $\frac{1}{2}$[a]

M	$G^{(1)}$	$G^{(2)}$	d_M	# paths
24	740424174	540424174	11	11
25	740415562	540415562	11	5
26	740424173	540424173	11	1
27	740424172	540424172	12	23
28	740424172	540424172	12	8
29	740424173	540424173	12	2
30	74042402074	54042402074	13	43
31	74042402072	54042402072	13	15
32	74042402071	54042402071	13	4
33	740424020714	540424020714	13	1
34	740424020712	540424020712	14	34
35	740424026637	540424026637	14	14
36	7404240266364	5404240266364	14	5
37	7404240266362	5404240266362	14	2
38	7404240207121	5404240207121	15	31
39	74042417136114	54042417136114	15	12
40	74042402071132	54042402071132	15	3
41	74042417136111	54042417136111	15	1
42	740424020712164	540424020712164	16	31
43	740424020712166	540424020712166	16	14
44	740424020713351	540424020713351	16	5
45	7404240207133514	5404240207133514	16	1
46	7404240207121636	5404240207121636	17	39
47	7404240207121635	5404240207121635	17	13
48	74042402071216354	54042402071216354	17	4
49	74042402071216356	54042402071216356	17	1
50	74042402071216357	54042402071216357	18	38

Source: Ref. 12: © 1976 IEEE; reprinted, with permission, from "Some Long Rate One-Half Binary Convolutional Codes with an Optimum Distance Profile", by R. Johannesson, from *IEEE Trans. Inf. Theory*, Vol. IT-22, Sept. 1976, p. 630.
[a]See Table A12.9 for an explanation of the QLI code.

Table A12.11 Optimal Distance Profile Systematic Codes with Rate $\frac{2}{3}$ [a]

ν	G_1^3	G_2^3	d_ν	#paths	d_∞	#paths
1	4	6	2	1	2	1
2	5	7	3	6	3	2
3	54	64	3	3	4	7
4	56	62	4	17	4	2
5	57	63	4	7	5	6
6	554	704	4	4	5	2
7	664	742	5	30	6	24
8	665	743	5	15	6	5
9	5734	6370	5	6	6	1
10	5736	6322	6	54	7	8
11	5736	6323	6	26	8	44
12	66414	74334	6	12	8	16
13	57372	63226	6	6	8	3
14	57371	63225	7	72	8	2
15	664150	743314	7	31	8	1
16	664072	743346	7	21	10	40
17	573713	632255	7	7	10	15
18	6640344	7431024	8	102	10	18
19	5514632	7023726	8	39	10	2
20	5514633	7023725	8	25	11	8
21	57361424	63235074	8	18	12	74
22	66415416	74311464	9	135	11	4
23	66415417	74311465	9	68	12	17

Source: Ref. 11: © 1978 IEEE; reprinted, with permission, from "Further Results on Binary Convolutional Codes with an Optimum Distance Profile," by R. Johannesson and E. Paaske, from *IEEE Trans. Inf. Theory*, Vol. IT-24, Mar. 1978, p. 266.

[a] The constraint length is $K = \nu + 1$.

Table A12.12 Optimum Distance Profile Nonsystematic Codes with Rate $\frac{2}{3}$ [a]

ν	G_1^1	G_1^2	G_1^3	G_2^1	G_2^2	G_2^3	Notes	d_ν	#paths	d_∞	#paths
3	6	2	4	1	4	7	1	3	2	4	1
4	6	3	7	1	5	5	1	4	17	5	7
5	60	30	70	34	74	40	1	4	7	6	9
6	50	24	54	24	70	54		4	2	6	1
7	54	30	64	00	46	66		5	30	7	6
8	64	12	52	26	66	44	1	5	15	8	8
9	54	16	66	25	71	60		5	9	8	1
10	53	23	51	36	53	67		6	54	9	9
11	710	260	670	320	404	714		6	29	10	29
12	740	260	520	367	414	515	2	6	27	10	4
13	710	260	670	140	545	533		6	5	11	9
14	676	046	704	256	470	442		7	65	12	58
15	722	054	642	302	457	435	2	7	38	12	25
16	7640	2460	7560	0724	5164	4260	3	7	14	12	7
17	5330	3250	5340	0600	7650	5434	3	7	7	13	18
18	6734	1734	4330	1574	5140	7014	3	8	106	14	?
19	5044	3570	4734	1024	5712	5622	3	8	43	14	?
20	7030	3452	7566	0012	6756	5100	3	8	23	14	?
21	6562	2316	4160	0431	4454	7225	3	8	11	15	?
22	57720	12140	63260	15244	70044	47730	3	9	144	16	?
23	51630	25240	42050	05460	61234	44334	3	9	60	16	?

Notes:
1. This code is OFD.
2. The search according to criterion 3) was not exhaustive, and hence a slightly better code might exist.
3. The search according to criterion 2) was not exhaustive, and hence a better code might exist.

Source: Ref. 11: © 1978 IEEE; reprinted, with permission, from "Further Results on Binary Convolutional Codes with an Optimum Distance Profile", by R. Johannesson and E. Paaske, from *IEEE Trans. Inf. Theory*, Vol. IT-24, Mar. 1978, p. 267.
[a] The constraint length is $K = \nu + 1$.

APPENDIX 12.2 DATA STRUCTURE OF A STACK DECODER

In this appendix we describe briefly the software implementation of the stack algorithm as we have programmed it in Fortran. It follows in essence the Z-J algorithm proposed by Jelinek [35].

An entry in the stack consists of the three items (metric value, node depth, encoder state) necessary for a possible further extension of a node, and of two pointers necessary for the stack ordering and path retrieval at the end of the decoding. This information is stored in five contiguous words of memory labelled, respectively, VALUE, DEPTH, STATE, STAKPT and PATHP. In addition, one auxiliary array called AUXPT is necessary to find the node to be extended next, as explained below.

Initially, all arrays are cleared, but the origin node (the first top node) is assigned an arbitrary positive metric value to avoid encountering negative total metric values during decoding. The information about new nodes is entered in the stack in consecutive addresses, and once entered, this information is never destroyed. All new entries whose accumulated metric, say Γ, yields the same integer part Q of Γ/H belong to the same substack Q. Moreover, the algorithm always keeps track of the highest substack Q_{top} ever reached.

In a given substack, say Q_m, all the nodes are linked in a continuous fashion by their pointers STAKPT. The STAKPT of the first node in Q_m contains 0, the STAKPT of the second entry contains the address of the first entry, and so on. The address of the last new entry in Q_m (i.e., the beginning of the chain) is contained in the Q_mth word of the auxiliary array AUXPT. Suppose that a new extension yielding metric value Γ and substack Q_m is about to be stored in the stack at address N. The metric value, state and depth are first properly stored in their respective registers in the stack at address N. Then the address contained in AUXPT (Q_m) is entered in STAKPT (N), and N is entered in AUXPT (Q_m). For any empty substack, Q_i, the corresponding contents of AUXPT (Q_i) is zero.

When a node is extended, its "elimination" from the stack consists simply in bypassing it in the chain of STAKPT pointers. Since the highest encountered substack Q_{top} is always known, to determine the top node of the stack, starting from Q_{top}, the first nonempty substack is found and, if necessary, scanned for the true maximum metric value. The chain of STAKPT pointers and the AUXPT array are then properly modified to exclude this top node. In the quantized Jelinek algorithm, no scanning is performed within the substacks and the top node is considered to be the last entry in the highest nonempty substack.

If the path extension cycle requires the extension of a number, say M nodes, these M nodes are extracted in sequence from the top node to the Mth, using the same technique.

As an example of entering and extracting nodes from the stack, let nodes stored at addresses 75, 90 and 100 have metric values 641, 647 and 645 respectively. Taking $H = 8$, the substack value corresponding to these metrics is

$641/8=80$; $647/8=80$ and $645/8=80$. The node at address 75 being the first entry and the node at address 100 being the last entry for this substack, the contents of the STAKPT pointers is then

$$STAKPT\ (75)=0$$
$$STAKPT\ (90)=75$$
$$STAKPT\ (100)=90$$

The starting point of this chain of pointers is the contents at address 80 of the auxiliary array AUXPT since the substack value is equal to 80. That is,

$$AUXPT\ (80)=100$$

Figure A12.1 gives a schematic description of this situation.

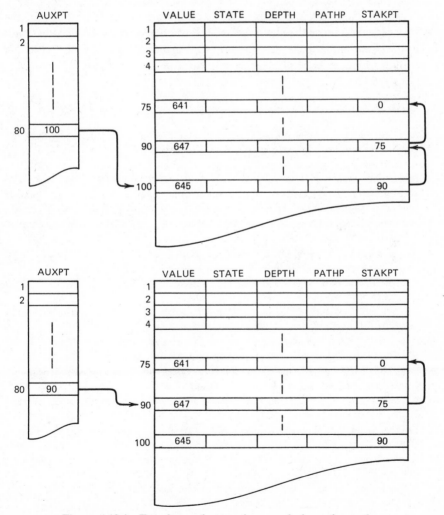

Figure A12.1   Entering and extracting a node from the stack.

Now suppose that a new node is to be extended and the highest substack is equal to 80. Following the last-in first-out procedure, we know that it is the node stored at address 100 in the stack. The program easily finds this node by looking at the contents of AUXPT (80). This node being extended, it must be "removed" from the stack. This "removal" from the stack is easily performed by bypassing this node in the chain of STAKPT pointers, that is, by setting the auxiliary array

$$\text{AUXPT }(80) = 90$$

With this simple procedure a node is never actually eliminated from the stack, but for the purpose of path extension it is in fact nonexistent.

When the top node reaches the end of the tree, the information bits on the finally decoded path must be recovered. This is accomplished by the chain of PATHP pointers, which are part of the data structure of the stack. For every new entry in the stack, the stack address of the parent node is stored in the PATHP register. Hence the chain of PATHP pointers specify the decoded path from the final node back to the origin. Further details on the software realization of the stack decoder may be found in Haccoun [29] or Geist [39].

REFERENCES

1   A. J. Viterbi, "Convolutional Codes and Their Performance in Communication Systems", *IEEE Trans. Commun. Technol.*, Vol. COM-19, Oct. 1971, pp. 751–772.

2   G. D. Forney, Jr., "Coding System Design for Advanced Solar Missions", Natl. Aeronaut. Space Adm., Final Rep., Contract NAS2-3637, Codex Corp., Watertown, Mass., Dec. 1967.

3   A. J. Viterbi, "Error Bounds for Convolutional Codes and an Asymptotically Optimum Decoding Algorithm", *IEEE Trans. Inf. Theory*, Vol. IT-13, Apr. 1967, pp. 260–269.

4   J. L. Massey, "Shift Register Synthesis and BCH Decoding", *IEEE Trans. Inf. Theory*, Vol. IT-15, Jan. 1969, pp. 122–127.

5   R. Johannesson, "Robustly Optimal Rate One-Half Binary Convolutional Codes", *IEEE Trans. Inf. Theory*, Vol. IT-21, July 1975, pp. 464–468.

6   P. Chevillat and D. Costello, "Distance and Computation in Sequential Decoding", *IEEE Trans. Commun.*, Vol. COM-24, Apr. 1976, pp. 440–447.

7   J. P. Odenwalder, "Optimal Decoding of Convolutional Codes", Ph. D. dissertation, Dept. of Elec. Eng., Univ. of Calif. at Los Angles, Jan. 1970.

8   K. J. Larsen, "Short Convolutional Codes with Maximal Free Distances for Rate $\frac{1}{2}$, $\frac{1}{3}$, and $\frac{1}{4}$", *IEEE Trans. Inf. Theory*, Vol. IT-19, May 1973, pp. 371–372.

9   L. R. Bahl and F. Jelinek, "Rate $\frac{1}{2}$ Convolutional Codes with Complementary Generators", *IEEE Trans. Inf. Theory*, Vol. IT-17, Nov. 1971, pp. 718–727.

10   E. Paaske, "Short Binary Convolutional Codes with Maximal Free Distances for Rates $\frac{2}{3}$ and $\frac{3}{4}$", *IEEE Trans. Inf. Theory*, Vol. IT-20, Sept. 1974, pp. 683–689.

11   R. Johannesson and E. Paaske, "Further Results on Binary Convolutional Codes with an Optimum Distance Profile", *IEEE Trans. Inf. Theory*, Vol. IT-24, Mar. 1978, pp. 264–268.

12   R. Johannesson, "Some Long Rate One-Half Binary Convolutional Codes with an Optimum Distance Profile", *IEEE Trans. Inf. Theory*, Vol. IT-22, Sept. 1976, pp. 629–631.

13   J. J. Bussgang, "Some Properties of Binary Convolutional Code Generators", *IEEE Trans. Inf. Theory*, Vol. IT-11, Jan. 1965, pp. 90–100.

14 J. L. Massey and D. J. Costello, Jr., "Nonsystematic Convolutional Codes for Sequential Decoding in Space Applications", *IEEE Trans. Commun. Technol.*, Vol. COM-19, Oct. 1971, pp. 806–813.

15 D. J. Costello, Jr., "A Construction Technique for Random-Error-Correcting Convolutional Codes", *IEEE Trans. Inf. Theory*, Vol. IT-15, Sept. 1969, pp. 631–636.

16 G. D. Forney, Jr., "Use of a Sequential Decoder to Analyze Convolutional Code Structure", *IEEE Trans. Inf. Theory*, Vol. IT-16, Nov. 1970, pp. 793–795.

17 J. M. Wozencraft and I. M. Jacobs, *Principles of Communication Engineering*, Wiley, New York, 1965.

18 R. G. Gallager, *Information Theory and Reliable Communication*, Wiley, New York, 1968.

19 J. K. Omura, "On the Viterbi Decoding Algorithm", *IEEE Trans. Inf. Theory*, Vol. IT-15, Jan. 1969, pp. 177–179.

20 J. A. Heller and I. M. Jacobs, "Viterbi Decoding for Satellite and Space Communication", *IEEE Trans. Commun. Technol.*, Vol. COM-19, Oct. 1971, pp. 835–848.

21 J. L. Massey, "Error Bounds for Tree Codes, Trellis Codes and Convolutional Codes with Encoding and Decoding Procedures", in *Coding and Complexity*, CISM Lecture Series No. 216, Springer-Verlag, New York, 1975.

22 J. Conan, "Implementation of Microprocessor-Based Viterbi Type Decoders", *Proc. Mini Microprocessors Symp.*, Zurich, June 1978, pp. 87–93.

23 G. D. Forney, Jr., "The Viterbi Algorithm", *Proc. IEEE*, Vol. 61, Mar. 1973, pp. 268–278.

24 J. G. Proakis, "Performance Capabilities of the Viterbi Algorithm for Combating Intersymbol Interference on Fading Multipath Channels", in *Proc. 1977 NATO Adv. Stud. Inst.*, Darlington, U.K..

25 I. M. Jacobs, "Practical Applications of Coding", *IEEE Trans. Inf. Theory*, Vol. IT-20, May 1974, pp. 305–310.

26 I. M. Jacobs, J. A. Heller *et al.*, "Performance Study of Viterbi Decoding as Related to Space Communications", Final Tech. Rep., Contract No. DAAB07-71-C-0148, prepared for the U.S. Army Satellite Commun. Agency, Linkabit Corp., San Diego, Calif., Aug. 1971.

27 L. B. Hofman and D. A. George, "Curriculum Sharing by Digital TV using HERMES", in *Proc. 20th Symp. R. Soc. Can.*, Vol. 1, Nov. 1977.

28 J. M. Wozencraft, "Sequential Decoding for Reliable Communication", Sc.D. thesis, Dept. of Elec. Eng., MIT, Cambridge, Mass., June 1957.

29 D. Haccoun, "Multiple-Path Stack Algorithms for Decoding Convolutional Codes", Ph.D. dissertation, Dept. of Elec. Eng., McGill Univ., Montréal, June 1974.

30 D. Haccoun and M. Ferguson, "Generalized Stack Algorithms for Decoding Convolutional Codes", *IEEE Trans. Inf. Theory*, Vol. IT-21, Nov. 1975, pp. 638–651.

31 J. L. Massey, *Threshold Decoding*, MIT Press, Cambridge, Mass., 1963.

32 I. M. Jacobs and E. R. Berlekamp, "A Lower Bound to the Distribution of Computation for Sequential Decoding", *IEEE Trans. Inf. Theory*, Vol. IT-13, Apr. 1967, pp. 167–174.

33 R. M. Fano, "A Heuristic Discussion of Probabilistic Decoding", *IEEE Trans. Inf. Theory*, Vol. IT-19, Apr. 1963, pp. 64–73.

34 J. L. Massey, "Variable-Length Codes and the Fano Metric", *IEEE Trans. Inf. Theory*, Vol. IT-18, Jan. 1972, pp. 196–198.

35 F. Jelinek, "A Fast Sequential Decoding Algorithm Using a Stack", *IBM J. Res. Dev.*, Vol. 13, Nov. 1969, pp. 675–685.

36 K. Zigangirov, "Some Sequential Decoding Procedures", *Probl. Peredachi Inf.*, Vol. 2, No. 4, 1966, pp. 13–25.

37 J. L. Massey, M. K. Sain and J. M. Geist, "Certain Infinite Markov Chains and Sequential Decoding", *Discrete Math.*, Vol. 3, Sept. 1972, pp. 163–175.

38 D. Haccoun, "A Markov Chain Analysis of the Sequential Decoding Metric", *IEEE Trans. Inf. Theory*, Vol. IT-26, Jan. 1980, pp. 109–113.

39 J. Geist, "Algorithmic Aspects of Sequential Decoding", Ph.D. dissertation, Dept. of Elec. Eng., Univ. of Notre Dame, Aug. 1970.

40 J. Geist, "An Empirical Comparison of Two Sequential Decoding Algorithms", *IEEE Trans. Commun. Technol.*, Vol. COM-19, Aug. 1971, pp. 415–419.

41 J. E. Savage, "The Computation Problem with Sequential Decoding", MIT Lincoln Lab. Tech. Rep. No. 371, Feb. 1965.

42 F. Jelinek, "An Upper Bound on Moments of Sequential Decoding Effort", *IEEE Trans. Inf. Theory*, Vol. IT-15, Jan. 1969, pp. 140–149.

43 G. D. Forney, Jr., "Convolutional Codes III: Sequential Decoding", *Inf. Control*, Vol. 25, July 1974, pp. 267–297.

44 D. Falconer, "A Hybrid Sequential and Algebraic Decoding Scheme", Ph.D. dissertation, Dept. of Elec. Eng., MIT, Cambridge, Mass., Feb. 1967.

45 G. D. Forney, Jr., and E. K. Bower, "A High-Speed Sequential Decoder, Prototype Design and Test", *IEEE Trans. Commun. Techol.*, Vol. COM-19, Oct. 1971, pp. 821–835.

46 D. Haccoun, "Simulated Communication with Sequential Decoding and Phase Estimation", S. M. thesis, Dept. of Elec. Eng., MIT, Cambridge, Mass., Sept. 1966.

47 J. A. Heller, "Sequential Decoding for Channels with Time Varying Phase", Ph.D. dissertation, Dept. of Elec. Eng., MIT, Cambridge, Mass., Aug. 1967.

48 D. Haccoun, "Branching Processes and Sequential Decoding", *1977 IEEE Int. Symp. Inf. Theory*, Ithaca, N.Y., Oct. 1977, p. 68.

49 R. Johannesson, "On the Distribution of Computation for Sequential Decoding Using the Stack Algorithm", *IEEE Trans. Inf. Theory*, Vol. IT-25, May 1980, pp. 323–331.

50 P. R. Chevillat and D. J. Costello, Jr., "An Analysis of Sequential Decoding for Specific Time-Invariant Convolutional Codes", *IEEE Trans. Inf. Theory*, Vol. IT-24, July 1978, pp. 443–451.

51 P. R. Chevillat and D. J. Costello, Jr., "A Multiple Stack Algorithm for Erasure Free Decoding of Convolutional Codes", *IEEE Trans. Commun.*, Vol. COM-25, Dec. 1977, pp. 1460–1470.

52 G. D. Forney, Jr., and R. M. Langelier, "A High-Speed Sequential Decoder for Satellite Communications", in *Proc. Int. Conf. Commun.*, Boulder, Colo., June 1969, pp. 33.9–33.17.

53 K. Gilhousen and D. R. Lumb, "A Very High Speed Hard Decision Sequential Decoder", in *Proc. Natl. Telecommun. Conf.*, 1972.

54 J. Hagenauer, "Sequential Decoding for Burst Error Channels in Communication Systems and Random Process Theory, in *Proc. 1977 NATO Adv. Stud. Inst.*, Darlington, U.K.

55 S. C. Lu, L. N. Lee and R. J. F. Fang, "An Integrated System for Secure and Reliable Communications over Noisy Channels", *COMSAT Tech. Rev.* Vol. 9, Spring 1979, pp. 49–60.

PROBLEMS

1 Write the encoding equation (12.14) for the encoder of Figure 12.2, and give the symbol by symbol diagram representation of this encoder as in Figure 12.5.

2 The rate-$\frac{3}{4}$ systematic code used in the INTELSAT SPADE system has the following generator polynomials:

$$G_1(D) = 1 + D^3 + D^{15} + D^{19}$$

$$G_2(D) = 1 + D^8 + D^{17} + D^{18}$$

$$G_3(D) = 1 + D^6 + D^{11} + D^{13}$$

Give a schematic diagram of the encoder. What is the constraint length of this code? What is the minimum distance d_{min}?

3 (a) Determine the generator matrix of the encoder of Figure 12.6 and find the encoded sequence corresponding to the input $U = (11, 01, 00, 11, 00)$.

(b) Show that the systematic rate-$\frac{2}{3}$ convolutional code

$$G_1^* = 101\ 001\ 000$$

$$G_2^* = 010\ 000\ 001$$

gives the same set of codewords of length-4 branches as does the code of part (a).

(c) The code above is fully specified by the codewords $X^{(1)}$ and $X^{(2)}$ corresponding to the information sequences $(100\ldots)$ and $(0100\ldots)$, respectively. Find the information sequences that would yield $X^{(1)}$ and $X^{(2)}$ when the code of part (a) is used.

4 Let the impulse response of a convolutional code by $X_I = (11, 10, 11, 00, \ldots)$.

(a) Expressing the encoder input $U = (110100\ldots)$ as a linear combination of the impulse input sequence $I = (1000\ldots)$, and using the linearity of the code, determine the output sequence of the encoder.

(b) Find the generator matrix $[G]$ of this code and evaluate by matrix multiplication the output coded sequence for the input of part (a).

(c) Using the polynomial representation of the code and the input sequence, determine the output polynomial corresponding to the input of part (a).

(d) Give a schematic diagram of this encoder and find its minimum distance d_{min}.

5 A constraint-length-3, rate-$\frac{1}{3}$, systematic convolutional code is described by the following connection vectors:

$$G_1 = 111, \qquad G_2 = 011, \qquad G_3 = 001$$

(a) Give a schematic diagram of the encoder and determine the generator matrix of the code.

(b) Draw the corresponding tree and trellis and show the path corresponding to the input sequence $U=(10110)$.

(c) Give the state diagram of this encoder and determine the generating function $T(D, B)$ of the code.

(d) Determine d_{min} and d_{free} for this code.

6 Specify a $K=3$ convolutional encoder that produces the coded tree shown in Figure P12.6. Extend the tree to level 4, and give the minimum distance for this code.

7 Consider the $K=5$, $r=\frac{1}{2}$ convolutional code defined by $G_1=11001$ and $G_2=11111$.

(a) Find the coded branch symbols pairs for the trellis corresponding to this code.

(b) Find the path yielding the minimum distance.

(c) Find the systematic convolutional code of constraint length 3 which would give the same set of codewords of length-5 branches as the original code.

8 Prove properties 1 and 2 in the subsection "Trellis Structure" of Section 12.1.

9 Find d_{free} and d_{min} for the rate-$\frac{1}{2}$ convolutional code specified by

$$G_1(D)=1+D+D^3, \qquad G_2(D)=1+D+D^2+D^3$$

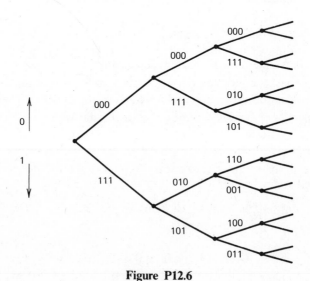

Figure P12.6

10 Explain why for the code of Figure 12.2, the paths corresponding to d_{free} are only three and four branches long, whereas the column distance function of the code indicates that d_{free} is reached for paths at least eight branches long.

11 (a) Use the Viterbi algorithm to determine the most likely transmitted sequence $X^{(U)}$ and the corresponding information sequence $(\hat{U})$, given the received sequence

$$Y = (11, 01, 00, 10, 00, 10, 01, 00, 00, 10, 00, 11, 00, 00, 00)$$

where the last two branches are tail branches. The code is the $K=3$, rate-$\frac{1}{2}$ code $G_1 = 111$, $G_2 = 101$; and assume the channel to be a BSC.

(b) What is the number of bit errors if the information sequence is

$$U = (0, \quad 0, \quad 1, \quad 1, \quad 1, \quad 1, \quad 0, \quad 1, \quad 1, \quad 0, \quad 1, \quad 0, \quad 0)$$

Discuss your results with respect to d_{free} of this code.

12 An additive white Gaussian noise channel is modelled as a discrete memoryless channel by the use of a modulator and a quantized receiver. The modulator generates 1 of A orthogonal signals of energy E_s and duration τ in successive intervals of length τ. For example, in the jth interval it generates one signal from the set $s_i(t-j\tau)$, $i=1,2,\ldots,A$. The receiver determines the output y_j for the jth interval by correlating the signal received during that interval with each of the A possible transmitted signals. It sets y_j equal to the number of the signal with the largest correlation if this correlation exceeds a given threshold value T; otherwise, it sets $y_j = 0$. Determine R_0 in bits per τ-second use of the channel.

13 Show that for d odd and for d even, the first-error event probability P_d of (12.55) can be upper bounded by (12.59), $P_d < 2^d [p(1-p)]^{d/2}$. [*Hint:* Observe that for $m > n$, $p^m < p^n$ and that $\sum_{i=0}^{d} \binom{d}{i} = 2^d$.]

14 (a) Determine the generating function $T(D, B)$ for the encoder of Figure 12.2. Identify the path(s) corresponding to d_{free}. Can you do the same for d_{min}? Explain.

(b) Evaluate the probability of first-error event $P(E)$, and the bit error probability for this code when used over a BSC with transition probability $p = 10^{-3}$, and over a Gaussian channel with bilateral noise spectral density $N_0/2$.

(c) What is the upper bound on the coding gain that can be achieved with this code?

15 This problem illustrates the problem of "catastrophic error propagation" which occurs with certain convolutional codes labelled "catastrophic".

(a) Draw the state diagram and attempt to determine the generating function $T(D)$ of the rate-$\frac{1}{2}$, $K=3$ convolutional code specified by $\mathbf{G}_1=110$ and $\mathbf{G}_2=101$. Explain why the attempt fails.

(b) Let the code used over a BSC with transition probability p. Suppose that the transmitted sequence is $\mathbf{X}=(000\ldots)$ and let the received sequence by $\mathbf{Y}=(1100000\ldots)$. Use the Viterbi algorithm to decode this sequence. What do you conclude?

(c) From observation of either the state diagram or the trellis, what is the property that makes this code propagate the errors, that is, to be catastrophic?

16 (a) Express as a function of p the probability assignments for the Fano branch metrics of a rate-$\frac{1}{2}$ code used over a BSC with transition probability p. Give the numerical results for $p=0.033$.

(b) Express as a function of p the possible Fano branch metrics and their probability assignments for a rate-$\frac{1}{3}$ code over a BSC with transition probability p. Give numerical results for $p=0.033$.

17 Consider some node U on the correct path and let its metric be Γ. Then a fundamental quantity related to its decoding is the metric difference Δ_j defined by $\Delta_j=\Gamma_j-\min_{i\geq j}\Gamma_i$.

(a) Show that $\Delta_j\geq 0$, $j=1,2,\ldots$, where equality holds only for breakout nodes.

(b) Verify that [37]

$$\Delta_j=\begin{cases}(\Delta_{j+1}-\gamma_{j+1}) & \text{if } (\Delta_{j+1}-\gamma_{j+1})\geq 0\\ 0 & \text{otherwise}\end{cases}$$

where $\Gamma_j=\Sigma_{i=0}^{j}\gamma_i$.

(c) Since the branch metrics γ are statistically independent random variables with a common distribution, the sequence $\Delta_{j+2},\Delta_{j+1},\Delta_j,\ldots$ induces a queuing process. This process is equivalent to a Markof chain where the states are the values of Δ_j and where the transition probabilities are given by

$$p_\ell=P\bigl(\gamma_j=-\ell\bigr)$$
$$q_k=P\bigl(\gamma_j=+k\bigr)$$

Give the transition probability matrix for the codes of Problem 16.

18 Show that over the ensemble of convolutional codes of rate r and for the general Q inputs J outputs DMC, the expected value of the symbol Fano metric is positive on the correct path, and negative on the incorrect paths.

19 Use the stack algorithm to perform the decoding of Problem 11. Compare your results (error performance and decoding effort) with those of Problem 11. Discuss.

20 Repeat Problem 19 by using the Fano algorithm. Compare your results with those of the stack algorithm.

21 (a) Devise a method whereby a stack decoder can recognize and exploit the remerging of the paths by keeping only the best path at a remerging node.

(b) Show that with such a modified stack algorithm, only very few computations will be repeated. Using typical received value trees, illustrate the duplicated computations that could and could not be eliminated with this algorithm.

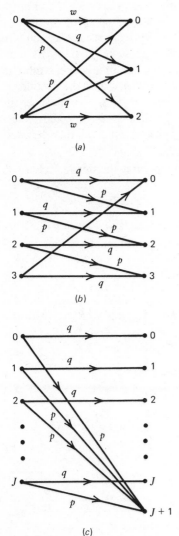

(a)

(b)

(c)

Figure P12.23 (a) $p+q+w=1$; (b) $p+q=1$; (c) $p+q=1$.

22 Consider the systematic $K=5$, rate-$\frac{1}{3}$ convolutional encoder specified by $G_1 = 10000$, $G_2 = 01110$, $G_3 = 00111$. Let the input sequence be $U = (10110)$, and let the corresponding coded sequence be $X^{(U)}$.

(a) Find the output coded sequence $X^{(U)}$.

(b) This coded sequence $X^{(U)}$ is transmitted over a BSC with transition probability p. The decoder observes the channel output sequence

$$Y = \left(y_1^{(1)} y_2^{(1)} y_3^{(1)}, \ y_1^{(2)} y_2^{(2)} y_3^{(2)}, \dots \right)$$

and operates as follows. The first symbol of each received branch is temporarily accepted as error-free when it is first received. Then based on that interim decision and the preceding ones, the decoder computes what it would expect to receive for the other symbols of the branch. If the sequence so computed differs too greatly from what was actually received, the decoder presumes it made a mistake, changes its temporary decision and tries again. Assume that the first coded symbol is received in error (i.e., $y_1^{(1)} = 0$). The interim decision of the decoder is thus $\hat{u}_1 = 0$. Assuming that no further symbol of $X^{(U)}$ is received in error, determine the 14 other symbols the decoder is expected to see. For equally likely inputs, what is the probability that this decoded sequence is correct?

(c) Assume the same received sequence except that the decoder changed its interim decision on $\hat{u}_1$ to a 1 and proceeded on. What would then be the probability that it has the correct sequence?

23 (a) Find R_{comp} for the discrete memoryless channels of Figure P12.23.

(b) What is the maximum average rate in bits per channel use that can be accommodated over the channel shown in Figure P12.23c with zero error probability when a noiseless feedback channel is available for retransmission requests?

24 Devise a graphical construction to determine R_{comp} and r/R_{comp} from the $E_0(\alpha)$ curves for $\alpha > 1$ and $\alpha < 1$.

CHAPTER 13

Error Detection and Correction Using Block Codes

In this chapter we discuss linear block codes, which were the first to be studied and have the best developed theory [1]. These codes can be used both to detect and correct errors in digital transmission. We first define the most significant algebraic structure necessary to understand the elementary properties of block codes. Further development in block codes depends upon the theory of finite fields. A brief treatment of finite fields is followed by a description of the most important block codes and their encoding and decoding procedures. Cyclic codes are treated in detail with emphasis on Bose–Chaudhuri–Hocquenghem (BCH) codes and the Reed–Solomon (RS) codes. For channels in which errors affect successive symbols independently, the BCH codes are the best codes. For burst channels the RS codes enjoy a similar position [2]. The chapter concludes with examples of applications of block codes in digital communications. The problems are carefully designed to enhance and augment the material presented in this chapter.

13.1 ALGEBRA OF CODING

A **group** G is a set of elements $a, b, c, \ldots$ for which an operation $*$ is defined and satisfies the following properties:

1 For a, b in G, $a*b$ is in G (axiom of closure).
2 For a, b, c in G, $(a*b)*c = a*(b*c)$ (associative law).
3 There is an element e in G such that for every a in G, $a*e = e*a = a$ (existence of an identity element).
4 For every a in G, there is an element a^{-1} in G such that $a*a^{-1} = a^{-1}*a = e$ (existence of an inverse).

In addition to the foregoing laws, a group may satisfy the commutative law: that is, $a*b = b*a$. Such a group is called **Abelian** or commutative. Some examples of group are:

1 G=set of all nonsingular $n \times n$ matrices with real entries.
  ∗ =matrix multiplication.

This group is not Abelian.

2 G=set of all real numbers
  ∗ =ordinary addition

This group is Abelian.

A **ring** R is a set of elements $a, b, c, \ldots$ for which two operations ∗ and ·
are defined and satisfies the following properties:

1 R is an Abelian group under ∗.
2 For a, b in R, $a \cdot b$ is in R (closure).
3 For a, b, c in R, $a \cdot (b \cdot c) = (a \cdot b) \cdot c$ (associative law).
4 For a, b, c in R, $a \cdot (b * c) = a \cdot b * a \cdot c$ and $(b * c) \cdot a = b \cdot a * c \cdot a$ (distributive
  law).

We may think of ∗ as addition and · as multiplication, even though these
operations may not be ordinary addition or multiplication of numbers.

A ring is called **commutative** if its multiplication operation is commutative;
that is, for a, b in R, $a \cdot b = b \cdot a$. Some examples of rings are:

1 R=set of all nonsingular $n \times n$ matrices with real entries
  ∗ =matrix addition
  · =matrix multiplication

This ring is noncommutative.

2 R=set of all real numbers
  ∗ =ordinary addition
  · =ordinary multiplication

This ring is commutative.

A **field** is a commutative ring with a unit element (multiplicative identity) in
which every nonzero element has a multiplicative inverse. Some examples of
fields are:

1 The set of all real numbers.
2 Binary field of two elements, with addition and multiplication.

Table 13.1 Modulo-5 Arithmetic

Addition						Multiplication					
+	0	1	2	3	4	·	0	1	2	3	4
0	0	1	2	3	4	0	0	0	0	0	0
1	1	2	3	4	0	1	0	1	2	3	4
2	2	3	4	0	1	2	0	2	4	1	3
3	3	4	0	1	2	3	0	3	1	4	2
4	4	0	1	2	3	4	0	4	3	2	1

A field with p elements can be formed by taking the integers modulo p, provided that p is a prime. (By integers modulo p we mean that we first divide the number in question by p and keep the remainder.)

In Table 13.1 we give addition and multiplication tables for a field with five elements ($p=5$ is a prime). We see that $4+4=8=3$ modulo 5, $3.3=9=4$ modulo 5 and so on.

Vector Spaces

A set V of elements is called a **vector space** over a field F if it satisfies the following axioms:

1 The set V is an Abelian group under addition.
2 For v in V and c in F, the product cv is in V.
3 For u,v in V and c in F, $c(u+v)=cu+cv$ (distributive law).
4 For v in V and c, d in F, $(c+d)v=cv+dv$ (distributive law).
5 For v in V and c, d in F, $(cd)v=c(dv)$ and $1v=v$ (associative law). (For convenience, we have omitted the symbol.)

Elements of F are called **scalars** and elements of V are called **vectors**.

An n-tuple is denoted as $(a_1, a_2, \ldots, a_n)$, where each a_i is an element of the field. Addition of n-tuples is defined as follows:

$$(a_1, a_2, \ldots, a_n) + (b_1, b_2, \ldots, b_n) = (a_1 + b_1, a_2 + b_2, \ldots, a_n + b_n)$$

Multiplication of an n-tuple by a field element is defined as

$$c(a_1, a_2, \ldots, a_n) = (ca_1, ca_2, \ldots, ca_n)$$

Multiplication of n tuples can be defined as follows:

$$(a_1, a_2, \ldots, a_n)(b_1, b_2, \ldots, b_n) = (a_1 b_1, a_2 b_2, \ldots, a_n b_n)$$

A linear combination of k vectors $v_1, v_2, \ldots, v_k$ is a sum of the form

$$u = a_1 v_1 + a_2 v_2 + \cdots + a_k v_k, \qquad a_i \text{ in } F \text{ (scalars)}$$

A set of vectors $v_1, v_2, \ldots, v_k$ is **linearly dependent** if and only if there are scalars $c_1, \ldots, c_k$, not all zero such that

$$c_1 v_1 + c_2 v_2 + \cdots + c_k v_k = 0$$

A set of vectors is **linearly independent** if it is not linearly dependent. A set of vectors is said to span a vector space if every vector in the vector space is a linear combination of the vectors in the set.

A subset of a vector space is called a **subspace** if it satisfies the axioms for a vector space. It is well known that the set of all linear combinations of a set of vectors $v_1, \ldots, v_k$ of a vector space V is a subspace of V [3].

In any space, the number of linearly independent vectors that span the space is called the dimension of the space. A set of k linearly independent vectors spanning a k-dimensional vector space is called a **basis**. If V is a k-dimensional vector space, any set of k linearly independent vectors in V is a basis of V [3].

An **inner product** or **dot product** of two n-tuples is a scalar and is defined as follows:

$$(a_1, \ldots, a_n) \cdot (b_1, \ldots, b_n) = a_1 b_1 + \cdots + a_n b_n$$

It is easily verified that $u \cdot v = v \cdot u$. If the inner product of two vectors is zero, they are said to be **orthogonal**.

Example 13.1

Consider the set of all 3-tuples over the binary field denoted as V_3:

$$v_1 = (0 \quad 0 \quad 0), \qquad v_5 = (1 \quad 0 \quad 0)$$
$$v_2 = (0 \quad 0 \quad 1), \qquad v_6 = (1 \quad 0 \quad 1)$$
$$v_3 = (0 \quad 1 \quad 0), \qquad v_7 = (1 \quad 1 \quad 0)$$
$$v_4 = (0 \quad 1 \quad 1), \qquad v_8 = (1 \quad 1 \quad 1)$$

Consider the following set of vectors from V_3:

$$v_2 = (0 \quad 0 \quad 1)$$
$$v_3 = (0 \quad 1 \quad 0)$$
$$v_5 = (1 \quad 0 \quad 0)$$

which are clearly independent. Also, any vector in V_3 can be written as a linear combination of this set of vectors. Hence the set $\{v_2, v_3, v_5\}$ span the vector space V_3. They constitute a basis for V_3. As concrete examples:

$$v_4 = (0 \quad 1 \quad 1) = v_2 + v_3$$

$$v_6 = (1 \quad 0 \quad 1) = v_2 + v_5$$

and so forth for any other vector in V_3.

Consider now the following set of vectors:

$$v_1 = (0 \quad 0 \quad 0)$$

$$v_3 = (0 \quad 1 \quad 0)$$

$$v_6 = (1 \quad 0 \quad 1)$$

$$v_8 = (1 \quad 1 \quad 1)$$

It is easy to verify that the set of 3-tuples above is a subspace of V_3.

An $n \times m$ matrix A is an ordered set of nm elements in a rectangular array of n rows and m columns:

$$A = \begin{bmatrix} a_{11} & a_{12} & \cdots & a_{1m} \\ a_{21} & a_{22} & \cdots & a_{2m} \\ \vdots & & & \\ a_{n1} & a_{n2} & \cdots & a_{nm} \end{bmatrix} = [a_{ij}]$$

The **row space** of an $n \times m$ matrix A is the set of all linear combinations of row vectors of A. They form a subspace of the vector space of m-tuples. Similarly, the set of all linear combinations of column vectors of the matrix form the **column space** whose dimension is called the **column rank**. It can be shown that row rank equals column rank; this value is referred to as rank of the matrix.

There is a set of **elementary row operations** defined for matrices:

1   Interchange of any two rows
2   Multiplication of any row by a nonzero field element
3   Addition of any multiple of one row to another

The following results can be readily established [3]:

R1   If one matrix is obtained from another by a succession of elementary row operations, both matrices have the same row space.

R2 The set of all n-tuples orthogonal to a subspace V_1 of n-tuples forms a subspace V_2 (called the null space) of n-tuples. Clearly, if a vector is orthogonal to every vector of a set that spans V_1, it is in the null space of V_1.

R3 If the dimension of a subspace of n-tuples is k, the dimension of the null space is $n-k$.

13.2 INTRODUCTION TO LINEAR BLOCK CODES

Next, we focus attention on linear block codes that are binary. A **linear block code** is a set of 2^k binary n-tuples. These n-tuples are called codewords of the code. The set of n-tuples over the binary field is a vector space. A set of these vectors of length n is called an (n, k) linear block code if and only if it is a k-dimensional subspace of the vector space of n-tuples. Here

$$k = \text{number of information digits}$$

$$n = \text{block length of the code}$$

$$n - k = \text{number of parity bits}$$

$$r = \frac{k}{n} = \text{code rate}$$

It is a common practice to encode the information blocks so that the first k bits of the codeword are identical to those of the information block and the last $n-k$ bits are parity bits. A code in this form is called a systematic code and is shown in Figure 13.1.

Some Fundamental Definitions

The **Hamming weight** of a vector $\mathbf{v}$, denoted $|\mathbf{v}|$, is defined to be the number of nonzero components of $\mathbf{v}$. For example, if $\mathbf{v} = (1 \quad 1 \quad 0 \quad 1 \quad 0 \quad 1)$, then $|\mathbf{v}| = 4$.

The **Hamming distance** between two vectors $\mathbf{v}_1$ and $\mathbf{v}_2$ denoted $d_H(\mathbf{v}_1, \mathbf{v}_2)$ is the number of positions in which they differ. For example, if

$$\mathbf{v}_1 = (1 \quad 1 \quad 0 \quad 1 \quad 0 \quad 1)$$

$$\mathbf{v}_2 = (1 \quad 1 \quad 1 \quad 0 \quad 0 \quad 0)$$

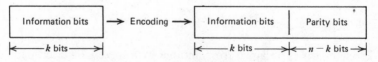

Figure 13.1 Systematic block code.

then $d_H(v_1,v_2)=3$. Clearly, $d_H(v_1,v_2)=|v_1+v_2|$; that is, the Hamming distance between v_1 and v_2 equals the Hamming weight of the sum of v_1+v_2. For the example given above, $v_1+v_2=001101$ and here $|v_1+v_2|=3=d_H(v_1,v_2)$. If v_1 and v_2 are both codewords of a linear block code, then v_1+v_2 must also be a codeword (since the set of all codewords is a vector space). Therefore, the distance between any two codewords equals the weight of some other codeword, and the **minimum distance** d_{min} for a linear block code equals the minimum weight of its nonzero codewords. This property is extremely helpful in analysing the error-correction capabilities of linear codes.

Example 13.2

Binary block code with $n=6$; the set of vectors (000000), (001101), (010011), (011110), (100110), (101011), (110101) and (111000) from a vector space V_1 of dimension three, and hence a $(6,3)$ linear block code. The minimum weight (of nonzero codewords) is 3, and hence the minimum distance is 3.

Error-Correction and Error-Detection Ability of Codes

It should be clear that a single error results in a Hamming distance 1 between transmitted codeword and the received codeword. If a code is used only for error detection and must detect all patterns of $d-1$ or fewer errors, it is necessary and sufficient for the minimum distance to be d. For if the minimum distance is d, no pattern of $d-1$ errors can change one codeword into another, whereas if the minimum distance is $d-1$ or less, there exists some pair of words at a distance less than d apart and then there is a pattern of fewer than d errors that will carry one codeword into another.

Similarly, it is possible to decode in such a way as to correct all patterns of t or fewer errors if and only if the minimum distance is at least $2t+1$. Then any received word with $t'\leq t$ errors differs from the transmitted codeword in t' symbols but from every other codeword in at least $2t+1-t'>t'$ symbols. On the other hand, if the minimum distance is less, there is at least one case where a t-fold error results in a received word at least as close to an incorrect codeword as to the transmitted codeword.

Example 13.3

The code in Example 13.2 has a minimum distance equal to three. Hence it is a single error correcting–double error detecting (SEC-DED) code.

From the preceding discussion, the basic problem is constructing block codes should be clear. For a specified codeword length n and code rate $(r=k/n)$, how do we generate a code with the largest possible d_{min}? There is no general answer, but coding theory does indicate when we can stop searching for a better code. This is in the form of Elias upper bound shown in Figure 13.2 [1]. In other words, for a given code rate $(r=k/n)$ we cannot find a code

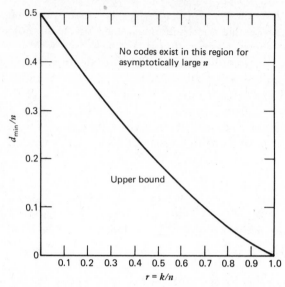

Figure 13.2 Elias upper bound on the minimum distance (the code length n is assumed to be very large).

for which the ratio d_{min}/n exceeds the Elias bound (for very large n). For example, consider a rate-$\frac{1}{2}$ code for which we must have $d_{min}/n < 0.2$ for large n. As a particular case, it is known that there is a binary $(128,64)$ code with d_{min} equal to 22. From the Elias upper bound, d_{min} for this code cannot be greater than $0.2 \times 128 = 26$. In general, the values of d_{min} achieved with known codes is less than what is theoretically achievable.

Description of Linear Block Codes by Matrices

Let $\mathbf{m} = (m_1, m_2, \ldots, m_k)$ be the k information bits and let $\mathbf{v} = (v_1, v_2, \ldots, v_n)$ be the encoded codeword. The "encoding rules" for an (n, k) code are conveniently expressed by the following matrix equation:

$$(v_1, v_2, \ldots, v_n) = (m_1, m_2, \ldots, m_k) \begin{bmatrix} g_{11} & g_{12} & \cdots & g_{1n} \\ g_{21} & g_{22} & \cdots & g_{2n} \\ \vdots & \vdots & & \vdots \\ g_{k1} & g_{k2} & \cdots & g_{kn} \end{bmatrix}$$

or more simply

$$\mathbf{v} = \mathbf{mG}$$

The matrix $\mathbf{G}$ is called **generator matrix** of the (n, k) code. Our requirement of 2^k distinct codewords is equivalent to the requirement that k rows of $\mathbf{G}$ be linearly independent, that is, that $\mathbf{G}$ be an $k \times n$ matrix with rank k. Moreover, it is obvious that every $k \times n$ matrix with rank k defines a valid (n, k) code. The code rate is $r = k/n$ and the set of codewords is just the row space of $\mathbf{G}$, that is, the set of all vectors formed as linear combinations of the rows of $\mathbf{G}$.

Alternative Description

Recall that if V is a subspace of dimension k, its null space is a vector space V^1 of dimension $n - k$. A matrix $\mathbf{H}$ of rank $n - k$ whose row space is V^1 can be made with a basis for V^1 as rows. Then V is the null space of V^1, and a vector $\mathbf{v}$ is in V if and only if it is orthogonal to every row of $\mathbf{H}$, that is, if and only if

$$\mathbf{v}\mathbf{H}^T = \mathbf{0} \tag{13.1}$$

If $\mathbf{H} = [h_{ij}]$, then (13.1) implies that for each row i of $\mathbf{H}$,

$$\sum_j v_j h_{ij} = 0 \tag{13.2}$$

that is, the components of $\mathbf{v}$ must satisfy a set of $n - k$ independent equations. [Of course, any linear combination of (13.2) also gives an equation that the components of $\mathbf{v}$ must satisfy, and this corresponds to the fact that $\mathbf{v}$ is orthogonal to every vector of V^1.] These equations are called **generalized parity checks**. The matrix $\mathbf{H}$ is called a **parity check matrix** of V.

Equation (13.1) holds for every vector in V. In particular, it holds for the k basis vectors of the matrix $\mathbf{G}$. These k equations can be expressed as

$$\mathbf{G}\mathbf{H}^T = \mathbf{0} \qquad \text{where } \mathbf{0} \text{ denotes the } k \times (n-k) \text{ all-zero matrix}$$

The row space of $\mathbf{H}$ is an $(n, n-k)$ linear block code V^1. V^1 is called the *dual code* of the (n, k) code V.

Example 13.4

For the code given in Example 13.2,

$$\mathbf{G} = \begin{bmatrix} 1 & 0 & 0 & 1 & 1 & 0 \\ 0 & 1 & 0 & 0 & 1 & 1 \\ 0 & 0 & 1 & 1 & 0 & 1 \end{bmatrix} \quad \text{and} \quad \mathbf{H} = \begin{bmatrix} 1 & 0 & 1 & 1 & 0 & 0 \\ 1 & 1 & 0 & 0 & 1 & 0 \\ 0 & 1 & 1 & 0 & 0 & 1 \end{bmatrix}$$

Note that $\mathbf{H}$ also generates a $(6, 3)$ code. This code is the dual of the $(6, 3)$ code generated by $\mathbf{G}$, and vice versa.

Systematic Block Codes

For channels in which errors occur independently, two codes that differ only in the arrangement of symbols have the same probability of error and in that sense are called **equivalent codes**. Thus performing column operations and elementary row operations on $\mathbf{G}$ results in $\mathbf{G}'$ which generate an equivalent code. Since $\mathbf{G}$ has rank k, by performing column permutations and elementary row operations, we can always put it in the form

$$\mathbf{G}'=[\mathbf{I}_k,\mathbf{P}], \qquad \mathbf{P}=[p_{ij}]$$

Consider an information block of k symbols $\mathbf{m}=[m_1, m_2,\ldots, m_k]$. The corresponding codeword $\mathbf{v}=[v_1, v_2,\ldots, v_n]$ is obtained as

$$\mathbf{v}=\mathbf{mG}'=\begin{bmatrix} 1 & 0 & \cdots & 0 & p_{1,1} & \cdots & p_{1,n-k} \\ 0 & 1 & \cdots & 0 & p_{2,1} & \cdots & p_{2,n-k} \\ \cdot & \cdot & \cdots & & \cdots & \cdots & \cdots \\ 0 & 0 & \cdots & 1 & p_{k,1} & \cdots & p_{k,n-k} \end{bmatrix}$$

Clearly, $v_i=m_i$, $i=1,2,\ldots, k$ (the code is in systematic form) and $v_{k+j}=p_{1,j}m_1+p_{2,j}m_2+\cdots+p_{k,j}m_k$ for $j=1,2,\ldots, n-k$. It can be shown (Problem 1) that every linear code is equivalent to a systematic code.

There is a simple way to find a parity check matrix for code if a generator matrix is in the systematic form, that is, $\mathbf{G}=[\mathbf{I}_k,\mathbf{P}]$ is given. If V is an (n, k) code generated by $\mathbf{G}=[\mathbf{I}_k,\mathbf{P}]$, where $\mathbf{I}_k$ is a $k\times k$ identity matrix and $\mathbf{P}$ is a $k\times(n-k)$ matrix, then V^1, the dual $(n, n-k)$ code, is generated by $\mathbf{H}=[\mathbf{P}^T,\mathbf{I}_{n-k}]$. The proof may be found in Problem 2.

Concept of the Syndrome

Recall that $\mathbf{v}$ is the codeword in the code generated by $\mathbf{G}$ if and only if

$$\mathbf{v}\mathbf{H}^T=\mathbf{0}$$

Here $\mathbf{0}=[0,0,\ldots,0]$ is a row of $n-k$ zeros. The received vector $\mathbf{r}$ is of the form

$$\mathbf{r}=\mathbf{v}+\mathbf{e}$$

where $\mathbf{e}$ is the noise introduced by the channel. The **syndrome** $\mathbf{s}$ is an $n-k$ component row vector defined as

$$\mathbf{s}=\mathbf{r}\mathbf{H}^T=(\mathbf{v}+\mathbf{e})\mathbf{H}^T=\mathbf{v}\mathbf{H}^T+\mathbf{e}\mathbf{H}^T=\mathbf{e}\mathbf{H}^T$$

Clearly, $\mathbf{s}=\mathbf{0}$ if and only if $\mathbf{e}=\mathbf{0}$ or $\mathbf{e}$ is a codeword itself. The fundamental importance of the syndrome stems from its use in error detection and correction.

Example 13.5

For the code in Example 13.4, let $\mathbf{r}=[1\ \ 0\ \ 1\ \ 1\ \ 1\ \ 0]$ be the received word. The syndrome is

$$\mathbf{s}=[1\ \ 0\ \ 1\ \ 1\ \ 1\ \ 0]\begin{bmatrix} 1 & 1 & 0 \\ 0 & 1 & 1 \\ 1 & 0 & 1 \\ 1 & 0 & 0 \\ 0 & 1 & 0 \\ 0 & 0 & 1 \end{bmatrix}=[1\ \ 0\ \ 1]$$

13.3 FINITE FIELDS

Building on the material of Section 13.1, we now present the minimum amount of the theory of finite fields necessary to understand cyclic codes. Results are stated as theorems without proof but with references and are illustrated by examples whenever possible. We denote a finite field with q elements, where $q=p^m$ is the power of a prime by GF(q), the **Galois field** of q elements.

Polynomials over a Finite Field

Let a_i be in GF(q), an expression

$$a_0+a_1x+\cdots+a_nx^n=a(x),\qquad a_n\neq 0$$

is called a **polynomial** of degree n over GF(q). We now state the Euclidean division algorithm for polynomials.

Theorem 13.1

For any polynomials $a(x)$ and $b(x)$ over GF(q) there exists unique polynomials $q(x)$ and $r(x)$ over GF(q) such that

$$a(x)=q(x)b(x)+r(x)$$

where the degree of $r(x)$ is less than the degree of $b(x)$.

Example 13.6

(a) Consider $a(x)=x^7+1$ and $b(x)=x^3+1$ over GF(2). Since $x^7+1=(x^4+x)(x^3+1)+(x+1)$, we see that $q(x)=x^4+x$ and $r(x)=x+1$.
(b) Consider $a(x)=x^7+1$ and $b(x)=x^3+x^2+1$ over GF(2). Since $x^7+1=(x^4+x^3+x^2+1)(x^3+x^2+1)$, we conclude that $q(x)=x^4+x^3+x^2+1$ and $r(x)=0$.

Using Example 13.6(b), we can define reducible and irreducible polynomials. A polynomial $p(x)$ of degree n over GF(q) is said to be **reducible** if it is divisible by some other polynomial of degree $k < n$ over GF(q); otherwise, $a(x)$ is said to be **irreducible** over the field GF(q).

Irreducible polynomials will be of great importance in the study of cyclic codes. In fact, just as we defined arithmetic modulo a prime number p (see Section 13.1), we can define operations modulo an irreducible polynomial. Let $r(x)$ be the unique remainder polynomial when the polynomial $a(x)$ is divided by an irreducible polynomial $p(x)$. Then by

$$a(x) = r(x) \text{ modulo } p(x)$$

we mean

$$a(x) = q(x)p(x) + r(x)$$

Example 13.7

Let $a(x) = x^5 + x^4 + x + 1$ and $p(x) = x^3 + x^2 + 1$ over GF(2). Then

$$(x^5 + x^4 + x + 1) = (x^2 + x + 1) \text{ modulo } (x^3 + x^2 + 1)$$

since

$$(x^5 + x^4 + x + 1) = x^2(x^3 + x^2 + 1) + (x^2 + x + 1)$$

Galois Fields GF(2^m)

Consider now a collection of polynomials of degree $m - 1$ or less over GF(2). We can generate this set of polynomials by performing (a) polynomial addition and (b) multiplication of polynomial modulo $p(x)$, where $p(x)$ is an irreducible polynomial of degree m over GF(2). The set of polynomials generated this way constitute a field of 2^m elements, which we designate GF(2^m).

Example 13.8

Let $p(x) = x^2 + x + 1$. This is an irreducible polynomial over GF(2). Considering the polynomial addition and multiplication [modulo $p(x)$] on $x + 1$ and $x^2 + 1$, we have

$$(x^2 + 1) + (x + 1) = (x^2 + x + 1) + 1 = 1 \qquad \left[\text{modulo } (x^2 + x + 1)\right]$$

$$(x^2 + 1)(x + 1) = x^3 + x^2 + x + 1 = 1 \qquad \left[\text{modulo } (x^2 + x + 1)\right]$$

Thus the elements of GF(2^2) may be represented as $1, x, x + 1, 0$.

Let α be in GF(2^m), and μ be the smallest integer such that

$$\alpha^\mu = 1$$

Then μ is called the **multiplicative order** of α. If $\mu = 2^m - 1$, then α is called a **primitive element** of GF(2^m).

Example 13.9

In Example 13.8, $\alpha_1 = x$ is a primitive element of GF(2^2), since $\mu = 2^2 - 1 = 3$. This can be seen as follows:

$$\alpha_1^2 = x^2 = x + 1 \qquad \left[\text{modulo } (x^2 + x + 1)\right]$$

$$\alpha_1^3 = \alpha_1 \cdot \alpha_1^2 = x \cdot (x + 1) = (x^2 + x) = 1 \qquad \left[\text{modulo } (x^2 + x + 1)\right]$$

It is an important point to observe that the powers of a primitive element of GF(2^m) will generate all nonzero elements of GF(2^m). Also, if $p(x)$ is an irreducible polynomial of degree m over GF(2) and α is a primitive element of GF(2^m) such that $p(\alpha) = 0$, then $p(x)$ is called a **primitive polynomial**.

The polynomial $m(x)$ of smallest degree with binary coefficients such that $m(\alpha) = 0$ is called the **minimum polynomial** of α. The minimum polynomial of α is irreducible.

We note the $\alpha, \alpha^2, \alpha^{2^2}, \ldots, \alpha^{2^l}$ are all roots of $m(x)$. Since $m(x)$ has finite degree, it must have a finite number of roots. Thus there must be a repetition in the sequence above. Let e be the degree of $m(x)$. It can be shown that $\alpha, \alpha^2, \ldots, \alpha^{2^{e-1}}$ are all the distinct roots of $m(x)$.

Example 13.10

With reference to Example 13.8, we have the following:

Minimum polynomial of 1: $m(x) = x + 1$.
Minimum polynomial of x: $m(x) = x^2 + x + 1$.
Minimum polynomial of $x + 1$: $m(x) = x^2 + x + 1$.

13.4 CYCLIC CODES

It is perhaps a remarkable fact that many of the important block codes found to date can be reformulated to be cyclic codes or closely related to cyclic codes. This class of codes can be easily encoded using linear shift registers with feedback. Further, because of their inherent algebraic structure, the decoding has been greatly simplified, both conceptually and in practice.

Fundamental Concepts of Cyclic Codes

For cyclic codes we shall number the n-tuples positions as 0 through $n-1$ rather than 1 through n. Thus we will let $\mathbf{v}=(v_0, v_1, \ldots, v_{n-1})$ denote a codeword. We shall also introduce a one-to-one correspondence between the codeword $\mathbf{v}$ and the code polynomial $v(x)$ of $\mathbf{v}$ as follows:

$$\mathbf{v}=(v_0, v_1, \ldots, v_{n-1}) \leftrightarrow v(x)=v_0+v_1 x+ \cdots +v_{n-1}x^{n-1}$$

A **cyclic code** is a linear code with the special property that if a vector is a codeword, so are all cyclic (end-around) shifts of the vector.

Example 13.11

The set of vectors (000), (011), (101) and (110) form a (3,2) cyclic code.

Let T denote the **cyclic shift** operation for n tuples, that is, $T\mathbf{v}= T(v_0, v_1, \ldots, v_{n-1}) = (v_{n-1}, v_1, \ldots, v_{n-2})$. It should be clear that $T^i(v_0, v_1, \ldots, v_{n-1})=(v_{n-i}, \ldots, v_{n-i-1})$ and represents the n-tuple obtained by shifting $\mathbf{v}$ cyclically i places.

We next investigate the polynomial operation corresponding to a cyclic shift. We have $T^i v(x)=v_{n-i} +v_{n-i+1}x+ \cdots +v_{n-1}x^{i-1} +v_0 x^i +v_1 x^{i+1} + \cdots +v_{n-i-1}x^{n-1}$. It can be shown that

$$x^i v(x)=q(x)(x^n+1)+T^i v(x)$$

that is, $x^i v(x)=T^i v(x)$ modulo $(x^n +1)$. Thus the ith shift of a codeword $\mathbf{v}$ is equivalent to multiplying the code polynomial $v(x)$ by x^i and reducing the expression $x^i v(x)$ modulo $(x^n +1)$.

Example 13.12

With reference to Example 13.11, $\mathbf{v}=(011) \leftrightarrow v(x)=x+x^2$. We have $T^2\mathbf{v}= (110)$. On the other hand, $x^2 v(x)=x^3 +x^4 =(1+x)(x^3 +1)+(1+x)$, that is, $x^2 v(x)=(1+x)$ modulo $(x^3 +1)$.

Let

$$\mathbf{m}=(m_0, m_1, \ldots, m_{k-1}) \leftrightarrow m(x)=m_0 +m_1 x+ \cdots +m_{k-1}x^{k-1}$$

denote the k information digits. The theory and practice of cyclic codes is based on the following important facts, which we summarise as theorems:

Theorem 13.2

An (n, k) binary cyclic code is said to be generated by a unique **generator polynomial**

$$g(x)=1+g_1x+g_2x^2+\cdots+g_{n-k-1}x^{n-k-1}+x^{n-k}; \qquad g_i=0 \text{ or } 1$$

such that code polynomial is a multiple of $g(x)$, that is,

$$v(x)=m(x)g(x)$$

Conversely, every polynomial of degree $n-1$ or less that is a multiple of $g(x)$ is a code polynomial [1].

Theorem 13.3

The generator polynomial $g(x)$ must divide x^n+1, that is,

$$x^n+1=g(x)h(x)$$

Here $h(x)$ is a polynomial of degree k and is referred to as the **parity polynomial** [1].

Example 13.13

Consider a $(7,3)$ cyclic code for which $g(x)=(1+x+x^2+x^4)$ is the generator polynomial. Since $x^7+1=(1+x+x^2+x^4)(1+x+x^3)$, we conclude that the parity polynomial is $h(x)=1+x+x^3$. The codewords are listed below.

Information	Code Polynomial	Codeword
(000)	$0 \cdot g(x)=0$	(0000000)
(100)	$1 \cdot g(x)=1+x+x^2+x^4$	(1110100)
(010)	$x \cdot g(x)=x+x^2+x^3+x^5$	(0111010)
(110)	$(1+x) \cdot g(x)=1+x^3+x^4+x^5$	(1001110)
(001)	$x^2 \cdot g(x)=x^2+x^3+x^4+x^6$	(0011101)
(101)	$(1+x^2) \cdot g(x)=1+x+x^3+x^6$	(1101001)
(011)	$(x+x^2) \cdot g(x)=x+x^4+x^5+x^6$	(0100111)
(111)	$(1+x+x^2) \cdot g(x)=1+x^2+x^5+x^6$	(1010011)

Systematic Cyclic Codes

In practical applications one uses systematic codes. In the procedure above (as illustrated in Example 13.13) a nonsystematic code is obtained. However, the following algorithm will yield a code in the systematic form:

1 For the information polynomial $m(x)$, form the product $x^{n-k}m(x)=m_0x^{n-k}+m_1x^{n-k+1}\cdots+m_{k-1}x^{n-1}$.

2   Divide $x^{n-k}m(x)$ by $g(x)$ to obtain

$$x^{n-k}m(x)=q(x)g(x)+p(x) \qquad (13.3)$$

where $p(x)$, the remainder, will have degree $n-k-1$ or less and thus can be written as

$$p(x)=p_0+p_1x+p_2x^2+ \cdots +p_{n-k-1}x^{n-k-1}$$

3   Rewrite (13.3) as

$$p(x)+x^{n-k}m(x)=q(x)g(x)$$

Since $p(x)+x^{n-k}m(x)$ is a polynomial of degree $n-1$ or less and is a multiple of the generator polynomial, it must be a codeword.

4   Writing out the polynomial $p(x)+x^{n-k}m(x)$ explicitly, we have

$$p(x)+x^{n-k}m(x)=p_0+p_1x+p_2x^2+ \cdots$$

$$+p_{n-k-1}x^{n-k-1}+m_0x^{n-k}+m_1x^{n-k-1}$$

$$+ \cdots +m_{k-1}x^{n-1}$$

This corresponds to the following codeword:

$$(p_0, p_1,..., p_{n-k-1}, m_0, m_1,..., m_{k-1})$$
$$|\leftarrow \text{parity digits} \rightarrow|\leftarrow \text{inform. digits} \rightarrow|$$

The codeword consists of $n-k$ parity check digits followed by k information digits and is clearly in the systematic form. Also note the difference in notation from that adopted in Section 13.2. This is due to the conventional practice when dealing with cyclic codes to identify the first $n-k$ digits as parity bits and the last k bits as the information bits.

Example 13.14

With reference to Example 13.12, let $\mathbf{m}=(101)$ be the information to be encoded in the systematic form. We have $m(x)=1+x^2$. Dividing $x^4m(x)$ by $g(x)=1+x+x^2+x^4$, we have

$$x^4+x^6=(x^2)(1+x+x^2+x^4)+x^2+x^3$$

that is, $p(x)=x^2+x^3$ and $q(x)=x^2$. The required codeword is $p(x)+x^{n-k}m(x)$, that is, $x^2+x^3+x^4+x^6$, and the corresponding codeword is (0011101), which is in the systematic form. All the codewords of this code in the systematic form are listed below.

Information	Codeword
(000)	(0000000)
(100)	(1110100)
(010)	(0111010)
(110)	(1001110)
(001)	(1101001)
(101)	(0011101)
(011)	(1010011)
(111)	(0100111)

To obtain the generator matrix for any cyclic code in the systematic form, we proceed as follows. Let

$$p_i(x)=p_{i,0}+p_{i,1}x+p_{i,2}x^2+\cdots+p_{i,n-k-1}x^{n-k-1}$$

be the remainder after dividing x^{n-k+i} by $g(x)$:

$$x^{n-k+i}=q_i(x)g(x)+p_i(x),\qquad i=0,1,\ldots,k-1$$

Clearly, $p_i(x)+x^{n-k+i}=q_i(x)g(x)$ is a code polynomial. If these polynomials, for $i=0,1,\ldots,k-1$ are taken as rows of the generator matrix, then

$$\mathbf{G}=\begin{bmatrix} p_{0,0} & p_{0,1} & \cdots & p_{0,n-k-1} & 1 & 0 & \cdots & 0 \\ p_{1,0} & p_{1,1} & \cdots & p_{1,n-k-1} & 0 & 1 & \cdots & 0 \\ & & & & & & & \\ p_{k-1,0} & p_{k-1,1} & \cdots & p_{k-1,n-k-1} & 0 & 0 & \cdots & 1 \end{bmatrix}=[\mathbf{P},\mathbf{I}_k]$$

From the above, we can obtain the parity check matrix

$$\mathbf{H}=\left[\mathbf{I}_{n-k},\mathbf{P}^T\right]$$

Example 13.15

For the $(7,3)$ code generated by

$$g(x)=1+x+x^2+x^4$$

$$x^4=(1)g(x)+1+x+x^2$$

$$x^5=(x)g(x)+x+x^2+x^3$$

$$x^6=(1+x^2)g(x)+1+x+x^3$$

and

$$G = \begin{bmatrix} 1 & 1 & 1 & 0 & 1 & 0 & 0 \\ 0 & 1 & 1 & 1 & 0 & 1 & 0 \\ 1 & 1 & 0 & 1 & 0 & 0 & 1 \end{bmatrix}$$

Encoding of Cyclic Codes

We have seen that the encoding of k bits involves computing the parity bits as the remainder $p(x)$ obtained by dividing $x^{n-k}m(x)$ by $g(x)$. The information bits are transmitted without alteration. This can be achieved by the circuit shown in Figure 13.3.

Encoding is accomplished as follows. With the gate turned on and the switch S in position 2, the k message digits are shifted into the register and simultaneously into the communication channel. As soon as the k information digits have entered the shift register, the $n-k$ digits in the register are the parity digits. The gate is then turned off, the switch S thrown to position 1 and the $n-k$ parity bits are shifted into the channel. This encoder requires an $(n-k)$-stage shift register and approximately $(n-k)/2$ modulo 2 adders.

Example 13.16

Consider again the $(7,3)$ binary cyclic code generated by $g(x)=1+x+x^2+x^4$. The encoding circuit is shown in Figure 13.4. Suppose that we wish to encode $\mathbf{m}=(0\ 1\ 0)$. As the information bits shift into register sequentially, the contents of the register are as follows:

	r_0	r_1	r_2	r_3
Initial state	0	0	0	0
First shift	0	0	0	0
Second shift	1	1	1	0
Third shift	0	1	1	1

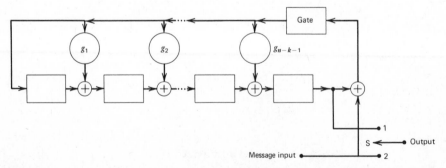

Figure 13.3 $(n-k)$-Stage shift register encoding circuit for the cyclic code generated by $g(x)=1+g_1x+g_2x^2+\cdots+g_{n-k-1}x^{n-k-1}$.

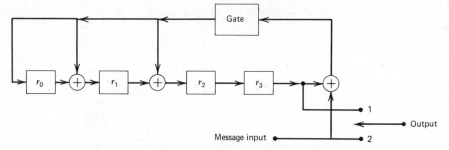

Figure 13.4 Encoder for the code of Example 13.16.

After three shifts, the contents of the register are (0 1 1 1), so that the encoded codeword is

$$(0 \quad \underbrace{1 \quad 1 \quad 1}_{\text{parity}} \quad \quad \underbrace{0 \quad 1 \quad 0}_{\text{inform.}})$$

The encoder shown in Figure 13.3 is used, in general, for higher-rate (say $r \geq \frac{1}{2}$) codes. For low-rate codes, a k-stage encoder, based on the parity check polynomial $h(x)$, is used. Such an encoder is shown in Figure 13.5. Here the k message bits are stored initially in the k-stage shift register. Now the register is shifted $(n-k)$ times to obtain the $(n-k)$ parity bits. This encoding circuit requires at most $(k-1)$ modulo 2 adders.

Example 13.17

For the $(7,3)$ code of Example 13.16, the parity check polynomial is

$$h(x) = \frac{x^7 + 1}{1 + x + x^2 + x^4} = 1 + x + x^3$$

The encoding circuit based on $h(x)$ is shown in Figure 13.6. Again we will encode the message $\mathbf{m} = (0 \quad 1 \quad 0)$. With s_0, s_1 and s_2 denoting the states of the

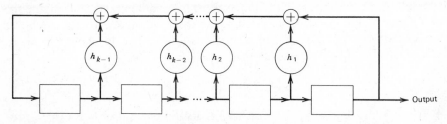

Figure 13.5 k-Stage shift-register encoding circuit for the cyclic code with parity polynomial $h(x) = h_0 + h_1 x + \cdots + h_k x^k$.

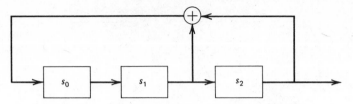

Figure 13.6 Encoder for the code of Example 13.17.

shift register, we have the following table:

	s_0	s_1	s_2	
Initial state	0	1	0	message bits
First shift	1	0	1	
Second shift	1	1	0	
Third shift	1	1	1	
Fourth shift	0	1	1	parity bits

Thus the encoded codeword is

$$(0 \quad 1 \quad 1 \quad 1 \qquad 0 \quad 1 \quad 0)$$
$$\underbrace{\qquad\qquad}_{\text{parity}} \qquad \underbrace{\qquad}_{\text{inform.}}$$

which is the same as before.

Syndrome Calculation and Error Detection with Cyclic Codes

Suppose that the received word in polynomial form is

$$r(x) = r_0 + r_1 x + r_2 x^2 + \cdots + r_{n-1} x^{n-1}$$

where according to our convention $r_0, r_1, \ldots, r_{n-k-1}$ are the received parity check bits and $r_{n-k}, \ldots, r_{n-1}$ are the received information bits. To compute the syndrome we must add the received parity check bits to the parity check bits calculated from the received information bits. If $e(x)$ denotes the error pattern introduced by the noisy channel, then

$$e(x) = q(x)g(x) + s(x) \tag{13.4}$$

If $s(x) = 0$, then the received word $r(x)$ is a valid codeword. If $s(x) \neq 0$, then errors have been detected. Further, as is evident from (13.4), the syndrome of the received word contains information about the error pattern, and thus can

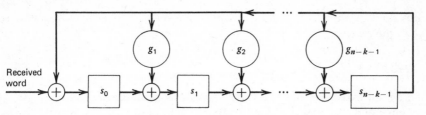

Figure 13.7 $(n-k)$-Stage shift-register syndrome calculator for cyclic code generated by $g(x)$.

be processed further to perform error correction. Thus error correction is substantially more involved than error detection.

The circuits of Figures 13.3 and 13.5 can be modified to perform syndrome calculation. In Figure 13.7, we show the syndrome calculator obtained by modifying Figure 13.3. The received word is shifted into the shift register with all stages initially set to zero. After the entire received word has entered the register, the contents will correspond to $s(x)$. *starting with highest order*

Example 13.18

We again consider the $(7, 3)$ binary, cyclic code generated by $g(x) = 1 + x + x^2 + x^4$. The syndrome calculator is shown in Figure 13.8. Suppose that we wish to compute the syndrome of the received word $(0\ 1\ 1\ 1\ 1\ 1\ 0)$. As the received bits shift into register sequentially, the contents of the register are as follows:

	s_0	s_1	s_2	s_3
Initial state	0	0	0	0
First shift	0	0	0	0
Second shift	1	0	0	0
Third shift	1	1	0	0
Fourth shift	1	1	1	0
Fifth shift	1	1	1	1
Sixth shift	0	0	0	1
Seventh shift	1	1	1	0

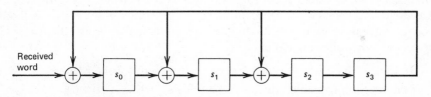

Figure 13.8 Syndrome calculator for the code of Example 13.18.

Thus $s(x) = 1 + x + x^2$, which is different from zero, and we have detected error (s) in the received word.

Listing of Cyclic Codes

There is an extensive listing of cyclic codes available in the literature. Chen [4] has tabulated n, k and d_{min} for all binary cyclic codes of odd length less than or equal to 65. This table has also been reproduced in Appendix D of Peterson and Weldon [5]. A similar listing was done by Promhouse and Tavares [6] for all binary cyclic codes of odd lengths from 69 to 99. MacWilliams and Sloane have provided tables of the best codes known, in Appendix A of Ref. 7. Many of the best cyclic codes can be found in their listing.

Modified Cyclic Codes

In many applications it is necessary to find values of n and k to accommodate required data format. Since cyclic codes exist only for certain values of n and k, it becomes necessary to modify them to fit the required format. We shall call such codes modified cyclic codes. For example, the (127, 112) BCH code (see Section 13.5) recommended in the INTELSAT V era [8] is first extended to a (128, 112) code, so that the parity and information bits are in multiples of four. Any (n, k) block code can be **extended** to an $(n+1, k)$ block code by adding an overall parity bit. Note that the weight of every codeword in such an extended code is even and if the original code has odd minimum distance, say $2t+1$, the extended code will have a minimum distance $2t+2$. Similarly, we can **shorten** the code by forcing a certain number, say k', of the leading information bits to obtain a $(n-k', k-k')$ shortened cyclic codes. The minimum distance of the shortened code will be at least the same as that of the original code. There are several other ways in which a code may be modified. For details, see Chapter 14 in Ref. 1.

13.5   BOSE–CHAUDHURI–HOCQUENGHEM (BCH) CODES

One of the most important class of linear codes discovered to date is that of BCH codes [1, 5, 7]. The BCH codes are the best constructive codes for channels in which errors affect successive symbols independently. These codes are cyclic and can be described in terms of generator polynomial, or more conveniently in terms of the roots of the generator polynomials. Binary BCH codes have the following parameters:

Block length: $n = 2^m - 1$, $m = 3, 4, 5, \ldots$ (or a factor of the number n)
Number of information digits: $k \geq n - mt$
Minimum distance: $d_{min} \geq 2t + 1$

The BCH codes are generated by polynomials of the form

$$g(x) = \text{LCM}(m_1(x), m_2(x), \ldots, m_{2t}(x)); \qquad \text{LCM} = \text{least common multiple}$$

which can be reduced (Problem 10) to

$$g(x) = \text{LCM}(m_1(x), m_3(x), \ldots, m_{2t-1}(x))$$

where $m_i(x)$ is the minimum polynomial of α^i and α is a primitive root of $GF(2^m)$. Clearly, $\alpha, \alpha^2, \alpha^3, \ldots, \alpha^{2t}$ are $2t$ consecutive roots of $g(x)$ and thus according to the **BCH bound**, the minimum distance $d \geq 2t+1$. (The BCH bound states that the minimum distance of the code generated by $g(x)$ must be greater than the largest number of consecutive roots. The bound applies to all cyclic codes and in the case of BCH codes is quite tight [1].)

Unfortunately, it is a somewhat difficult task to compute $g(x)$. However, the generator polynomials of BCH codes of practical interest have been tabulated in the literature [9]. See also Table 9.1 in Ref. 5 and Table 6.1 in Ref. 10.

Example 13.19

Consider what is now virtually a classic example of a BCH code. It was used in the original paper on BCH codes as well as in some standard textbooks [1, 5, 7].

Let $m = 4$ so that $n = 15$ and $t = 2$ and consider the BCH code generated by

$$g(x) = \text{LCM}(m_1(x)m_2(x)m_3(x)m_4(x)) = \text{LCM}(m_1(x)m_3(x))$$

so that $\alpha, \alpha^2, \alpha^3$ and α^4 are consecutive roots of $g(x)$. After carrying out the relevant computations or by looking up in Appendix C of Ref. 5, we find that

$$m_1(x) = 1 + x + x^4$$

and

$$m_3(x) = 1 + x + x^2 + x^3 + x^4$$

so that

$$g(x) = \text{LCM}(m_1(x)m_3(x)) = m_1(x)m_3(x) = 1 + x^4 + x^6 + x^7 + x^8$$

which then is the generator polynomial of a $(15,7)$ double-error-correcting BCH code.

Since BCH codes are cyclic, the encoding is fairly straightforward and can be done by using the circuits of Figures 13.3 and 13.5.

Decoding Binary BCH Codes

The decoding procedure of the t-error-correcting BCH codes was first given by Peterson [11]. This was modified and generalized by Zierler [12], Chien [13], Massey [14, 15], Berlekamp [16] and Burton [17]. Bartee and Schneider [18] constructed an electronic decoder for a binary five-error-correcting (127,92) BCH code based on the Peterson procedure. The circuitry consists of a 127-stage shift register for codeword storage, and a special-purpose digital computer for the error-correction procedure. Pehlert [19] designed a decoder to implement the Berlekamp procedure for a (144,96) shortened BCH code that can correct up to four errors per codeword. The design includes the use of several feedback registers per syndrome calculations and the technique for finding the error-locator polynomial involves time sharing a single $GF(2^8)$ multiplier.

These decoders have disadvantages in hardware complexity decoding delay, size and cost. On the other hand, a software implementation cannot achieve the speed of a hardware implementation but may be more economical in some cases.

Recently, it has been proposed that an optimum implementation of a BCH decoder might involve a combination of specially designed hardware for some operations and a software implementation on a computer for the remaining operations [20]. A microprocessor-based decoder can meet this requirement with many advantages.

The decoding of BCH codes consists of the following main steps:

Computing the Syndrome

Recall that the received word $r(x)$ can be written as the sum of the valid codeword $v(x)$ and an error polynomial $e(x)$:

$$r(x) = v(x) + e(x)$$

Since for $j = 0, 1, 2, \ldots, 2t-1$, the codeword $v(x)$ is a multiple of the minimum polynomial of the roots α^i, we have

$$r(\alpha^j) = \sum_{i=0}^{n-1} v_i(\alpha^j)^i + \sum_{i=0}^{n-1} e_i(\alpha^j)^i$$

$$= 0 + \sum_{i=0}^{n-1} e_i(\alpha^j)^i$$

Note that $e_i = 0$ or 1 and $e_i = 1$ denotes an error in the ith bit. If we consider at most t errors and denote them as $x_0, x_1, \ldots, x_{t-1}$, we have

$$r(\alpha^j) = \sum_{k=0}^{t-1} x_k^j = s_j$$

The s_j's are called power sums. Also, it can be shown that

$$[s_1, s_3, \ldots, s_{2t+1}] = \mathbf{r}\mathbf{H}^T$$

where the parity check matrix $\mathbf{H}$ is given by [5]

$$\mathbf{H} = \begin{bmatrix} 1 & \alpha_1 & \alpha_1^2 & \cdots & \alpha_1^{n-1} \\ 1 & \alpha_2 & \alpha_2^2 & \cdots & \alpha_2^{n-1} \\ \vdots & \vdots & \vdots & & \vdots \\ 1 & \alpha_{n-k} & \alpha_{n-k}^2 & \cdots & \alpha_{n-k}^{n-1} \end{bmatrix}$$

Computing the Error-Locator Polynomial

After the set of power sums is obtained, the next step to establish an error-locator polynomial $\sigma(x)$ with coefficients σ_k, $k = 1, 2, \ldots, t$, which are called elementary symmetric functions:

$$\sigma(x) = x^t + \sigma_1 x^{t-1} + \sigma_2 x^{t-2} + \cdots + \sigma_{t-1} x + \sigma_t$$

$$= \prod_{i=1}^{t} (x - \beta_t), \qquad \text{where } \beta_t\text{'s are the roots of } \sigma(x)$$

The elementary symmetric functions are obtained by solving a set of Newton's identities:

$$s_1 - \sigma_1 = 0$$

$$s_3 - \sigma_1 s_2 + \sigma_2 s_1 - 3\sigma_3 = 0$$

$$s_5 - \sigma_1 s_4 + \sigma_2 s_3 - \sigma_3 s_2 + \sigma_4 s_1 - 5\sigma_5 = 0$$

Originally, the Peterson's algorithm [11] was used as follows. Assume that there are t or $t-1$ errors, and the first t equations are solved. If the determinant is zero, the last two equations are dropped and the process is tried again to find σ_1. The process is repeated in this manner until a solution is found.

The Peterson's algorithm requires a large number of computations, even for fairly modest values of t. Berlekamp [16] developed an iterative algorithm avoiding the direct computation of the set of identities as shown below:

(a) Let $1 + s = 1 + s_1 x + s_2 x^2 + \cdots + s_{2t-1} x^{2t-1}$

(b) Set $\sigma(0) = \tau(0) = D(0) = 1$.

Proceed recursively as follows.

(c) If s_{2k+1} is unknown, then *stop*.

Otherwise,

(d) Define $\Delta(2k)$ as the coefficient of x^{2k+1} in the product polynomial $(1+s)\sigma(2k)$.

(e) Let $\sigma(2k+2)=\sigma(2k)+\Delta(2k)\tau(2k)x$.

(f) Let

$$\tau(2k+2)=\begin{cases} \dfrac{\sigma(2k)}{\Delta(2k)}x, & \text{if } \Delta(2k)\neq 0 \text{ and } D(2k)\leq k+1 \\[2ex] x^2\tau(2k), & \text{otherwise.} \end{cases}$$

The last polynomial $\sigma(2k)$ before a stop occurs is the error-locator polynomial of the codeword.

Error-Location Procedure

After the error-locator polynomial, $\sigma(x)$, is found, the reciprocal roots of $\sigma(x)$ can be found by the cyclic error location procedure developed by Chien [13] discussed below.

The error-locator polynomial can be written as

$$\sigma(x)=x^t+\sum_{i=1}^{t}\sigma_i x^{t-i}=\prod_{i=1}^{t}(x-\beta_i)$$

so that

$$\sigma_1=\sum_{j=1}^{t}\beta_j, \qquad \sigma_2=\sum_{\substack{j,k=1\\j<k}}^{t}\beta_j\beta_k, \qquad \sigma_3=\sum_{\substack{i,j,k=1\\i<j<k}}^{t}\beta_i\beta_j\beta_k\cdots \qquad (13.5)$$

If $\bar{\beta}_j=\alpha\beta_j$, $\bar{\beta}_j^{(\tau)}=\alpha^\tau\beta_j$, $j=1,2,\ldots,t$ are the set of transformed roots of

$$\bar{\sigma}(x)=x^t+\sum_{i=1}^{t}\bar{\sigma}_1 x^{t-i}=\prod_{i=0}^{t}(x-\bar{\beta}_i)$$

and

$$\bar{\sigma}^{(\tau)}(x)=x^t+\sum_{i=1}^{t}\bar{\sigma}_i^{(\tau)}x^{t-i}=\prod_{i=0}^{t}(x-\bar{\beta}_i^{(\tau)})$$

Then from (13.5) it can be shown that

$$\bar{\sigma}_1=\alpha^i\sigma_i \qquad \text{and} \qquad \bar{\sigma}_i^{(\tau)}=\alpha^{\tau i}\sigma_i$$

Assume that $\alpha = 1$ is the root of $\sigma(x)$,

$$\sigma(1) = 0 \quad \text{or} \quad \sum_{i=1}^{t} \sigma_i = 1$$

Thus, if

$$\sum_{i=1}^{t} \bar{\sigma}_i^{(\tau)} = 1$$

after $s_1, s_2, \ldots, s_t$ shifts the roots of $\sigma(x)$ are $\alpha^{n-\tau_1}, \alpha^{n-\tau_2}, \alpha^{n-\tau_3}, \ldots, \alpha^{n-\tau_t}$, which are the error locations.

To summarize, the decoding of BCH codes is done in a manner that computes directly the locations and values of the individual bit errors and consists of three main steps:

1 Compute the syndrome: $s = (s_1, s_2, \ldots, s_{n-k})$.
2 Calculate the so-called error-locator polynomial from the syndrome.
3 Solve the roots of the error-locator polynomial and correct the erroneous bits.

Step 2 is by far the most difficult part of BCH decoding in terms of analysis and explanation. However, from an implementation point of view, it is no more difficult than any of the other steps.

Steps 1 and 3 require t identical m-stage shift registers. In addition, step 3 requires m identical t-input modulo 2 adders and a m-input OR gate. Step 2 is best done on a software and requires approximately $2t^2$ additions and $2t^2$ multiplications. With a hardware implementation of steps 1 and 3 and software implementation of step 2, data rates of up to 200 kbps can be achieved. A complete software implementation [with step 1 requiring $(n-1)$ t additions and nt multiplications, step 3 requiring $n(t-1)$ multiplications and nt additions] will achieve data rates that are acceptable for HF and telephone data transmission [10].

Performance of BCH Codes

In a recent paper, Weng [21] has computed the performance of a number of BCH codes (with CPSK modulation) over the Gaussian channel with both soft- and hard-decision decoding. His results are reproduced in Figures 13.9 and 13.10. Recall that the soft-decision decoding (or channel measurement decoding [22]) offers up to 2 dB additional coding gain over the hard-decision decoding for the same code when employed over the Gaussian channel. However, the complexity of a hard-decision decoder is substantially less than

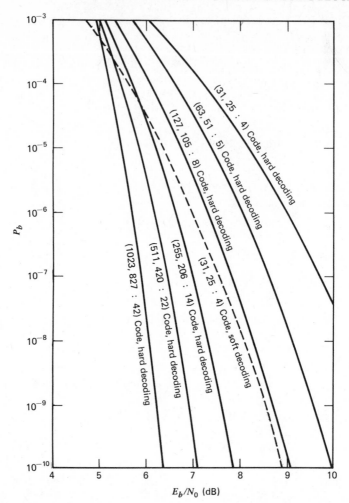

Figure 13.9 Performance of BCH codes with code rate $\geq \frac{25}{31}$ over the Gaussian channel. CPSK modulation. $(n, k: d_{min})$ code: n, block length; k, number of information digits; d_{min}, minimum distance. [From Ref. 21: ©1979 IEEE; reprinted, with permission, from "Soft and Hard Decoding Performance Comparisons for BCH Codes", by L.-J. Weng, from *Conf. Rec., 1979 Int. Conf. Commun.* (Vols. 1–4), 10–14 June 1979, Boston, Mass. (79CH1435-7CSCB).]

that of the corresponding soft-decision decoder. Further, for a given code rate, the decoded bit-error rate is known to decrease with increasing block length. Thus for a given code rate it is of interest to compare the required code lengths for the hard-decision decoding performance to be comparable to the soft-decision decoding. Several conclusions can be drawn from Figures 13.9 and 13.10 [21].

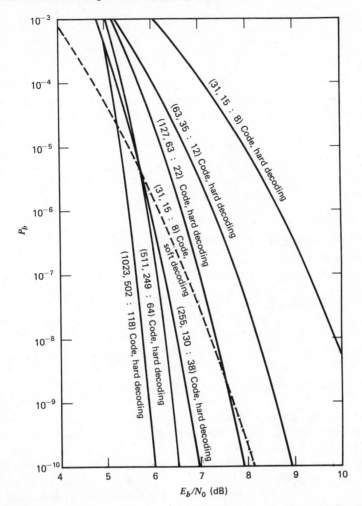

Figure 13.10 Performance of BCH codes with code rate $\geq \frac{15}{31}$ over the Gaussian channel. CPSK modulation. [From Ref. 21: © 1979 IEEE; reprinted from "Soft and Hard Decoding Performance Comparisons for BCH Codes", L.-J. Weng, from *Conf. Rec. 1979 Int. Conf. Commun.* (Vols. 1–4), 10–14 June 1979, Boston, Mass. (79CH1435-7CSCB).]

1 For a fixed code rate, a hard-decision decoded BCH code of length $4n$ or longer has a better performance than a soft-decision decoded BCH code of length n.

2 For low code rates, low decoded bit-error rates and limited decoding delay (associated with long codes), soft-decision decoded BCH codes of shorter length will prevail.

Weng's method of computing the performance of soft-decision decoding is applicable to any block code whose weight distribution is known [23].

13.6 REED–SOLOMON CODES

Historically, the discovery of RS codes preceded the BCH codes. The Reed–Solomon codes may be viewed as a subclass of the general BCH codes. This class of codes is useful principally for burst error correction. (We treat burst correction much more generally in Section 13.8.)

The Reed–Solomon codes are generated by polynomials of the form

$$g(x) = (x - \alpha)(x - \alpha^2) \cdots (x - \alpha^{2t})$$

for correcting t symbols (bytes). Here α is a primitive root of $GF(2^m)$. The parameters of an RS code are the following:

Symbol: m bits per symbol
Block length: $n = (2^m - 1)$ symbols $= m(2^m - 1)$ bits
Number of parity symbols: $(n - k) = 2t$ symbols $= m \cdot 2t$ bits
Minimum distance: $d_{min} = 2t + 1$

Recall that a symbol from $GF(2^m)$ can be represented by a binary m-tuple. From this observation we can deduce the multiple-burst-correcting capability of the RS codes; the code can correct t in-phase bursts of length m bits in each codeword. To see this, divide the received binary vector into $2^m - 1$ symbols of $GF(2^m)$. Thus if an error pattern affects mt or fewer bits, it affects t or fewer symbols of a t-error-correcting RS code.

Since $d_{min} = n - k + 1$ for the RS codes, they have the greatest minimum distance possible for any given rate. Such codes are also called **maximum-distance separable** [5]. The encoding for RS codes is simple since they are cyclic. For decoding, the same three steps used for decoding a binary BCH code are required; in addition, a fourth step involving calculation of t or fewer error values is also required. RS decoders are now commercially available, with development spurred by military tactical communications requirements. In fact, a powerful 100-Mbps RS decoder can be constructed on one or two inexpensive cards [2]. A 3.8- to 5.1-dB coding gain can be achieved without soft decisions with only a slight increase in bandwidth.

Additional coding gain can be achieved by using a concatenated coding system due to Forney [24]. The basic idea of concatenation is to factor the channel encoder and decoder in a way shown in Figure 13.11. By choosing an inner code (block or convolutional) randomly and taking an RS code as the outer code, lower decoding complexity and larger coding gains are possible

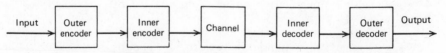

Figure 13.11 Concatenated coding system.

compared to an unfactored system. In summary, RS codes are extremely well suited for burst error correction and for use as outer codes in concatenated coding systems.

13.7 OTHER IMPORTANT CLASSES OF BLOCK CODES

In this section we discuss the characteristics of some important classes of block codes. Most of them are again cyclic (or related to cyclic) codes.

Golay Code

The Golay $(23, 12)$ code is a cyclic code generated by either

$$g_1(x) = 1 + x^2 + x^4 + x^5 + x^6 + x^{10} + x^{11}$$

or

$$g_2(x) = 1 + x + x^5 + x^6 + x^7 + x^9 + x^{11}$$

Note that

$$x^{23} + 1 = (1 + x)g_1(x)g_2(x)$$

This code is capable of correcting any combination of three or fewer random errors in a block of 23 digits. The encoding of the Golay code can be performed with an 11-stage feedback shift register with feedback connections selected according to the coefficients of either $g_1(x)$ or $g_2(x)$. From a coding theory point of view, the Golay code is the single most important error-correcting code and has inspired much research. Unfortunately, the code cannot be generalized to a family of codes.

There are several methods to decode the Golay code. Among these, the best perhaps is Kasami's modification of the error-trapping technique [25]. However, with new technologies, the "table-lookup" algorithm discussed in Problem 6 can be practical. (There are $2^{11} = 2048$ syndromes, which correspond one-to-one with all the 23-bit error patterns.)

Quasi-Cyclic Codes

The practical known codes in this class have rates $\frac{1}{2}$ and $\frac{2}{3}$ [26, 27]. A $(2m, m)$ rate-$\frac{1}{2}$ quasi-cyclic code is generated by a generator matrix of the form

$$G = [I_m, A] \tag{13.6}$$

where

$I_m = m \times m$ identity matrix

$$A = \begin{bmatrix} a_0 & a_1 & \cdots & a_{m-1} \\ a_{m-1} & a_0 & \cdots & a_{m-2} \\ a_1 & a_2 & \cdots & a_0 \end{bmatrix}$$ is an $m \times m$ binary circulant matrix

Clearly, to specify the code we only need to know the first row of the matrix A. Some good rate-$\frac{1}{2}$ quasi-cyclic codes are listed in Table 13.2. These codes were first described in Ref. 28.

If the matrix product

$$AA^T = I_m \text{ modulo } 2$$

the code generated by (13.6) is said to be a self-dual code [29, 30]. The coding gains for these codes are extremely attractive (in excess of 5 dB) and have been detailed in Ref. 23.

A $(3m, 2m)$ rate-$\frac{2}{3}$ code is specified by a generator matrix of the form

$$G = \begin{bmatrix} I_{2m} & \begin{matrix} A_1 \\ A_2 \end{matrix} \end{bmatrix} \tag{13.7}$$

where I_{2m} is an $2m \times 2m$ identity matrix and A_1, A_2 are $m \times m$ binary circulant matrices. Some good rate-$\frac{2}{3}$ quasi-cyclic codes are listed in Table 13.3.

Table 13.2 Parameters of Some $(2m, m)$ Quasi-cyclic Codes

$(2m, m)$	d_{min}	First Row of the Circulant A (in Octal)
(6,3)	3	3
(8,4)	4	7
(10,5)	4	7
(12,6)	4	43
(14,7)	4	414
(16,8)	5	426
(18,9)	6	362
(20,10)	6	5244
(22,11)	7	3426
(24,12)	8	4336
(26,13)	7	7442
(28,14)	8	76614
(30,15)	8	2167
(32,16)	8	40372
(34,17)	8	557
(36,18)	8	573
(38,19)	8	557
(40,20)	9	5723
(42,21)	10	145730

Table 13.3 Parameters of Some $(3m, 2m)$ Quasi-cyclic Codes

$(3m, 2m)$	d_{min}	First Row of the Circulants (in Octal)	
		A_1	A_2
$(12, 8)$	3	4	6
$(15, 10)$	4	54	7
$(18, 12)$	4	64	54
$(21, 14)$	4	52	62
$(24, 16)$	4	57	76
$(27, 18)$	4	75	76
$(30, 20)$	5	57	726
$(33, 22)$	6	4762	4574
$(36, 24)$	6	4566	4754
$(39, 26)$	6	657	7363
$(42, 28)$	6	30676	727
$(45, 30)$	6	7342	75
$(48, 32)$	6	66301	557
$(51, 34)$	6	557	2167
$(54, 36)$	7	400166	475271

Majority-Logic Decodable Codes

These are a class of codes that because of the special form of their parity check equations are majority-logic decodable. Majority-logic decoding is the simplest form of threshold decoding which is applicable to both block codes and convolutional codes. We discuss the decoding algorithm in a somewhat general framework in Chapter 14. Unfortunately, the construction of the majority-logic decodable cyclic codes rely heavily on the properties of finite geometries, a treatment of which is beyond the scope of this presentation. Thus we confine ourselves to stating the parameters of these codes, with the hope that the reader will be able to get some idea of their usefulness and performance in digital communications system. Also note that the codes discussed may also be decodable by other known decoding procedures.

Maximum-Length Codes

These are cyclic codes with the following parameters:

Block length: $n = 2^m - 1$

Number of information digits: $k = m$

Minimum distance: $d_{min} = 2^{m-1}$

The codes are generated by polynomials of the form

$$g(x) = \frac{x^n + 1}{h(x)}$$

where $h(x)$ is any primitive polynomial of degree m. The codes are called maximum-length codes because they can be encoded using an m-stage linear feedback shift registers which also generate the maximal-length sequences described in Section 9.2. The $(7, 3)$ cyclic code generated by $1 + x + x^2 + x^4$ and used extensively as an example in previous sections is also a maximum-length code with $m = 3$.

Hamming Codes

These are cyclic codes whose generator polynomial is a primitive polynomial of degree m. Thus these codes are the dual of maximum-length codes and thus related to m-sequences. Indeed, the polynomials listed in Table 9.1 can be used as the generator polynomials for Hamming codes. The code has the following parameters:

Block length: $n = 2^m - 1$

Number of information digits: $k = m$

Minimum distance: $d_{min} = 3$

Thus the codes can correct any single error in the block of length n digits. Syndrome decoding is especially suited for Hamming codes. In fact, the syndrome directly identifies the error location [1].

Difference-Set Codes

These are cyclic codes with the following parameters:

Block length: $n = 2^{2s} + 2^s + 1$; $s = 1, 2, \ldots$

Number of parity check digits: $n - k = 3^s + 1$

Minimum distance: $d_{min} = 2^s + 2$

These codes compare well with the best known cyclic codes. Unfortunately, there are relatively few codes with useful parameters in this class. A list of the first few codes is given in Table 13.4.

Table 13.4   Parameters of Some Binary Difference-Set Cyclic Codes

s	n	k	d_{min}
1	7	3	4
2	21	11	6
3	73	45	10
4	273	191	18
5	1057	813	34

Finite Projective Geometry Codes (PG Codes)

These are cyclic codes that contain difference-set codes and maximal-length codes as subclasses. These codes have the following parameters:

Block length: $n = \dfrac{2^{ms} - 1}{2^s - 1}$, $s = 1, 2, \ldots$; $m = 1, 2, \ldots$

Number of information digits: $k =$ extremely complicated expression [see (7.18) in Ref. 10].

Minimum distance: $d_{\min} = \dfrac{2^{(m-L)} - 1}{2^s + 1} + 1$, $L = 1, 2, \ldots$ and $L < M$

A 105-Mbps codec incorporating the (85,68) double-error-correcting PG code has been constructed [31]. A list of some finite projective geometry codes is given in Table 13.5.

Euclidean Geometry Codes (EG Codes)

These are cyclic codes with the following parameters:

Block length: $n = 2^{ms} - 1$, $s = 1, 2, \ldots$; $m = 1, 2, \ldots$

Number of information digits: $k =$ extremely complicated expression [see (7.10) in Ref. 10].

Minimum distance: $d_{\min} = \dfrac{2^{(m-L+1)s} - 1}{2^s - 1} - 1$, $L = 1, 2, \ldots$ and $L < m$

For moderate n, the error-correcting capability of EG codes is slightly inferior to BCH codes of comparable parameters. However, they are much easier to decode. In Table 13.6 we list the parameters of some EG codes.

Table 13.5 Parameters of Some Finite Projective Geometry Codes

n	k	$d_{\min}$	L
22	11	5	1
73	45	9	1
85	24	21	1
85	68	5	2
273	191	17	1
341	45	85	1
341	195	21	2
341	315	5	3
585	184	73	1
585	520	9	2
1057	813	33	1
1365	78	341	1

Table 13.6 Parameters of Some Euclidean Geometry Codes

n	k	d_{min}	L
63	13	21	1
63	37	9	1
255	19	85	1
255	127	21	2
255	175	17	1
255	231	5	3

Reed–Müller Codes

These are a subclass of EG codes ($s=1$). These codes are of practical interest and have the following parameters:

Block length: $n = 2^m - 1$, $m = 1, 2, \ldots$

Number of information digits: $k = \sum_{i=0}^{L-1} \binom{m}{i}$

Minimum distance: $d_{min} = 2^{m-L+1} - 1$

Certain extended Reed–Müller codes with parameters $(2^m, m+1)$ are extremely well suited for decoding by an algorithm that is based on fast Fourier transform (FFT) over the elementary Abelian group of order 2^m (see Chapter 14 in Ref. 7). The unique feature of this algorithm is that a soft-decision decoding is inherent in it. From 1969 to 1977 all of NASA's Mariner class deep-space probes were equipped with a (32,6) Reed–Müller code [32].

13.8 CODING TO COMBAT BURST ERROR

In previous sections we were concerned primarily with coding techniques for channels in which each transmitted digit is affected independently by noise. This section discusses methods for dealing with a burst of noise. In certain communication channels errors are due primarily to impulsive noise having large amplitude and relatively short duration. At very high transmission rates, even a short noise pulse can affect a large number of bits, causing errors to cluster in bursts. Therefore, it is desirable to design coding schemes to combat burst errors. A burst of length b is a vector whose nonzero components are confined to b consecutive digit positions, the first and last of which are nonzero. Thus the vector

$$\mathbf{v} = (00111001100)$$

is a burst of length 7. The following theorem relates code redundancy to the ability of the code to detect a single burst of length b or less [10].

Theorem 13.4

For detecting any burst of length b or less with a linear code of length n, the necessary and sufficient number of parity check digits is b.

The following theorem relates code redundancy to the ability of the code to correct a single burst of length b or less [10].

Theorem 13.5

For correcting any burst of length b or less with a code of length n, the necessary and sufficient number of parity check digits is $2b$.

We now present several coding schemes to combat burst errors.

Fire Codes

Fire codes constitute the most versatile class of codes for correcting single bursts. These codes are generated by the generator polynomials of the form

$$g(x) = (x^c - 1)p(x)$$

where $p(x)$ is an irreducible polynomial of degree m. This class of codes corrects all burst of length $\leq b$ and detects all burst of length $\leq d$ provided that (1) $c \geq d + b - 1$, (2) $m \geq b$, and (3) the period of $p(x)$ denoted e, does not divide c.

The Fire code can be decoded by very simple logic circuitry. These codes are also very efficient in terms of redundancy. The length of this code is equal to the least common multiple of e and c, that is, $n = \text{LCM}(e, c)$.

Example 13.20

Let $g(x) = (x^{16} - 1)(x^7 + x^2 + 1)$; thus $m = 7$, $c = 16$ and it can be verified that $e = 127$. Hence $n = 16 \times 127 = 2032$ and $n - k = 7 + 16 = 23$, so that $k = 2032 - 23 = 2009$. Thus we have a $(2032, 2009)$ Fire code capable of correcting any burst of length 7 or less. Note the high code rate (0.99) and the ease of implementation, requiring only $n - k = 23$ stages in the encoder shift register. The only obvious disadvantage is that very long codes are required for modest burst-correcting capability.

Fire codes are extremely well suited for high-speed burst correction and detection. For details, the reader is referred to Ref. 33.

Interleaved Block Codes

Interleaving can best be explained by examining the interleaved encoder shown in Figure 13.12. We assume that an (n, k) block code is the code being interleaved. As shown in the figure, the input message stream is separated into a number of parallel streams. The encoded codewords are then interleaved for serial transmission over the channel. At the receiver the stream is de-interleaved and decoded separately according to appropriate algorithm. The decoder outputs are recombined to give the original message stream. The parameter i is called the **degree of interleaving**. The success of interleaving is based on the fact that if a burst of errors spans no more than $i \times t$ digits of the transmitted sequence, it can only affect t digits of any given codeword. Hence by using a code with a moderate error-correcting capability, it is possible to correct very long bursts of errors by using a sufficiently large value of i.

The obvious way to implement the codec is in terms of i identical encoders and i identical decoders, as shown in Figure 13.12. The simplest implementation results when the code interleaved is an (n, k) cyclic code. This is due to the fact that an (n, k) cyclic code with generator polynomial $g(x)$ interleaved i times results in an (ni, ki) cyclic code with generator polynomial $g(x^i)$.

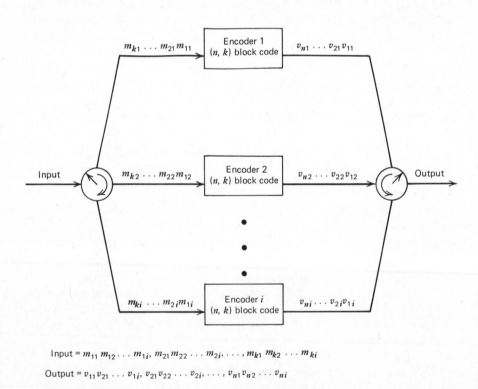

Input $= m_{11} m_{12} \cdots m_{1i}, m_{21} m_{22} \cdots m_{2i}, \ldots, m_{k1} m_{k2} \cdots m_{ki}$

Output $= v_{11} v_{21} \cdots v_{1i}, v_{21} v_{22} \cdots v_{2i}, \ldots, v_{n1} v_{n2} \cdots v_{ni}$

Figure 13.12 Interleaved encoder.

Further, the decoder of the interleaved code can be derived from the decoder of the (n, k) code simply by replacing each register stage of the original decoder by i stages without changing the other connections. Therefore, if the decoder of the original code is simple, so is the decoder for the interleaved code. Also, for cyclic codes, an error pattern with errors confined to the first j high-order positions and $(b-j)$ low-order positions are considered as a burst of length b. This is called an end-around burst.

The interleaved block coding to combat burst errors has the advantage of "spreading out" the effects of noise bursts by distributing over separately coded data streams. Further, it is easy to implement. The disadvantage of this scheme is that it does not fully exploit the dependence of errors ("memory") in bursts. Interleaving is an information-destroying process which increases the entropy of the noise. Hence an interleaved coding system will often perform inferiorly on a burst channel compared to a noninterleaved coding system specifically designed to combat burst errors [34].

Reed–Solomon Codes Revisited for Burst Error Correction

In Section 13.6 it was pointed out how RS codes can be used for burst error correction. The key point is to view each element of $GF(2^m)$ as a binary m-tuple.

Reed–Solomon codes compare very favorably, from the standpoint of code efficiency and burst-correcting power, with the Fire codes. The decoding is complicated since it requires a special-purpose computer to carry out Galois field manipulations. However, recent advances in theory as well as technology have resulted in substantial reductions in decoder complexity [35]. Hardware for Reed–Solomon codes is now commercially available.

Cyclic Product Codes

Cyclic product codes [36, 37] are a small subclass of the class of cyclic codes. These codes are capable of correcting both random and burst errors. Unfortunately, these codes either have low rates or are very long. Details of encoding and decoding such codes may be found in Ref. 10. See also Problem 16.

We have by no means covered the entire spectrum of coding to combat burst error. The codes we have described, however, do represent the most basic type of block codes for correcting burst errors.

13.9 SOFT-DECISION DECODING OF BLOCK CODES

We have noted in Chapter 12 and Section 13.5 that soft-decision decoding offers a substantial increase in performance (approximately 2 dB) over the hard-decision decoding for the same code when employed over the Gaussian channel. In non-Gaussian noise the improvement is even more [38]. This improvement in performance is due to the extended error-correcting capabilities obtained when soft-decision information is used effectively. Until recently soft-decision decoding was much easier to use for convolutional codes than for block codes. With new developments in semiconductor devices, soft-decision decoding of block codes is both possible and practical. This section presents a brief survey of soft-decision techniques proposed for various block codes [39]. Such techniques can also be used for ARQ systems [40].

The "generalized-minimum-distance" (GMD) decoding as suggested by Forney [41] resulted in the application of soft-decision decoding techniques to numerous important classes of block codes [22, 42, 43]. Among these codes are the product codes, iterated and concatenated coding systems. Weldon [44] introduced a soft-decision technique that is applicable to any block code with a known decoding procedure. His method has been extended to include burst errors as well [43, 45]. Hard-decision threshold decoders are described in Chapter 14. Analog soft-decision threshold decoding [46] combines the benefit of soft decisions with the simple structure of the binary quantized threshold decoder. An important technique for codes with high rate has been described in Ref. 47. The technique may be called soft-decision dual-code domain and minimises the symbol error probability. More generally, it has been shown [48] that any linear block code can be soft-decision-decoded using the Viterbi algorithm described in Section 12.4 [48]. Finally, it may be noted that for an (n, k) block code with minimum distance d, a maximum likelihood decoder consisting of 2^k matched filters can always be considered [49]. With the availability of cheap integrated circuits and microprocessors, such an implementation should be possible for a large number of codes. From the preceding discussion, we see that soft-decision decoding techniques have been found for a wide variety of block codes.

13.10   APPLICATIONS OF BLOCK CODING TO DIGITAL COMMUNICATIONS BY SATELLITES

Most of the previous applications of forward error correction techniques to digital communications by satellites have involved convolutional codes. This was mainly because of difficulties in implementing decoders for powerful block codes. With the recent developments in IC technology, the situation has changed. In fact, from the viewpoint of applying FEC techniques to a TDMA system, the block code is superior to the convolutional code. The primary reason is that the convolutional code needs overhead bits to terminate encoding/decoding at the end of each burst. Furthermore, for certain high-speed TDMA applications BCH codes have been found to yield greater coding improvements [8, 50].

In this section we present examples of the application of error control coding techniques (using block codes) to digital communications by satellites. Since, in general, there is no systematic procedure for selecting codes for a particular application, we first present a discussion on how to select codes for error control. We then present examples and summaries of the tradeoffs for proposed or already implemented systems. Since 1960 the cost of digital electronics have been decreasing at a phenomenal rate. Thus conclusions on factors such as speed, cost, weight or power based on the technology of the 1960s may with new technology be either irrelevant or no longer valid. This coupled with improvements in decoding algorithms has made sophisticated block coding systems increasingly attractive from an economic viewpoint [2].

How to Select Codes for Error Control

The first step is to determine user requirements. These may include all or some of the following system parameters [51, 52]:

1 Decoded bit-error rate
2 Data rate in bits per second which must be transmitted by the system
3 Available ratio of energy per bit to noise power density, E_b/N_0
4 Communication channel characteristics—random errors, burst error or a combination of both
5 Processing (decoding speed) compared to data rate, possible need for buffering
6 Decoder implementation (hardware versus software, general- versus special-purpose computer)
7 Availability and characteristics of a feedback link
8 Required information throughput rate
9 Decoder cost and complexity
10 In case of jamming, jamming-to-signal ratio, duty cycle, frequency coverage and so on

Based on the foregoing considerations, we select one of the three basic error-control techniques listed below:

1 Error detection and retransmission (ARQ) (this also includes employing error detection without retransmission; i.e., the user may only require that erroneous blocks be identified).
2 Forward error-correction technique (FEC), block or convolutional.
3 A hybrid, FEC and ARQ, with joint correction and detection with retransmission.

Having selected an error-control scheme, we proceed to choose all candidate codes that can be employed with this scheme.

The next step is to estimate the performance of various codes, analytically or otherwise, using channel error statistics and determine suitable range for the code parameters. This may conclude the code selection procedure in some cases unless more accuracy is required.

Finally, if necessary, simulate the encoder, channel and decoder to obtain accurate performance estimations. Select the coding scheme with parameter ranges optimized that meets user requirements having the least decoder cost and complexity.

By performing the above-mentioned steps for a large number of techniques, it is possible to select codes for error control.

BCH Codes for the INTELSAT TDMA System

The INTELSAT V satellite of the 1980s will utilize frequencies in the 14/12-GHz bands, as well as those in the 6/4-GHz transmission bands. Investigations by the researchers at KDD and COMSAT [8, 50] of the 120-Mbps TDMA system performance using QPSK modulation over an INTELSAT V satellite resulted in the choice of a (128,112) double error correcting–triple error detecting extended BCH code to maximize data throughput at a given carrier power, while maintaining bit-error-rate objectives (of $<10^{-3}$ for the interference-limited cases and for the thermal-noise-limited cases).

Laboratory tests were conducted to determine the error statistics of a QPSK signal transmitted through simulated TDMA satellite links using actual hardware. The tests indicated that error patterns caused by noise and co-channel interference were sufficiently random to justify the use of random-error-correcting codes as opposed to burst-error-correcting codes.

The authors then evaluated the performances of a large number of block and convolutional random-error-correcting codes. Their study covered all available coding/decoding techniques. Soft decision decoding methods were not considered since the resulting hardware is very complicated and in addition inappropriate for high-speed operation. Viterbi decoding and sequential decoding are known to be very powerful in the decoding of convolutional codes; however, they are only practical for low code rates.

After these considerations the (127,112) BCH code was selected. The code is generated by

$$g(x) = 1 + x^2 + x^7 + x^{10} + x^{11} + x^{12} + x^{13} + x^{14} + x^{15}$$

$$= (1+x)(1 + x^2 + x^3 + x^4 + x^5 + x^6 + x^{10} + x^{12} + x^{14})$$

This was then extended to a (128,112) code to provide system compatibility since all timing controlled by the MUX/DEMUX is a multiple of four symbols in the INTELSAT TDMA system. The preferred point of FEC application is between the MUX/DEMUX and the scrambler/descrambler [50]. This enables the shared use of a single codec for all bursts.

Encoding

Figure 13.13 illustrates the data burst structure using (128, 112) extended BCH code. Each burst is encoded as a whole by segmenting it into blocks of 112 bits. A shorter block is encoded by assuming hypothetical zeros. The shortened block also contains the normal complement of 15 parity bits plus one dummy bit to fill the parity symbol time. Any polarity is acceptable for the dummy bit. Encoding is accomplished by modifying the circuit shown in Figure 13.3.

Decoding

In principle, the code can be decoded using the decoding procedure outlined in Section 13.5. However, due to its high-speed requirement, the 120-Mbps decoder calculates a syndrome pattern and employs a read-only memory (ROM) table lookup to determine error location. Recall that each syndrome pattern corresponds to a distinct error pattern. Since the feedback from the decoder output is not needed, the bit-error-rate analysis is not complicated. The expected performance of this decoder based upon computer simulation is plotted in Figure 13.14. The performance predicted by computer simulation has been experimentally verified [8].

Error-Control Coding for the Electronic Message System Using a Satellite

Digital communication via satellite has been considered for the U.S. Postal Service (USPS) Electronic Message System (EMS). Binary error probabilities as low as 10^{-12} have been stated as goals to be achieved in the 1980s with the appropriate error-control techniques [53].

Numerous error control alternatives were considered [53] and a hybrid, FEC and ARQ, with joint correction and detection roles emerged as the preferred overall schemes for reasons of flexibility and error-correcting performance. A simplified block diagram of this scheme is shown in Figure 1.12. Convolutional codes with Viterbi decoding are effective for FEC uses on one-way channels. The ARQ schemes are nominally content with a code of any reasonable block length. The code rate can be quite high, such as 0.95 to 0.99. Except for storage, the ARQ codes permit very simple implementation. In his study, Nesenbergs concluded that a block code with block length $n \geq 10,000$ and not less than 40 parity checks (which amounts to 0.4% redundancy) will provide error control to keep the BER at the desired level [53].

To the best of our knowledge, no one has experimentally verified signal-to-noise ratios required for coded or uncoded 10^{-12} error rates. Conservative estimates indicate that more than 20 dB is needed for PSK modems. However, it is shown in Ref. 53 that such an extremely low output-error-rate operation appears possible with the hybrid coding scheme. If one is willing to accommodate bandwidth expansion, then unimpeded throughput seems possible down to signal-to-noise ratios as low as 6 dB. The biggest drawback of the ARQ operation is the need for large data storage facilities.

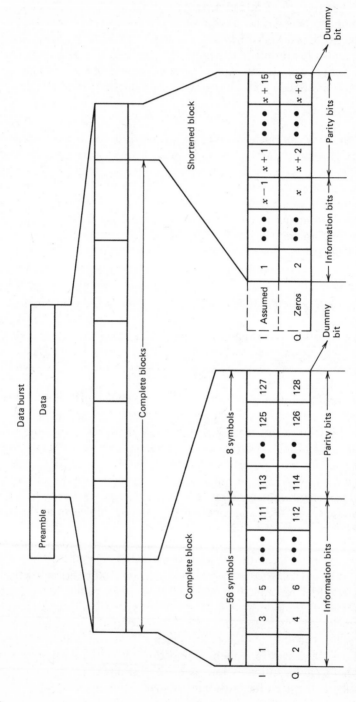

Figure 13.13 Data burst structure using the (128,112) extended BCH code. I, in-phase channel; Q, quadrature channel. (From ref. 8: © 1980 AIAA; reprinted from "Forward Error Correction for Satellite TDMA in the INTELSAT V Era", by J. S. Snyder and T. Muratani, from *A Collection of Technical Papers*, AIAA 8th Commun. Satellite Syst. Conf., 1980.)

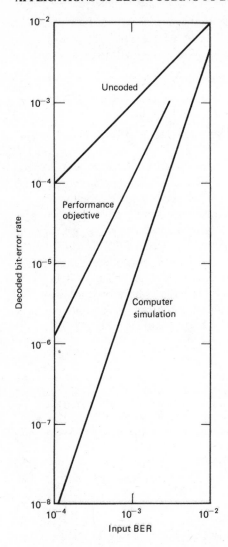

Figure 13.14 BER performance of the (138, 112) extended BCH code. (From Ref. 8: © 1980 AIAA. Reprinted from "Forward Error Correction for Satellite TDMA in the INTELSAT V Era", by J. S. Snyder and T. Muratani, from *A Collection of Technical Papers*, AIAA 8th Commun. Satellite Syst. Conf., 1980.)

Block Coding for a UHF Satellite System

In a very sketchy paper [54], Berlekamp suggests that convolutional codes with conventional decoding methods are ill suited to the correction of erasure bursts (even with interleaving!). This case occurs, for example, in the UHF DAMA environment, where the erasures are due to RFI interference with pulse length significantly longer than the bit transmission time. As an alternative Berlekamp suggests long cyclic block codes with soft-decision decoding, which, rather than dispersing the bursts by interleaving, recognizes and exploits the burst character of errors to improve performance.

As to how long the codes should be—the author suggests that a block length no greater than 128 will suffice for Navy's UHF DAMA system. There have been investigations of many binary cyclic codes in this region. Several codes have been found which have very good immunity to the combination of erasure bursts and Gaussian noise that characterizes the shipboard noise environment [2, 35].

Block Coding for Satellite Communications to Mobile Platforms

Many of the applications described so far are based on hard- or soft-decision demodulation with coherent BPSK or QPSK modulation techniques. With mobile airborne terminals operating in the presence of large doppler offset and doppler rate conditions, as well as multipath and fading environments, noncoherent FSK and M-ary FSK modulation techniques are more appropriate. The first application of coding to this channel involved a (7,2) Reed–Solomon code using octal symbols, based on octal M-ary FSK noncoherent modulation and utilizing a soft-decision decoder [55]. This technique, called TATS (TActical Transmission System), was developed by Lincoln Laboratory and originally used in TACSAT (U.S. Defense Department TACtical SATellite); it is now becoming operational in AFSATCOM wideband channels [56].

REFERENCES

1 E. R. Berlekamp, *Algebraic Coding Theory*, McGraw-Hill, New York, 1968.

2 E. R. Berlekamp, "The Technology of Error-Correcting Codes", *Proc. IEEE*, Vol. 68, May 1980, pp. 564–593.

3 G. Birkhoff and T. C. Bartee, *Modern Applied Algebra*, McGraw-Hill, New York, 1970.

4 C. L. Chen, "Computer Results on the Minimum Distance of Some Binary Cyclic Codes", *IEEE Trans. Inf. Theory*, Vol. IT-16, 1970, pp. 359–360.

5 W. W. Peterson and E. J. Weldon, Jr., *Error-Correcting Codes*, 2nd ed., MIT Press, Cambridge, Mass., 1972.

6 G. Promhouse and S. E. Tavares, "The Minimum Distance of All Binary Cyclic Codes of Odd Lengths from 69 to 99", *IEEE Trans. Inf. Theory*, Vol. IT-24, July 1978, pp. 438–442.

7 F. J. MacWilliams and N. J. A. Sloane, *The Theory of Error-Correcting Codes, I and II*, North-Holland, New York, 1977.

8 J. S. Snyder and T. Muratani, "Forward Error Correction for Satellite TDMA in the INTELSAT V Era", in *Proc. AIAA 8th Commun. Satellite Syst. Conf.*, Orlando, Fla., Apr. 1980, pp. 674–683.

9 J. P. Stenbit, "Tables of Generators for Bose–Chaudhuri–Hocquenghem Codes", *IEEE Trans. Inf. Theory*, Vol. IT-10, Oct. 1964, pp. 390–391.

10 S. Lin, *An Introduction to Error-Correcting Codes*, Prentice-Hall, Englewood Cliffs, N.J., 1970.

11 W. W. Peterson, "Encoding and Error-Correction Procedures for the Bose–Chaudhuri Codes", *IRE Trans. Inf. Theory*, Vol. IT-6, Sept. 1960, pp. 459–470.

12 N. Zierler, "A Complete Theory for Generalized BCH Codes", in H. B. Mann (Ed.), in *Error Correcting Code*, Wiley, New York, 1968.

13 R. T. Chien, "Cyclic Decoding Procedures for Bose–Chaudhuri–Hocquenghem Codes", *IEEE Trans. Inf. Theory*, Vol. IT-10, Oct. 1964, pp. 357–363.

14 J. L. Massey, "Step-by-Step Decoding of the Bose–Chaudhuri–Hocquenghem Codes", *IEEE Trans. Inf. Theory*, Vol. IT-11, Oct. 1965, pp. 580–585.

15 J. L. Massey, "Shift-Register Synthesis and BCH Codes", *IEEE Trans. Inf. Theory*, Vol. IT-15, Jan. 1969, pp. 122–127.

16 E. R. Berlekamp, "On Decoding Binary Bose–Chaudhuri–Hocquenghem Codes", *IEEE Trans. Inf. Theory*, Vol. IT-11, Oct. 1965, pp. 577–580.

17 H. O. Burton, "Inversionless Decoding of BCH Codes", *IEEE Trans. Inf. Theory*, Vol. IT-17, July 1971, pp. 464–466.

18 T. C. Bartee and D. I. Schneider, "An Electronic Decoder for Bose–Chaudhuri–Hocquenghem Error Correction Codes", *IEEE Trans. Inf. Theory*, Vol. IT-8, Sept. 1962, pp. 17–24.

19 W. K. Pehlert, Jr., "Design and Evaluation of a Generalized Burst-Trapping Error Control System", *IEEE Trans. Commun. Technol.*, Vol. COM-19, Oct. 1971, pp. 863–868.

20 T. Le-Ngoc and V. K. Bhargava, "A Microprocessor Based Decoder for the BCH Codes", *Can. Elec. Eng. J.*, Vol. 4, 1979, pp. 29–32.

21 L.-J. Weng, "Soft and Hard Decoding Performance Comparisons for BCH Codes", in *Proc. Int. Conf. Commun.*, 1979, pp. 25.5.1–25.5.5.

22 D. Chase, "Class of Algorithms for Decoding Block Codes with Channel Measurement Information", *IEEE Trans. Inf. Theory*, Vol. IT-18, Jan. 1972, pp. 170–182.

23 M. Avni and V. K. Bhargava, "On Improving the Bit Error Rate of Various Modulation Systems by Block Coding", in *Proc. Natl. Telecommun. Conf.*, Dec. 1980, pp. 52.3.1–52.3.5.

24 G. D. Forney, Jr., *Concatenated Codes*, Res. Monogr. No. 37, MIT Press, Cambridge, Mass., 1966.

25 T. Kasami, "A Decoding Procedure for Multiple-Error Correcting Cyclic Codes", *IEEE Trans. Inf. Theory*, Vol. IT-10, Apr. 1964, pp. 134–139.

26 J. M. Stein and V. K. Bhargava, "Equivalent Rate 1/2 Quasi-cyclic Codes", *IEEE Trans. Inf. Theory*, Vol. IT-21, Sept. 1975, pp. 588–589.

27 J. M. Stein, V. K. Bhargava and S. E. Tavares, "Weight Distribution of Some 'Best' $(3m, 2m)$ Binary Quasi-cyclic Codes", *IEEE Trans. Inf. Theory*, Vol. IT-21, Nov. 1975, pp. 708–711.

28 C. L. Chen, W. W. Peterson and E. J. Weldon, Jr., "Some Results on Quasi-cyclic Codes", *Inf. Control*, Vol. 15, 1969, pp. 407–423.

29 V. K. Bhargava and J. M. Stein, "(v, k, λ) Configurations and Self-Dual Codes", *Inf. Control*, Vol. 28, Aug. 1975, pp. 352–355.

30 M. Karlin, V. K. Bhargava and S. E. Tavares, "A Note on Extended Quaternary Quadratic Residue Codes and Their Binary Images", *Inf. Control*, Vol. 38, Aug. 1978, pp. 148–153.

31 J. E. Medlin and N. J. Bryg, "A 105 Mbit/s Error Correcting Codec", *IEEE Trans. Commun.*, Vol. COM-26, Oct. 1978, pp. 1425–1432.

32 R. J. McEliece, *The Theory of Information and Coding*, Addison-Wesley, Reading, Mass., 1977.

33 R. T. Chien, "Block-Coding Techniques for Reliable Data Transmission", *IEEE Trans. Commun. Technol.*, Vol. COM-19, Oct. 1971, pp. 743–751.

34 J. L. Massey, "Coding for Everyday Channels", Short course presented at the *Int. Conf. Commun.*, June 1972.

35 E. R. Berlekamp and R. J. McEliece, "Implementation and Performance of Encoders and Decoders", and Short course notes, Cyclotomics, Inc., Berkeley, Calif., Sept. 1978.

36 H. O. Burton and E. J. Weldon, Jr., "Cyclic Product Codes", *IEEE Trans. Inf. Theory*, Vol. IT-11, July 1965, pp. 433–439.

37 N. Abramson, "Cascade Decoding of Cyclic Product Codes", *IEEE Trans. Commun. Technol.*, Vol. COM-16, 1968, pp. 398–402.

38   D. Chase, "Digital Signal Design Concepts for a Time-Varying Rician Channel", *IEEE Trans. Commun.*, Vol. COM-24, Feb. 1976, pp. 164–172.

39   P. G. Farrell, "Soft-Decision Minimum-Distance Decoding", *Proc. NATO ASI Commun. Syst. Random Process Theory*, Darlington, U.K., Oct. 1977.

40   C. E. Sundberg, "A Class of Soft-Decision Error Detectors for the Gaussian Channel", *IEEE Trans. Commun.*, Vol. COM-24, Jan. 1976, pp. 106–112.

41   G. D. Forney, Jr., "Generalized Minimum Distance Decoding", *IEEE Trans. Inf. Theory*, Vol. IT-12, Apr. 1966, pp. 125–131. See also ref. 24.

42   D. Chase, "A Combined Coding and Modulation Approach for Communication over Dispersive Channels", *IEEE Trans. Commun.*, Vol. COM-21, Mar. 1973, pp. 159–174.

43   S. M. Reddy, "Further Results on Decoders for Q-ary Output Channels", *IEEE Trans. Inf. Theory*, Vol. IT-20, July 1974, pp. 552–554.

44   E. J. Weldon, Jr., "Decoding Binary Block Codes on Q-ary Output Channels", *IEEE Trans. Inf. Theory*, Vol. IT-17, Nov. 1971, pp. 713–718.

45   S. Wainberg and J. K. Wolf, "Burst Decoding of Binary Block Codes on Q-ary Output Channels", *IEEE Trans. Inf. Theory*, Vol. IT-18, Sept. 1972, pp. 684–686.

46   C. E. Sundberg, "One-Step Majority Logic Decoding with Symbol Reliability Information", *IEEE Trans. Inf. Theory*, Vol. IT-21, Mar. 1975, pp. 236–242.

47   C. R. P. Hartmann and L. D. Rudolph, "An Optimum Symbol-by-Symbol Decoding Rule for Linear Codes", *IEEE Trans. Inf. Theory*, Vol. IT-22, Sept. 1976, pp. 514–517.

48   J. K. Wolf, "Efficient Maximum Likelihood Decoding of Linear Block Codes Using a Trellis", *IEEE Trans. Inf. Theory*, Vol. IT-24, Jan. 1978, pp. 76–80.

49   J. M. Wozencraft and I. M. Jacobs, *Principles of Communication Engineering*, Wiley, New York, 1965.

50   T. Muratani, H. Saitoh, K. Koga, T. Mizuno, Y. Yasuda and J. S. Snyder, "Application of FEC Coding to the INTELSAT TDMA Systems", in *Proc. ICDSC-4*, Montreal, Oct. 1978, pp. 108–115.

51   *Error Protection Manual*, Computer Sciences Corp., Falls Church, Va., Nov. 1972.

52   D. Wiggert, *Error Control Coding and Applications*, Artech House, Dedham, Mass., 1978.

53   M. Nesenbergs, "Study of Error Control Coding for the U.S. Postal Service Electronic Message System", U.S. Dept. of Commerce, Report prepared for Advanced Mail Systems Directorate, U.S. Postal Service, Rockville, Md., May 1975.

54   E. R. Berlekamp, "Long Block Codes Which Use Soft-Decisions and Correct Erasure Bursts without Interleaving", in *Proc. Natl. Telecommun. Conf.*, Dec. 1977, pp. 36.1.1–36.1.2.

55   I. L. Lebow, K. L. Jordan, Jr. and P. R. Drouilhet, Jr., "Satellite Communications to Mobile Platforms", *Proc. IEEE*, Vol. 59, Feb. 1971, pp. 139–159.

56   J. P. Odenwalder and A. J. Viterbi, "Overview of Existing and Projected Uses of Coding in Military Satellite Communications", in *Proc. Natl. Telecommun. Conf.*, Dec. 1977, pp. 36.4.1–36.4.2.

PROBLEMS

1   Prove that every linear block code is equivalent to a systematic block code.

2   If a systematic (n, k) linear block code is generated by

$$\mathbf{G} = [\mathbf{I}_k, \mathbf{P}]$$

show that its parity check matrix is given by

$$H = \left[P^T, I_{n-k} \right]$$

which generates the $(n, n-k)$ dual code.

3 Show that every combination of $d-1$ or fewer columns of the parity check matrix H of an (n, k) block code is linearly independent. Here d denotes the minimum distance of the code. Also, prove that there exists at least one combination of d columns of H which is linearly independent.

4 Use Problem 3 to compute d_{min} for the code of Example 3.4.

5 This problem illustrates the **standard array** for an (n, k) linear code. We place the 2^k codewords in a row with the all-zero codeword v_1 as the leftmost element. From the remaining $(2^n - 2^k)$ n-tuples, select an n-tuple e_2 (called a **coset leader**) and place it under v_1. Complete the row by adding e_2 to each v_i and placing the sum $(e_2 + v_i)$ under v_i. The row thus obtained is called a **coset**. We now pick another unused n-tuple e_3 and place it under v_1. The third row is now completed by adding e_3 to each codeword v_i and placing the sum $(e_3 + v_i)$ under v_i. This process is continued until all n-tuples are used up. The array thus obtained is called a standard array. The coset leader are normally chosen to be the error patterns that are most likely to occur for a given channel.

(a) Show that no two-tuples in the same row of the standard array are identical.

(b) Show that no n-tuples appear in different rows.

(c) Show that there are precisely 2^{n-k} cosets, each containing 2^k n-tuples.

6 This problem illustrates the **table-lookup decoding** of an (n, k) block code using the standard array.

(a) Show that the 2^k n-tuples of a coset have the same syndrome.

(b) Show that syndromes for different cosets are different.

(c) Show that there is a one-to-one correspondence between a coset leader and a syndrome. Hence the **decoding table** need only the 2^{n-k} coset leaders and their corresponding syndromes.

(d) Show that the following steps will result in the correct decoding of the received n-tuple r.

(1) Calculate the syndrome $s = rH^T$.

(2) Identify the coset leader e_i whose syndrome is equal to s.

(3) $v_i = r + e_i$ is the decoded codeword.

7 How many bits of memory would a table-lookup decoder require to decode the $(128, 112)$ extended BCH code recommended for the INTELSAT V era?

8 Construct the decoding table for the code in Example 13.4. Decode the received words (101110), (111000) and (110111).

9 Given an (n, k) block code with minimum distance d_{min}, generator matrix **G** and parity check matrix **H**, give d_{min} and the form of **G** and **H** for the following related codes:

   (a) The $(n+1, k)$ extended block code formed by adding an overall parity check digit to each codeword.

   (b) The $(n-k', k-k')$ shortened block code obtained by deleting k' information digits.

10 Let $m_i(x)$ denote the minimum polynomial of α^i. Show that
$$LCM(m_1(x), m_2(x), \ldots, m_{2t}(x)) = LCM(m_1(x), m_3(x), \ldots, m_{2t-1}(x))$$

11 An (15,5) cyclic code is generated by

$$g(x) = 1 + x + x^2 + x^4 + x^5 + x^8 + x^{10}$$

   (a) Find the parity check polynomial of this code.

   (b) Find the code polynomial in the systematic form for the message polynomial

$$m(x) = 1 + x + x^4$$

   (c) Construct a five-stage shift register encoder for this code.

   (d) Find the generator matrix and the parity check matrix for this code.

   (e) Is the received polynomial

$$r(x) = 1 + x^6 + x^7 + x^9 + x^{13} + x^{14}$$

   a code polynomial? If not, find the syndrome.

12 Suppose that we use the (15,7) double-error-correcting BCH code of Example 13.19 for error-control purposes. Decode the following received words:

$$r_1(x) = 1 + x^2 + x^6 + x^7 + x^8$$

$$r_2(x) = 1 + x^4 + x^7 + x^8 + x^9 + x^{11} + x^{12}$$

13 Consider the (7,3) RS code correcting two symbol errors. Let $GF(2^3)$ be represented as 0 and the powers of a primitive root α such that $\alpha^3 = \alpha + 1$.

   (a) Construct a table of $GF(2^3)$.

   (b) Show that the generator polynomial of the (7,3) RS code is given by

$$g(x) = \alpha^3 + \alpha x + x^2 + \alpha^3 x^3 + x^4$$

(c) Decode the received word

$$r(x) = 1 + \alpha^3 x + \alpha^2 x^3 + x^4 + \alpha x^5 + \alpha^3 x^6$$

14 A source produces binary digits at a rate of 32 kbps. The channel is subjected to error bursts lasting up to 1 ms. Design an interleaved block coding scheme using (n, k) single-error-correcting codes that will allow full correction of a burst. What is the minimum time between bursts if the system is to operate properly?

15 The burst-error-correcting efficiency of an (n, k) block code is defined as

$$\eta = \frac{2b}{n-k}$$

Show that η is at most equal to $\frac{2}{3}$ for Fire codes.

16 Consider an (n_1, k_1) block code with burst-error-correcting capability b_1 minimum distance d_1 and an (n_2, k_2) block code with burst-error-correcting capability b_2 minimum distance d_2. Then the so-called product code is an $(n_1 n_2, k_1 k_2)$ code consisting of a rectangular array of n_1 columns and n_2 rows in which every row is a codeword of the (n_1, k_1) code and every column is a codeword of the (n_2, k_2) code.

(a) Show that the minimum distance of the product code is $d_1 d_2$.

(b) Show that the burst-error-correcting capability b of the product code is lower-bounded by

$$b \geq \max(n_1 b_1, n_2 b_2)$$

(c) Discuss the suitability of using a product code for both burst- and random-error correction.

17 Soft-decision decoding is often rejected because it requires a modification or replacement of the hard-decision demodulator in a receiver. Discuss the advantages and disadvantages of a simple modification in the form of an additional output terminal for the demodulated signal (before hard-decision, limiting or pulse regeneration).

18 In Section 5.1 we saw that noisy carrier phase reference causes a loss in received E_b/N_0. Show that the effect of noisy carrier phase reference is always worse for a coded system than for an uncoded system.

19 An error-correcting code is said to be transparent if $x + 1$ (where 1 is the all 1's n-tuple) is a valid codeword whenever x is a valid codeword. That is, the complement of every codeword is also a codeword.

(a) Show that a block code is transparent whenever the all 1's n-tuple 1 is a valid codeword.

(b) Show that all the rate-$\frac{1}{2}$ self-dual codes are transparent.

(c) Show that a rate-$\frac{1}{2}$ convolutional code is transparent if each of the encoded sequence $x_i^{(1)}$, $x_i^{(2)}$ is formed by tapping an odd number of encoder stages. Can you generalize this observation for convolutional codes of arbitrary rates?

20 In the discussion of source and channel coding in Section 1.4, it was remarked that differential coding internal to or external to error-control coding impacts on the effective coding gain of the FEC. The two possible configurations are depicted in Figure P13.20.

(a) Show that transparent or nontransparent codes may be used in the configuration of Figure P13.20*a*, but the input bit error rate of the FEC decoder is doubled.

(b) Show that only transparent codes may be used in the configuration of Figure P13.20*b*.

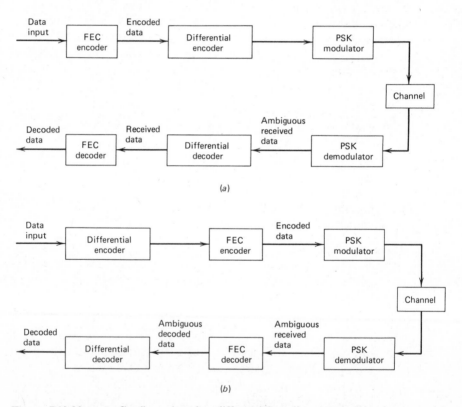

Figure P13.20 (*a*) Configuration for differential coding internal to error-control coding; (*b*) configuration for differential coding external to transparent error-control coding.

(c) Show that the configuration of Figure P13.20*a* requires greater E_b/N_0 to achieve same bit-error rate as the configuration of Figure P13.20*b*. Hence transparent codes are clearly desirable.

21 Figure P13.21 depicts a simplified data path in a TDMA terminal. Discuss the relative advantages and disadvantages of applying FEC at point *A*, *B*, *C* or *D*.

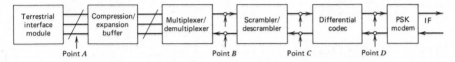

Figure P13.21 Simplified data path in a TDMA terminal.

22 Let S_u and S_c denote the SNR per bit for the uncoded and coded system (using a code with rate *r*), respectively. Clearly,

$$S_c = S_u - 10 \log_{10}\left(\frac{1}{r}\right) \quad \text{dB}$$

(a) Compute the loss (in dB) due to redundancy $[=10\log_{10}(1/r)]$ for codes with rate $\frac{7}{8}$, $\frac{3}{4}$, $\frac{1}{2}$, $\frac{1}{3}$ and $\frac{1}{4}$, respectively.

(b) Explain qualitatively why you would use a high rate code for very low bit-error-rate requirements.

(c) Show that the decoded bit-error rate p_0 for an (n, k) code with error-correcting capability *t* can be approximated by

$$p_0 = \sum_{i=t+1}^{n} \frac{e_i}{n}\binom{n}{i} p^i (1-p)^{n-i}$$

where e_i = average number of output errors resulting from *i* input errors

p = channel crossover probability (corresponding to S_c)

23 For the BCH code proposed for the INTELSAT V era, show that

$$p_0 \simeq 7875 p^3$$

Use $n = 127$ as the block length of the code (excluding the unused dummy bit). How well does this easy-to-use approximation compare with the computer simulation result presented in Figure 13.14?

24 Discuss why with a rate-$\frac{1}{2}$ coded system, the automatic pairing of data and parity bits possible with QPSK is an important consideration.

CHAPTER 14

Threshold Decoding of Block and Convolutional Codes

Threshold decoding is a decoding method applicable to both block and convolutional codes. Much of the basic work on threshold decoding is due to the pioneering work of Massey and our presentation will follow his monumental writing [1]. Threshold decoders require only one decoding cycle per decoded information bit and can be implemented using simple threshold logic gates. However, their performance is decidedly suboptimum.

In this chapter we first consider the simplest form of threshold decoding, **majority-logic decoding**. We then consider in Section 14.2 its generalization to L-step majority logic decoding. Convolutional codes eminently suitable for threshold decoding are considered in Section 14.3. Diffuse threshold decoding and the Gallager adaptive burst-finding scheme, two important variations of threshold decoding, are considered in Sections 14.4 and 14.5. These algorithms can deal with burst errors as well.

14.1 MAJORITY-LOGIC DECODING

Threshold decoding was developed as a simple method of implementing logic circuitry that operates on a syndrome to produce a likely estimate of some selected error digit. The main point is that any syndrome digit, being a linear combination of error digits, represents a known sum of error digits. Further, any linear combination of syndrome digits is thus also a known sum of error digits. Hence all 2^{n-k} such possible combinations of syndrome digits are all of the known sums of error digits available at the receiver. We call such a sum a parity check equation and denote it by A_i (the ith parity check equation). Thus each A_i is a syndrome digit or a known sum of syndrome digits. A parity check equation A_i is said to check an error digit e_j if e_j appears in A_i. A set (A_i) of parity check equations is said to be orthogonal on e_m if each A_i checks e_m but no other error digits is checked by more than one A_i. For example, the

498

following set is orthogonal on e_3 (all additions are modulo 2):

$$A_1 = e_1 + e_2 + e_3$$

$$A_2 = \qquad\quad e_3 + e_4 + e_5$$

$$A_3 = \qquad\quad e_3 \qquad\quad + e_6 + e_7$$

As we see, e_3 appears in each A_i, but each of the other error digits appears in only a single A_i. Majority-logic decoding is a technique of solving for a specific error digit given an orthogonal set of parity check equations for that error digit and is characterized by the following [1]:

Theorem 14.1

Given a set of $J = 2T + S$ parity checks orthogonal on e_m, then any pattern of T or fewer errors in the digits checked by the set (A_i) will cause no decoding error (i.e., are correctable) and patterns of $T+1, \ldots, T+S$ errors are detectable if e_m is decoded by the rule

$$\hat{e}_m = \begin{cases} 1 & \text{if more than } \dfrac{J+S}{2} \text{ of the } A_i \text{ have value 1} \\ 0 & \text{if } \dfrac{J-S}{2} \text{ or fewer have values 1} \end{cases}$$

error detection otherwise. Here $\hat{e}_m$ denotes the estimate of e_m.

Thus $J+1$ corresponds to the effective minimum distance for majority-logic decoding. Further, it can be shown that the code must have a minimum distance of at least $J+1$. We define a code to be completely orthogonalized if $d_{\min} - 1$ orthogonal parity check equations can be found for each error digit. At this point, an example would be helpful in clarifying the concepts developed so far.

Example 14.1

Consider the $(7,3)$ maximum-length code generated by $g(x) = 1 + x + x^2 + x^4$. The generator matrix in the systematic form for this code was derived in Example 13.15. The parity check matrix then is

$$\mathbf{H} = \begin{bmatrix} 1 & 0 & 0 & 0 & 1 & 0 & 1 \\ 0 & 1 & 0 & 0 & 1 & 1 & 1 \\ 0 & 0 & 1 & 0 & 1 & 1 & 0 \\ 0 & 0 & 0 & 1 & 0 & 1 & 1 \end{bmatrix}$$

Let $\mathbf{e} = (e_0, e_1, e_2, e_3, e_4, e_5, e_6)$ be an error vector. The syndrome corresponding to $\mathbf{e}$ is

$$\mathbf{s}=(s_0, s_1, s_2, s_3)=\mathbf{rH}^T=\mathbf{eH}^T$$

which gives

$$s_0 = e_0 + e_4 + e_6$$
$$s_1 = e_1 + e_4 + e_5 + e_6$$
$$s_2 = e_2 + e_4 + e_5$$
$$s_3 = e_4 + e_5 + e_6$$

From these four syndrome digits, three (and no more) parity check sums orthogonal on e_6 can be formed, as follows:

$$A_1 = s_0 \qquad = e_0 \qquad + e_4 \qquad + e_6$$

$$A_2 = s_1 + s_2 = e_1 + e_2 \qquad\qquad + e_6$$

$$A_3 = s_3 \qquad = \qquad e_3 \quad + e_5 + e_6$$

From the preceding discussion we can construct the decoder as shown in Figure 14.1 [1].

The operation of the decoder can be described as follows:

1   The syndrome is calculated as usual by feeding the received word $\mathbf{r}$ into the shift register.

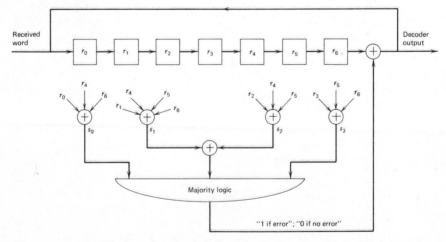

Figure 14.1   Majority-logic decoder for the $(7,3)$ cyclic code generated by $g(x)=1+x+x^2+x^4$.

2   The set of three parity check sums orthogonal on e_6 is formed and fed into a three-input majority logic. If the output of the majority logic is "1", the decoder assumes that the first received digit is erroneous. The received word is shifted again and r_6 is corrected and fed back to the first stage of the shift register. If the majority logic output is "0", the decoder assumes the first received digit to be correct and the received word is shifted again.

3   Repeat steps 1 and 2 to decide if $e_5 = 1$ or 0.

4   The procedure is repeated until a total of seven received digits have been decoded.

It is, of course, possible to calculate the syndrome directly by a four-stage shift register, resulting in the decoder of Figure 14.2. This reduces the number of modulo 2 adders required. This decoder operates in a similar fashion to the decoder of Figure 14.1.

Generalizing the discussion of Example 14.1 to an (n, k) code, we see that both realizations of the decoder require an n-stage shift register. Thus this criterion cannot be used to decide which can be implemented more economically. As a general rule, the implementation given in Figure 14.1 is preferred for low-rate codes, while the implementation given in Figure 14.2 is preferred for high-rate codes.

The decoders described up to this point are referred to as one-step majority-logic decoding. The following classes of cyclic and noncyclic block codes described in Section 13.7 are one-step-majority-logic-decodable:

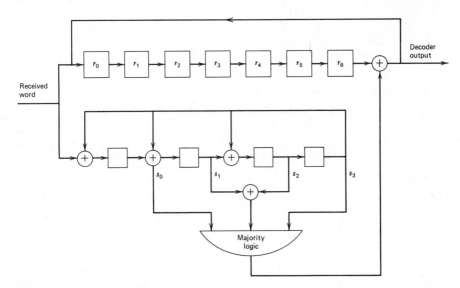

Figure 14.2   Majority-logic decoder for the (7, 3) cyclic code using a syndrome calculator.

1 Maximum-length codes.
2 Difference-set codes [2, 3].
3 Finite projective geometry codes with $L=1$ [4, 5].
4 Euclidean geometry codes with $L=1$ [4, 5].
5 Reed–Muller codes with $L=1$ [5, 6].
6 The $(15,7)$ double-error-correcting BCH code of Example 13.19.
7 A subset of the class of quasi-cyclic codes. Details of one-step majority-logic decodable quasi-cyclic codes may be found in Refs. 7 to 9.

14.2 L-STEP MAJORITY-LOGIC DECODING

The concept of a set of check sums orthogonal on a single bit generalizes without undue complication to sum of bits. A set of check sums is said to be orthogonal on a sum of bits if every bit in the sum is checked by every check sum and no other bit is involved in more than one check sum. It should be clear that the value of any on which it is possible to construct $d_{min}-1$ orthogonal check sums can be determined correctly if $t=\lceil(d_{min}-1)/2\rceil$ or fewer errors occurred. This sum can be considered as an additional check sum and used for decoding, and the process is continued. This method of employing a succession of L majority decisions to decode a bit is called L-step majority-logic decoding. The implementation is obviously more complex but applies to a larger class of cyclic codes.

Example 14.2

Consider the $(7,4)$ cyclic code generated by $g(x)=1+x+x^3$. This is also an Hamming code. The generator matrix **G** and the parity check matrix **H** can be found by applying the method outlined in Section 13.4. Omitting details of computation, we have

$$\mathbf{G}=\begin{bmatrix} 1 & 1 & 0 & 1 & 0 & 0 & 0 \\ 0 & 1 & 1 & 0 & 1 & 0 & 0 \\ 1 & 1 & 1 & 0 & 0 & 1 & 0 \\ 1 & 0 & 1 & 0 & 0 & 0 & 1 \end{bmatrix} \quad \text{and} \quad \mathbf{H}=\begin{bmatrix} 1 & 0 & 0 & 1 & 0 & 1 & 1 \\ 0 & 1 & 0 & 1 & 1 & 1 & 0 \\ 0 & 0 & 1 & 0 & 1 & 1 & 1 \end{bmatrix}.$$

The syndrome vector $\mathbf{s}=(s_0, s_1, s_2)$ of the received vector $\mathbf{r}$ is given as usual by

$$\mathbf{s}=\mathbf{r}\mathbf{H}^T=\mathbf{e}\mathbf{H}^T$$

Thus

$$s_0 = e_0 \quad +e_3 \quad +e_5+e_6$$
$$s_1 = \quad e_1 \quad +e_3+e_4+e_5$$
$$s_2 = \quad e_2 \quad +e_4+e_5+e_6$$

We see that the syndrome equations s_0 and s_2 are orthogonal on the sum $B_1 = e_5 + e_6$, while the check sums $s_0 + s_1$ and s_2 are orthogonal on the sum $B_2 = e_4 + e_6$. Thus, majority-logic decoding can give estimates of B_1 and B_2. However, note that

$$B_1 = \quad e_5 + e_6$$

$$B_2 = e_4 \quad + e_6$$

are orthogonal on e_6. Thus e_6 can be estimated by majority-logic decoding on B_1 and B_2. If a single error has occurred in the error vector, then the estimates of B_1 and B_2 are correct and consequently e_6 is also correctly estimated. Thus we have a two-step majority-logic decoding procedure for this code.

Figure 14.3 depicts an obvious implementation of the decoder. If, however, a syndrome calculator is available to compute s_0, s_1 and s_2, the implementation of Figure 14.4 is desirable.

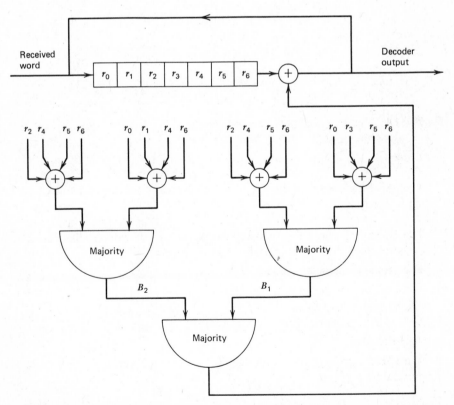

Figure 14.3 Two-step majority-logic decoder for the (7,4) cyclic code generated by $g(x) = 1 + x + x^3$.

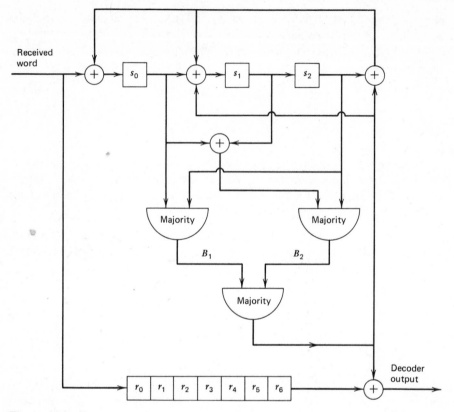

Figure 14.4 Two-step majority logic decoder for the (7,4) cyclic code using a syndrome calculator.

The following classes of block codes described in Sections 13.4 and 13.7 are L-step-majority-logic-decodable:

1. The Hamming codes are two-step majority-logic decodable. Several other decoding methods are applicable and are often simpler [5].
2. The projective geometry codes.
3. The Euclidean projective codes.
4. The Reed–Muller codes.
5. $(2^m - 1, m + 1)$ BCH codes correcting $2^{m-2} - 1$ error are two-step majority-logic decodable.
6. The (31, 16) triple-error-correcting BCH code is three-step-majority-logic decodable.

It can be shown [1] that an L-step majority-logic decoder will require L levels of majority logic, no more than R majority elements and at most $(d) \cdot (k)$

modulo 2 adders. Hence a number of majority elements growing exponentially with L is not required. L-step majority-logic decoding is not applicable to convolutional codes [1].

14.3 THRESHOLD DECODING OF CONVOLUTIONAL CODES

The concept of parity check sums orthogonal on an error digit e_m is readily applicable to convolutional codes. However, as the information and the encoded output occur as sequences for convolutional codes, the syndrome is also a sequence. Following the polynomial form representation developed in Section 12.1, we define the syndrome sequence $S(D)$ to be

$$S_j(D) = G_j(D)[E_I(D) + E_P(D)], \qquad j = 1, 2, \ldots, v$$

$$= s_{j0} + s_{j1}D + s_{j2}D^2 + \cdots \tag{14.1}$$

where

$G_j(D) = $ generator polynomials

$E_I(D) = e_{I0} + e_{I1}D + e_{I2}D^2 + \cdots$ is the information error sequence

$E_P(D) = e_{P0} + e_{P1}D + e_{P2}D^2 + \cdots$ is the parity error sequence

Observe that e_{I0} can affect only the first constraint length of the syndrome digits.

Example 14.3

Consider the rate-$\frac{1}{2}$ code due to Massey [1], who constructed a collection of orthogonalizable codes by a trial-and-error method. The generator polynomials are $G_1(D) = 1$, and $G_2(D) = 1 + D^3 + D^4 + D^5$. The code is systematic [thus we can drop subscripts in (14.1)] with constraint length 6. To compute the parity check sums orthogonal on e_{I0}, the syndrome digits are computed using (14.1). That is, we simply reencode the received information digits and add them to the received check digits. This requires only a copy of the encoding circuit, which is normally a very simple device. The first six syndrome digits are as follows:

$$s_0 = e_{I0} \quad + e_{P0}$$

$$s_1 = \quad\quad e_{I1} + e_{P1}$$

$$s_2 = \quad\quad e_{I2} + e_{P2}$$

$$s_3 = e_{I0} \qquad\qquad\qquad + e_{I3} + e_{P3}$$

$$s_4 = e_{I0} \qquad\qquad\qquad + e_{I4} + e_{P4}$$

$$s_5 = e_{I0} + e_{I1} + \quad e_{I2} \qquad\qquad + e_{I5} + e_{P5}$$

From these six syndrome digits, four (and no more) parity check sums orthogonal on e_{I0} can be formed as follows:

$$A_1 = s_0 \quad = e_{I0} \qquad\qquad\qquad\qquad\qquad + e_{P0}$$

$$A_2 = s_1 + s_5 = e_{I0} + e_{I2} \qquad + e_{I5} \qquad + e_{P1} \qquad + e_{P5}$$

$$A_3 = s_3 \quad = e_{I0} \quad + e_{I3} \qquad\qquad\qquad\qquad + e_{P3} \qquad (14.2)$$

$$A_4 = s_4 \quad = e_{I0} \qquad + e_{I4} \qquad\qquad\qquad + e_{P4}$$

We can now obtain $\hat{e}_{I0}$ an estimate of e_{I0} correctly using Theorem 14.1 whenever two or fewer errors occur among the 10 (in general, $2K$) digits checked by these four syndrome digits.

Decision Feedback Decoder and Error Propagation

Since decoding is likely to have been correct, we can use the estimate $\hat{e}_{I0}$ to cancel the contribution of e_{I0} from each syndrome bit it affects, as shown in Figure 14.5. Decoders, which make use of past decisions, are **called decision feedback decoders**. It can be seen from Figure 14.5 that after one shift, the decision feedback decoder is ready to decode e_{I1} and so on. The simplicity of the majority-logic decoder is again apparent.

If the error estimate is wrong, that is, $\hat{e}_{I0} \neq e_{I0}$, additional errors in subsequent decodings will occur. This is called the error propagation effect and is a serious flaw for threshold decoders using decision feedback. Fortunately, the error progation is a solvable problem.

One solution is to select codes that have the automatic recovery property that limits error propagation. For such codes, the decoder recovers from error propagation upon receipt of a relatively short span of error-free channel digits. Further details of convolutional codes having the automatic recovery property may be found in Refs. 10 to 12.

Definite Decoding

The second solution to eliminate error propagation is simply to break the feedback connection in a decision feedback threshold decoder. This is called **definite decoding**. The disadvantage is the increased probability of error; that is, the error-correcting capability of the code is reduced [13]. We shall illustrate

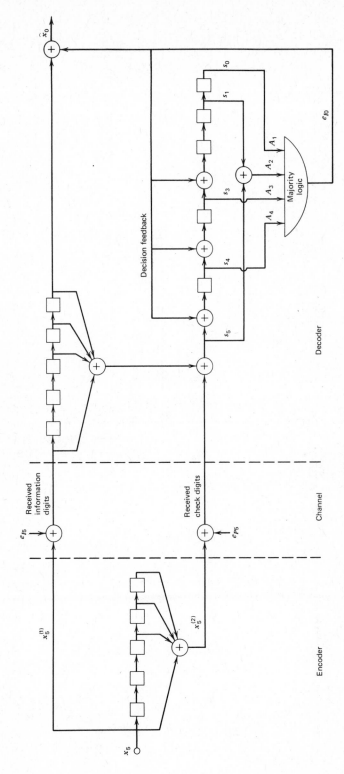

Figure 14.5 Threshold decoder for the $r = \frac{1}{2}$ systematic convolutional code generated by $G_2(D) = 1 + D^3 + D^4 + D^5$.

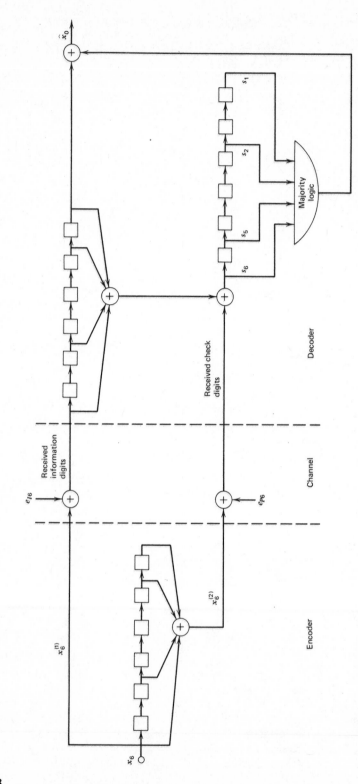

Figure 14.6 Threshold decoder for the $r = \frac{1}{2}$ systematic self. Orthogonal code generated by $G_2(D) = 1 + D^2 + D^5 + D^6$.

definite decoding by considering a code from the class of self-orthogonal codes first studied by Massey [1]. A code is said to be self-orthogonal if the set of syndrome digits checking each of the information error digits forms a set of parity check sums orthogonal on that digits. Self-orthogonal codes have been extensively tabulated in Refs. 1 and 11 and reproduced in Ref. 14, 25 and 26.

Example 14.4

Consider the rate-$\frac{1}{2}$ self-orthogonal code due to Robinson and Bernstein [11]. The generator polynomials are $G_1(D)=1$ and $G_2(D)=1+D^2+D^5+D^6$. The first seven syndrome digits are computed as follows:

$$s_0 = e_{I0} + e_{p0}$$

$$s_1 = \quad e_{I1} + e_{P1}$$

$$s_2 = e_{I0} \quad + e_{I2} + e_{P2}$$

$$s_3 = \quad e_{I1} \quad + e_{I3} + e_{P3}$$

$$s_4 = \quad + e_{I2} \quad + e_{I4} + e_{P4}$$

$$s_5 = e_{I0} \quad + e_{I3} \quad + e_{I5} + e_{P5}$$

$$s_6 = e_{I0} + e_{I1} \quad + e_{I4} \quad + e_{I6} + e_{P6}$$

We recognize that the four syndrome digits (and no more) s_0, s_2, s_5 and s_6 which check e_{I0} are orthogonal on e_{I0}. In other words, the parity check sums are syndrome digits themselves and not sums of syndrome digits. Again using Theorem 14.1, whenever two or fewer errors occur in the H digits checked by these syndrome digits, e_{I0} can be correctly decoded. The threshold decoder in the definite decoding mode is shown in Figure 14.6.

It is a simple matter to show that for decoding subsequent digits, the number of error digits included in the orthogonal parity check equations will increase from 11 to 17 (see Problem 6), thereby reducing the error-correcting capability of the code. The advantage gained has been the absence of any error propagation. Nevertheless, it has been shown that decision feedback decoders (with the associated error propagation) generally outperform definite decoders [15].

14.4 DIFFUSE THRESHOLD-DECODABLE CONVOLUTIONAL CODES

Diffuse convolutional codes are very effective on channels where both random and burst disturbances appear, such as the Aerosat data channel [16]. A rate-$\frac{1}{2}$ code from this class can provide a coding gain of about 8 dB at a bit error rate of 10^{-5} for a 1200-bps information rate over a simulated DPSK Aerosat channel with an $S/I = 10$ dB and a channel fading bandwidth of 120 Hz [17, 18]. The concept of diffuse convolutional codes is due to Kohlenberg [19] and is perhaps best illustrated by an example.

Example 14.5

Consider the rate-$\frac{1}{2}$ systematic diffuse convolutional code for which the encoder and decoder are shown in Figure 14.7. Encoding is similar to any convolutional code, except that the encoding constraint length is made long to "diffuse" the information over a span of digits. The length of the shift register is denoted by β, which also determines the burst-error-correcting capability of the code. Note that at least β digits separate each tap used to provide the check digits. As before, $x_i^{(1)}$ are the information digits and $x_i^{(2)}$ are the parity digits of the code.

The syndrome can always be formed at the receiver by reencoding the received information digits and adding them to the received check digits. The syndrome digits are stored in a register of length $3\beta + 1$.

Calculating the syndromes shown in the decoder diagram, we have:

$$
\begin{aligned}
s_0 \quad &= e_{I0} && + e_{P0} \\[4pt]
s_\beta \quad &= e_{I0} + e_{I\beta} && \quad + e_{P\beta} \\[4pt]
s_{3\beta} + s_{2\beta} &= e_{I0} && + e_{I(3\beta)} && + e_{P(2\beta)} + e_{P(3\beta)} \\[4pt]
s_{3\beta+1} &= e_{I0} + e_{I(\beta+1)} + e_{I(2\beta+1)} && + e_{I(3\beta+1)} && + e_{P(3\beta+1)}
\end{aligned}
\tag{14.3}
$$

For $\beta > 1$ (and $\beta \gg 1$ will be the case of interest), (14.3) constitute a set of parity check sums orthogonal on e_{I0}. Thus we can use Theorem 14.1 to obtain $\hat{e}_{I0}$, an estimate of e_{I0}, whenever two or fewer errors occur among the 11 digits checked by (14.3).

Suppose, on the other hand, that a burst of length 2β or less on the channel has contributed to check sums in (14.3). It is clear that such a burst can cause at most two of the error digits to be 1. Therefore, a burst of length 2β or less is correctable in the absence of other errors, which implies an error-free space (called the guard space) on either side of length $6\beta + 2$ digits. Note that this code has the automatic recovery property.

Several hardware implementations of diffuse convolutional encoders–decoders are currently available. On the theoretical side Ferguson [20] has constructed a family of $r = \frac{1}{2}$ diffuse threshold decodable codes. Tong [21] has constructed a class of diffuse codes which are self-orthogonal. Iwadare [22] and Massey [14] have found a simple $r = (n-1)/n$ class of diffuse single-error-correcting codes, which are very attractive for use on telephone channels.

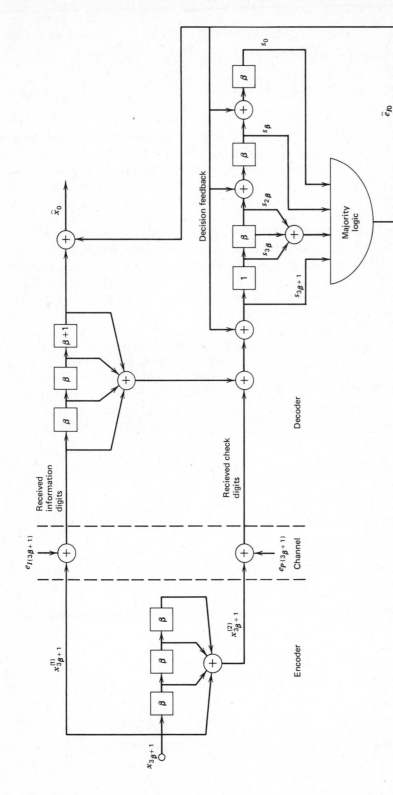

Figure 14.7 Encoder and decoder for a $r = \frac{1}{2}$ systematic diffuse threshold decodable convolutional code.

511

14.5   GALLAGER ADAPTIVE BURST-FINDING SCHEME

This technique relates to threshold decoding and is a very effective scheme for corrections of burst errors [23]. We shall illustrate the basic concepts by considering a simplified block diagram of an $r=1/2$ Gallager encoder. As illustrated in Figure 14.8, the scheme employs a conventional convolutional code (which is threshold-decodable), modified by adding the information digits, delayed by B time units directly into the parity digits. Normally, B will be quite large, say several hundred. In the random mode a decision feedback threshold decoder can be used except that the decoding devisions will be delayed.

The delay and add feature gives the code a burst-error-correcting capability as well. For when the channel errors become too numerous, implying a burst of errors in the received digits, the decoder enters the burst mode. In this mode of correction the decoder attempts to recover the erroneous information digits from the delayed digits which were superimposed on the parity digits. This scheme is capable of correcting bursts of length $2B$ provided that an error-free guard space of $2B$ digits follows the burst.

A major drawback of the Gallager scheme is its vulnerability to errors occurring in the guard space of the code. Sullivan [24] has generalized this scheme to cope with errors in the guard space at some sacrifice in data rate. Kohlenberg and Forney [19] have concluded that Gallager's scheme performs better than diffuse convolutional coding schemes on "dense-burst" (e.g., the troposcatter) channels but performs worse on "diffuse-burst" channels.

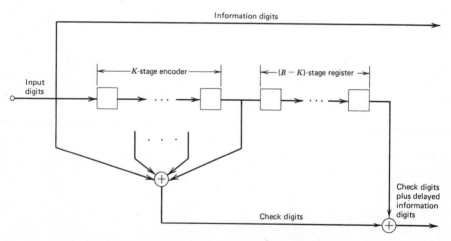

Figure 14.8   Encoder for a $r=\frac{1}{2}$ Gallager adaptive burst-finding code.

14.6 CONCLUDING REMARKS

It is evident from Theorem 14.1 that only certain block and convolutional codes are suitable for threshold decoding. For block codes one-step majority-logic decoding is most efficient or optimum when J is equal to or close to $d_{min} - 1$. Unfortunately, for most powerful block codes, this is not the case. For convolutional codes also, the threshold decoding algorithm is suboptimum (Problem 8). However, since threshold decoders are much simpler to implement, they are competitive for practical applications [25, 26].

Finally, we mention that by extending the work of Massey [1] and Robinson [11, 13], a large number of threshold decodable convolutional codes from rates $\frac{2}{3}$ to $\frac{49}{50}$ have been constructed by Wu [25, 26]. Actual applications for digital communications by satellite include the INTELSAT SPADE system (Problem 9), the DITEC system (Problem 10) and many others [25, 26].

REFERENCES

1   J. L. Massey, *Threshold Decoding*, MIT Press, Cambridge, Mass., 1963.

2   E. J. Weldon, Jr., "Difference-Set Cyclic Codes", *Bell Syst. Tech. J.*, Vol. 45, Sept. 1966, pp. 1045–1055.

3   R. L. Graham and F. J. MacWilliams, "On the Number of Parity Checks in Difference-Set Cyclic Codes", *Bell Syst. Tech. J.*, Vol. 45, Sept. 1966, pp. 1056–1070.

4   K. J. C. Smith, "Majority Decodable Codes Derived from Finite Geometries", *Institute of Statistics Mimeo Series No. 561*, Univ. of North Carolina, Chapel Hill, N.C., 1967.

5   W. W. Peterson and E. J. Weldon, Jr., *Error-Correcting Codes*, 2nd ed., MIT Press, Cambridge, Mass., 1972.

6   I. S. Reed, "A Class of Multiple-Error Correcting Codes and the Decoding Scheme", *IRE Trans. Inf. Theory*, Vol. IT-4, Sept. 1954, pp. 38–49.

7   E. J. Weldon, Jr., "Quasi-cyclic Codes", *IEEE Trans. Inf. Theory*, Vol. IT-12, Apr. 1967, pp. 183–195.

8   V. K. Bhargava, "One-Step Majority Logic Decodable Quasi-cyclic Codes", in *Proc. Int. Symp. Inf. Theory*, Tbilisi, USSR, Vol. 111, July 1979, pp. 22–24.

9   V. K. Bhargava, "Some Codes of Rahman and Blake for Computer Applications", *IEEE Trans. Comp.*, Vol. C-27, Aug. 1978, pp. 765–767.

10   D. D. Sullivan, "Control of Error Propagation in Convolutional Codes", Tech. Rept. EE-667, Dept. of Elec. Eng., Univ. of Notre Dame, Notre Dame, Ind., Nov. 1966.

11   J. P. Robinson and A. J. Bernstein, "A Class of Binary Recurrent Codes with Limited Error Propagation", *IEEE Trans. Inf. Theory*, Vol. IT-13, Jan. 1967, pp. 106–113.

12   J. L. Massey and M. K. Sain, "Inverses of Linear Sequential Circuits", *IEEE Trans. Comp.*, Vol. C-17, Apr. 1968, pp. 330–337.

13   J. P. Robinson, "Error Propagation and Definite Decoding of Convolutional Codes", *IEEE Trans. Inf. Theory*, Vol. IT-14, Jan. 1968, pp. 121–128.

14   S. Lin, *An Introduction to Error-Correcting Codes*, Prentice-Hall, Englewood Cliffs, N.J., 1970.

15   T. N. Morrissey, Jr., "A Unified Markovian Analysis of Decoders for Convolutional Codes", Tech. Rept. EE-687, Dept. of Elec. Eng., Univ. of Notre Dame, Notre Dame, Ind., Oct. 1968.

16 R. Lyons and L. Beaudet, "Application of Forward Error Correction over Aeronautical Satellite Data Links", *Commun. Res. Centre Rep. No. 1314*, Dept. of Commun., Ottawa, Canada, Apr. 1978.

17 S. G. Wilson, R. W. Sutton and E. H. Schroeder, "Differential Phase-Shift Keying Performance on L-Band Aeronautical Satellite Channels: Test Results and a Coding Evaluation", *IEEE Trans. Commun.*, Mar. 1976, pp. 374–380.

18 H. Salwen, "Differential Phase-Shift Keying Performance under Time Selective Multipath Fading", *IEEE Trans. Commun.*, Mar. 1975, pp. 383–385.

19 A. K. Kohlenberg and G. D. Forney, Jr., "Convolutional Coding for Channels with Memory", *IEEE Trans. Inf. Theory*, Vol. IT-14, Sept. 1968, pp. 618–626.

20 M. J. Ferguson, "Diffuse Threshold Decodable Rate 1/2 Convolutional Codes", *IEEE Trans. Inf. Theory*, Vol. IT-17, Mar. 1971, pp. 171–180.

21 S. Y. Tong, "Systematic Construction of Self-Orthogonal Diffuse Codes", *IEEE Trans. Inf. Theory*, Vol. IT-16, Sept. 1970, pp. 594–604.

22 Y. Iwadare, "On Type B1 Burst-Error-Correcting Convolutional Codes", *IEEE Trans. Inf. Theory*, Vol. IT-14, July 1968, pp. 577–583.

23 R. G. Gallager, *Information Theory and Reliable Communication*, Wiley, New York, 1968.

24 D. D. Sullivan, "A Generalization of Gallager's Adaptive Error Control Scheme", *IEEE Trans. Inf. Theory*, Vol. IT-17, Nov. 1971, pp. 727–735.

25 W. W. Wu, "New Convolutional Codes—Part I", *IEEE Trans. Commun.*, Vol. COM-23, Sept. 1975, pp. 942–956.

26 W. W. Wu, "New Convolutional Codes—Part II", *IEEE Trans. Commun.*, Vol. COM-24, Jan. 1976, pp. 19–33.

PROBLEMS

1 Prove Theorem 14.1.

2 Show that the $(15, 7)$ double-error-correcting BCH code of Example 13.19 can be completely orthogonalized in one step. Give a schematic diagram of the decoder.

3 The simplest way to characterize a "(v, k, λ) cyclic difference set" is in terms of its "incidence matrix" **A**. The matrix **A** is a $v \times v$ binary circulant matrix with each row containing exactly k 1's such that any two rows have precisely λ 1's in common. Consider now a rate-$\frac{1}{2}$ quasi-cyclic code with parity check matrix of the form

$$\mathbf{H} = [\mathbf{I}_v, \mathbf{A}]$$

(a) For $\lambda = 1$, show that k parity check sums orthogonal on any message error digit can be formed.

(b) For $\lambda = 1$, show that the minimum distance of this code is exactly $k + 1$.

4 Consider a $(14, 7)$ quasi-cyclic code whose parity check matrix is

$$
\mathbf{H} = \begin{bmatrix}
1 & 0 & 0 & 0 & 0 & 0 & 0 & 0 & 1 & 1 & 0 & 1 & 0 & 0 \\
0 & 1 & 0 & 0 & 0 & 0 & 0 & 0 & 0 & 1 & 1 & 0 & 1 & 0 \\
0 & 0 & 1 & 0 & 0 & 0 & 0 & 0 & 0 & 0 & 1 & 1 & 0 & 1 \\
0 & 0 & 0 & 1 & 0 & 0 & 0 & 1 & 0 & 0 & 0 & 1 & 1 & 0 \\
0 & 0 & 0 & 0 & 1 & 0 & 0 & 0 & 1 & 0 & 0 & 0 & 1 & 1 \\
0 & 0 & 0 & 0 & 0 & 1 & 0 & 1 & 0 & 1 & 0 & 0 & 0 & 1 \\
0 & 0 & 0 & 0 & 0 & 0 & 1 & 1 & 1 & 0 & 1 & 0 & 0 & 0
\end{bmatrix}
$$

(a) Show that this code can be completely orthogonalized in one step.

(b) Show that the minimum distance of this code is exactly 4.

(c) Construct a one-step majority-logic decoder for this code.

5 The $(15, 5)$ triple-error-correcting BCH code is generated by

$$ g(x) = 1 + x + x^2 + x^4 + x^5 + x^8 + x^{10} $$

Show that this code can be completely orthogonalized in two steps. Give a schematic diagram of the decoder.

6 For the code of Example 14.4, write the parity check orthogonal on the digit e_{j_1} after e_{j_0} has been decoded.

7 A serious flaw for threshold decoders using decision feedback is the catastrophic error propagation illustrated in Problem 15 of Chapter 12. In this situation bits can be decoded in error indefinitely.

(a) Show that a convolutional code is catastrophic if and only if the code generator polynomials contain a common factor of degree 1 or greater.

(b) Show that all systematic convolutional codes are noncatastrophic.

8 (a) Let K' denote the number of error digits included in the orthogonal parity check equations of a threshold decodable convolutional code. (K' is often called the decoding constraint length.) Assuming BPSK modulation and an AWGN channel, show that for large E_b/N_0, the decoded bit error rate is

$$ P_0 = \left(\frac{K'}{\dfrac{d_{min} - 1}{2}} \right) \left(\frac{1}{2} \right)^{(d_{min} + 1)/2} \exp \left[-\frac{r(d_{min} + 1)}{2} \cdot \frac{E_b}{N_0} \right] $$

(b) By using (12.30) and (12.75) or otherwise, show that threshold decoding is a suboptimum decoding technique.

9 (a) Show that the rate-$\frac{3}{4}$ systematic convolutional code used in the INTELSAT SPADE system whose generator polynomials are given in Problem 2 of Chapter 12 can be completely orthogonalized.

(b) Give a schematic diagram of the decoder.

(c) By using the performance curve for this code given in Figure 11.5 or otherwise, show that an error-rate improvement from 10^{-4} to about 10^{-9} is possible.

10 DITEC is a digital television system for transmitting color TV signals at an information rate of 33.6 Mbps. In this system the DPCM coded video signal is very susceptible to channel errors. To meet the recommended quality a rate-$\frac{7}{8}$ systematic convolutional code having the following generator polynomials has been proposed.

$$G_1(D)=1$$

$$G_2(D)=1+D^2+D^8+D^{32}+D^{88}+D^{142}$$

$$G_3(D)=1+D^3+D^{19}+D^{52}+D^{78}+D^{146}$$

$$G_4(D)=1+D^{11}+D^{12}+D^{62}+D^{85}+D^{131}$$

$$G_5(D)=1+D^{21}+D^{25}+D^{39}+D^{82}+D^{126}$$

$$G_6(D)=1+D^5+D^{20}+D^{47}+D^{84}+D^{144}$$

$$G_7(D)=1+D^{58}+D^{96}+D^{106}+D^{113}+D^{141}$$

$$G_8(D)=1+D^{41}+D^{77}+D^{108}+D^{117}+D^{130}$$

The constraint length and minimum distance of this code can be computed to be 7 and 1176, respectively.

(a) Show that the code can be completely orthogonalized.

(b) Show that using this code an error-rate improvement from 10^{-4} to below 10^{-8} is possible.

11 A rate-$\frac{1}{2}$ convolutional code is generated by the following generator polynomials:

$$G_1(D)=1$$

$$G_2(D)=1+D+D^4+D^6$$

The code has minimum distance 5. Is this code self-orthogonal? If not, can this code be completely orthogonalized?

CHAPTER 15

Related Topics and Technological Trends

This chapter identifies and discusses briefly some of the related technologies and other factors that influence the main themes of this book: modulation, multiple access and coding. Some trends in technological developments are pointed out and preferred directions are cited from the literature as of 1980. Of course, it should be clear that these views are either being reported or are the opinions and estimates of one or more of the authors. These topics are only introduced here, with references for follow-up reading.

We introduce some regulatory and administrative aspects of satellite communications, some institutional considerations and cost implications. We also identify some new technologies and systems, such as on-board processing and intersatellite links, and outline some broad systems aspects.

15.1 TRENDS IN MODULATION TECHNIQUES

This section addresses the importance of efficient spectrum utilization, the tradeoffs between bandwidth and power and some options for modulation techniques.

Utilization of the Spectrum

Spectrum management is an important factor in the design of a communication system. The allocation of a finite frequency band to a particular service is a regulatory function that serves to conserve the frequency spectrum. Given a certain band within which to work, operating companies must eventually contend with maximizing their system throughput. As the needs increase, they will also have to convince regulators to agree to and themselves be willing to pay for the use of new frequencies through system revision.

It is within the interests of all parties to be concerned with bandwidth efficiency. It is for this reason that the trend will be more and more to the use of bandwidth-efficient modulation techniques [1–3]. Although future systems may well provide greater satellite output power, efficient use of signal power will continue to be an important factor.

Bandwidth/Power Considerations

In the assessment of various modulation techniques for use in satellite communications, bandwidth and power limitations and their tradeoffs are significant.

From the Shannon capacity formula

$$R = B \log_2\left(1 + \frac{S}{N}\right)$$

discussed in Chapter 11 and from

$$\frac{E_b}{N_0} = \frac{S}{N} \times \frac{B}{R}$$

it can be shown that

$$\frac{E_b}{N_0} = 10 \log_{10}\left[\frac{B}{R}(2^{R/B} - 1)\right] \qquad \text{dB}$$

It is instructive to determine the limiting value of E_b/N_0 as B/R grows. This is found by expanding $2^{R/B}$ in a series giving

$$\frac{E_b}{N_0} = 10 \log_{10}\frac{B}{R}\left[1 + \frac{R}{B}\ln 2 + \frac{1}{2!}\left(\frac{R}{B}\ln 2\right)^2 + \cdots - 1\right]$$

$$= -1.6 \text{ dB} \qquad \text{as } B/R \to \infty$$

The limit exists and the convergence is monotonic. Below this value of -1.6 dB there can be no error-free performance, even if B is arbitrarily large or if R is vanishingly small. Above this level but near it, good performance is likely to be expensive. In fact, real systems operate well above -1.6 dB.

A modulation scheme involves the transmission of one of a set of symbols in every symbol period T_s seconds. The mapping of these symbols into a modulated output can be made to be spectrally efficient either by (a) signal shaping or by (b) using a representation where 2 or more input bits are taken during each interval T_s to specify a symbol. (For example, in QPSK 2 bits are used to specify one of four phases.) Bandwidth is controlled by the symbol signalling rate. The symbol rate and consequently the occupied bandwidth decrease since each symbol represents more of the input bits.

It should be noted that the definition of bandwidth has not been generally standardised. A number of definitions exist, including:

1 Frequency band occupied by the signal's main lobe
2 92% or 99% power bandwidth
3 Minimum bandwidth occupied by equally spaced modulated signals where the error rate degradation between adjacent signals does not exceed a specified value
4 U.S. government FCC Docket Number 19311 [4]

Care should be exercised because of the ambiguity that does arise in comparison of the spectral efficiency of modulation techniques.

Several modulation techniques are compared with the Shannon capacity curve in Figure 15.1. The channel capacities of M-ary PSK, M-ary DPSK (differential PSK) and quadrature amplitude modulation suppressed carrier (QAM-SC) are plotted. The channel capacity is given for $P_b = 10^{-4}$ and is plotted against average E_b/N_0. Thus the comparison of various modulation techniques is made for the same bit rate. It should be noted that the curves shown represent theoretical performance for ideal conditions: optimum signal shaping and detection. Implicit in these conditions is that signalling occurs in

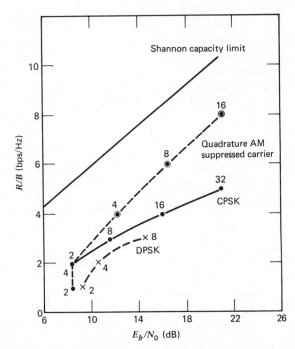

Figure 15.1 Channel capacity in bps/Hz versus E_b/N_o, $P_b = 10^{-4}$.

the Nyquist bandwidth which is equal to the symbol rate for carrier modulation. As can be expected, real systems fall short of these objectives by 1 or more dB.

The ideal results show that the best performance is attained with modulations that combine amplitude and phase to represent input symbols. For values of $R/B>2$, these techniques are more appropriate. Note, however, that an increase in E_b/N_0 is also required. For $R/B=2$ bps/Hz, 4-$\emptyset$ CPSK and 2-QAM are equivalent and require an $E_b/N_0=8.4$ dB. DPSK does not perform as well as CPSK as previously noted.

Modulation Options

Of the various modulation techniques that are available, some exhibit very high spectral efficiency. Included among these are hybrid techniques such as amplitude–phase shift keying (APK), which uses several amplitude and phase values to define a set of symbols. A special case of APK is quadrature partial response [5, 6], in which a controlled amount of intersymbol interference is introduced to PSK signals to achieve bandwidth efficiency. Generally, these techniques are applicable to situations where the channel is linear or where significant SNR is available to compensate for the inherent E_b/N_0 penalty associated with the techniques. For example, in terrestrial microwave systems margins of 40 dB are included in the system design. These margins have not been possible in satellite systems, where margins of 3 to 6 dB are typical. In broadband terrestrial systems employing a series of regenerative repeaters, TWT nonlinearities are controllable as to spectral spreading and performance degradation by means of postamplifier filtering. In such situations the TWT input is the unfiltered constant envelope signal and the TWT output is distortionless because of the absence of envelope fluctuations on the input. The TWT output is then filtered as required. With the advant of powerful regenerative satellites, such techniques may eventually be feasible in satellite communications.

Studies into the application of APK signals in filtered and nonlinear channels [7, 8] show that APK is several dB worse than PSK signals, even though the APK signals may have been as much as 5 dB better in the linear channel case. As a result of these findings, constant envelope techniques are of greater interest in the satellite environment. By constant envelope what is meant are techniques where phase or frequency is the variable, such as PSK and FSK. Especially in the power-efficient operating region near satellite TWT saturation, techniques that are robust to nonlinear distortion are of primary interest.

In satellite systems to date, available E_b/N_0 at the receive station has been of the order of 20 dB and less; consequently, the choices among modulation techniques are restricted. For digital communications, the choice is often a sinusoidal roll-off QPSK which utilizes a bandwidth of $1.2T_s$. For example, a 60-Mbps serial data rate is equivalent to a 30-Msymbol/s rate. The bandwidth requirement is then 36 MHz and the bandwidth efficiency is given by $R/B=$ 1.67 bps/Hz. Typical of the performance attainable with such a QPSK modem are measurements made over the TELESAT 61 Mbps TDMA system [9]. In a laboratory modem back-to-back configuration the E_b/N_0 degradation from theoretical was 2.2 dB for $P_b = 10^{-5}$. The Earth station equipment and link introduced additional degradation for a total of 3.2 dB. An additional 5 to 6 dB was allocated as margin for rain attenuation and other effects.

A number of FSK-related modulation techniques which are constant envelope and spectrally efficient have been of interest recently. These include [10] M-ary CPFSK employing pulse shaping, CPFSK formats using a number of modulation indices and M-ary techniques using correlative encoding. MSK (FFSK) has been described as a binary CPFSK having a modulation index $h=0.5$. The phase trellis is piecewise linear. Extensions to M-ary systems are considered in Refs. 11 and 12. Modifications to the piecewise-linear MSK phase trellis have been proposed by a number of authors [13–15]. In these schemes the piecewise-linear phase transitions of MSK are modified so that the phase function and one or more of its derivatives are continuous at the data transition.

Another approach is one where the modulation index can assume a number of distinct values. In multimode [16] and multi-h [17] modulations, a set of modulation indices are used cyclically by the transmitter. Multi-h phase-coded modulation can provide bandwidth and SNR advantages relative to MSK and has been described in detail in Chapter 6.

In previous descriptions of CPFSK signals it has been implicitly assumed that the phase pulse function is rectangular and of duration equal to the symbol interval. In partial response (correlative) systems having a single modulation index, pulse waveforms are employed that extend beyond the basic symbol interval and better spectral efficiencies are possible [18]. One example of such a technique is *tamed frequency modulation* [19]. Significant spectral roll-off improvements relative to MSK can be attained with an E_b/N_0 penalty of 1 dB relative to ideal CPSK (differential encoding effects ignored). The implementation of the technique can be relatively straightforward [20].

Filtering Options

Filtering a signal invariably causes some degradation in performance in a satellite link. Proper filter design and filter compensation (equalization) at the receiver can reduce this degradation. For example, Mesiya et al. [21] specified an optimal receiver filter for BPSK transmission for a channel consisting of the tandem connection of a linear filter, a bandpass nonlinearity and additive white Gaussian noise. The transmitter filter in this case was a 0.05-dB-ripple, eighth-order pseudo-elliptic function filter with $BT=0.5$, where the one-sided ripple bandwidth B was equal to the Nyquist bandwidth of $(2T)^{-1}$. The receiver filter was constrained to be linear. The authors showed that a 2-dB

SNR improvement compared with a standard choice for the receiver filter could be attained. Fredricson [22] dealt with the same problem for QPSK and specified an optimum receiver filter.

Maximum Likelihood Sequence Estimation

A sophisticated approach to recovering system losses is to employ maximum likelihood sequence estimation (MLSE). A channel that introduces interference over a finite number of transmitted symbols can be thought of as creating a number of states corresponding to the various combinations of symbols sent. These states can be used with the Viterbi algorithm to provide an estimate of the transmitted symbols on a maximum likelihood basis. This represents an approach with good potential. It requires a significant increase in modem complexity. The possibilities of such an approach will be seen in the following.

Bandwidth compaction can be attained by deliberately introducing inter-symbol interference at the transmitter by severe filtering. This improves the R/B efficiency. At the receiver, maximum likelihood demodulation and decoding are used to reduce the error rate. Within this context modulation and coding are no longer separated as two distinct operations.

Often, a modulation technique is fixed for a given system with coding added to achieve a reduction in E_b/N_0 for a given P_b [23,24]. This approach often assumes a hard decision made by the demodulator. From a coding standpoint, the demodulator output should be quantized to several levels to allow the likeliness of a decision (soft decision) to be determined. For example, 3-bit quantization can yield a 2-dB improvement for BPSK in the linear additive white Gaussian noise channel [25]. Section 13.9 discusses such soft decisions followed by decoding.

The maximum likelihood receiver, which uses the Viterbi algorithm, is most effective for large E_b/N_0 and has application where satellites become bandwidth-limited rather than power-limited. In future satellite systems this may very well be the case.

A study [26] of the throughput achievable for satellite-type channels showed that much higher R/B efficiencies are achievable than was previously thought. Given a fixed bandwidth, it has been seen that there is a limit to the maximum symbol rate that can be handled without intersymbol interference (ISI). Beyond this limit, a degradation due to ISI results. Coding can be used to compensate for the ISI degradation but at the expense of more bits transmitted. This study showed that there is an optimum signalling rate at which Viterbi demodulation and decoding can effectively compensate for ISI. An example was cited wherein a QPSK satellite link, downlink-limited by additive noise, could provide an information rate of 3 bps/Hz as compared with conventional results of 1.5 to 1.7 bps/Hz. This means that up to twice as many bits per hertz can be handled by the system. These results show that careful selection of modulation and coding results in increased R/B efficiency.

The increased efficiency is attained at a penalty of processing complexity at the receiver.

For a linear channel with ISI, the BPSK maximum likelihood receiver consists of a matched filter followed by a Viterbi algorithm to compensate for the ISI [27,28]. In the nonlinear channel with no uplink noise, a bank of filters (or correlators) followed by a Viterbi algorithm is best.

Chakraborty et al. [29] are considering the use of a maximum likelihood sequence estimation (MLSE) processor using dynamic programming (Viterbi algorithm) as a probable solution for obtaining performance close to the ideal for the nonlinear satellite channel. The authors cite an example where a decision is based upon observing the demodulator outputs for four symbol intervals for QPSK over a simulated INTELSAT channel. The authors estimate a degradation from the ideal CPSK response of 0.5 dB for $P_b = 10^{-5}$. This represents an improvement of 9.5 dB for the case of saturated nonlinearities.

15.2 DIRECTIONS IN MULTIPLE ACCESS

Although there are now many viable systems, multiple-access techniques are considered to be yet in the formative and developmental stages. It is expected that all classes and many types of multiple-access schemes will be used in the near future, since each has particular merits. There will continue to be ample scope and opportunity for systems analysis and design as these schemes are compared, improved and extended in applications. Some trends in the four types of multiple access are discussed here.

Frequency-Division Multiple Access

We have already noted that not all FDMA systems are digital and that digital applications in FDMA and SCPC are expected to increase in many subsystems, such as voice processing modulation, coding and monitoring of operations; we can expect ever more automation in demand assignment, control, billing and in other end-user interactions.

A typical system specification for the NATO digital satellite communications system is outlined in Ref. 30. Indonesia is planning to extend its digital services, Australia has specified a new domestic satellite system, Brazil is considering a system and the Andean nations of South America are studying a regional system. The previous chapters have already outlined developments in Europe, the United States, Canada and Japan.

As indicated in Chapter 7, trunking applications are well suited to digital FDMA and such will continue and expand. SCPC can be expected to continue in thin-route applications, in situations in which low cost and portability are predominant, and in applications where minimal control is important. The present wide availability and proven performance are strong attributes of SCPC systems.

Time-Division Multiple Access

An overview of TDMA technologies and systems is presented in Chapter 8. *High-capacity* TDMA will be in commercial service by SBS, followed closely by INTELSAT, Japan and Europe. SBS will make use of highly flexible networking and provide a variety of services; INTELSAT's TDMA/DSI at 120 Mbps is to start in 1984 to achieve increased capacity, which is growing at about 17 to 20% per year. Future systems operating at 10 Gbps ($G = 10^9$) in TDMA/FDMA mode at 30/20 GHz to serve multiple-rate users are described in Ref. 31.

Medium-capacity (1 to 20 Mbps) systems are now in operation or undergoing field trials; such systems are expected to be a high-growth area, because of their flexibility, responsiveness to specific requirements and lower cost. A very significant aspect is that many functions are readily subject to LSI or MSI by cost-effective means.

Low-capacity TDMA, such as for maritime applications or for a SPADE control system, can be anticipated in more future applications. The authors are not aware of new developments in this area as of 1980.

Synchronization methods for high- and medium-capacity TDMA systems are discussed in Chapter 8. There seems to be a trend toward hybrid and composite techniques, wherein the features of closed loop and open loop are advantageously used. Continued research and more applications are expected to yield practical results. There is also now a melding of methods from terrestrial and satellite systems, made necessary and possible by interconnection aspects. The 1980 Special Issue on Synchronization [32] of the *IEEE Transactions on Communications* presents a comprehensive overview of problems, solutions and trends.

Code-Division Multiple Access

Chapter 9 presents principles of CDMA systems. Several CDMA techniques, particularly direct-sequence SSMA (spread spectrum), can be expected in commercial and military applications. We have already noted some applications of CDMA techniques in synchronization, signalling and control. CDMA is likely to be used also in the command channels for on-board processing.

Packet Satellite Networks

Packet satellite networks (PSN) are in relatively early stages of development, field trials and operations. There appears to be enormous interest in analysis and optimization of many varieties; the area has high potential for wider applications.

15.3   TRENDS IN CODING

Sophisticated coding systems are now used almost routinely in digital satellite systems and the trend is destined to continue [33]. There are several reasons for this trend:

1   Phenomenal decrease in the cost of digital electronics
2   Significant improvement in various decoding algorithms [34–37]
3   Much slower or no decrease in the cost of analog components, such as power amplifier, antenna and so on

Although the coding systems have been separated into block and convolutional error-correcting techniques, the trend is toward the gradual blurring of this separation. There are two reasons for this trend [38]: first, binary algebraic decoders (which dominate the research and application of block coding techniques) can be used directly with soft-decision information to yield up to 2-dB improvement on a Guassian channel due to a factor of 2 increase in the error-correcting capability of the code [39]; and second, block codes have now been shown to be closely related to convolutional codes [34]. Indeed, the issues may now really be as to the performance of and implementations of the decoding algorithms, (and not the type of codes) in a given application.

As presented in Chapter 12, Viterbi decoding achieves optimum performance in Gaussian channels, but the storage requirements indicate a constraint-length limit of 10. Thus methods (such as reduced-state Viterbi decoding) must be found for Viterbi decoding of larger-constraint-length codes. Soft-decision Viterbi decoders have achieved coding gains of up to 6 dB at speeds of 10 Mbps [40]. With analog implementation similar gains at speeds around 200 Mbps may be achieved [35]. At moderate speeds, say between tens of kbps and a few Mbps, convolutional codes with Viterbi decoding appear to be the most attractive technique for digital satellite links. This assumes that there is no appreciable interference other than Gaussian noise; that a decoded bit error rate of 10^{-5} is satisfactory; that code rates between $\frac{1}{2}$ and $\frac{5}{6}$ are to be used due to bandwidth considerations; and that the overall system transmits long sequences of bit streams [33].

For larger coding gains at high speeds and low bit-error rates (such as 10^{-8} to 10^{-12}, which are needed for computer data files, electronic mail, etc.), sequential decoding appears to be more attractive. As seen in Chapter 12, the performance of sequential decoding is nearly optimum. However, because of buffer size requirements, metric sensitivity and computation speed, soft-decision sequential decoding is not practical for data transmitted at rates above 100 kbps. Thus the trend here will be the development of high-data-rate soft-decision sequential decoders. Hard-decision sequential decoders have been built yielding a coding gain of up to 6 dB at speeds of 60 Mbps [40].

In situations where the system protocols require the transmission of blocks of data (such as TDMA), all convolutional systems require flushouts and restarts. This can result in less than expected performance, which is actually given for infinite-length messages. For such systems block codes appear to be more attractive [41]. For high-data-rate TDMA application, if sufficient bandwidth is available, high performance with lower power can be achieved by a concatenated system [42]. Concatenated coding technique is very effective in implementing long block codes for a low bit-error rate whose complexity depends only on the much shorter component codes used in the concatenated systems.

It is shown in Chapter 14 that threshold decoding is a very attractive technique at high speeds with small complexity. In digital satellite systems a potential use of this techniques will be in digital telephony, where the user requires a smaller bit error rate than that of the uncoded system but desires extremely high bandwidth efficiency [43]. It may be noted that threshold decoding is a suboptimum decoding algorithm.

For mobile terminals operating in the presence of large doppler offset and doppler rate, as well as multipath and fading, the use of Reed–Solomon codes with soft-decision decoding appears extremely attractive to combat the bursty nature of noise [44]. In fact, the Joint Tactical Information Distribution System (JTIDS) program has adopted the (31, 15) Reed–Solomon code, which may now be a standard for all tactical military communication links in the United States and perhaps for all the NATO countries [33]. With the existing and future technology, the amount of hardware needed to implement RS decoders is rather modest compared with the hardware required for other system functions, such as timing, synchronizing, buffering and controlling. Alternative coding techniques applicable are hard-decision feedback decoder [45] with interleaving or soft-decision Viterbi decoding with interleaving [46]. However, interleaving requires a significant increase in the encoding delay. In packet networks where decoding, reencoding and retransmission at several intermediate nodes is required, the delay associated with interleaved convolutional codes may not be acceptable. In such situations the solution might be to avoid interleaving by using one of the block codes suited for this purpose, which rather than dispersing the bursts by interleaving, exploits it for improved error performance [47].

As noted in Chapter 11, Automatic Repeat for Request (ARQ) may be useful in many applications, such as computer-to-computer communications, provided that large transmit buffers for high-speed circuits are available. It has been shown that Continuous ARQ schemes, whether go-back-N or Selective-Repeat, are the most efficient versions of ARQ techniques. For satellite links Selective-Repeat ARQ is the most attractive, although the most complex to implement.

For bit error rates as low as 10^{-12} a hybrid, FEC and ARQ, with joint correction and detection, is an attractive technique for reasons of performance and flexibility [48, 49].

15.4 NEW TECHNOLOGIES

This section introduces some new technologies that are expected to be applied widely in the future and that have major impact on systems.

On-Board Processing

Communications payloads will include intersatellite links (ISL); satellite switches, which are treated in Chapter 8 and in a survey article [50]; and *on-board processing*. SS/TDMA will be operational in the near future; on-board processing is now being actively analyzed and developed for likely post-1990 systems. The type of processing considered is regeneration of digital signals, remodulation, baseband switching, store-and-forward, and spectral transform systems.

Regeneration in a satellite consists of receiving and detecting uplink wideband signals, followed by modulation of a satellite-borne carrier and transmission in the downlink. Usually, hard-decisions are used and in systems studied to date, PSK modulation is preferred, and the modulation type is the same in both links. *Remodulation* may be regarded as regeneration with the additional feature that uplink and downlink formats and modulations may be different. Only a few papers treat these subjects [51]. The major advantage of regenerative systems is the separation of the uplink and downlink, so that only errors add, but not noise; the two links can be designed separately under their very different constraints of power, G/T and interference. A further step in on-board processing is to reconstruct and *switch the baseband signals* as required—a true switchboard in the sky. Yet other features can be considered, such as store-and-forward systems, which are already in use in low-capacity systems for data retransmission; computer data acquisition and processing are also space-tested.

There are also proposals for discrete spectral analysis of uplink signals and transmission of the results in the downlink [52]; analysis shows that this Frequency Hopping/Spectral Transform approach is more robust than FDMA and SSMA and more readily implemented than coherent processing systems.

Regeneration, remodulation and baseband switching are expected to be applied to wideband signals. However, for high-capacity systems, store-and-forward and spectral transforms appear to be far in the future for wide applications.

Intersatellite Links

Satellites have a wide coverage area; in the so-called global beams, more than one-third of the Earth can communicate through a single satellite. For reasons already explained, there are limited-area beams in use and planned. Then, in general, an Earth station does not have single-hop access to the whole world. The **intersatellite link** (ISL) (crosslink) is to provide this connectivity by providing a direct communication link *between satellites*. Microwave links are now in use, and in the future laser links with several-Gbps capacity will be available [53]; systems aspects are under study [54].

Land-Mobile Satellite Systems

In the United States and Canada, and to some degree in other countries, there is keen interest in the *cellular* concept for mobile communications systems. A coverage area with limited capacity is divided into cells and operation within each cell is self-contained. When contiguous cells are connected by a communications link, it is called a cellular system. In terrestrial systems the cells are smaller than are otherwise possible by line-of-sight paths, and the transmitters are usually power-constrained; in a satellite system the cell sizes are controlled by spot beam coverage and power considerations. In both forms, the frequency spectrum can be reused many times in FDM form or the whole available spectrum can be shared in SSMA mode [55]. There is considerable controversy over which is more spectrally efficient. The United States and Canada are in the program definition phase of a project in a Land-Mobile Satellite System (LMSS) and a satellite called MSAT [55].

15.5   MACRO-SYSTEMS ASPECTS

The main emphasis of this book is the use of modulation, multiple access and coding within a repeater, whereas here some systems aspects of the whole satellite, both within a satellite and among satellites, are mentioned; we call this the macro-systems aspects.

Interference

A satellite has many channels and repeaters, so there are *adjacent channel interference* (ACI) effects; within a given channel this is the total interference from other channels in the repeater, the satellite, nearby satellites and, in some cases, distant satellites. ACI is closely connected with the frequency plan, the spacing of satellites and the antenna patterns, as well as modulation methods, power levels and such factors. For the *Fixed Satellite Service* (FSS), a CCIR term meaning all services to fixed Earth stations offered via satellites, up to 10,000 pWOp (picowatts of interference) is allowed in a budget of 50,000 pWOp. There is at present (1980) much discussion in CCIR meetings on whether to change the interference allocation and how to apportion it [56]. This group is working on the broader problems of improving the efficient use of the geostationary satellite orbit.

Orbit Utilization

Clearly, the synchronous orbit is a unique and valuable resource which consists of 360° to be used by all nations for a variety of services. From interference and operational considerations, synchronous satellites should have minimum spacings; this is now normally at 5°, with possibilities of negotiating different spacings and with the trend being to smaller values. The World Administrative Radio Council (WARC '79) in 1979 adopted resolution BP; the main resolution is quoted here [57].

WARC, Geneva, 1979 *resolves*

That a World Administrative Radio Conference shall be convened not later than 1984 to guarantee in practice for all countries equitable access to the geostationary-satellite orbit and the frequency bands allocated to space services.

This WARC is intended to establish the principles, criteria and technical parameters for planning and to provide guidelines for regulatory procedures. CCIR and its national members are carrying out preparatory studies and will report to the WARC. The significance of the space WARC in 1984/1986 is evident since conclusions and recommendations will be in effect for several decades.

The potential capacity of the synchronous orbit has been estimated to be higher than foreseen demand. However, the realizations of such capacities depend on development and application of ever-more-sophisticated techniques and systems; there is a rising need for coordination and adoption of spectrum/orbit conservation measures. These scenarios include the use of spot beams, limited-area and regional coverage systems, dual polarization, antennas within the radiation recommendations, control of inhomogeneities and use of 6/4-, 14/12- and 30/18-GHz bands. There remains much work to be done on computer models to include specific constraints, such as realistic geographic features. Effective measures of the orbit/spectrum efficiency are being proposed.

Earth Station Antenna Patterns

At present, for parabolic antennas operating with synchronous satellites, the recommendation is for side-lobe levels not to exceed 32 to 25 log θ dB, where θ is the off-axis angle from the main lobe. Present large antennas do not meet this curve, but smaller antennas can. The IWP [56] recommends that after 1985 all new antennas should meet a 3-dB reduction, to 29 to 25 log θ. CCIR Reports 466-2, 488-1 and 523 discuss the matter further. There is a vigorous program in progress collecting data on the performance of existing Earth station antennas. Similar studies are under way for satellite antennas.

These and other macro-systems topics are recommended for further intensive work by the international community. A COMSAT paper for the INTELSAT system constructs and analyses planning scenarios [58].

15.6 REGULATORY AND ADMINISTRATIVE ASPECTS

This book emphasizes the technical aspects of digital communications by satellite. The regulatory and administrative aspects are recognized as being of equal, or perhaps greater, importance. The practitioner and student must take such factors into account to get a complete picture of satellite communications; brief glimpses of specific topics are offered here.

ITU, CCIR and CCITT provide the international forums for discussion of global and regional aspects on most topics in communication, information dissemination and influencing policy decisions. The Recommendations of CCIR have the equivalent effects of regulations and the resolutions of its meetings have the force of policy decisions. For example, WARC '79 is already having a major impact on satellite communications [59, 60].

Regional agreements and domestic communications regulatory agencies usually have a profound effect on many aspects of the business of satellite communications. Although many rulings are on nondigital services, there are often spillover and extensions to digital services. Policies in neighboring countries can be very different: in the United States, private ownership of Earth stations for reception is allowed and several operating companies own or lease satellite capacity; in Canada, individuals may not operate Earth stations and TELESAT has the charter for all domestic satellites.

The administration and management of the operating companies must also be considered. Again, in contrast, transponder leasing is very flexible with U.S. companies, whereas lease arrangements are reported to be difficult to establish in Canada.

INTELSAT provides very different basic services (voice and data), with only occasional broadband (TV), and provides only the satellite segment. Its policies have long permitted point-to-point voice-band services; when the PACSAT experiment was conducted, long delays were reported in establishing a tariff. New services must be guided over such administrative hurdles as well as the technical difficulties.

REFERENCES

1 E. Bedrosian, "Spectrum Conservation by Efficient Channel Utilization", *IEEE Commun. Soc. Mag.*, Vol. 15, Mar. 1977, pp. 20–27.

2 C. L. Cuccia, "Bandwidth Conservation Is Essential", *Microwave Syst. News Mag.*, Oct. 1978, p. 67.

3 J. G. Smith, "Spectrally Efficient Modulation", in *Proc. Int. Conf. Commun.*, Chicago, 12–15, June 1977, pp. 3.1.37–3.1.41.

4 FCC Report on Docket No. 19311, adopted 19 Sept. 1974 and released 27 Sept. 1974.

5 P. Kabal and S. Pasupathy, "Partial Response Signalling", *IEEE Trans. Commun.*, Vol. COM-23, Sept. 1975, pp. 921–934.

6 S. Barber, "Implementation of Partial Response in a 91 Mbps Digital Radio", in *Proc. Int. Conf. Commun.*, Chicago, 12–15 June 1977, pp. 5.6.111–5.6.115.

7 C. M. Thomas, M. Y. Weidner and S. H. Durrani, "Digital Amplitude-Phase Keying with *M*-ary Alphabets", *IEEE Trans. Commun.*, Vol. COM-22, Feb. 1974, pp. 168–180.

8 D. P. Taylor, H. C. Chan and S. S. Haykin, "Comparative Evaluation of Digital Modulation Techniques", Commun. Res. Lab., Rep. CRL-18, McMaster Univ., Hamilton, Ontario, Apr. 1974.

9 D. A. Gray, "Telesat's Sixty-One Mbps TDMA System Operational Experience", in *Proc. Natl. Telecommun. Conf.*, Birmingham, Ala., Dec. 1978, pp. 11.3.1–11.3.6.

10 N. Rydbeck and C-E. Sundberg, "Recent Results on Spectrally Efficient Constant Envelope Digital Modulation Methods", in *Proc. Int. Conf. Commun.*, Boston, June 1979, pp. 42.1.1–42.1.6.

11 W. J. Weber, P. H. Stanton and J. T. Sumida, "A Bandwidth Compressive Modulation System Using Multiple Amplitude Minimum Shift Keying (MAMSK)", *IEEE Trans. Commun.*, Vol. COM-26, May 1978, pp. 543–551.

12 T. Aulin and C-E. Sundberg, "*M*-ary CPFSK Type of Signalling with Input Data Symbol Pulse Shaping—Minimum Distance and Spectrum", Tech. Rep. TR-111, Telecommun. Theory, Univ. of Lund, Sweden, Aug. 1978.

13 F. Amoroso, "Pulse and Spectrum Manipulation in the Minimum (Frequency) Shift Keying (MSK) Format", *IEEE Trans. Commun.*, Vol. COM-24, Mar. 1976, pp. 381–384.

14 M. K. Simon, "A Generalization of Minimum-Shift Keying (MSK)-Type Signalling Based upon Input Data Symbol Pulse Shaping", *IEEE Trans. Commun.*, Vol. COM-24, Aug. 1976, pp. 845–856.

15 M. Rabzel and S. Pasupathy, "Spectral Shaping in Minimum Shift Keying (MSK) Type Signals", *IEEE Trans. Commun.*, Vol. COM-26, Jan. 1978, pp. 189–195.

16 H. Miyakawa, H. Harashima and Y. Tanaka, "A New Digital Modulation Scheme—Multimode Binary CPFSK", in *Proc. ICDSC-3*, Kyoto, Nov. 1975, pp. 105–112.

17 J. B. Anderson and D. P. Taylor, "A Bandwidth-Efficient Class of Signal Space Codes", *IEEE Trans. Inf. Theory*, Vol. IT-24, Nov. 1978, pp. 703–712.

18 G. S. Deshpande and P. H. Wittke, "The Spectrum of Correlative Encoded FSK" in *Proc. Int. Conf. Commun.*, Toronto, June 1978, pp. 25.3.1–25.3.5.

19 F. de Jager and C. B. Dekker, "Tamed Frequency Modulation, a Novel Method to Achieve Spectrum Economy in Digital Transmission", *IEEE Trans. Commun.*, Vol. COM-26, May 1978, pp. 534–541.

20 C. B. Dekker, "On the Application of Tamed Frequency Modulation to Various Fields of Digital Transmission via Radio", in *Proc. Int. Zurich Semin. Digital Commun.*, 1980, pp. A1.1–A1.10.

21 M. F. Mesiya, P. J. McLane and L. L. Campbell, "Optimal Receiver Filters for BPSK Transmission over a Bandlimited Nonlinear Channel," *IEEE Trans. Commun.*, Vol. COM-26, Jan. 1978, pp. 12–22.

22 S. A. Fredricson, "Optimum Receiver Filters in Digital Quadrature Phase-Shift-Keyed Systems with a Nonlinear Repeater", *IEEE Trans. Commun.*, Vol. COM-23, Dec. 1975, pp. 1389–1399.

23 L. Golding, "DITEC—A Digital Television Communications System for Satellite Links", in *Proc. ICDSC-2*, Paris, Nov. 1972, pp. 384–397.

24 W. W. Wu, "Applications of Error-Correcting Techniques to Satellite Communications", *COMSAT Tech. Rev.*, Vol. 1, Fall 1971, pp. 183–219.

25 J. L. Massey, "Coding and Modulation in Digital Communications", in *Proc. Int. Zurich Semin. Digital Commun.*, Switzerland, 12–15 Mar. 1974, pp. E2.1–E.2.4.

26 Lin Com Corporation, "Final Report on Investigation of Modulation/Coding Trade-off for Military Satellite Communications", performed for the Defense Commun. Agency, Contract No. DCA 100-76-00062; presented at *Natl. Telecommun. Conf.*, Los Angeles, 1977.

27 G. D. Forney, "Maximum Likelihood Sequence Estimation of Digital Sequences in Intersymbol Interference", *IEEE Trans. Inf. Theory*, Vol. IT-18, May 1972, pp. 363–378.

28 M. F. Mesiya, P. J. McLane and L. L. Campbell, "Maximum Likelihood Sequence Estimation of Binary Sequences Transmitted over Bandlimited Non-linear Channels", *IEEE Trans. Commun.*, Vol. COM-25, July 1977, pp. 633–643.

29 D. Chakraborty, T. Noguchi, S. J. Campanella and C. Wolejsza, "Digital Modem Design for Non-linear Channel", in *Proc. ICDSC-4*, Montreal, Oct. 1978, pp. 123–130.

30 M. Celebiler, J. Munns and E. Turner, "The NATO Digital Satellite Communications System", in *Proc. AIAA 8th Commun. Satellite Conf.*, Orlando, Fla., Apr. 1980, pp. 140–147.

31 H. G. Raymond and W. M. Holmes, Jr., "An Advanced Mixer User Domestic Satellite System Architecture," in *Proc. AIAA 8th Commun. Satellite Conf.*, Orlando, Fla., Apr. 1980, pp. 148–153.

32 Special Issue on Synchronization, *IEEE Trans. Commun.*, Vol. COM-28, Aug. 1980.

33 E. R. Berlekamp, "The Technology of Error-Correcting Codes", *Proc. IEEE*, Vol. 68, May 1980, pp. 564–593.

34 J. K. Wolf, "Efficient Maximum Likelihood Decoding of Linear Block Codes Using a Trellis", *IEEE Trans. Inf. Theory*, Vol. IT-24, Jan. 1978, pp. 76–80.

35 A. S. Acampora and R. P. Gilmore, "Analog Viterbi Decoding for High Speed Digital Satellite Channels", *IEEE Trans. Commun.*, Vol. COM-26, Oct. 1978, pp. 1463–1470.

36 M. I. Weng and L. N. Lee, "Weighted Erasure Decoding for Binary Block Codes with Two-Bit Soft-Decision Demodulation", in *Proc. Natl. Telecommun. Conf.*, 1979, p. 23.6.1.

37 "Analog Soft-Decision Threshold Decoding", INTELSAT Attachment No. 1 to BG-42-23E, B/6180, p. 14.

38 J. P. Odenwalder and A. J. Viterbi, "Overview of Existing and Projected Uses of Coding in Military Satellite Communications", in *Proc. Natl. Telecommun. Conf.*, 1977, pp. 36:4.1–36:4.2.

39 D. Chase, "A Class of Algorithms for Decoding Block Codes with Channel Measurement Information", *IEEE Trans. Inf. Theory*, Vol. IT-18, Jan. 1972, pp. 170–182.

40 Error Control Products Brochure, Linkabit Corporation, San Diego, Calif., 1978.

41 J. S. Snyder and T. Muratani, "Forward Error Correction for Satellite TDMA in the INTELSAT V Era", in *Proc. AIAA 8th Commun. Satellite Syst. Conf.*, Orlando, Fla., Apr. 1980, pp. 674–683.

42 R. W. Boyd, J. Bibb Cain and G. C. Clark, "A Concatenated Coding Approach for High Data Rate Applications", in *Proc. Natl. Telecommun. Conf.*, 1977, pp. 36.2.1–36.2.7.

43 J. L. Massey, "Convolutional Coding Techniques in Digital Communications," Tutorial notes, *Int. Commun. Conf.*, 1976.

44 I. L. Lebow, K. J. Jordon, Jr. and P. R. Drouilhet, Jr., "Satellite Communications to Mobile Platforms", *Proc. IEEE*, Vol. 59, Feb. 1971, pp. 139–159.

45 J. A. Heller, "Feedback Decoding of Convolutional Codes", in A. J. Viterbi (Ed.,), *Advances in Communications Systems*, Vol. 4, Academic Press, New York, 1975, pp. 261–278.

46 A. J. Viterbi and I. M. Jacobs, "Advances in Coding and Modulation for Noncoherent Channels Affected by Fading, Partial Band and Multiple-Access Interference", in A. J. Viterbi (Ed.), *Advances in Communications Systems*, Vol. 4, Academic Press, New York, 1975, pp. 279–308.

47 E. R. Berlekamp, "Long Block Coes Which Use Soft Decisions and Correct Erasure Bursts without Interleaving", in *Proc. Natl. Telecommun. Conf.*, 1977, pp. 36.1.1–36.1.2.

48 M. Nesenbergs, "Study of Error Control Coding for the U.S. Postal Service Electronic Message System", U.S. Dept. of Commerce, Report prepared for Advanced Mail Systems Directorate, U.S. Postal Service, Rockville, Md., May 1975.

49 J. Conan and D. Haccoun, "High-Speed Transmission of Reliable Data on Satellite Channels", in *Proc. ICDSC-4*, Montreal, Oct. 1978, pp. 269–274.

50 C. R. Carter, "Survey of Synchronization Techniques for a TDMA Satellite-Switched System", *IEEE Trans. Commun.*, Vol. COM-28, Aug. 1980, pp. 1291–1301.

51 Session XI, in *Proc. ICDSC-5*, Genoa, Italy, Mar. 1981.

52 A. J. Viterbi, "A Processing Satellite Transponder for Multiple Access by Low-Rate Mobile Users", in *Proc. ICDSC-4*, Montreal, Oct. 1978, pp. 166–174.

53 J. A. Maynard, M. Ross and J. D. Wolf, "Multi-gigabit Laser Communications for Satellite Cross-Links", in *Proc. ICDSC-4*, Montreal, Oct. 1978, pp. 155–159.

54 J. Deal, "Digital Transmission Involving Intersatellite Links", in *Proc. ICDSC-4*, Montreal, Oct. 1978, pp. 160–165.

55 G. H. Knouse, "Terrestrial/Land Mobile Satellite Considerations, NASA Plans and Critical issues", in *Proc. Natl. Telecommun. Conf.*, Nov. 1979, pp. 30.1.1–30.1.6.

56 CCIR Interim Working Party 4/1 (IWP), Report to Study Group 4 by CCIR IWP 4/1 on its 9th Meeting, Paris, 5–10 May 1980.

57 *WARC '79*, Resolution BP, Geneva, 1979.

58 H. L. Van Trees, J. C. Fuenzalida, M. Mohajeri and J. E. Martin, "Planning for the Post-1985 INTELSAT System", in *Proc. AIAA 7th Commun. Satellite Conf.*, Apr. 1978, pp. 43–54.

59 R. R. Bowen, "Satellite Broadcasting after WARC '79", in *Proc. AIAA 8th Commun. Satellite Conf.*, Apr. 1980, pp. 156–158.

60 "WARC-79: Three Perspectives", *IEEE Commun. Mag.*, Vol. 17, Sept. 1979, pp. 10–24.

APPENDIX A

Mathematical Relationships

A.1   TRIGONOMETRIC IDENTITIES

$$\sin(A+B) = \sin A \cos B + \cos A \sin B$$

$$\sin(A-B) = \sin A \cos B - \cos A \sin B$$

$$\cos(A+B) = \cos A \cos B - \sin A \sin B$$

$$\cos(A-B) = \cos A \cos B + \sin A \sin B$$

$$\cos A \cos B = \tfrac{1}{2}[\cos(A+B) + \cos(A-B)]$$

$$\sin A \sin B = \tfrac{1}{2}[\cos(A-B) - \cos(A+B)]$$

$$\sin A \cos B = \tfrac{1}{2}[\sin(A+B) + \sin(A-B)]$$

$$\sin 2A = 2 \sin A \cos A$$

$$\cos 2A = \cos^2 A - \sin^2 A$$

$$\sin nA = \sin A \left[(2\cos A)^{n-1} - \binom{n-2}{1}(2\cos A)^{n-3} \right.$$
$$\left. + \binom{n-3}{2}(2\cos A)^{n-5} - \cdots \right]$$

$$\cos nA = \frac{1}{2} \left[(2\cos A)^{n} - \frac{n}{1}(2\cos A)^{n-2} + \frac{n}{2}\binom{n-3}{1}(2\cos A)^{n-4} \right.$$
$$\left. - \frac{n}{3}\binom{n-4}{2}(2\cos A)^{n-6} + \cdots \right]$$

$$\sin^{2n-1}A = \frac{(-1)^{n-1}}{2^{2n-2}} \left[\sin(2n-1)A - \binom{2n-1}{1} \right.$$

$$\cdot \sin(2n-3)A + \cdots + (-1)^{n-1}\binom{2n-1}{n-1}\sin A\Bigg]$$

$$\cos^{2n-1}A = \frac{1}{2^{2n-2}}\Bigg[\cos(2n-1)A + \binom{2n-1}{1}\cos(2n-3)A + \cdots$$

$$+\binom{2n-1}{n-1}\cos A\Bigg]$$

$$\sin^{2n}A = \frac{1}{2^{2n}}\binom{2n}{n} + \frac{(-1)^n}{2^{2n-1}}\Bigg[\cos 2nA - \binom{2n}{1}\cos(2n-2)A + \cdots$$

$$+(-1)^{n-1}\binom{2n}{n-1}\cos 2A\Bigg]$$

$$\cos^{2n}A = \frac{1}{2^{2n}}\binom{2n}{n} + \frac{1}{2^{2n-1}}\Bigg[\cos 2nA - \binom{2n}{1}\cos(2n-2)A + \cdots$$

$$+\binom{2n}{n-1}\cos 2A\Bigg]$$

$$\sin^2 A + \cos^2 A = 1$$

$$\cos^2 A = \tfrac{1}{2}(1 + \cos 2A)$$

$$\sin^2 A = \tfrac{1}{2}(1 - \cos 2A)$$

$$\cos^3 A = \tfrac{1}{4}(3\cos A + \cos 3A)$$

$$\sin^3 A = \tfrac{1}{4}(3\sin A - \sin 3A)$$

$$\cos(\pm A) = \cos A$$

$$\sin(\pm A) = \pm \sin A$$

$$e^{\pm jA} = \cos A \pm j\sin A$$

$$\cos A = \tfrac{1}{2}(e^{jA} + e^{-jA})$$

$$\sin A = \frac{j}{2}(e^{jA} - e^{-jA})$$

A.2   INDEFINITE INTEGRALS

$$\int u^n du = \frac{u^{n+1}}{n+1}, \quad n \neq -1$$

$$\int \frac{du}{u} = \ln u \quad \text{if } u > 0 \qquad \text{or} \qquad \ln(-u) \quad \text{if } u < 0$$

$$= \ln|u|$$

$$\int e^u du = e^u$$

$$\int a^u du = \int e^{u \ln a} du = \frac{e^{u \ln a}}{\ln a} = \frac{a^u}{\ln a}, \quad a > 0, \quad a \neq 1$$

$$\int \sin u \, du = -\cos u$$

$$\int \cos u \, du = \sin u$$

$$\int \tan u \, du = \ln \sec u = -\ln \cos u$$

$$\int \sin^2 u \, du = \frac{u}{2} - \frac{\sin 2u}{4}$$

$$\int \cos^2 u \, du = \frac{u}{2} + \frac{\sin 2u}{4}$$

$$\int \tan^2 u \, du = \tan u - u$$

A.3   DEFINITE INTEGRALS

$$\int_0^\infty \frac{dx}{x^2 + a^2} = \frac{\pi}{2a}$$

$$\int_0^a \frac{dx}{\sqrt{a^2 - x^2}} = \frac{\pi}{2}$$

$$\int_0^a \sqrt{a^2 - x^2} \, dx = \frac{\pi a^2}{4}$$

$$\int_0^\pi \sin mx \sin nx \, dx = \begin{cases} 0 & m, n \text{ integers and } m \neq n \\ \pi/2 & m, n \text{ integers and } m = n \end{cases}$$

$$\int_0^\pi \cos mx \cos nx \, dx = \begin{cases} 0 & m, n \text{ integers and } m \neq n \\ \pi/2 & m, n \text{ integers and } m = n \end{cases}$$

$$\int_0^\pi \sin mx \cos nx \, dx = \begin{cases} 0 & m, n \text{ integers and } m+n \text{ odd} \\ 2m/(m^2 - n^2) & m, n \text{ integers and } m+n \text{ even} \end{cases}$$

$$\int_0^{\pi/2} \sin^2 x \, dx = \int_0^{\pi/2} \cos^2 x \, dx = \frac{\pi}{4}$$

$$\int_0^{\pi/2} \sin^{2m} x \, dx = \int_0^{\pi/2} \cos^{2m} x \, dx = \frac{1 \cdot 3 \cdot 5 \cdots (2m-1)}{2 \cdot 4 \cdot 6 \cdots 2m} \frac{\pi}{2}, \quad m = 1, 2, \ldots$$

$$\int_0^{\pi/2} \sin^{2m+1} x \, dx = \int_0^{\pi/2} \cos^{2m+1} x \, dx = \frac{2 \cdot 4 \cdot 6 \cdots 2m}{1 \cdot 3 \cdot 5 \cdots (2m+1)}, \quad m = 1, 2, \ldots$$

$$\int_0^\infty \frac{\sin px}{x} dx = \begin{cases} \pi/2 & p > 0 \\ 0 & p = 0 \\ -\pi/2 & p < 0 \end{cases}$$

A.4 SERIES EXPANSIONS

$$(1+x)^n = 1 + nx + \frac{n(n-1)}{2!} x^2 + \cdots, \quad |nx| \ll 1$$

$$a^x = 1 + x \ln a + \tfrac{1}{2!}(x \ln a)^2 + \cdots$$

$$e^x = 1 + x + \tfrac{1}{2!} x^2 + \cdots$$

$$\ln(1+x) = x - \frac{x^2}{2} + \frac{x^3}{3} + \cdots$$

$$\sin x = x - \frac{x^3}{3!} + \frac{x^5}{5!} - \cdots$$

$$\cos x = 1 - \frac{x^2}{2!} + \frac{x^4}{4!} - \cdots$$

$$\text{sinc } x = 1 - \frac{(\pi x)^2}{3!} + \frac{(\pi x)^4}{5!} - \cdots$$

A.5   FOURIER TRANSFORM THEOREMS

$$G(f) = \int_{-\infty}^{\infty} g(t)e^{-j\omega t}dt, \quad \omega = 2\pi f$$

Property	Expression	Transform
Superposition	$a_1 g_1(t) + a_2 g_2(t)$	$a_1 G_1(f) + a_2 G_2(f)$
Time scaling	$g(at)$	$\dfrac{1}{\|a\|} G\left(\dfrac{f}{a}\right)$
Time delay	$g(t-\tau)$	$G(f)e^{-j\omega\tau}$
Differentiation	$\dfrac{dg(t)}{dt}$	$(j\omega)G(f)$
Integration	$\displaystyle\int_{-\infty}^{t} g(t')dt'$	$(j\omega)^{-1}G(f)$
Time convolution	$g_1(t)*g_2(t)$	$G_1(f)G_2(f)$
Time multiplication	$g_1(t)g_2(t)$	$G_1(f)*G_2(f)$
Frequency shift	$g(t)e^{j\omega_c t}$	$G(f-f_c)$
Modulation	$g(t)\cos\omega_c t$	$\frac{1}{2}[G(f+f_c)+G(f-f_c)]$
	$g(t)\sin\omega_c t$	$\frac{j}{2}[G(f+f_c)-G(f-f_c)]$

Baseband Data Formats

Figure B.1 depicts the various baseband data formats. These are some of the possible methods of representing digital information.

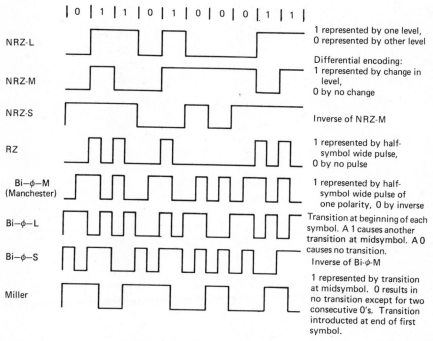

Figure B.1 *Baseband data formats.*

APPENDIX C

Error Function Tabulation

$$\operatorname{erf} x = \frac{2}{\sqrt{\pi}} \int_0^x e^{-u^2} du = 1 - \operatorname{erfc} x$$

x	erf x	x	erf x
0.00	0.00000	0.29	0.31828
0.01	0.01128	0.30	0.32863
0.02	0.02256	0.31	0.33891
0.03	0.03384	0.32	0.34915
0.04	0.04511	0.33	0.35928
0.05	0.05637	0.34	0.36936
0.06	0.06762	0.35	0.37938
0.07	0.07886	0.36	0.38933
0.08	0.09008	0.37	0.39921
0.09	0.10128	0.38	0.40901
0.10	0.11246	0.39	0.41874
0.11	0.12362	0.40	0.42839
0.12	0.13476	0.41	0.43797
0.13	0.14587	0.42	0.44747
0.14	0.15695	0.43	0.45689
0.15	0.16800	0.44	0.46623
0.16	0.17901	0.45	0.47548
0.17	0.18999	0.46	0.48466
0.18	0.20094	0.47	0.49375
0.19	0.21184	0.48	0.50275
0.20	0.22270	0.49	0.51167
0.21	0.23352	0.50	0.52050
0.22	0.24430	0.51	0.52924
0.23	0.25502	0.52	0.53790
0.24	0.26570	0.53	0.54646
0.25	0.27633	0.54	0.55494
0.26	0.28690	0.55	0.56332
0.27	0.29742	0.56	0.57162
0.28	0.30788	0.57	0.57982

x	erf x	x	erf x
0.58	0.58792	1.06	0.86614
0.59	0.59594	1.07	0.86977
0.60	0.60386	1.08	0.87333
0.61	0.61168	1.09	0.87680
0.62	0.61941	1.10	0.88021
0.63	0.62705	1.11	0.88353
0.64	0.63459	1.12	0.88679
0.65	0.64203	1.13	0.88997
0.66	0.64938	1.14	0.89308
0.67	0.65663	1.15	0.89612
0.68	0.66378	1.16	0.89910
0.69	0.67084	1.17	0.90200
0.70	0.67780	1.18	0.90484
0.71	0.68467	1.19	0.90761
0.72	0.69143	1.20	0.91031
0.73	0.69810	1.21	0.91296
0.74	0.70468	1.22	0.91553
0.75	0.71116	1.23	0.91805
0.76	0.71754	1.24	0.92051
0.77	0.72382	1.25	0.92290
0.78	0.73001	1.26	0.92524
0.79	0.73610	1.27	0.92751
0.80	0.74210	1.28	0.92973
0.81	0.74800	1.29	0.93190
0.82	0.75381	1.30	0.93401
0.83	0.75952	1.31	0.93606
0.84	0.76514	1.32	0.93807
0.85	0.77067	1.33	0.94002
0.86	0.77610	1.34	0.94191
0.87	0.78144	1.35	0.94376
0.88	0.78669	1.36	0.94556
0.89	0.79184	1.37	0.94731
0.90	0.79691	1.38	0.94902
0.91	0.80188	1.39	0.95067
0.92	0.80677	1.40	0.95229
0.93	0.81156	1.41	0.95385
0.94	0.81627	1.42	0.95538
0.95	0.82089	1.43	0.95686
0.96	0.82542	1.44	0.95830
0.97	0.82987	1.45	0.95970
0.98	0.83423	1.46	0.96105
0.99	0.83851	1.47	0.96237
1.00	0.84270	1.48	0.96365
1.01	0.84681	1.49	0.96490
1.02	0.85084	1.50	0.96611
1.03	0.85478	1.51	0.96728
1.04	0.85865	1.52	0.96841
1.05	0.86244	1.53	0.96952

x	erf x		
1.54	0.97059	1.78	0.98817
1.55	0.97162	1.79	0.98864
1.56	0.97263	1.80	0.98909
1.57	0.97360	1.81	0.98952
1.58	0.97455	1.82	0.98994
1.59	0.97546	1.83	0.99035
1.60	0.97635	1.84	0.99074
1.61	0.97721	1.85	0.99111
1.62	0.97804	1.86	0.99147
1.63	0.97884	1.87	0.99182
1.64	0.97962	1.88	0.99216
1.65	0.98034	1.89	0.99248
1.66	0.98110	1.90	0.99279
1.67	0.98181	1.91	0.99309
1.68	0.98249	1.92	0.99338
1.69	0.98315	1.93	0.99366
1.70	0.98379	1.94	0.99392
1.71	0.98441	1.95	0.99418
1.72	0.98500	1.96	0.99443
1.73	0.98558	1.97	0.99466
1.74	0.98613	1.98	0.99489
1.75	0.98667	1.99	0.99511
1.76	0.98719	2.00	0.99532
1.77	0.98769		

‖‖‖

Frequency Designations and Bands

Table D.1

Band Number[a]	Symbols	Frequency Range (Lower Limit Exclusive[b], Upper Limit Inclusive)	Corresponding Metric Subdivision	Metric Abbreviations for the Bands
4	VLF	3 to 30 kHz	Myriametric waves	B.Mam
5	LF	30 to 300 kHz	Kilometric waves	B.km
6	MF	300 to 3000 kHz	Hectometric waves	B.hm
7	HF	3 to 30 MHz	Decametric waves	B.dam
8	VHF	30 to 300 MHz	Metric waves	B.m
9	UHF	300 to 3000 MHz	Decimetric waves	B.dm
10	SHF	3 to 30 GHz	Centimetric waves	B.dm
11	EHF	30 to 300 GHz	Millimetric waves	
12		300 to 3000 GHz (or 3T Hz)	Decimillimetric waves	

Source: Radio Regulations. Resolutions and Recommendations. International Telecommunications Union (ITU). Art. 2, Para. 112, 1976 ed., rev. 1979, p. RR2-7.
[a] "Band number N" (N=band number) extends from 0.3×10^N Hz to 3×10^N Hz.
[b] Prefix: k=kilo (10^3), M=mega (10^6), G=giga (10^9), T=tera (10^{12}).

Table D.2

Band Symbol	Frequency Range
L	1000–2000 MHz
S	2000–4000 MHz
C	4000–8000 MHz
X	8000–12,500 MHz
K_u	12.5–18 GHz
K	18–26.5 GHz
K_a	26.5–40 GHz
Millimeter	>40 GHz

Source: M. I. Skolnick, *Introduction to Radar Systems*, McGraw-Hill, New York, 1962, p. 8.

APPENDIX E

Digital Speech Interpolation Techniques

E.1 SYSTEM FUNCTION AND TRANSMISSION MODES

The basic functions necessary for the operation of a digital speech interpolation (DSI) system can be grouped under two broad categories corresponding to the two practical systems so far developed.

In the first system, specified by INTELSAT, at the transmission side the activity of each input trunk is detected periodically and only the active input trunks are connected to the available transmission channels. At the reception side, the received signals from the transmission channels are connected to the proper output trunks according to the received DSI assignment message, which provides information on the status of the connections performed at the transmitting side.

In the second kind of DSI system, called SPEC (Speech Predictive Encoding Communication System), the speech is detected at the transmitting terminal, and the speech sample is transmitted only when the difference in level between the actual sample and the preceding one exceeds a given amplitude. The transmitted sample is stored in a memory at the receiving terminal. Depending on the information contained in the assignment signalling message, this stored sample is used for the reconstruction of the succeeding untransmitted samples [1].

There are two ways of incorporating a digital speech interpolation (DSI) facility into a digital system:

1 The DSI technique is applied individually to each bundle of circuits all having the same destination. This mode is referred to as "point-to-point" DSI.

2 The DSI technique is applied globally to several bundles of circuits having different destinations. This mode is referred as "multidestination" DSI.

544

E.2　DSI GAIN AND SPEECH QUALITY

In a DSI system, the theoretical capacity gain has its limit set by the inverse of the mean activity factor of the input trunks. DSI, however, may introduce some degradation of the voice signal of the systematic or sporadic type.

The systematic degradations are due to the voice detection process and to the internal signalling message and can be reduced or even partially eliminated by employing suitable hardware solutions [2, 3].

The sporadic degradations are due to the statistical behaviour of traffic, which causes system overload to occur when the number of active input trunks is higher than the number of available transmission channels. The degradations introduced in this case are directly related to the DSI gain and therefore to the system capacity.

When the system goes into overload, clipping is introduced whose average value is referred to as freeze-out (Φ). Practical tests on users' reactions to the presence of speech degradation introduced by DSI show that the speech degradation is unnoticeable by an average listener as long as the freeze-out

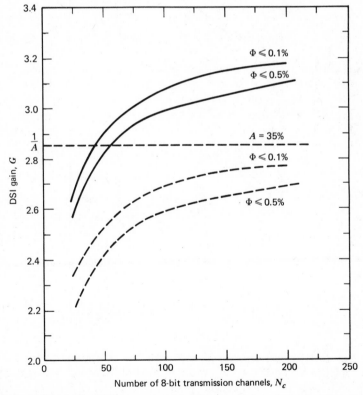

Figure E.1　DSI gain for different numbers of transmission channels (N_0). Solid line, bit-reduction strategy; dashed line, normal DSI.

fraction is less than 0.5%. Theoretically, the DSI gain G and the freeze-out fraction Φ are related to the mean voice activity A, the number of input trunks N_t and the number of transmission channels N_c [4].

Figure E.1 shows G as a function of N_c, when $A = 35\%$ and $\Phi \leq 0.5\%$ and $\leq 1.0\%$. From this curve it appears that the DSI advantage increases as the number of transmission channels increases.

A bit-reduction strategy can be employed to reduce the freeze-out fraction further. With this strategy the eighth bit of the various normal channels is used to yield additional transmission channels when all the normal channels of 8 bits are overloaded.

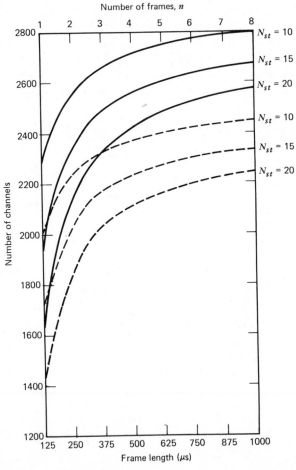

Figure E.2 Net channel capacity of a 36-MHz satellite transponder for a TDMA system equipped with a DSI device when adopting different DSI strategies, assuming that $A = 35\%$. Solid line, bit-reduction strategy ($\phi \leq -1\%$); dashed line, concentrated freeze-out strategy ($\phi \leq 1\%$); N_{st}, number of stations in the system.

In this case the degradation introduced consists of a decrease of the S/N ratio of channels involved, which is limited to the overload time, and a possible residual freeze-out, which is qualitatively similar to that which occurs without the bit-reduction strategy.

Subjective tests performed in Italy on equipment with and without the bit-reduction strategy under the same operational conditions (i.e. with N_t, A and G) revealed that the voice quality obtained by means of the bit-reduction strategy is better than that achieved without the bit-reduction strategy [2]. For example, for the system with bit reduction, a freeze-out fraction of less than 0.05% can be expected, compared with 0.5% for the system without bit reduction but of the same DSI gain. Moreover, a higher DSI gain and therefore a greater total system capacity is achieved through the bit-reduction strategy as compared with the normal DSI for the same voice quality, other conditions being equal [5]. Figure E.2 compares, as an example, the channel capacities using TDMA multiframe and DSI, for the two strategies, assuming equivalent subjective quality of service.

E.3 DATA TRANSMISSION VIA DSI

The effects of DSI on data signals in overload conditions depend on the SI philosophy adopted. If a bit-reduction philosophy is adopted, the data signals are always transmitted even if over a lower-resolution channel. Tests carried out on a 1200-bps voice-band data signal have ascertained that the increase in quantizing noise due to bit reduction limited to the overload condition periods does not impair the transmission of data signals, in practice, even in the extreme-overload condition. When the bit-reduction strategy is not employed, as in systems like the SPEC, the data signals are noticeably degraded during the overload intervals. Laboratory tests [2] provided evidence on the different behaviour of the two systems with respect to data transmission and on the different performances achieved. High-speed data up to 64 kbps is exclusively used on dedicated channels. Therefore, they will not undergo the interpolation process and will use 8-bit time slots, assigned on a permanent basis.

REFERENCES

1 J. A. Sciulli and S. J. Campanella, "A Speech Predictive Encoding Communication System for Multichannel Telephony", *IEEE Trans. Commun.*, Vol. COM-21, July 1973.

2 I. Poretti, G. C. Monti and A. Bagnoli, "Speech Interpolation Systems and Their Application in TDM-TDMA Systems", *Conf. Rec.*, *2nd Int. Conf. Digital Satellite Commun.*, Paris, Nov. 1972.

3 G. Quaglione, L. Ruspantini and D. Lembo, "Ottimizzazione della capacita di un sistema di accesso multiplo a divisione de tempo per satellite di telecommunicazioni" ("Maximization of

the Capacity of the Time-Division Multiple Access Satellite Communication System"), *Elettron. Telecommun.*, Vol. 5, 1972.

4 E. Guinand and J. Penicaud, "Qualité des conversations transmises sur des circuits téléphoniques concentrés par interpolation de la parole" ("Quality of Conversations Transmitted on Telephone Circuits with Concentration by Speech Interpolation"), *Onde Elec.*, Vol. 53, Mar. 1973.

5 I. Poretti, G. C. Monti, A. Bagnoli and G. Bernasconi, "Use of Digital Speech Interpolation Equipment to Increase the Capacity of Telecommunication Systems", *Alta Freq.*, Vol. 7, July 1974.

APPENDIX F

Demand-Assignment Multiple Access

The fundamental multiple-access techniques of FDMA, TDMA and CDMA have been presented in Chapters 7 through 9. These techniques have been analysed as a function of the division of the time-frequency plane. In this appendix an introduction to the principles of the dynamic assignment of the channels according to the demand of the users is presented. The so-called **demand-assignment multiple access** (DAMA) technique may be used with either the division of time or frequency of the time-frequency plane.

In general, when most of the users in a communication system do not communicate continuously, the channels need not be allocated permanently. Therefore, it could be very advantageous to dynamically assign the different channels according to the traffic requirements imposed on the system.

To illustrate the potential benefit and improvement brought about by demand assignment, consider the telephone network. It has been observed that only about 10% of the total number of subscribers want to utilize the telephone facilities during peak time. Therefore, if a dynamic allocation of the circuits were implemented, then only about 10% of the facilities would be needed by the telephone companies.

As another example, consider a satellite system in which an Earth station is required to communicate to some N destinations with a given traffic α (Erlangs) at a required blocking probability. A comparison of the number of channels required to each destination can be calculated when the channels are all preassigned and when they are allocated according to a demand-assignment mode of operation [1, 2]. Table F.1 compares the required number of channels for various traffic loads when the blocking probability is 0.01. Table F.1 shows that a dynamic allocation scheme can substantially increase the use of the available bandwidth and therefore expand the individual services. In general, overall system capacity improvement is the principal motivation for demand-assignment techniques.

Table F.1 Comparison of Satellite Channel Requirements for Preassignment and Demand-Assignment Schemes for a Blocking Probability of 0.01

Number of Destinations per Station N	Channel Requirement								
	$\alpha = 0.1$ Erlang			$\alpha = 0.5$ Erlang			$\alpha = 1.0$ Erlang		
	Pre-assign-ment	Demand assign-ment	Im-prove-ment	Pre-assign-ment	Demand assign-ment	Im-prove-ment	Pre-assign-ment	Demand assign-ment	Im-prove-ment
1	2	2	—	4	4	—	5	5	—
2	4	3	1.3	8	5	1.6	10	7	1.4
4	8	3	2.7	16	7	2.9	20	10	2.0
8	16	4	4.0	32	10	3.2	40	15	2.7
10	20	5	4.0	40	11	3.6	50	18	2.7
20	40	7	5.7	80	18	4.4	100	30	3.0
40	80	10	8.0	160	30	5.3	200	53	3.8

Source: Ref. 2.

A large variety of multiple-access assignment schemes have been developed in the last several years and they may be classified according to the technique with which the channels are allocated. We thus have the following techniques: **polling, preassignment** and **demand assignment**.

Polling is an assignment technique frequently used in computer communications whereby the entire population of users is periodically interrogated in order to know which user requires access to the communication system. The operating system then assigns to that user the requested channel for its exclusive utilization during the entire duration of the message. Although such a system exploits very efficiently the available bandwidth, the idle time experienced by the users waiting to gain access to the system may be excessive if the messages are very long. The problem is somewhat alleviated if several channels can be accessed simultaneously. But then the simplicity of the system is lost.

Perhaps the simplest assignment technique consists of allocating the channel *a priori* to the different users. The channels are thus said to be **preassigned** and this preassignment may be **fixed** or **programmed**. In a fixed preassignment scheme each user will always use the same channels and no other user may use those channels. Such a technique is clearly indicated when all users have a high traffic. In satellite communications there is always the need to accommodate the heavy traffic between certain Earth stations on a fixed preassignment scheme. However, if all the links were preassigned this way, the capacity of the satellite will be exhausted quickly. Therefore, in order to share the channel utilization among users with less than 100% duty cycle, the preassignment of the channels may be programmed rather than fixed.

In a **demand-assignment** multiple-access scheme the channels are allocated upon the user's demand. The assignment may be **ordered**, whereby the assigned signals are kept separated and do not interfere with one another, or the assignment may be **random**, that is, where interferences and collisions are possible. Random-access techniques are presented in Chapter 10.

In an ordered demand assignment scheme a central control unit is necessary, and all requests, signalling, system status and so on are conveyed to the central control via special service channels. When the demand-assignment scheme is random, all users attempt to use the same channels regardless of the status of the system. There is no central control and no service channels. Although such a procedure simplifies the system somewhat, collisions and interferences are bound to occur, which results in a drastic reduction of the overall capacity.

Finally, to satisfy a variety of services and traffics, an efficient system will often consist of a mixture of both preassignment and demand-assignment multiple-access schemes. It is worthwhile to note that contrary to the preassignment scheme, in demand-assignment the channels are never permanently assigned between any pair of stations. Consequently, the demand-assigned channels have a larger traffic-carrying capability than do the preassigned channels, and hence can serve a larger number of circuits. This advantage of demand assignment is warranted only for light to medium traffic to many destinations.

In satellite communication FDMA and TDMA multiple-access techniques are particularly suited for demand-assignment schemes. Therefore, in such systems the usual multiple-access equipment (FDMA or TDMA) is needed together with their coding and modulation equipment. In addition, **demand-assignment signalling and switching** (DASS) equipment is necessary to select the channels from the available pool, to perform the connections to the terrestrial equipment and in general to control the entire system. All service signals (channel requests, busy signals, system status report, etc.) are handled through a routing channel under the control of DASS.

Demand-assignment techniques enhance the usefulness of satellite communications networks and have been implemented by INTELSAT with the following objectives [3].

1 Provide efficient service to light traffic links.
2 Handle overflow traffic from preassigned links.
3 Establish communication links between stations on demand.
4 Optimize the use of both the satellite capacity and the existing Earth station equipment.

One of the most publicized examples of a demand-assignment scheme implemented in conjunction with FDMA is the SPADE system of INTELSAT [4], introduced in Chapter 7. The objective of SPADE is to provide efficient service by sharing a pool of satellite voice circuits among several Earth stations. The circuits are assigned on demand and whenever no longer required they are returned to the common pool. This pool consists of some 800 individual carriers (each representing an access to the satellite) within the 36-MHz bandwidth of a single transponder. PCM source coding is used together with four-phase coherent PSK modulation on each carrier [4]. An interesting feature of the system is that since a separate carrier is assigned to each one-way voice channel, the carrier need not be transmitted unless voice is actually present (voice activation) on the channel. Hence although the channel allocation remains fixed for the entire duration of the call, satellite power is saved by turning the carrier on and off according to the talker activity. Thus with this scheme there is sufficient satellite power to support the entire pool of 800 channels. The SPADE system has proved to be more efficient of power and bandwidth per channel than is conventional FDMA or TDMA.

REFERENCES

1 H. L. Van Trees (Ed.), *Satellite Communications*, IEEE Press Book, New York, 1979. Also distributed by Wiley, New York.
2 G. D. Dill, "Comparison of Circuit Call Capacity of Demand Assignment and Preassignment Operation," *COMSAT Tech. Rev.*, Vol. 2, Spring 1972, pp. 243–256.
3 J. G. Puente and A. M. Werth, "Demand-Assignment Service for the INTELSAT Global Network," *IEEE Spectrum*, Vol. 8, Jan. 1971, pp. 59–69.
4 A. M. Werth, "SPADE: A PCM FDMA Demand Assignment System for Satellite Communications," in *Proc. ICDSC-1*, London, Nov. 1969.

Glossary of Satellite Communications Terms

Satellite

A body that revolves around another body of preponderant mass and that has a motion primarily and permanently determined by the force of attraction of this body.

Orbit

1 The path, relative to a specified frame of reference, described by the centre of mass of a satellite or other object in space, subjected solely to natural forces, principally gravitational attraction.
2 By extension, the path described by the centre of mass of an object in space subjected to natural forces and occasional low-energy corrective forces exerted by a propulsive device in order to achieve and maintain a desired path.

Orbital Plane

The plane containing the centre of mass of the primary body and the velocity vector of a satellite, the frame of reference being that specified for defining the orbital elements.

Direct Orbit

A satellite orbit such that the projection of the centre of mass of the satellite on the reference plane revolves about the axis of the primary body in the same direction as that in which the primary body rotates.

Inclination

The angle between the place of the orbit of a satellite and the reference plane. By convention, the inclination of a direct orbit of a satellite is an acute angle and the inclination of a retrograde orbit is an obtuse angle.

Circular Orbit

A satellite orbit in which the distance between the centres of mass of the satellite and of the primary body is constant.

Elliptical Orbit

A closed satellite orbit in which the distance between the centres of mass of the satellite and of the primary body in not constant.

Equatorial Orbit

A satellite orbit in the plane of which coincides with that of the equator of the primary body.

Period of Revolution

The time between two consecutive passages of a satellite through a characteristic point on its orbit.

Sidereal Period of Revolution

The time elapsing between two consecutive intersections of the projection of a satellite on a reference plane that passes through the centre of mass of the primary body with a line in that plane extending from the centre of mass to infinity, both the normal to the reference plane and the direction of the line being fixed in relation to the stars.

Sidereal Period of Rotation (of an Object in Space)

Period of rotation, around its own axis, of an object in space, such as a natural satellite or a spacecraft, in a frame of reference fixed in relation to the stars. The sidereal period of rotation of an Earth satellite is about 23 h 56 min.

Station-Keeping Satellite

A satellite, the position of the centre of mass of which is controlled to follow a specified law, either in relation to the position of other satellites belonging to the same system or in relation to a point on Earth that is fixed or moves in a specified way.

Synchronous Satellite

A satellite for which the mean sidereal period of revolution is equal to the sidereal period of rotation of the primary body about its own axis.

Geosynchronous Satellite

Synchronous Earth satellite.

Stationary Satellite

A synchronous satellite with an equatorial, circular and direct orbit. A stationary satellite remains fixed in relation to the surface of the primary body.

Geostationary Satellite

A stationary satellite having the Earth as its primary body. The orbit on which a satellite should be placed to be a geostationary satellite is called the geostationary satellite orbit.

Frequency Reuse Satellite Networks

A satellite network in which the satellite utilizes a frequency band more than once, by means of antenna polarization discrimination, or by multiple satellite antenna beams or both.

Glossary of Notation

Symbol	Definition	Typical Equation or Section Reference
A	Semimajor axis of the ellipse	(1.1)
ACI	Adjacent channel interference	Sec. 1.4
AM/AM	Amplitude-to amplitude conversion	Sec. 1.4
AM/PM	Amplitude-to phase conversion	Sec. 1.4
BCW	Burst codeword	Sec. 8.1
B_s	Receiver noise bandwidth	(1.12)
BO_i	Input backoff	(1.5)
BO_o	Output backoff	(1.13)
BPSK	Binary phase shift keying	Sec. 2.4
c	Velocity of light	(1.3)
C	Channel capacity	(8.5)
C/N_0	Carrier power/noise density	(1.5)
CCI	Co-channel interference	Sec. 1.4
CDM	Code-division multiplexing	Sec. 1.5
CDMA	Code-division multiple access	Sec. 1.5
D	Time unit delay	(12.16)
dB	Decibel	Sec. 1.3
$d_c(n)$	Column distance function	(12.27)
d_{free}	Free distance	(12.29)
$d_{\min}$	Minimum distance	(12.28)
DS	Direct sequence	Sec. 9.1
e	Number 2.71828	Sec. 10.2
E	Expected-value operator	Sec. 2.3
E_b/N_0	Energy per bit/noise density	Sec. 1.3
eff	Frame efficiency	(8.1)
e.i.r.p.	Effective isotropic radiated transmitted power	(1.6)
$\text{erfc}(x)$	Complementary error function	(2.2)
FDM	Frequency-division multiplexing	Sec. 1.5
FDMA	Frequency-division multiple access	Sec. 1.5
FFSK	Fast frequency shift keying	Sec. 2.4

Symbol	Definition	Typical Equation or Section Reference
FH	Frequency hopping	Sec. 9.1
G	Generator matrix of the code	Sec. 13.2
$GF(q)$	Galois field of q elements	Sec. 13.3
$\mathbf{G}_j$	Connection vector	(12.5)
$G_j(D)$	Generator polynomial	(12.17)
G_p	Processing gain	(9.1)
G_r	Antenna gain of the receiver	(1.7)
$\left.\dfrac{G}{T}\right)_s$	Figure of merit for a satellite	(1.5)
$g(x)$	Generator polynomial of a cyclic code	Sec. 13.4
h	Modulation index	Sec. 2.4
H	Parity check matrix of the code	Sec. 13.2
HPA	High-power amplifier	Sec. 1.1
$h(t)$	Filter impulse response	Sec. 2.3
$h(x)$	Parity polynomial of a cyclic code	Sec. 13.4
ICDSC	International Conference on Digital Satellite Communication	Sec. 4.3
IF	Intermediate frequency	Sec. 1.4
k	Boltzmann's constant	(1.5)
k	Number of information symbols in a block code	Sec. 11.1
K	Constraint length of a convolutional code	(12.28)
LNA	Low-noise amplifier	Sec. 1.1
$L_{p,u}$	Free-space path loss, uplink	(1.6)
$L_{r,u}$	Uplink margin for rainfall attenuation	(1.5)
m	Number of accesses	(7.29)
M_j	Jamming margin	(9.2)
MSK	Minimum shift keying	Sec. 2.3
n	Number of coded symbols in a block code	Sec. 11.1
$N=2^m-1$	Period of m-stage maximal-length shift-register sequence	Sec. 9.2
$N_0/2$	Double-sided noise density	Sec. 2.2
NRZ	Non-return-to-zero	Sec. 1.4
$n(t)$	Additive noise	Sec. 2.2
P	Power in the waveform	Sec. 2.4
P_b	Probability of bit error	Sec. 1.3
P_B	Probability of block error	(11.17)
PCM	Pulse code modulation	Sec. 7.3
$P(E)$	Probability of first error event	(12.60)
P_s	Probability of symbol error	(2.6)

Symbol	Definition	Typical Equation or Section Reference
PSN	Packet satellite networks	Sec. 1.5
$p(t)$	Rectangular pulse with unit amplitude	(5.1)
P_u	Probability of undetected error	Sec. 1.6
$p(x)$	Probability density function	Sec. 2.3
QPSK	Quadrature phase shift keying	Sec. 2.4
r	Code rate	(1.16)
R	Source rate	(1.16)
R_{comp}	Computation cutoff rate	(12.97)
RF	Radio frequency	Sec. 1.4
R_0	Two-codeword exponent for the DMC	(12.43)
R_s	Transmission (symbol) rate	Sec. 1.6
$r(t)$	Received waveform	Sec. 2.2
$\mathcal{R}(\tau)$	Autocorrelation	Sec. 2.3
$\mathbf{s}$	Syndrome of the code	Sec. 13.2
S	States of a convolutional encoder	(12.25)
SCPC	Single channel per carrier	Sec. 7.3
$S(D)$	Syndrome sequence	(14.1)
SDMA	Space-division multiple access	Sec. 1.5
SIC	Station identification code	Sec. 8.1
SPADE	Single-channel-per-carrier access on demand	Sec. 7.3
$s(t)$	Transmitted waveform	Sec. 2.2
T	(Receiver) noise temperature	(1.7)
T_c	Chip time	Sec. 9.1
t_d	Error-detecting capability of the code	(11.39)
$T(D)$	Generating function	(12.50)
TDM	Time-division multiplexing	Sec. 1.5
TDMA	Time-division multiple access	Sec. 1.5
TRMA	Time-random multiple access	Sec. 1.5
TWTA	Travelling-wave-tube amplifier	Sec. 1.1
U	Information sequence	(12.1)
U/C	Up-converter	Sec. 1.4
UW	Unique word	Sec. 8.1
X	Output coded sequence	(12.2)
$\delta(t)$	Dirac delta function	Sec. 2.3
ΔT	Overall round-trip delay	(11.16)
η	Efficiency of the antenna	(1.8)
η	Transmission efficiency of a burst	(8.4)
η	Throughput of ARQ systems	(11.16)
$\theta(t)$	Signal phase	(5.5)
λ	Wavelength	(1.8)
μ	Gravitational constant	(1.1)
ρ	Correlation coefficient	Sec. 2.3

Index